PROBABILITY

PROBABILITY

With Applications and R

ROBERT P. DOBROW
Department of Mathematics
Carleton College
Northfield, MN

Published by John Wiley & Sons, Inc., Hoboken, New Jersey
Published simultaneously in Canada

Library of Congress Cataloging-in-Publication Data:

Dobrow, Robert P.
Probability: with applications and R / Robert P. Dobrow, Department of Mathematics, Carleton College.
pages cm
Includes bibliographical references and index.
ISBN 978-1-118-24125-7 (hardback)
1. Probabilities–Data processing. 2. R (Computer program language) I. Title.
QA276.45.R3D63 2013
519.20285′5133–dc23

2013013972

Printed in the United States of America

ISBN: 9781118241257

10 9 8 7 6 5 4 3 2 1

To my wonderful family
Angel, Joe, Danny, Tom

CONTENTS

PREFACE

Probability: With Applications and R is a probability textbook for undergraduates. It assumes knowledge of multivariable calculus. While the material in this book stands on its own as a "terminal" course, it also prepares students planning to take upper level courses in statistics, stochastic processes, and actuarial sciences.

There are several excellent probability textbooks available at the undergraduate level, and we are indebted to many, starting with the classic *Introduction to Probability Theory and Its Applications* by William Feller.

Our approach is tailored to our students and based on the experience of teaching probability at a liberal arts college. Our students are not only math majors but come from disciplines throughout the natural and social sciences, especially biology, physics, computer science, and economics. Sometimes we will even get a philosophy, English, or arts history major. They tend to be sophomores and juniors. These students love to see connections with "real-life" problems, with applications that are "cool" and compelling. They are fairly computer literate. Their mathematical coursework may not be extensive, but they like problem solving and they respond well to the many games, simulations, paradoxes, and challenges that the subject offers.

Several features of our textbook set it apart from others. First is the emphasis on simulation. We find that the use of simulation, both with "hands-on" activities in the classroom and with the computer, is an invaluable tool for teaching probability. We use the free software R and provide an appendix for getting students up to speed in using and understanding the language. We recommend that students work through this appendix and the accompanying worksheet. The book is not meant to be an instruction manual in R; we do not teach programming. But the book does have numerous examples where a theoretical concept or exact calculation is reinforced by a computer simulation. The R language offers simple commands for generating samples from probability distributions. The book includes pointers to numerous R script files, that are available for download, as well as many short R "one-liners" that are easily shown in the classroom and that students can quickly and easily duplicate on their computer. Throughout the book are numerous "R:" display boxes that contain these code and scripts.

In addition to simulation, another emphasis of the book is on applications. We try to motivate the use of probability throughout the sciences and find examples from subjects as diverse as homelessness, genetics, meteorology, and cryptography. At the same time, the book does not forget its roots, and there are many classical chestnuts like the problem of points, Buffon's needle, coupon collecting, and Montmort's problem of coincidences.

Following is a synopsis of the book's 10 chapters.

Chapter 1 begins with basics and general principles: random experiment, sample space, and event. Probability functions are defined and important properties derived. Counting, including the multiplication principle and permutations, are introduced in the context of equally likely outcomes. Random variables are introduced in this chapter. A first look at simulation gives accessible examples of simulating several of the probability calculations from the chapter.

Chapter 2 emphasizes conditional probability, along with the law of total probability and Bayes formula. There is substantial discussion of the birthday problem.

Independence and sequences of independent random variables are the focus of Chapter 3. The most important discrete distributions—binomial, Poisson, and uniform—are introduced early and serve as a regular source of examples for the concepts to come. Binomial coefficients are introduced in the context of deriving the binomial distribution.

Chapter 4 contains extensive material on discrete random variables, including expectation, functions of random variables, and variance. Joint discrete distributions are introduced. Properties of expectation, such as linearity, are presented, as well as the method of indicator functions. Covariance and correlation are first introduced here.

Chapter 5 highlights several families of discrete distributions: geometric, negative binomial, hypergeometric, multinomial, and Benford's law.

Continuous probability begins with Chapter 6. The chapter introduces the uniform and exponential distributions. Functions of random variables are emphasized, including maximums, minimums, and sums of independent random variables. There is material on geometric probability.

Chapter 7 highlights several important continuous distributions starting with the normal distribution. There is substantial material on the Poisson process, constructing the process by means of probabilistic arguments from i.i.d. exponential inter-arrival times. The gamma and beta distributions are presented. There is also a section on the Pareto distribution with discussion of power law and scale invariant distributions.

Chapter 8 is devoted to conditional distributions, both in the discrete and continuous settings. Conditional expectation and variance are emphasized as well as computing probabilities by conditioning.

The important limit theorems of probability—law of large numbers and central limit theorem—are the topics of Chapter 9. There is a section on moment-generating functions, which are used to prove the central limit theorem.

Chapter 10 has optional material for supplementary discussion and/or projects. The first section covers the bivariate normal distribution. Transformations of two or more random variables are presented next. Another section introduces the method of moments. The last three sections center on random walk on graphs and Markov chains, culminating in an introduction to Markov chain Monte Carlo. The treatment does not assume linear algebra and is meant as a broad strokes introduction.

There is more than enough material in this book for a one-semester course. The range of topics allows much latitude for the instructor. We feel that essential material for a first course would include Chapters 1–4, 6, and parts of 8 and 9.

Additional features of the book include the following:

- Over 200 examples throughout the text and some 800 end-of-chapter exercises. Includes short numerical solutions for most odd-numbered exercises.
- End-of-chapter reviews and summaries highlight the main ideas and results from each chapter for easy access.
- Starred subsections are optional and contain more challenging material, assuming a higher mathematical level.
- Appendix A: *Getting up to speed in R* introduces students to the basics of R. Includes worksheet for practice.
- Appendix E: *More problems to practice* contains a representative sample of 25 exercises with fully worked out solutions. The problems offer a good source of additional problems for students preparing for midterm and/or final exams.
- A website containing relevant files and errata has been established. All the R code and script files used throughout the book, including the code for generating the book's graphics, are available at this site. The URL is

  `http://www.people.carleton.edu/~rdobrow/Probability`
- An instructor's solutions manual with detailed solutions to all the exercises is available for instructors who teach from this book.

ACKNOWLEDGMENTS

We are indebted to friends and colleagues who encouraged and supported this project. The students of my Fall 2012 Probability class were real troopers for using an early manuscript that had an embarrassing number of typos and mistakes and offering a volume of excellent advice. We also thank Marty Erickson, Jack Goldfeather, Matthew Rathkey, Wenli Rui, and Zach Wood-Doughty. Professor Laura Chihara field-tested an early version of the text in her class and has made many helpful suggestions. Thank you to Jack O'Brien at Bowdoin College for a detailed reading of the manuscript and for many suggestions that led to numerous improvements.

Carleton College and the Department of Mathematics were enormously supportive, and I am grateful for a college grant and additional funding that supported this work. Thank you to Mike Tie, the Department's Technical Director, and Sue Jandro, the Department's Administrative Assistant, for help throughout the past year.

The staff at Wiley, including Steve Quigley, Amy Hendrickson, and Sari Friedman, provided encouragement and valuable assistance in preparing this book.

INTRODUCTION

All theory, dear friend, is gray, but the golden tree of life springs ever green.
—Johann Wolfgang von Goethe

Probability began by first considering games of chance. But today it is the practical applications in areas as diverse as astronomy, economics, social networks, and zoology that enrich the theory and give the subject its unique appeal.

In this book, we will flip coins, roll dice, and pick balls from urns, all the standard fare of a probability course. But we have also tried to make connections with real-life applications and illustrate the theory with examples that are current and engaging.

You will see some of the following case studies again throughout the text. They are meant to whet your appetite for what is to come.

I.1 WALKING THE WEB

There are about one trillion websites on the Internet. When you google a phrase like "Can Chuck Norris divide by zero?," a remarkable algorithm called PageRank searches these sites and returns a list ranked by importance and relevance, all in the blink of an eye. PageRank is the heart of the Google search engine. The algorithm assigns an "importance value" to each web page and gives it a rank to determine how useful it is.

PageRank is a significant accomplishment of mathematics and linear algebra. It can be understood using probability, particularly a branch of probability called Markov chains and random walk. Imagine a web surfer who starts at some web page and clicks on a link at random to find a new site. At each page, the surfer chooses

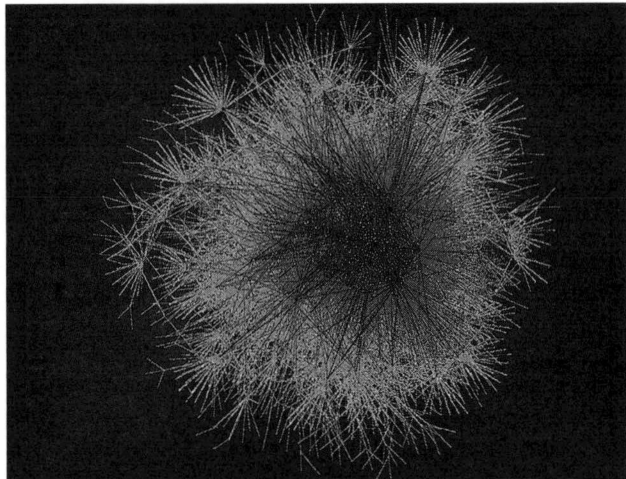

FIGURE I.1: A subgraph of the BGP (Gateway Protocol) web graph consisting of major Internet routers. It has about 6400 vertices and 13,000 edges. Image produced by Ross Richardson and rendered by Fan Chung Graham.

from one of the available hypertext links equally at random. If there are two links, it is a coin toss, heads or tails, to decide which one to pick. If there are 100 links, each one has a 1% chance of being chosen. As the web surfer moves from page to random page, they are performing a random walk on the web (Fig. I.1).

What is the PageRank of site x? Suppose the web surfer has been randomly walking the web for a very long time (infinitely long in theory). The probability that they visit site x is precisely the PageRank of that site. Sites that have lots of incoming links will have a higher PageRank value than sites with fewer links.

The PageRank algorithm is actually best understood as an assignment of *probabilities* to each site on the web. Such a list of numbers is called a *probability distribution.* And since it comes as the result of a theoretically infinitely long random walk, it is known as the *limiting distribution* of the random walk. Remarkably, the PageRank values for billions of websites can be computed quickly and in real time.

I.2 BENFORD'S LAW

Turn to a random page in this book. Look in the middle of the page and point to the first number you see. Write down the first digit of that number.

You might think that such first digits are equally likely to be any integer from 1 to 9. But a remarkable probability rule known as Benford's law predicts that most of your first digits will be 1 or 2; the chances are almost 50%. The probabilities go down as the numbers get bigger, with the chance that the first digit is 9 being less than 5% (Fig. I.2).

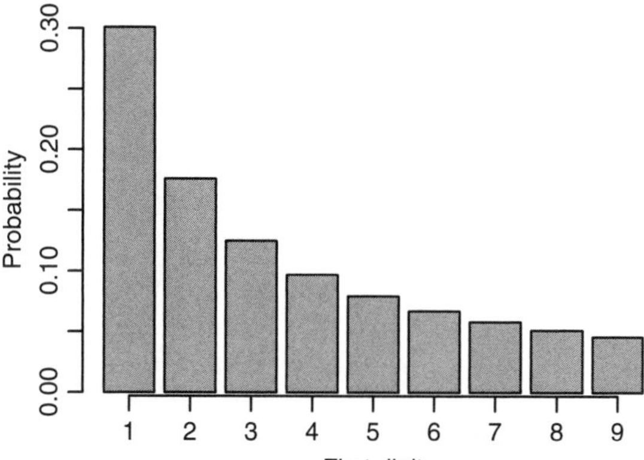

FIGURE I.2: Benford's law describes the frequencies of first digits for many real-life datasets.

Benford's law, also known as the "first-digit phenomenon," was discovered over 100 years ago, but it has generated new interest in the past 10 years. There are a huge number of datasets that exhibit Benford's law, including street addresses, populations of cities, stock prices, mathematical constants, birth rates, heights of mountains, and line items on tax returns. The last example, in particular, caught the eye of business Professor Mark Nigrini (2012), who showed that Benford's law can be used in forensic accounting and auditing as an indicator of fraud.

Durtschi et al. (2004) describe an investigation of a large medical center in the western United States. The distribution of first digits of check amounts differed significantly from Benford's law. A subsequent investigation uncovered that the financial officer had created bogus shell insurance companies in her own name and was writing large refund checks to those companies.

I.3 SEARCHING THE GENOME

Few areas of modern science employ probability more than biology and genetics. A strand of DNA, with its four nucleotide bases adenine, cytosine, guanine, and thymine, abbreviated by their first letters, presents itself as a sequence of outcomes of a four-sided die. The enormity of the data—about three billion "letters" per strand of human DNA—makes randomized methods relevant and viable.

Restriction sites are locations on the DNA that contain a specific sequence of nucleotides, such as G-A-A-T-T-C. Such sites are important to identify because they are locations where the DNA can be cut and studied. Finding all these locations is akin to finding patterns of heads and tails in a long sequence of coin tosses. Theoretical limit theorems for idealized sequences of coin tosses become practically relevant for exploring the genome. The locations for such restriction sites are well described by the Poisson process, a fundamental class of random processes that model locations

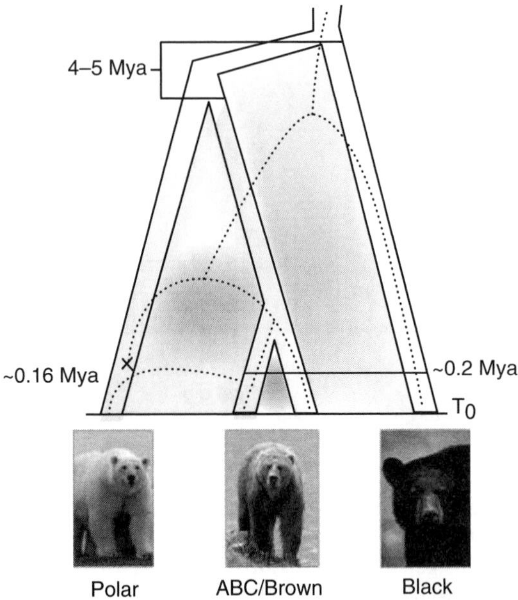

FIGURE I.3: Reconstructed evolutionary tree for bears. Polar bears may have evolved from brown bears four to five million years ago with occasional exchange of genes between the two species (shaded gray areas) fluctuating with key climactic events. Credit: Penn State University.

of restriction sites on a chromosome, as well as car accidents on the highway, service times at a fast food chain, and when you get your text messages.

On the macro level, random processes are used to study the evolution of DNA over time in order to construct evolutionary trees showing the divergence of species. DNA sequences change over time as a result of mutation and natural selection. Models for sequence evolution, called Markov processes, are continuous time analogues of the type of random walk models introduced earlier.

Miller et al. (2012) analyze the newly sequenced polar bear genome and give evidence that the size of the bear population fluctuated with key climactic events over the past million years, growing in periods of cooling and shrinking in periods of warming. Their paper, published in the *Proceedings of the National Academy of Sciences*, is all biology and genetics. But the appendix of supporting information is all probability and statistics (Fig. I.3).

I.4 BIG DATA

The search for the Higgs boson, the so-called "God particle," at the Large Hadron Collider in Geneva, Switzerland, generated 200 petabytes of data (1 petabyte = 10^{15} bytes). That is as much data as the total amount of printed material in the world! In physics, genomics, climate science, marketing, even online gaming and film, the

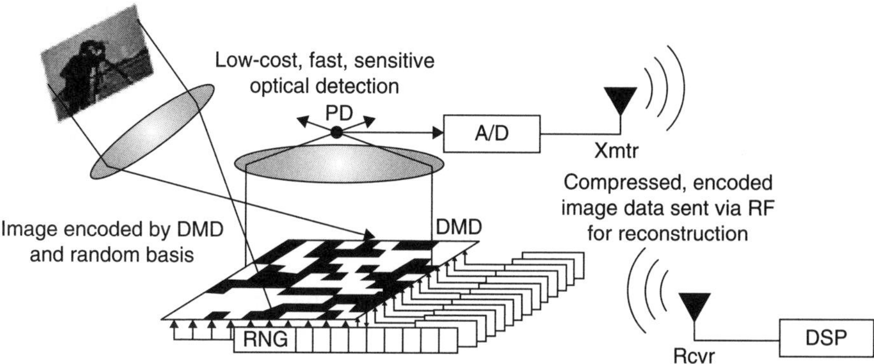

FIGURE I.4: Researchers at Rice University have developed a one-pixel camera based on compressed sensing that randomly alters where the light hitting the single pixel originates from within the camera's field of view as it builds the image. Image courtesy: Rice University.

sizes of datasets are staggering. How to store, transmit, visualize, and process such data is one the great challenges of science.

Probability is being used in a central way for such problems in a new methodology called *compressed sensing*.

In the average hospital, many terabytes (1 terabyte = 10^{12} bytes) of digital magnetic resonance imaging (MRI) data are generated each year. A half-hour MRI scan might collect 100 Mb of data. These data are then compressed to a smaller image, say 5 Mb, with little loss of clarity or detail. Medical and most natural images are compressible since lots of pixels have similar values. Compression algorithms work by essentially representing the image as a sum of simple functions (such as sine waves) and then discarding those terms that have low information content. This is a fundamental idea in signal processing, and essentially what is done when you take a picture on your cell phone and then convert it to a JPEG file for sending to a friend or downloading to the web.

Compressed sensing asks: If the data are ultimately compressible, is it really necessary to acquire all the data in the first place? Can just the final compressed data be what is initially gathered? And the startling answer is that by *randomly* sampling the object of interest, the final image can be reconstructed with similar results as if the object had been fully sampled. Random sampling of MRI scans produces an image of similar quality as when the entire object is scanned. The new technique has reduced MRI scan time to one-seventh the original time, from about half an hour to less than 5 minutes, and shows enormous promise for many other applied areas (Fig. I.4).

I.5 FROM APPLICATION TO THEORY

Having sung the praises of applications and case studies, we come back to the importance of theory.

Probability has been called the science of uncertainty. "Mathematical probability" may seem an oxymoron like jumbo shrimp or civil war. If any discipline can profess a claim of "certainty," surely it is mathematics with its adherence to rigorous proof and timeless results.

One of the great achievements of modern mathematics was putting the study of probability on a solid scientific foundation. This was done in the 1930s, when the Russian mathematician Andrey Nikolaevich Kolmogorov built up probability theory in a rigorous way similarly to how Euclid built up geometry. Much of his work is the material of a graduate-level course, but the basic framework of axiom, definition, theory, and proof sets the framework for the modern treatment of the subject.

One of the joys of learning probability is the compelling imagery we can exploit. Geometers draw circles and squares; probabilists toss coins and roll dice. There is no perfect circle in the physical universe. And the "fair coin" is an idealized model. Yet when you take *real* pennies and toss them repeatedly, the results conform so beautifully to the theory.

In this book, we use the computer package R. R is free software and an interactive computing environment available for download at `http://www.r-project.org/`. If you have never used R before, work through Appendix A: *Getting Started with R*.

Simulation plays a significant role in this book. Simulation is the use of random numbers to generate samples from a random experiment. Today it is a bedrock tool in the sciences and data analysis. Many problems that were for all practical purposes impossible to solve before the computer age are now easily handled with simulation.

There are many compelling reasons for including simulation in a probability course. Simulation helps build invaluable intuition for how random phenomena behave. It will also give you a flexible platform to test how changes in assumptions and parameters can affect outcomes. And the exercise of translating theoretical models into usable simulation code (easy to do in R) will make the subject more concrete and hopefully easier to understand.

And, most importantly, it is fun! Students enjoy the hands-on approach to the subject that simulation offers. This author still gets a thrill from seeing some complex theoretical calculation "magically" verified by a simulation.

To succeed in this subject, read carefully, work through the examples, and do as many problems as you can. But most of all, enjoy the ride!

> The results concerning fluctuations in coin tossing show that widely held beliefs ... are fallacious. They are so amazing and so at variance with common intuition that even sophisticated colleagues doubted that coins actually misbehave as theory predicts. The record of a simulated experiment is therefore included. . . .
>
> —William Feller, *An Introduction to Probability Theory and Its Applications*

1

FIRST PRINCIPLES

> First principles, Clarice. Read Marcus Aurelius. Of each particular thing ask: what is it in itself? What is its nature?
>
> —Hannibal Lecter, *Silence of the Lambs*

1.1 RANDOM EXPERIMENT, SAMPLE SPACE, EVENT

Probability begins with some activity, process, or experiment whose outcome is uncertain. This can be as simple as throwing dice or as complicated as tomorrow's weather.

Given such a "random experiment," the set of all possible outcomes is called the *sample space*. We will use the Greek capital letter Ω (omega) to represent the sample space.

Perhaps the quintessential random experiment is flipping a coin. Suppose a coin is tossed three times. Let H represent heads and T represent tails. The sample space is

$$\Omega = \{\text{HHH, HHT, HTH, HTT, THH, THT, TTH, TTT}\},$$

consisting of eight outcomes. The Greek lowercase omega ω will be used to denote these outcomes, the elements of Ω.

Probability: With Applications and R, First Edition. Robert P. Dobrow.

An *event* is a set of outcomes. The event of getting all heads in three coin tosses can be written as

$$A = \{\text{Three heads}\} = \{HHH\}.$$

The event of getting at least two tails is

$$B = \{\text{At least two tails}\} = \{\text{HTT, THT, TTH, TTT}\}.$$

We take probabilities of events. But before learning how to find probabilities, first learn to identify the sample space and relevant event for a given problem.

Example 1.1 The weather forecast for tomorrow says rain. The amount of rainfall can be considered a random experiment. If at most 24 inches of rain will fall, then the sample space is the interval $\Omega = [0, 24]$. The event that we get between 2 and 4 inches of rain is $A = [2, 4]$. ■

Example 1.2 Roll a pair of dice. Find the sample space and identify the event that the sum of the two dice is equal to 7.

The random experiment is rolling two dice. Keeping track of the roll of each die gives the sample space

$$\Omega = \{(1,1),(1,2),(1,3),(1,4),(1,5),(1,6),(2,1),(2,2),\ldots,(6,5),(6,6)\}.$$

The event is $A = \{\text{Sum is } 7\} = \{(1,6),(2,5),(3,4),(4,3),(5,2),(6,1)\}$. ■

Example 1.3 Yolanda and Zach are running for president of the student association. One thousand students will be voting. We will eventually ask questions like, What is the probability that Yolanda beats Zach by at least 100 votes? But before actually finding this probability, first identify (i) the sample space and (ii) the event that Yolanda beats Zach by at least 100 votes.

(i) The outcome of the vote can be denoted as $(x, 1000-x)$, where x is the number of votes for Yolanda, and $1000 - x$ is the number of votes for Zach. Then the sample space of all voting outcomes is

$$\Omega = \{(0, 1000), (1, 999), (2, 998), \ldots, (999, 1), (1000, 0)\}.$$

(ii) Let A be the event that Yolanda beats Zach by at least 100 votes. The event A consists of all outcomes in which $x - (1000 - x) \geq 100$, or $550 \leq x \leq 1000$. That is, $A = \{(550, 450), (551, 449), \ldots, (999, 1), (1000, 0)\}$. ■

Example 1.4 Joe will continue to flip a coin until heads appears. Identify the sample space and the event that it will take Joe at least three coin flips to get a head.

The sample space is the set of all sequences of coin flips with one head preceded by some number of tails. That is,

$$\Omega = \{\text{H, TH, TTH, TTTH, TTTTH, TTTTTH}, \ldots\}.$$

The desired event is $A = \{\text{TTH, TTTH, TTTTH}, \ldots\}$. Note that in this case both the sample space and the event A are infinite. ∎

1.2 WHAT IS A PROBABILITY?

What does it mean to say that *the probability that* A *occurs* is equal to x?

From a formal, purely mathematical point of view, a probability is a number between 0 and 1 that satisfies certain properties, which we will describe later. From a practical, empirical point of view, a probability matches up with our intuition of the likelihood or "chance" that an event occurs. An event that has probability 0 "never" happens. An event that has probability 1 is "certain" to happen. In repeated coin flips, a coin comes up heads about half the time, and the probability of heads is equal to one-half.

Let A be an event associated with some random experiment. One way to understand the probability of A is to perform the following thought exercise: imagine conducting the experiment over and over, infinitely often, keeping track of how often A occurs. Each experiment is called a *trial.* If the event A occurs when the experiment is performed, that is a *success.* The proportion of successes is the probability of A, written $P(A)$.

This is the *relative frequency* interpretation of probability, which says that the probability of an event is equal to its relative frequency in a large number of trials.

When the weather forecaster tells us that tomorrow there is a 20% chance of rain, we understand that to mean that if we could repeat today's conditions—the air pressure, temperature, wind speed, etc.—over and over again, then 20% of the resulting "tomorrows" will result in rain. Closer to what weather forecasters actually do in coming up with that 20% number, together with using satellite and radar information along with sophisticated computational models, is to go back in the historical record and find other days that match up closely with today's conditions and see what proportion of those days resulted in rain on the following day.

There are definite limitations to constructing a rigorous mathematical theory out of this intuitive and empirical view of probability. One cannot actually repeat an experiment infinitely many times. To define probability carefully, we need to take a formal, axiomatic, mathematical approach. Nevertheless, the relative frequency viewpoint will still be useful in order to gain intuitive understanding. And by the end of the book, we will actually derive the relative frequency viewpoint as a consequence of the mathematical theory.

1.3 PROBABILITY FUNCTION

We assume for the next several chapters that the sample space is *discrete*. This means that the sample space is either finite or countably infinite.

A set is *countably infinite* if the elements of the set can be arranged as a sequence. The natural numbers $1, 2, 3, \ldots$ is the classic example of a countably infinite set. And all countably infinite sets can be put in one-to-one correspondence with the natural numbers.

If the sample space is finite, it can be written as $\Omega = \{\omega_1, \ldots, \omega_k\}$. If the sample space is countably infinite, it can be written as $\Omega = \{\omega_1, \omega_2, \ldots\}$.

The set of all real numbers is an infinite set that is not countably infinite. It is called *uncountable*. An interval of real numbers, such as (0,1), the numbers between 0 and 1, is also uncountable. Probability on uncountable spaces will require differential and integral calculus, and will be discussed in the second half of this book.

A *probability function* assigns numbers between 0 and 1 to events according to three defining properties.

PROBABILITY FUNCTION

Definition 1.1. Given a random experiment with discrete sample space Ω, a *probability function* P is a function on Ω with the following properties:

1. $$P(\omega) \geq 0, \text{ for all } \omega \in \Omega$$

2. $$\sum_{\omega \in \Omega} P(\omega) = 1 \tag{1.1}$$

3. For all events $A \subseteq \Omega$,

$$P(A) = \sum_{\omega \in A} P(\omega). \tag{1.2}$$

You may not be familiar with some of the notations in this definition. The symbol $\in$ means "is an element of." So $\omega \in \Omega$ means ω is an element of Ω. We are also using a generalized Σ-notation in Equation 1.1 and Equation 1.2, writing a condition under the Σ to specify the summation. The notation $\sum_{\omega \in \Omega}$ means that the sum is over all ω that are elements of the sample space, that is, all outcomes in the sample space.

In the case of a finite sample space $\Omega = \{\omega_1, \ldots, \omega_k\}$, Equation 1.1 becomes

$$\sum_{\omega \in \Omega} P(\omega) = P(\omega_1) + \cdots + P(\omega_k) = 1.$$

And in the case of a countably infinite sample space $\Omega = \{\omega_1, \omega_2, \ldots\}$, this gives

$$\sum_{\omega \in \Omega} P(\omega) = P(\omega_1) + P(\omega_2) + \cdots = \sum_{i=1}^{\infty} P(\omega_i) = 1.$$

In simple language, probabilities sum to 1.

The third defining property of a probability function says that the probability of an event is the sum of the probabilities of all the outcomes contained in that event.

We might describe a probability function with a table, function, graph, or qualitative description.

Example 1.5 A type of candy comes in red, yellow, orange, green, and purple colors. Choose a candy at random. The sample space is $\Omega = \{\text{R, Y, O, G, P}\}$. Here are three equivalent ways of describing the probability function corresponding to equally likely outcomes:

1.

R	Y	O	G	P
0.20	0.20	0.20	0.20	0.20

2. $P(\omega) = 1/5$, for all $\omega \in \Omega$.
3. The five colors are equally likely. ■

In the discrete setting, we will often use *probability model* and *probability distribution* interchangeably with probability function. In all cases, to specify a probability function requires identifying (i) the outcomes of the sample space and (ii) the probabilities associated with those outcomes.

Letting H denote heads and T denote tails, an obvious model for a simple coin toss is

$$P(\text{H}) = P(\text{T}) = 0.50.$$

Actually, there is some extremely small, but nonzero, probability that a coin will land on its side. So perhaps a better model would be

$$P(\text{H}) = P(\text{T}) = 0.49999999995 \quad \text{and} \quad P(\text{Side}) = 0.0000000001.$$

Ignoring the possibility of the coin landing on its side, a more general model is

$$P(\text{H}) = p \quad \text{and} \quad P(\text{T}) = 1 - p,$$

where $0 \leq p \leq 1$. If $p = 1/2$, we say the coin is *fair*. If $p \neq 1/2$, we say that the coin is *biased*.

In a mathematical sense, all of these coin tossing models are "correct" in that they are consistent with the definition of what a probability is. However, we might debate which model most accurately reflects reality and which is most useful for modeling actual coin tosses.

Example 1.6 A college has six majors: biology, geology, physics, dance, art, and music. The proportion of students taking these majors are 20, 20, 5, 10, 10, and 35, respectively. Choose a random student. What is the probability they are a science major?

The random experiment is choosing a student. The sample space is

$$\Omega = \{\text{Bio, Geo, Phy, Dan, Art, Mus}\}.$$

The probability function is given in Table 1.1. The event in question is

$$A = \{\text{Science major}\} = \{\text{Bio, Geo, Phy}\}.$$

Finally,

$$\begin{aligned} P(A) &= P(\{\text{Bio, Geo, Phy}\}) = P(\text{Bio}) + P(\text{Geo}) + P(\text{Phy}) \\ &= 0.20 + 0.20 + 0.05 = 0.45. \end{aligned}$$

■

TABLE 1.1: Probabilities for majors.

Bio	Geo	Phy	Dan	Art	Mus
0.20	0.20	0.05	0.10	0.10	0.35

This example is probably fairly clear and may seem like a lot of work for a simple result. However, when starting out, it is good preparation for the more complicated problems to come to clearly identify the sample space, event and probability model before actually computing the final probability.

Example 1.7 In three coin tosses, what is the probability of getting at least two tails?

Although the probability model here is not explicitly stated, the simplest and most intuitive model for fair coin tosses is that every outcome is equally likely. Since the sample space

$$\Omega = \{\text{HHH, HHT, HTH, THH, HTT, THT, TTH, TTT}\}$$

has eight outcomes, the model assigns to each outcome the probability $1/8$.

The event of getting at least two tails can be written as $A = \{\text{HTT, THT, TTH, TTT}\}$. This gives

$$\begin{aligned} P(A) &= P(\{\text{HTT, THT, TTH, TTTT}\}) \\ &= P(\text{HTT}) + P(\text{THT}) + P(\text{TTH}) + P(\text{TTT}) \\ &= \frac{1}{8} + \frac{1}{8} + \frac{1}{8} + \frac{1}{8} = \frac{1}{2}. \end{aligned}$$ ∎

1.4 PROPERTIES OF PROBABILITIES

Events can be combined together to create new events using the connectives "or," "and," and "not." These correspond to the set operations union, intersection, and complement.

For sets $A, B \subseteq \Omega$, the *union* $A \cup B$ is the set of all elements of Ω that are in either A or B or both. The *intersection* AB is the set of all elements of Ω that are in both A and B. (Another common notation for the intersection of two events is $A \cap B$.) The *complement* A^c is the set of all elements of Ω that are not in A.

In probability word problems, descriptive phrases are typically used rather than set notation. See Table 1.2 for some equivalences.

TABLE 1.2: Events and sets.

Description	Set notation
Either A or B or both occur	$A \cup B$
A and B	AB
Not A	A^c
A implies B	$A \subseteq B$
A but not B	AB^c
Neither A nor B	A^cB^c
At least one of the two events occurs	$A \cup B$
At most one of the two events occurs	$(AB)^c = A^c \cup B^c$

A Venn diagram is a useful tool for working with events and subsets. A rectangular box denotes the sample space Ω, and circles are used to denote events. See Figure 1.1 for examples of Venn diagrams for the most common combined events obtained from two events A and B.

One of the most basic, and important, properties of a probability function is the simple addition rule for mutually exclusive events. We say that two events are *mutually exclusive,* or *disjoint,* if they have no outcomes in common. That is, A and B are mutually exclusive if $AB = \emptyset$, the empty set.

ADDITION RULE FOR MUTUALLY EXCLUSIVE EVENTS

If A and B are mutually exclusive events, then

$$P(A \text{ or } B) = P(A \cup B) = P(A) + P(B).$$

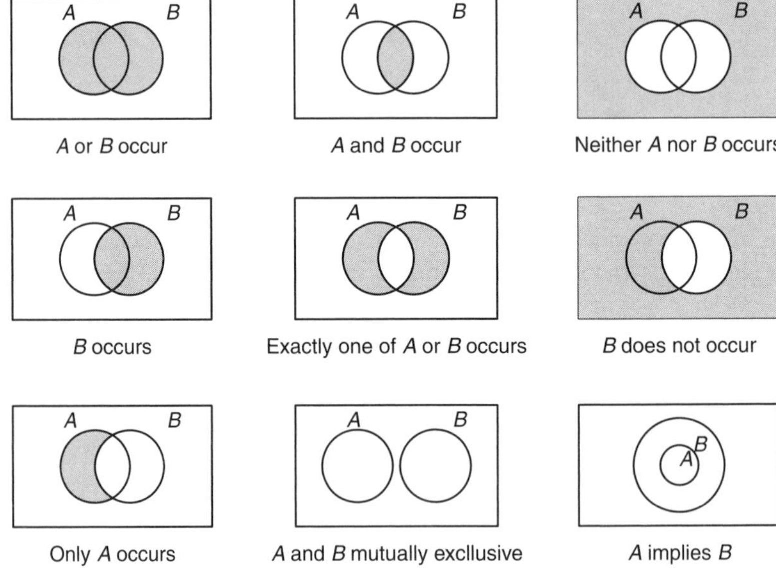

FIGURE 1.1: Venn diagrams.

The addition rule is a consequence of the third defining property of a probability function. We have that

$$\begin{aligned} P(A \text{ or } B) = P(A \cup B) &= \sum_{\omega \in A \cup B} P(\omega) \\ &= \sum_{\omega \in A} P(\omega) + \sum_{\omega \in B} P(\omega) \\ &= P(A) + P(B), \end{aligned}$$

where the third equality follows since the events are disjoint. The addition rule for mutually exclusive events extends to more than two events.

ADDITION RULE FOR MUTUALLY EXCLUSIVE EVENTS

Suppose $A_1, A_2, \ldots$ is a sequence of pairwise mutually exclusive events. That is, A_i and A_j are mutually exclusive for all $i \neq j$. Then

$$P(\text{at least one of the } A_i\text{'s occurs}) = P\left(\bigcup_{i=1}^{\infty} A_i\right) = \sum_{i=1}^{\infty} P(A_i).$$

Here we highlight other key properties that are consequences of the defining properties of a probability function and the addition rule for disjoint events.

PROPERTIES OF PROBABILITIES

1. If A implies B, that is, if $A \subseteq B$, then $P(A) \leq P(B)$.
2. $P(A \text{ does not occur}) = P(A^c) = 1 - P(A)$.
3. For all events A and B,

$$P(A \text{ or } B) = P(A \cup B) = P(A) + P(B) - P(AB). \tag{1.3}$$

Each property is derived next.

1. As $A \subseteq B$, write B as the disjoint union of A and BA^c. By the addition rule for disjoint events,

   $$P(B) = P(A \cup BA^c) = P(A) + P(BA^c) \geq P(A),$$

   since probabilities are nonnegative.
2. The sample space Ω can be written as the disjoint union of any event A and its complement A^c. Thus,

   $$1 = P(\Omega) = P(A \cup A^c) = P(A) + P(A^c).$$

   Rearranging gives the result.
3. Write $A \cup B$ as the disjoint union of A and A^cB. Also write B as the disjoint union of AB and A^cB. Then $P(B) = P(AB) + P(A^cB)$ and thus

   $$P(A \cup B) = P(A) + P(BA^c) = P(A) + P(B) - P(AB).$$

   Observe that the addition rule for mutually exclusive events follows from Property 3 since if A and B are disjoint, then $P(AB) = 0$.

Example 1.8 In a city, 75% of the population have brown hair, 40% have brown eyes, and 25% have both brown hair and brown eyes. A person is chosen at random. What is the probability that they

1. have brown eyes or brown hair?
2. have neither brown eyes nor brown hair?

To gain intuition, draw a Venn diagram, as in Figure 1.2. Let H be the event of having brown hair; let E denote brown eyes.

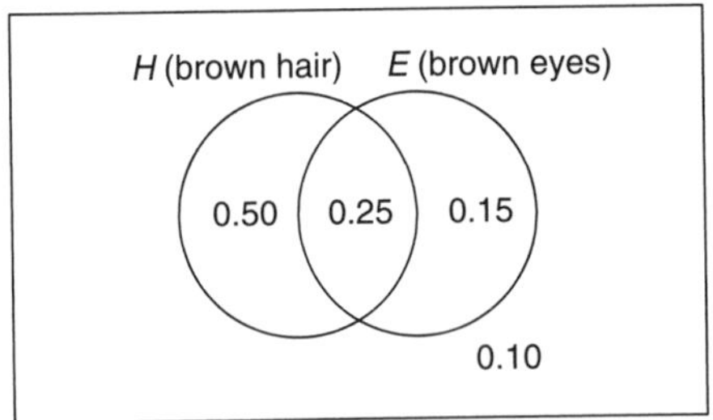

FIGURE 1.2: Venn diagram.

1. The probability of having brown eyes or brown hair is

$$P(E \text{ or } H) = P(E) + P(H) - P(EH) = 0.75 + 0.40 - 0.25 = 0.90.$$

   Notice that E and H are not mutually exclusive. If we made a mistake and used the simple addition rule $P(E \text{ or } H) = P(E) + P(H)$, we would mistakenly get $0.70 + 0.40 = 1.10 > 1$.
2. The complement of having neither brown eyes nor brown hair is having brown eyes or brown hair. Thus

$$P(E^c H^c) = P((E \text{ or } H)^c) = 1 - P(E \text{ or } H) = 1 - 0.90 = 0.10. \quad \blacksquare$$

1.5 EQUALLY LIKELY OUTCOMES

The simplest probability model for a finite sample space is that all outcomes are equally likely. If Ω has k elements, then the probability of each outcome is $1/k$, since probabilities sum to 1. That is, $P(\omega) = 1/k$, for all $\omega \in \Omega$.

Computing probabilities for equally likely outcomes takes a fairly simple form. Suppose A is an event with s elements, with $s \leq k$. Since $P(A)$ is the sum of the probabilities of all the outcomes contained in A,

$$P(A) = \sum_{\omega \in A} P(\omega) = \sum_{\omega \in A} \frac{1}{k} = \frac{s}{k} = \frac{\text{Number of elements of } A}{\text{Number of elements of } \Omega}.$$

Probability with equally likely outcomes reduces to *counting*.

Example 1.9 A *palindrome* is a word that reads the same forward or backward. Examples include mom, civic, and rotator. Pick a three-letter "word" at random choosing from D, O, or G for each letter. What is the probability that the resulting word is a palindrome?

There are 27 possible words (three possibilities for each letter). List them and count the palindromes: DDD, OOO, GGG, DOD, DGD, ODO, OGO, GDG, and GGG. The probability of getting a palindrome is $9/27 = 1/3$. ■

Example 1.10 A bowl has r red balls and b blue balls. What is the probability of selecting a red ball?

The sample space consists of $r+b$ balls. The event $A = \{\text{Red ball}\}$ has r elements. Therefore, $P(A) = r/(r+b)$. ■

Example 1.11 A field has 36 trees planted in six equally spaced rows and columns as in Figure 1.3. A tree is chosen at random so that each tree is equally likely to be chosen. What is the probability that the tree that is picked is within the wedge at the lower left-hand corner of the field?

There are eight points in the wedge, so the desired probability is $8/36 = 2/9$. ■

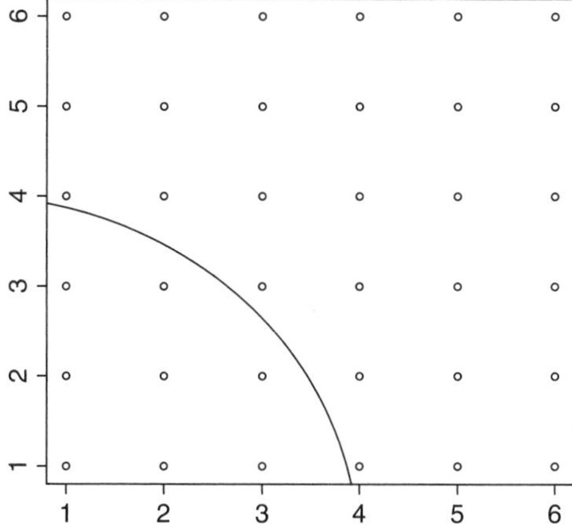

FIGURE 1.3: Trees in a field.

A model for equally likely outcomes assumes a finite sample space. Interestingly, it is impossible to have a probability model of equally likely outcomes on an infinite sample space. To see why, suppose $\Omega = \{\omega_1, \omega_2, \ldots\}$ and $P(\omega_i) = c$ for all i, where c is a nonzero constant. Then summing the probabilities gives

$$\sum_{i=1}^{\infty} P(\omega_i) = \sum_{i=1}^{\infty} c = \infty \neq 1.$$

While equally likely outcomes are not possible in the infinite case, there are many ways to assign probabilities for an infinite sample space where outcomes are not equally likely. For instance, let $\Omega = \{\omega_1, \omega_2, \ldots\}$ with $P(\omega_i) = (1/2)^i$, for $i = 1, 2, \ldots$ Then

$$\sum_{i=1}^{\infty} P(\omega_i) = \sum_{i=1}^{\infty} \left(\frac{1}{2}\right)^i = \left(\frac{1}{2}\right) \frac{1}{1-(1/2)} = 1.$$

Since counting plays a fundamental role in probability when outcomes are equally likely, we introduce some basic counting principles.

1.6 COUNTING I

Counting sets is sometimes not as easy as 1, 2, 3. . . . But a basic counting principle known as the *multiplication principle* allows for tackling a wide range of problems.

> **MULTIPLICATION PRINCIPLE**
>
> If there are m ways for one thing to happen, and n ways for a second thing to happen, there are $m \times n$ ways for both things to happen.
>
> More generally—and more formally—consider an n-element sequence $(a_1, a_2, \ldots, a_n)$. If there are k_1 possible values for the first element, k_2 possible values for the second element, ..., and k_n possible values for the nth element, there are $k_1 \times k_2 \times \cdots \times k_n$ possible sequences.

For instance, in tossing a coin three times, there are $2 \times 2 \times 2 = 2^3 = 8$ possible outcomes. Rolling a die four times gives $6 \times 6 \times 6 \times 6 = 6^4 = 1296$ possible rolls.

Example 1.12 License plates in Minnesota are issued with three letters from A to Z followed by three digits from 0 to 9. If each license plate is equally likely, what is the probability that a random license plate starts with G-Z-N?

The solution will be equal to the number of license plates that start with G-Z-N divided by the total number of license plates. By the multiplication principle, there are $26 \times 26 \times 26 \times 10 \times 10 \times 10 =$ 17,576,000 possible license plates.

For the number of plates that start with G-Z-N, think of a 6-element plate of the form G-Z-N-_-_-_. For the three blanks, there are $10 \times 10 \times 10$ possibilities. Thus the desired probability is $10^3/(26^3 \times 10^3) = 1/26^3 = 0.0000569$.

Aside: The author has the sense that everywhere he goes he sees license plates with the same initial three letters as his own car. This can be explained either by an interesting psychological phenomenon involving coincidences or by the fact that license plates in Minnesota are not, in fact, equally likely. ■

Example 1.13 A DNA strand is a long polymer string made up of four nucleotides—adenine, cytosine, guanine, and thymine. It can be thought of as a sequence of As, Cs, Gs, and Ts. DNA is structured as a double helix with two paired strands running in opposite directions on the chromosome. Nucleotides always pair the same way: A with T and C with G. A *palindromic sequence* is equal to its "reverse complement." For instance, the sequences ACGT and GTTAGCTAAC are palindromic sequences, but ACCA is not. Such sequences play a significant role in molecular biology.

Suppose the nucleotides on a DNA strand of length six are generated in such a way so that all strands are equally likely. What is the probability that the DNA sequence is a palindromic sequence?

By the multiplication principle, the number of DNA strands is 4^6 since there are four possibilities for each site. A palindromic sequence of length six is completely determined by the first three sites. There are 4^3 palindromic sequences. The desired probability is $4^3/4^6 = 1/64$. ■

Example 1.14 Mark is taking four final exams next week. His studying was erratic and all scores A, B, C, D, and F are equally likely for each exam. What is the probability that Mark will get at least one A?

Take complements. The complementary event of getting at least one A is getting no As. Since outcomes are equally likely, by the multiplication principle there are 4^4 exam outcomes with no As (four grade choices for each of four exams). And there are 5^4 possible outcomes in all. The desired probability is $1 - 4^4/5^4 = 0.5904$. ■

Given a set of distinct objects, a *permutation* is an ordering of the elements of the set. For the set $\{a, b, c\}$, there are six permutations:

$$(a, b, c), (a, c, b), (b, a, c), (b, c, a), (c, a, b), \text{ and } (c, b, a).$$

How many permutations are there of an n-element set? There are n possibilities for the first element of the permutation, $n - 1$ for the second, and so on. The result follows by the multiplication principle.

COUNTING PERMUTATIONS

There are $n \times (n-1) \times \cdots \times 1 = n!$ permutations of an n-element set.

The factorial function $n!$ grows very large very fast. In a classroom of 10 people with 10 chairs, there are $10! = 3{,}628{,}800$ ways to seat the students. There are $52! \approx 8 \times 10^{67}$ orderings of a standard deck of cards, which is "almost" as big as the number of atoms in the observable universe, which is estimated to be about 10^{80}.

Functions of the form c^n, where c is a constant, are said to exhibit *exponential* growth. The factorial function $n!$ grows like n^n, which is sometimes called *super-exponential* growth.

Example 1.15 Bob has three bookshelves in his office and 15 books—5 are math books and 10 are novels. If each shelf holds exactly five books and books are placed randomly on the shelves (all orderings are equally likely), what is the probability that the bottom shelf contains all the math books?

There are $15!$ ways to permute all the books on the shelves. There are $5!$ ways to put the math books on the bottom shelf and $10!$ ways to put the remaining novels on the other two shelves. Thus by the multiplication principle, the desired probability is $(5!10!)/15! = 1/3003 = 0.000333$. ■

Example 1.16 A bag contains six Scrabble tiles with the letters A-D-M-N-O-R. You reach into the bag and take out tiles one at a time. What is the probability that you will spell the word R-A-N-D-O-M?

How many possible words can be formed? All the letters are distinct and a "word" is a permutation of the set of six letters. There are $6! = 720$ possible words. Only one of them spells R-A-N-D-O-M, so the desired probability is $1/720 = 0.001389$. ■

Example 1.17 Scrabble continued. Change the previous problem. After you pick a tile from the bag, write down that letter and then return the tile to the bag. So every time you reach into the bag it contains the six original letters.

Now there are $6 \times \cdots \times 6 = 6^6 = 46,656$ possible words, and the desired probability is $1/46,656 = 0.0000214$. ■

SAMPLING WITH AND WITHOUT REPLACEMENT

The last examples highlight two different sampling methods called *sampling without replacement* and *sampling with replacement*. When sampling with replacement, a unit that is selected from a population is returned to the population before another unit is selected. When sampling without replacement, the unit is not returned to the population after being selected. When solving a probability problem involving sampling (such as selecting cards or picking balls from urns), make sure you know the sampling method before computing the related probability.

Example 1.18 When national polling organizations conduct nationwide surveys, they often select about 1000 people sampling without replacement. If N is the number of people in a target population, then by the multiplication principle there are $N \times (N-1) \times (N-2) \times \cdots \times (N-999)$ possible ordered samples. For national polls in the United States, where N, the number of people age 18 or over, is about 250 million, that gives about $(250{,}000{,}000)^{1000}$ possible ordered samples, which is a mind-boggling 2.5 with 8000 zeros after it. ■

1.7 PROBLEM-SOLVING STRATEGIES: COMPLEMENTS, INCLUSION–EXCLUSION

Consider a sequence of events $A_1, A_2, \ldots$. In this section, we consider strategies to find the probability that *at least* one of the events occurs, which is the probability of the union $\cup_i A_i$.

Sometimes the complement of an event can be easier to work with than the event itself. The complement of the event that at least one of the A_is occurs is the event that none of the A_is occur, which is the intersection $\cap_i A_i^c$.

Check with a Venn diagram (and if you are comfortable working with sets prove it yourself) that

$$(A \cup B)^c = A^c B^c \quad \text{and} \quad (AB)^c = A^c \cup B^c.$$

Complements turn unions into intersections, and vice versa. These set-theoretic results are known as DeMorgan's laws. The results extend to infinite sequences. Given events $A_1, A_2, \ldots$,

$$\left(\bigcup_{i=1}^{\infty} A_i \right)^c = \bigcap_{i=1}^{\infty} A_1^c \quad \text{and} \quad \left(\bigcap_{i=1}^{\infty} A_i \right)^c = \bigcup_{i=1}^{\infty} A_i^c.$$

Example 1.19 Four dice are rolled. Find the probabilty of getting at least one 6.

The sample space is the set of all outcomes of four dice rolls

$$\Omega = \{(1,1,1,1), (1,1,1,2), \ldots, (6,6,6,6)\}.$$

By the multiplication principle, there are $6^4 = 1296$ elements. If the dice are fair, each of these outcomes is equally likely. It is not obvious, without some new tools, how to count the number of outcomes that have at least one 6.

Let A be the event of getting at least one 6. Then the complement A^c is the event of getting no sixes in four rolls. An outcome has no sixes if the dice rolls a 1, 2, 3, 4, or 5 on every roll. By the multiplication principle, there are $5^4 = 625$ possibilities. Thus $P(A^c) = 5^4/6^4 = 625/1296$ and

$$P(A) = 1 - P(A^c) = 1 - \frac{625}{1296} = 0.5177. \quad \blacksquare$$

Recall the formula in Equation 1.3 for the probability of a union of two events. We generalize for three or more events using the principle of inclusion–exclusion.

Proposition 1.2. For events A, B, and C,

$$P(A \cup B \cup C) = P(A) + P(B) + P(C) - P(AB) - P(AC) - P(BC) + P(ABC).$$

Since we first *include* the sets, then *exclude* the pairwise intersections, then *include* the triple intersection, this is called the *inclusion–exclusion principle.* The proof is intuitive with the help of a Venn diagram. Write

$$A \cup B \cup C = \left[A \cup B\right] \cup \left[C(AC \cup BC)^c\right].$$

The bracketed sets $A \cup B$ and $C(AC \cup BC)^c$ are disjoint. Thus,

$$\begin{aligned} P(A \cup B \cup C) &= P(A \cup B) + P\left(C(AC \cup BC)^c\right) \\ &= P(A) + P(B) - P(AB) + P\left(C(AC \cup BC)^c\right). \end{aligned} \qquad (1.4)$$

Write C as the disjoint union

$$C = [C(AC \cup BC)] \cup [C(AC \cup BC)^c] = [AC \cup BC] \cup [C(AC \cup BC)^c].$$

This gives

$$P(C(AC \cup BC)^c) = P(C) - P(AC \cup BC).$$

Together with Equation 1.4,

$$\begin{aligned} P(A \cup B \cup C) = & \, P(A) + P(B) - P(AB) \\ & + P(C) - [P(AC) + P(BC) - P(ABC)]. \end{aligned}$$

Extending further to more than three events gives the general principle of inclusion–exclusion. We will not prove it, but if you know how to use mathematical induction, give it a try.

INCLUSION–EXCLUSION

Given events $A_1, \ldots, A_n$, the probability that at least one event occurs is

$$\begin{aligned} P(A_1 \cup \cdots \cup A_n) = & \sum_i P(A_i) - \sum_{i<j} P(A_i A_j) \\ & + \sum_{i<j<k} P(A_i A_j A_k) - \cdots + (-1)^{n+1} P(A_1 \cdots A_n). \end{aligned}$$

■ **Example 1.20** An integer is drawn uniformly at random from $\{1, \ldots, 1000\}$ such that each number is equally likely. What is the probability that the number drawn is divisible by 3, 5, or 6?

Let D_3, D_5, and D_6 denote the events that the number drawn is divisible by 3, 5, and 6, respectively. The problem asks for $P(D_3 \cup D_5 \cup D_6)$. By inclusion–exclusion,

$$\begin{aligned} P(D_3 \cup D_5 \cup D_6) = & \, P(D_3) + P(D_5) + P(D_6) - P(D_3 D_5) - P(D_3 D_6) \\ & - P(D_5 D_6) + P(D_3 D_5 D_6). \end{aligned}$$

Let $\lfloor x \rfloor$ denote the integer part of x. There are $\lfloor 1000/x \rfloor$ numbers from 1 to 1000 that are divisible by x. Since all selections are equally likely,

$$P(D_3) = \lfloor 1000/3 \rfloor / 1000 = 0.333.$$

$$P(D_5) = \lfloor 1000/5 \rfloor / 1000 = 0.20.$$

$$P(D_6) = \lfloor 1000/6 \rfloor / 1000 = 0.166.$$

A number is divisible by 3 and 5 if and only if it is divisible by 15. Thus, $D_3 D_5 = D_{15}$. If a number is divisible by 6, it is also divisible by 3, so $D_3 D_6 = D_6$. Also, $D_5 D_6 = D_{30}$. And $D_3 D_5 D_6 = D_{30}$. This gives

$$P(D_3 D_5) = \lfloor 1000/15 \rfloor / 1000 = 0.066.$$

$$P(D_3 D_6) = 0.166.$$

$$P(D_5 D_6) = \lfloor 1000/30 \rfloor / 1000 = 0.033.$$

$$P(D_3 D_5 D_6) = 0.033.$$

Putting it all together gives

$$P(D_3 \cup D_5 \cup D_6)$$

$$= 0.333 + 0.2 + 0.166 - 0.066 - 0.166 - 0.033 + 0.033 = 0.467. \quad \blacksquare$$

We have presented two different ways of computing the probability that at least one of several events occurs: (i) a "back-door" approach of taking complements and working with the resulting "and" probabilities; and (ii) a direct "frontal-attack" by inclusion-exclusion. Here is a third way, which illustrates decomposing an event into a union of mutually exclusive subsets.

Example 1.21 Consider a random experiment that has k equally likely outcomes, one of which we call *success*. Repeat the experiment n times. Let A be the event that at least one of the n outcomes is a success. For instance, consider rolling a die 10 times, where success means rolling a three. Here $n = 10$, $k = 6$, and A is the event of rolling at least one 3.

Define a sequence of events $A_1, \ldots, A_n$, where A_i is the event that the ith trial is a success. Then $A = A_1 \cup \cdots \cup A_n$ and $P(A) = P(A_1 \cup \cdots \cup A_n)$. We cannot use the addition rule on this probability since the A_is are not mutually exclusive.

To define a sequence of mutually exclusive events, let B_i be the event that the *first* success occurs on the ith trial. Then the B_is are mutually exclusive. Furthermore,

$$B_1 \cup \cdots \cup B_n = A_1 \cup \cdots \cup A_n = A.$$

Thus,

$$P(A) = P(B_1 \cup \cdots \cup B_n) = P(B_1) + \cdots + P(B_n).$$

To find $P(B_i)$, observe that if the first success occurs on the ith trial, then the first $i-1$ trials are necessarily not successes and the ith trial is a success. There are $k-1$ possible outcomes for each of the first $i-1$ trials, one outcome for the ith trial, and k possible outcomes for each of the remaining $n-i$ trials. By the multiplication principle, there are $(k-1)^{i-1}k^{n-i}$ outcomes where the first success occurs on the ith trial. And there are k^n possible outcomes in all. Thus,

$$P(B_i) = \frac{(k-1)^{i-1}k^{n-i}}{k^n} = \frac{1}{k}\left(\frac{k-1}{k}\right)^{i-1} = \frac{1}{k}\left(1-\frac{1}{k}\right)^{i-1},$$

for $i = 1, \ldots, n$. The desired probability is

$$\begin{aligned} P(A) = P(B_1) + \cdots + P(B_n) &= \sum_{i=1}^{n} \frac{1}{k}\left(1-\frac{1}{k}\right)^{i-1} \\ &= \frac{1}{k}\left(\frac{1-(1-1/k)^n}{1-(1-1/k)}\right) \\ &= 1-\left(1-\frac{1}{k}\right)^n. \end{aligned}$$

For instance, the probability of rolling at least one 3 in 10 rolls of a die is

$$1-\left(1-\frac{1}{6}\right)^{10} = 1-\left(\frac{5}{6}\right)^{10} = 0.8385.$$

■

1.8 RANDOM VARIABLES

Often the outcomes of a random experiment take on numerical values. For instance, we might be interested in how many heads occur in three coin tosses. Let X be the number of heads. Then X is equal to 0, 1, 2, or 3, depending on the outcome of the coin tosses. The object X is called a *random variable*. The possible *values* of X are 0, 1, 2, and 3.

RANDOM VARIABLE

A random variable assigns numerical values to the outcomes of a random experiment.

Random variables are enormously useful and allow us to use algebraic expressions, equalities, and inequalities when manipulating events. In many of the previous examples, we have been working with random variables without using the name, for example, the number of threes in rolls of a die, the number of votes received, the number of palindromes, the number of heads in repeated coin tosses.

Example 1.22 In tossing three coins, let X be the number of heads. Then the event of getting two heads can be written as $\{X = 2\}$. The probability of getting two heads is thus

$$P(X = 2) = P(\{\text{HHT, HTH, THH}\}) = \frac{3}{8}. \qquad \blacksquare$$

Example 1.23 If we throw two dice, what is the probability that the sum of the dice is greater than four?

We can, of course, find the probability by direct counting. But we will use random variables. Let Y be the sum of two dice rolls. Then Y is a random variable whose possible values are $2, 3, \ldots, 12$. The event that the sum is greater than 4 can be written as $\{Y > 4\}$. Observe that the complementary event is $\{Y \leq 3\}$. By taking complements,

$$\begin{aligned} P(Y > 4) &= 1 - P(Y \leq 3) \\ &= P(Y = 2 \text{ or } Y = 3) = P(Y = 2) + P(Y = 3) \\ &= P(\{(1,1)\}) + P(\{(1,2),(2,1)\}) = \frac{1}{36} + \frac{2}{36} = \frac{1}{12}. \end{aligned} \qquad \blacksquare$$

Example 1.24 Recall Example 1.3. One thousand students are voting. Suppose the number of votes that Yolanda receives is equally likely to be any number from 0 to 1000. What is the probability that Yolanda beats Zach by at least 100 votes?

We approach the problem using random variables. Let Y be the number of votes for Yolanda. Let Z be the number of votes for Zach. Then the total number of votes is $Y + Z = 1000$. Thus, $Z = 1000 - Y$. The event that Yolanda beats Zach by at least 100 votes is $\{Y - Z \geq 100\} = \{Y - (1000 - Y) \geq 100\} = \{2Y \geq 1100\} = \{Y \geq 550\}$. The desired probability is

$$P(Y - Z \geq 100) = P(Y \geq 550) = 451/1001,$$

since there are 1001 possible votes for Yolanda and 451 of them are greater than 550.

$\blacksquare$

If a random variable X takes values in a finite set all of whose elements are equally likely, we say that X is *uniformly distributed* on that set.

UNIFORM RANDOM VARIABLE

Let $S = \{s_1, \ldots, s_k\}$ be a finite set. A random variable X is *uniformly distributed on* S if

$$P(X = s_i) = \frac{1}{k}, \quad \text{for } i = 1, \ldots, k.$$

We write $X \sim \text{Unif}(S)$.

Example 1.25 Rachel picks an integer "at random" between 1 and 50. (i) Find the probability that she picks 13. (ii) Find the probability that her number is between 10 and 20. (iii) Find the probability that her number is prime.

Let X be Rachel's number. Then X is uniformly distributed on $\{1, \ldots, 50\}$.

(i) The probability that Rachel picks 13 is $P(X = 13) = 1/50$.

(ii) There are 11 numbers between 10 and 20 (including 10 and 20). The desired probability is

$$P(10 \leq X \leq 20) = \frac{11}{50} = 0.22.$$

(iii) One counts 15 prime numbers between 1 and 50. Thus,

$$P(X \text{ is prime}) = \frac{15}{50} = 0.3. \quad \blacksquare$$

We write $\{X = 2\}$ for the event that the random variable takes the value 2. More generally, we write $\{X = x\}$ for the event that the random variable X takes the value x, where x is a specific number. The difference between the uppercase X (a random variable) and the lowercase x (a number) can be confusing but is extremely important to clarify.

Example 1.26 Roll a pair of dice. What is the probability of getting a 4? Of getting each number between 2 and 12?

Assuming the dice fair, each die number is equally likely. There are six possibilities for the first roll, six possibilities for the second roll, so $6 \times 6 = 36$ possible rolls. We thus assign the probability of 1/36 to each dice pair. Let X be the sum of the two dice. Then

$$\begin{aligned} P(X = 4) &= P(\{(1,3), (3,1), (2,2)\}) \\ &= P((1,3)) + P((3,1)) + P((2,2)) = 3\left(\frac{1}{36}\right) = \frac{1}{12}. \end{aligned}$$

Consider $P(X = x)$ for $x = 2, 3, \ldots, 12$. By counting all the possible combinations, verify the probabilities in Table 1.3.

TABLE 1.3: Probability distribution for the sum of two dice.

x	2	3	4	5	6	7	8	9	10	11	12
$P(X = x)$	$\frac{1}{36}$	$\frac{2}{36}$	$\frac{3}{36}$	$\frac{4}{36}$	$\frac{5}{36}$	$\frac{6}{36}$	$\frac{5}{36}$	$\frac{4}{36}$	$\frac{3}{36}$	$\frac{2}{36}$	$\frac{1}{36}$

Observe that while the outcomes of each individual die are equally likely, the values of the *sum* of two dice are not. ■

1.9 A CLOSER LOOK AT RANDOM VARIABLES

> While writing my book [Stochastic Processes] I had an argument with Feller. He asserted that everyone said "random variable" and I asserted that everyone said "chance variable." We obviously had to use the same name in our books, so we decided the issue by a [random] procedure. That is, we tossed for it and he won.
>
> —Joe Doob, quoted in *Statistical Science*

Random variables are central objects in probability. The name can be confusing since they are really neither "random" nor a "variable" in the way that that word is used in algebra or calculus. A random variable is actually a *function*, a function whose domain is the sample space.

A random variable assigns every outcome of the sample space a real number. Consider the three coins example, letting X be the number of heads in three coin tosses. Depending upon the outcome of the experiment, X takes on different values. To emphasize the dependency of X on the outcome ω, we can write $X(\omega)$, rather than just X. In particular,

$$X(\omega) = \begin{cases} 0, & \text{if } \omega = \text{TTT} \\ 1, & \text{if } \omega = \text{HTT, THT, or TTH} \\ 2, & \text{if } \omega = \text{HHT, HTH, or THH} \\ 3, & \text{if } \omega = \text{HHH.} \end{cases}$$

The probability of getting exactly two heads is written as $P(X = 2)$, which is shorthand for $P(\{\omega : X(\omega) = 2\})$.

You may be unfamiliar with this last notation, used for describing sets. The notation $\{\omega : \text{Property}\}$ describes the set of all ω that satisfies some property. So $\{\omega : X(\omega) = 2\}$ is the set of all ω with the property that $X(\omega) = 2$. That is, the set of all outcomes that result in exactly two heads, which is $\{HHT, HTH, THH\}$.

Similarly, the probability of getting at most one head in three coin tosses is

$$P(X \leq 1) = P(\{\omega : X(\omega) \leq 1\}) = P(\{\text{TTT, HTT, THT, TTH}\}).$$

Because of simplicity and ease of notation, authors typically use the shorthand X in writing random variables instead of the more verbose $X(\omega)$.

1.10 A FIRST LOOK AT SIMULATION

Using random numbers on a computer to simulate probabilities is called the Monte Carlo method. Today, Monte Carlo tools are used extensively in statistics, physics, engineering, and across many disciplines. The name was coined in the 1940s by mathematicians John von Neumann and Stanislaw Ulam working on the Manhattan Project. It was named after the famous Monte Carlo casino in Monaco.

Ulam's description of his inspiration to use random numbers to simulate complicated problems in physics is quoted in Eckhardt (1987):

> The first thoughts and attempts I made to practice [the Monte Carlo method] were suggested by a question which occurred to me in 1946 as I was convalescing from an illness and playing solitaires.
>
> The question was what are the chances that a Canfield solitaire laid out with 52 cards will come out successfully? After spending a lot of time trying to estimate them by pure combinatorial calculations, I wondered whether a more practical method than "abstract thinking" might not be to lay it out say one hundred times and simply observe and count the number of successful plays. This was already possible to envisage with the beginning of the new era of fast computers, and I immediately thought of problems of neutron diffusion and other questions of mathematical physics, and more generally how to change processes described by certain differential equations into an equivalent form interpretable as a succession of random operations. Later [in 1946], I described the idea to John von Neumann, and we began to plan actual calculations.

The Monte Carlo simulation approach is based on the relative frequency model for probabilities. Given a random experiment and some event A, the probability $P(A)$ is estimated by repeating the random experiment many times and computing the proportion of times that A occurs.

More formally, define a sequence $X_1, X_2, \ldots,$ where

$$X_k = \begin{cases} 1, & \text{if } A \text{ occurs on the } k\text{th trial} \\ 0, & \text{if } A \text{ does not occur on the } k\text{th trial,} \end{cases}$$

for $k = 1, 2, \ldots$. Then

$$\frac{X_1 + \cdots + X_n}{n}$$

is the proportion of times in which A occurs in n trials. For large n, the Monte Carlo method estimates $P(A)$ by

$$P(A) \approx \frac{X_1 + \cdots + X_n}{n}. \tag{1.5}$$

MONTE CARLO SIMULATION

Implementing a Monte Carlo simulation of $P(A)$ requires three steps:

1. **Simulate a trial.** Model, or translate, the random experiment using random numbers on the computer. One iteration of the experiment is called a "trial."
2. **Determine success.** Based on the outcome of the trial, determine whether or not the event A occurrs. If yes, call that a "success."
3. **Replication.** Repeat the aforementioned two steps many times. The proportion of successful trials is the simulated estimate of $P(A)$.

Monte Carlo simulation is intuitive and matches up with our sense of how probabilities "should" behave. We give a theoretical justification for the method and Equation 1.5 in Chapter 9, where we study limit theorems and the law of large numbers.

Here is a most simple, even trivial, starting example. Consider simulating the probability that an ideal fair coin comes up heads. One could do a *physical* simulation by just flipping a coin many times and taking the proportion of heads to estimate $P(\text{Heads})$.

Using a computer, choose the number of trials n (the larger the better) and type the R command

```
> sample(0:1,n,replace=T)
```

The command samples with replacement from $\{0, 1\}$ n times such that outcomes are equally likely. Let 0 represent tails and 1 represent heads. The output is a sequence of n ones and zeros corresponding to heads and tails. The average, or mean, of the list is precisely the proportion of ones. To simulate $P(\text{Heads})$, type

```
> mean(sample(0:1,n,replace=T))
```

Repeat the command several times (use the up arrow key). These give repeated Monte Carlo estimates of the desired probability. Observe the accuracy in the estimate with one million trials:

```
> mean(sample(0:1,1000000,replace=T))
[1] 0.500376
> mean(sample(0:1,1000000,replace=T))
[1] 0.499869
> mean(sample(0:1,1000000,replace=T))
```

```
[1] 0.498946
> mean(sample(0:1,1000000,replace=T))
[1] 0.500115
```

The R script **CoinFlip.R** simulates a familiar probability—the probability of getting three heads in three coin tosses.

R: SIMULATING THE PROBABILITY OF THREE HEADS IN THREE COIN TOSSES

```
# CoinFlip.R

# Trial
> trial <- sample(0:1,3,replace=TRUE)

# Success
> if (sum(trial)==3) 1 else 0

# Replication
> n <- 10000   # Number of repetitions
> simlist <- numeric(n) # Initialize vector
> for (i in 1:n) {
    trial <- sample(0:1, 3, replace=TRUE)
    success <- if (sum(trial)==3) 1 else 0
    simlist[i] <- success }
> mean(simlist) # Proportion of trials with 3 heads
[1] 0.1231
```

The script is divided into three parts to illustrate (i) coding the trial, (ii) determining success, and (iii) implementing the replication.

To simulate three coin flips, use the `sample` command. Again letting 1 represent heads and 0 represent tails, the command

```
> trial <- sample(0:1, 3, replace=TRUE)
```

chooses a head or tails three times. The three results are stored as a three-element list (called a vector in R) in the variable `trial`.

After flipping three coins, the routine must decide whether or not they are all heads. This is done by summing the outcomes. The sum will equal three if and only if all flips are heads. This is checked with the command

```
> if (sum(trial)==3) 1 else 0
```

which returns a 1 for success, and 0, otherwise.

For the actual simulation, the commands are repeated n times in a loop. The output from each trial is stored in the vector `simlist`. This vector will consist of n ones and zeros corresponding to success or failure for each trial, where success is flipping three heads.

Finally, after repeating n trials, we find the proportion of successes in all the trials, which is the proportion of ones in `simlist`. Given a list of zeros and ones, the average, or mean, of the list is precisely the proportion of ones in the list. The command `mean(simlist)` finds this average giving the simulated probability of getting three heads.

Run the script file and see that the resulting estimate is fairly close to the exact solution $1/8 = 0.125$. Increase n to 100,000 or even a million to get more precise estimates.

The script **Divisible356.R** simulates the divisibility problem in Example 1.20 that a random integer from $\{1, \ldots, 1000\}$ is divisible by 3, 5, or 6. The problem, and the resulting code, is more complex.

The function `simdivis()` simulates one trial. Inside the function, the expression `num%%x==0` checks whether `num` is divisible by x. The `if` statement checks whether `num` is divisible by 3, 5, or 6, returning 1 if it is, and 0, otherwise.

After defining the function, typing `simdivis()` will simulate one trial. By repeatedly typing `simdivis()` on your computer, you get a feel for how this random experiment behaves over repeated trials.

In this script, instead of writing a loop, we use the `replicate` command. This powerful R command is an alternative to writing loops for simple expressions. The syntax is `replicate(n,expr)`. The expression `expr` is repeated n times creating an n-element vector. Thus the result of typing

```
> simlist <- replicate(1000, simdivis())
```

is a vector of 1000 ones and zeros stored in the variable `simlist` corresponding to success or failure in the divisibility experiment. The average `mean(simlist)` gives the simulated probability.

Play with this script. Based on 1000 trials, you might guess that the true probability is between 0.45 and 0.49. Increase the number of trials to 10,000 and the estimates are roughly between 0.46 and 0.48. At 100,000, the estimates become even more precise, between 0.465 and 0.468.

We can actually quantify this increase in precision in Monte Carlo simulation as n gets large. But that is a topic that will to have to wait until Chapter 9.

R: SIMULATING THE DIVISIBILITY PROBABILITY

```
# Divisible356.R
# simdivis() simulates one trial
> simdivis <- function()  {
    num <- sample(1:1000,1)
    if (num%%3==0 || num%%5==0 || num%%6==0) 1 else 0
    }
> simlist <- replicate(1000, simdivis())
mean(simlist)
```

1.11 SUMMARY

In this chapter, the first principles of probability were introduced: from random experiment and sample space to the properties of probability functions. We start with discrete sample spaces—sets are either finite or countably infinite. The simplest probability model is when outcomes in a finite sample space are equally likely. In that case, probability reduces to "counting." Basic counting principles are presented. General properties of probabilities are derived from the three defining properties of a probability function. Random variables are introduced. The chapter ends with a first look at simulation.

- **Random experiment:** An activity, process, or experiment in which the outcome is uncertain.
- **Sample space Ω:** Set of all possible outcomes of a random experiment.
- **Outcome ω:** The elements of a sample space.
- **Event:** A subset of the sample space; a collection of outcomes.
- **Random variable X:** Assigns numbers to the outcomes of a random experiment. A real-valued function defined on the sample space.
- **Probability function:** A function P that assigns numbers to the elements $\omega \in \Omega$ such that
  1. $P(\omega) \geq 0$
  2. $\sum_\omega P(\omega) = 1$
  3. For events A, $P(A) = \sum_{\omega \in A} P(\omega)$
- **Equally likely outcomes:** Probability model for a finite sample space in which all elements have the same probability.
- **Uniform distribution:** Let $S = \{s_1, \ldots, s_k\}$. Then X is uniformly distributed on S if $P(X = s_i) = 1/k$, for $i = 1 \ldots, k$.
- **Counting**
  1. **Multiplication principle:** If there are m ways for one thing to happen, and n ways for a second thing to happen, then there are mn ways for both things to happen.
  2. **Permutations:** A permutation of $\{1, \ldots, n\}$ is an n-element ordering of the n numbers. There are $n!$ permutations of an n-element set.
- **Sampling:** When sampling from a population, *sampling with replacement* is when objects are returned to the population after they are sampled; *sampling without replacement* is when objects are not returned to the population after they are sampled.
- **Properties of probabilities**
  1. **Simple addition rule:** If A and B are mutually exclusive, that is, disjoint, then $P(A \text{ or } B) = P(A \cup B) = P(A) + P(B)$.

2. **Implication:** If A implies B, that is, if $A \subseteq B$, then $P(A) \leq P(B)$.
3. **Complement:** The probability that A does not occur $P(A^c) = 1 - P(A)$.
4. **General addition rule:** For all events A and B, $P(A \text{ or } B) = P(A \cup B) = P(A) + P(B) - P(AB)$.

- **Monte Carlo simulation** is based on the relative frequency interpretation of probability. Given a random experiment and an event A, $P(A)$ is approximately the fraction of times in which A occurs in n repetitions of the random experiment. A Monte Carlo simulation of $P(A)$ is based on three principles:
  1. Trials—Simulate the random experiment, typically on a computer using the computer's random numbers.
  2. Success—Based on the outcome of each trial, determine whether or not A occurs.
  3. Replication—Repeat the aforementioned steps n times. The proportion of successful trials is the simulated estimate of $P(A)$.
- **Problem–solving strategies**
  1. **Taking complements:** Finding $P(A^c)$, the probability of the complement of an event, might be easier in some cases than finding $P(A)$, the probability of the event. This arises in "at least" problems. For instance, the complement of the event that "at least one of several things occur" is the event that "none of those things occur." In the former case, the event involves a union. In the latter case, the event involves an intersection.
  2. **Inclusion–exclusion:** This is another method for tackling "at least" problems. For three events, inclusion–exclusion gives

$$P(A \cup B \cup C)$$
$$= P(A) + P(B) + P(C) - P(AB) - P(AC) - P(BC) + P(ABC).$$

EXERCISES

Sample space, event, random variable

1.1 Your friend was sick and unable to make today's class. Explain to your friend, using your own words, the meaning of the terms (i) random experiment, (ii) sample space, (iii) event, and (iv) random variable.

For the following problems 1.2–1.5, identify (i) the random experiment, (ii) sample space, (iii) event, and (iv) random variable. Express the probability in question in terms of the defined random variable, but do not compute the probability.

1.2 Roll four dice. Consider the probability of getting all fives.

1.3 A pizza shop offers three toppings: pineapple, peppers, and pepperoni. A pizza can have 0, 1, 2, or 3 toppings. Consider the probability that a random customer asks for two toppings.

1.4 Bored one day, you decide to play the video game Angry Birds until you win. Every time you lose, you start over. Consider the probability that you win in less than 1000 tries.

1.5 In Angel's garden, there is a 3% chance that a tomato will be bad. Angel harvests 100 tomatoes and wants to know the probability that at most five tomatoes are bad.

1.6 In two dice rolls, let X be the outcome of the first die, and Y the outcome of the second die. Then $X+Y$ is the sum of the two dice. Describe the following events in terms of simple outcomes of the random experiment:

(a) $\{X+Y=4\}$. (Solution: $\{13, 22, 31\}$.)

(b) $\{X+Y=9\}$.

(c) $\{Y=3\}$.

(d) $\{X=Y\}$.

(e) $\{X>2Y\}$.

1.7 A bag contains r red and b blue balls. You reach into the bag and take k balls. Let R be the number of red balls you take. Let B be the number of blue balls. Express the following events in terms of the random variables R and B:

(a) You pick no red balls. (Solution: $\{R=0\}$.)

(b) You pick one red and two blue balls.

(c) You pick four balls.

(d) You pick twice as many red balls as blue balls.

1.8 A couple plans to continue having children until they have a girl or until they have six children, whichever comes first. Describe a sample space and a reasonable random variable for this random experiment.

Probability functions

1.9 A sample space has four elements $\omega_1, \ldots, \omega_4$ such that ω_1 is twice as likely as ω_2, which is three times as likely as ω_3, which is four times as likely as ω_4. Find the probability function.

1.10 A random experiment has three possible outcomes a, b, and c, with

$$P(a)=p, \quad P(b)=p^2, \quad \text{and} \quad P(c)=p.$$

What choice(s) of p makes this a valid probability model?

1.11 Let P_1 and P_2 be two probability functions on Ω. Define a new function P such that $P(A)=(P_1(A)+P_2(A))/2$. Show the P is a probability function.

1.12 Suppose $P_1, \ldots, P_k$ are probability functions on Ω. Let $a_1, \ldots, a_k$ be a sequence of numbers. Under what conditions on the a_is will

$$P = a_1 P_1 + \cdots + a_k P_k$$

be a probability function?

1.13 Let P be a probability function on $\Omega = \{a, b\}$ such that $P(a) = p$ and $P(b) = 1 - p$ for $0 \leq p \leq 1$. Let Q be a function on Ω defined by $Q(\omega) = [P(\omega)]^2$. For what value(s) of p will Q be a valid probability function?

Equally likely outcomes and counting

1.14 A club has 10 members and is choosing a president, vice-president, and treasurer. All selections are equally likely.

(a) What is the probability that Tom is selected president?

(b) What is the probability that Brenda is chosen president and Liz is chosen treasurer?

1.15 A fair coin is flipped six times. What is the probability that the first two flips are heads and the last two flips are tails? Use the multiplication principle.

1.16 Suppose that license plates can be two, three, four, or five letters long, taken from the alphabets A to Z. All letters are possible, including repeats. A license plate is chosen at random in such a way so that all plates are equally likely.

(a) What is the probability that the plate is "A-R-R?"

(b) What is the probability that the plate is four letters long?

(c) What is the probability that the plate is a palindrome?

(d) What is the probability that the plate has at least one "R?"

1.17 Suppose you throw five dice and all outcomes are equally likely.

(a) What is the probability that all dice are the same? (In the game of Yahtzee, this is known as a *yahtzee*.)

(b) What is the probability of getting at least one 4?

(c) What is the probability that all the dice are different?

1.18 Amy is picking her fall term classes. She needs to fill three time slots, and there are 20 distinct courses to choose from, including probability 101, 102, and 103. She will pick her classes at random so that all outcomes are equally likely.

(a) What is the probability that she will get probability 101?

(b) What is the probability that she will get probability 101 and Probability 102?

(c) What is the probability she will get all three probability courses?

1.19 Suppose k numbers are chosen from $\{1, \ldots, n\}$, where $k < n$, sampling without replacement. All outcomes are equally likely. What is the probability that the numbers chosen are in increasing order?

Properties of probabilities

1.20 Suppose $P(A) = 0.40$, $P(B) = 0.60$, and $P(A \text{ or } B) = 0.80$. Find

(a) $P(\text{neither } A \text{ nor } B \text{ occur})$.

(b) $P(AB)$.

(c) $P(\text{one of the two events occurs, and the other does not})$.

1.21 Suppose A and B are mutually exclusive, with $P(A) = 0.30$ and $P(B) = 0.60$. Find the probability that

(a) At least one of the two events occurs

(b) Both of the events occur

(c) Neither event occurs

(d) Exactly one of the two events occur

1.22 Suppose $P(A \cup B) = 0.6$ and $P(A \cup B^c) = 0.8$. Find $P(A)$.

1.23 Suppose X is a random variable that takes values on all positive integers. Let $A = \{2 \leq X \leq 4\}$ and $B = \{X \geq 4\}$. Describe the events (i) A^c; (ii) B^c; (iii) AB; and (iv) $A \cup B$.

1.24 Suppose X is a random variable that takes values on $\{0, 0.01, 0.02, \ldots, 0.99, 1\}$. If each outcome is equally likely, find

(a) $P(X \leq 0.33)$.

(b) $P(0.55 \leq X \leq .66)$.

1.25 Let A, B, C, be three events. At least one event always occurs. But it never happens that exactly one event occurs. Nor does it ever happen that all three events occur. If $P(AB) = 0.10$ and $P(AC) = 0.20$, find $P(B)$.

1.26 See the assignment of probabilities to the Venn diagram in Figure 1.4. Find the following:

(a) P(No events occur).

(b) P(Exactly one event occurs).

(c) P(Exactly two events occur).

(d) P(Exactly three events occur).

(e) P(At least one event occurs).

(f) P(At least two events occur).

(g) P(At most one event occurs).

(h) P(At most two events occur).

1.27 Four coins are tossed. Let A be the event that the first two coins comes up heads. Let B be the event that the number of heads is odd. Assume that all

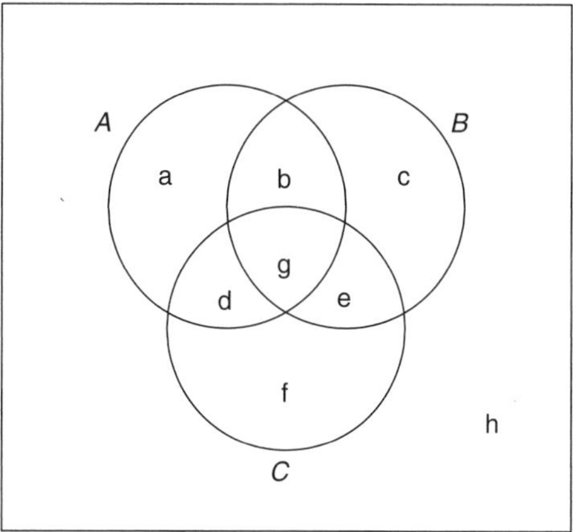

FIGURE 1.4: Venn diagram.

16 elements of the sample space are equally likely. Describe and find the probabilities of (i) AB, (ii) $A \cup B$, and (iii) AB^c.

1.28 Two dice are rolled. Let X be the maximum number obtained. (Thus, if 1 and 2 are rolled, $X = 2$; if 5 and 5 are rolled, $X = 5$.) Assume that all 36 elements of the sample space are equally likely. Find the probability function for X. That is, find $P(X = x)$, for $x = 1, 2, 3, 4, 5, 6$.

1.29 Judith has a penny, nickel, dime, and quarter in her pocket. So does Joe. They both reach into their pockets and choose a coin. Let X be the greater (in cents) of the two.

(a) Construct a sample space and describe the events $\{X = k\}$ for $k = 1, 5, 10, 25$.

(b) Assume that coin selections are equally likely. Find the probabilities for each of the aforementioned four events.

(c) What is the probability that Judith's coin is worth more than Joe's? (It is not $1/2$.)

1.30 A tetrahedron dice is four-sided and labeled with 1, 2, 3, and 4. When rolled it lands on the base of a pyramid and the number rolled is the number on the base. In five rolls, what is the probability of rolling at least one 2?

1.31 Let

$$Q(k) = \frac{2}{3^{k+1}}, \text{ for } k = 0, 1, 2, \ldots.$$

(a) Show that Q is a probability function. That is, show that the terms are nonnegative and sum to 1.

(b) Let X be a random variable such that $P(X = k) = Q(k)$, for $k = 0, 1, 2, \ldots$. Find $P(X > 2)$ without summing an infinite series.

1.32 The function

$$P(k) = c\frac{3^k}{k!}, \text{ for } k = 0, 1, 2, \ldots,$$

is a probability function for some choice of c. Find c.

1.33 Let A, B, C be three events. Find expressions for the events:

(a) At least one of the events occurs.

(b) Only B occurs.

(c) At most one of the events occurs.

(d) All of the events occur.

(e) None of the events occur.

1.34 The *odds in favor* of an event is the ratio of the probability that the event occurs to the probability that it will not occur. For example, the odds that you were born on a Friday, assuming birth days are equally likely, is 1 to 6, often written 16 or 1 to 6.

(a) In Texas Hold'em Poker, the odds of being dealt a pair (two cards of the same denomination) is 116. What is the chance of not being dealt a pair?

(b) For sporting events, bookies usually quote odds as odds against, as opposed to odds in favor. In the Kentucky Derby horse race, our horse Daddy Long Legs was given 2–9 odds. What is the chance that Daddy Long Legs wins the race?

1.35 An exam had three questions. One-fifth of the students answered the first question correctly; one-fourth answered the second question correctly; and one-third answered the third question correctly. For each pair of questions, one-tenth of the students got that pair correct. No one got all three questions right. Find the probability that a randomly chosen student did not get any of the questions correct.

1.36 Suppose $P(ABC) = 0.05$, $P(AB) = 0.15$, $P(AC) = 0.2$, $P(BC) = 0.25$, $P(A) = P(B) = P(C) = 0.5$. For each of the events given next, write the event using set notation in terms of A, B, and C, and compute the corresponding probability.

(a) At least one of the three events A, B, C occur.

(b) At most one of the three events occurs.

(c) All of the three events occurs.

(d) None of the three events occurs.

(e) At least two of the three events occurs.

(f) At most two of the three events occurs.

1.37 Find the probability that a random integer between 1 and 5000 is divisible by 4, 7 or 10.

1.38

(a) Each of the four squares of a two-by-two checkerboard is randomly colored red or black. Find the probability that at least one of the two columns of the checkerboard is all red.

(b) Each of the six squares of a two-by-three checkerboard is randomly colored red or black. Find the probability that at least one of the three columns of the checkerboard is all red.

1.39 Given events A and B, show that the probability that exactly one of the events occurs equals

$$P(A) + P(B) - 2P(AB).$$

1.40 Given events A, B, C, show that the probability that exactly one of the events occurs equals

$$P(A) + P(B) + P(C) - P(AB) - P(AC) - P(BC) + 3P(ABC).$$

Simulation and R

1.41 Modify the code in the R script **CoinFlip.R** to simulate the probability of getting exactly one head in four coin tosses.

1.42 Modify the code in the R script **Divisible356.R** to simulate the probability that a random integer between 1 and 5000 is divisible by 4, 7, or 10. Compare with your answer in Exercise 1.37.

1.43 Use R to simulate the probability of getting at least one 8 in the sum of two dice rolls.

1.44 Use R to simulate the probability in Exercise 1.30.

1.45 See the help file for the `sample` command (type `?sample`). Let X be a random variable taking values 1, 4, 8, and 16 with respective probabilities $0.1, 0.2, 0.3, 0.4$. Show how to simulate X.

1.46 Write a function `dice(`k`)` for generating k throws of a fair die. Use your function and R's `sum` function to generate the sum of two dice throws.

1.47 Make up your own random experiment and write an R script to simulate it.

2

CONDITIONAL PROBABILITY

In the fields of observation chance favors only the prepared mind.

—Louis Pasteur

2.1 CONDITIONAL PROBABILITY

Sixty students were asked, "Would you rather be attacked by a big bear or swarming bees?" Their answers, along with gender, are collected in the following *contingency table*, a common way to present data for two variables, in this case gender and attack preference. The table includes row and column totals, called marginals or marginal totals, and the overall total surveyed.

	Big bear	**Swarming bees**	
F	27	9	36
M	10	14	24
	37	23	60

The table of counts is the basis for creating a probability model for selecting a student at random and asking their gender and attack preference. The sample space consists of the four possible responses

$$\Omega = \{(F, B), (F, S), (M, B), (M, S)\},$$

Probability: With Applications and R, First Edition. Robert P. Dobrow.

where M is male, F is female, B is big bear, and S is swarming bees. The probability function is constructed from the contingency table so that the probability of each outcome is the corresponding proportion of responses. That is,

	Big bear	**Swarming bees**
F	27/60 = 0.450	9/60 = 0.150
M	10/60 = 0.167	14/60 = 0.233

Some questions of interest are

1. What is the probability that a student is male and would rather be attacked by a big bear?
2. What is the probability that a male student would rather be attacked by a big bear?

These questions are worded similarly but ask different things. The proportion of students who are male and prefer a big bear is $10/60 = 0.167$. That is, $P(M \text{ and } B) = 0.167$. But the proportion of male students who prefer a big bear is $10/24 = 0.4167$ since there are 24 males and 10 of them prefer a big bear.

The second probability is an example of a *conditional probability*. In a conditional probability, some information about the outcome of the random experiment is known—in this case that the selected student is male. The probability is *conditional* on that knowledge.

For events A and B, the conditional probability of A given that B occurs is written $P(A|B)$. We also read this as "the probability of A conditional on B." Hence, the probability the student would rather be attacked by a big bear conditional on being male is $P(B|M) = 0.4167$.

The probability of preferring a big bear conditional on being male is computed from the table by taking the number of students who are both male and prefer a big bear as a proportion of the total number of males. That is,

$$P(B|M) = \frac{\text{Number of males who prefer big bear}}{\text{Number of males}}.$$

Dividing numerator and denominator by the total number of students, this is equivalent to

$$P(B|M) = \frac{P(B \text{ and } M)}{P(M)}.$$

This suggests how to define the general conditional probability $P(A|B)$.

CONDITIONAL PROBABILITY

For events A and B such that $P(B) > 0$, the *conditional probability of* A *given* B is

$$P(A|B) = \frac{P(AB)}{P(B)}. \quad (2.1)$$

Example 2.1 In a population, 60% of the people have brown hair (H), 40% have brown eyes (E), and 30% have both (H and E). The probability that someone has brown eyes given that they have brown hair is

$$P(\text{E}|\text{H}) = \frac{P(\text{EH})}{P(\text{H})} = \frac{0.30}{0.60} = 0.50.$$

■

Example 2.2 Consider the Venn diagram in Figure 2.1. If each outcome x is equally likely, then $P(A) = 5/14$, $P(B) = 7/14$, $P(AB) = 2/14$, $P(A|B) = 2/7$, and $P(B|A) = 2/5$.

■

Example 2.3 At a college, 5% of the students are math majors. Of the math majors, 10% are double majors. Collegewide, 20% of the students are double majors. What is the probability (i) that a math major is a double major and (ii) that a double major is a math major?

Let D and M denote being a double major and math major, respectively.
(i) The problem asks for

$$P(D|M) = \frac{P(DM)}{P(M)} = \frac{(0.10)(0.05)}{0.05} = 0.10.$$

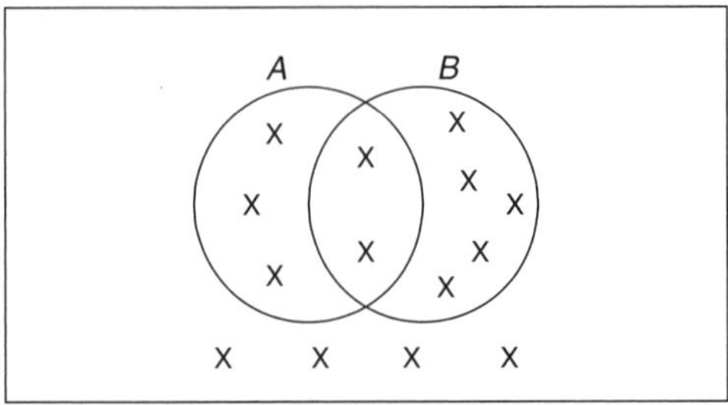

FIGURE 2.1

(ii) Here we want

$$P(M|D) = \frac{P(DM)}{P(D)} = \frac{(0.10)(0.05)}{0.20} = 0.025. \qquad \blacksquare$$

Example 2.4 Two dice are rolled. What is the probability that the first die is a 2 given that the sum of the dice is 7?

We use random variables to notate the problem. Let X_1 and X_2 be the outcomes of the first and second die, respectively. Then the sum of the dice is $X_1 + X_2$. The problem asks for

$$\begin{aligned}
P(X_1 = 2|X_1 + X_2 = 7) &= \frac{P(X_1 = 2 \text{ and } X_1 + X_2 = 7)}{P(X_1 + X_2 = 7)} \\
&= \frac{P(X_1 = 2 \text{ and } 2 + X_2 = 7)}{P(X_1 + X_2 = 7)} \\
&= \frac{P(X_1 = 2 \text{ and } X_2 = 5)}{P(X_1 + X_2 = 7)} \\
&= \frac{P(\{25\})}{P(\{16, 25, 34, 43, 52, 61\})} \\
&= \frac{1/36}{6/36} = \frac{1}{6}.
\end{aligned}$$

Observe that for the second equality the variable X_1 is replaced with its value 2.

It is interesting to observe that the *unconditional* probability $P(X_1 = 2)$ that the first die is 2 is also equal to 1/6. In other words, the information that the sum of the dice is 7 did not affect the probability that the first die is 2.

On the other hand, if we are given that the sum is 6, then

$$P(X_1 = 2|X_1 + X_2 = 6) = \frac{P(\{24\})}{P(\{15, 24, 33, 42, 51\})} = \frac{1}{5} > \frac{1}{6}.$$

Information that the sum of the dice is 6 makes the probability that first die is 2 a little more likely than if nothing was known about the sum of the two dice. ■

R: SIMULATING A CONDITIONAL PROBABILITY

Simulating the conditional probability $P(A|B)$ requires repeated simulations of the underlying random experiment, but restricting to trials in which B occurs. To simulate the conditional probability in the last example requires simulating repeated pairs of dice tosses, but the only data that are relevant are those pairs that result in a sum of 7.

See **ConditionalDice.R**. Every time a pair of dice is rolled, the routine checks whether the sum is 7 or not. If not, the dice are rolled again until a 7 occurs. Once a 7 is rolled, success is recorded if the first die is 2, and failure if it is not. The proportion of successes is taken just for those pairs of dice rolls that sum to 7.

A counter `ctr` is used that is iterated every time a 7 is rolled. This keeps track of the actual trials. We set the number of trials at $n = 10{,}000$. However, more than 10,000 random numbers will be generated. Since the probability of getting a 7 is $1/6$, the 10,000 trials are about one-sixth of the total number of random numbers generated, which we expect to be about 60,000.

```
# ConditionalDice.R
> n <- 10000
> ctr <- 0
> simlist <- numeric(n)
> while (ctr < n) {
    trial <- sample(1:6,2,replace=TRUE)
    if (sum(trial) == 7)    { # Check if sum is 7
    success <- if (trial[1] == 2) 1 else 0
    ctr <- ctr + 1
    simlist[ctr] <- success }        }
> mean(simlist)
[1] 0.1611
```

Example 2.5 John flips three coins. The probability of getting all heads is $(1/2)^3 = 1/8$. Suppose Amy peeks and sees that the first coin came up heads. For Amy, what is the probability that John gets all heads?

Amy's probability is conditional on the first coin coming up heads. This gives

$$P(\text{HHH}|\text{ First coin is H}) = \frac{P(\text{HHH and First coin is H})}{P(\text{ First coin is H})} = \frac{P(\text{HHH})}{P(\text{ First coin is H})} = \frac{1/8}{1/2} = \frac{1}{4}. \quad \blacksquare$$

Warning: A common mistake when first working with conditional probability is to write $P(A|B) = P(A)/P(B)$. In general, this is just wrong. However, in the special case when A implies B, that is, $A \subseteq B$, then it is correct since $A \cap B = A$ and

$$P(A|B) = \frac{P(A \cap B)}{P(B)} = \frac{P(A)}{P(B)}.$$

That is what happens in the last example, since getting all heads implies that the first coin is heads.

2.2 NEW INFORMATION CHANGES THE SAMPLE SPACE

In Example 2.5 the fact that Amy's probability of getting three heads is different from John's highlights the fact that probability is not an intrinsic "physical" property of a random experiment. It changes based on information and context. When we ask what is the probability of getting all heads in three coin tosses, implicit in that question is that you have not seen the outcome of the experiment. If you see the outcome, then you know that either all heads came up or they did not, so the probability is either 1 or 0. On the other hand, when some part of the experiment is observed, then that partial information becomes relevant in the probability calculation.

Partial information about the outcome of a random experiment actually changes the set of possible outcomes, that is, it changes the sample space of the original experiment and reduces it based on new information. For the three coin tosses, and before Amy peeks, the sample space is

$$\Omega = \{\text{HHH, HHT, HTH, THH, HTT, THT, TTH, TTT}\}.$$

But after she looks and sees that the first coin is heads, the sample space reduces to

$$\Omega' = \{\text{HHH, HHT, HTH, HTT}\}.$$

The resulting conditional probability is a probability function computed on the restricted sample space.

Conditional probability is a probability function. In Chapter 1, a probability function is defined based on Definition 1.1. Here we show that for a fixed event B, the conditional probability $P(A|B)$—as a function of A—is itself a similarly defined probability function, but on a reduced sample space.

Starting from a random experiment and sample space Ω, let $B \subseteq \Omega$ be an event such that $P(B) > 0$. We use lowercase letters b to denote the elements of B. Consider $P(A|B)$ as a function of A. Better to write this as $P(\cdot|B)$ to emphasize that the conditional probability is a function of its first argument. This function is itself a probability function on the restricted sample space B and satisfies the defining conditions. In particular,

1. $P(b|B) \geq 0$, for all $b \in B$.
2. $\sum_{b \in B} P(b|B) = 1$.
3. For all $A \subseteq B$, $P(A|B) = \sum_{b \in A} P(b|B)$.

The properties are verified subsequently.

1. This is true since $P(b|B)$ is defined as a ratio of two probabilities, which are both nonnegative.

2.
$$\sum_{b\in B} P(b|B) = \sum_{b\in B} \frac{P(\{b\} \text{ and } B)}{P(B)} = \sum_{b\in B} \frac{P(b)}{P(B)}$$
$$= \frac{1}{P(B)} \sum_{b\in B} P(b) = \frac{P(B)}{P(B)} = 1.$$

3. Let $A \subseteq \Omega$. Then

$$P(A|B) = \frac{P(A \cap B)}{P(B)} = \sum_{b\in A\cap B} \frac{P(b)}{P(B)}$$
$$= \sum_{b\in A} \frac{P(\{b\} \cap B)}{P(B)} = \sum_{b\in A} P(b|B)$$

In summary, conditional probabilities are themselves probability functions defined on the restricted sample space of the conditioning event.

Example 2.6 Let X be a random integer picked uniformly from 0 to 6. The probability function for X is

$$P(X = 0) = \cdots = P(X = 6) = \frac{1}{7}.$$

We are told that X is odd. Then for $k = 1, 3, 5,$

$$P(X = k|X \text{ is odd}) = \frac{P(X = k, X \text{ is odd})}{P(X \text{ is odd})} = \frac{P(X = k)}{P(X \text{ is odd})} = \frac{1/7}{3/7} = \frac{1}{3}.$$

This gives the new probability function

$$\tilde{P}(X = 1) = \tilde{P}(X = 3) = \tilde{P}(X = 5) = \frac{1}{3},$$

with reduced sample space $\{1, 3, 5\}$, where

$$\tilde{P}(X = k) = P(X = k|X \text{ is odd}), \quad \text{for } k = 1, 3, 5.$$ ■

2.3 FINDING $P(A$ AND $B)$

So far we have focused on finding the conditional probability $P(A|B)$, which requires knowledge of $P(A \text{ and } B)$. But sometimes what is unknown is precisely $P(A \text{ and } B)$. Rearranging the formula $P(A|B) = P(A \text{ and } B)/P(B)$ gives the following.

GENERAL FORMULA FOR $P(A$ AND $B)$

$$P(A \text{ and } B) = P(AB) = P(A|B)P(B), \tag{2.2}$$

This is a general formula for working with "and" probabilities. Observe that by switching the roles of A and B, this also gives

$$P(A \text{ and } B) = P(B|A)P(A).$$

Example 2.7 Draw two cards from a standard deck. What is the probability of getting two aces?

The probability of drawing an ace is $4/52 = 1/13$. If we have already drawn an ace, the probability of drawing a second ace is $3/51$, since there are three aces left in a reduced deck of 51. Let A_1 be the event of drawing an ace on the first card, and A_2 the event of drawing an ace on the second card. Then

$$P(A_1A_2) = P(A_2|A_1)P(A_1) = \left(\frac{3}{51}\right)\left(\frac{1}{13}\right) = \frac{1}{221} = 0.0045. \quad \blacksquare$$

What is the probability of being dealt three aces? Intuitively, you might guess the answer $(4/52)(3/51)(2/50) = 0.000178$. The three factors in the probability are (i) the probability of getting an ace on the first card, (ii) the probability of getting an ace on the second card given that the first card is an ace, and (iii) the probability of getting an ace on the third card given that the first two cards are aces.

This intuition is correct and suggests the extension of Equation 2.2 for more than two events. Given events A_1, A_2, and A_3,

$$\boxed{P(A_1A_2A_3) = P(A_3|A_1A_2)P(A_2|A_1)P(A_1).} \tag{2.3}$$

To see this, let $A = A_1$ and $B = A_2A_3$. Then

$$\begin{aligned} P(A_1A_2A_3) &= P(AB) = P(A|B)P(B) \\ &= P(A_1|A_2A_3)P(A_2A_3) \\ &= P(A_1|A_2A_3)P(A_2|A_3)P(A_3). \end{aligned}$$

More generally, for k events $A_1, \ldots, A_k$,

$$\boxed{P(A_1 \cdots A_k) = P(A_k|A_1 \cdots A_{k-1})P(A_{k-1}|A_1 \cdots A_{k-2}) \cdots P(A_2|A_1)P(A_1).} \tag{2.4}$$

The general result is similarly proven using mathematical induction.

Example 2.8 A subject in an experiment is given three tries to complete a task. On the first try, the probability of success is 0.30. If they fail, the chance of success on the second attempt is 0.50. And if they fail that, the chance of success on the third try is 0.65. What is the probability that they complete the task?

Let S_1, S_2, S_3 denote the events that the task is completed on the first, second, and third tries, respectively. The desired probability is

$$\begin{aligned} P(S_1 \cup S_2 \cup S_3) &= 1 - P(S_1^c S_2^c S_3^c) \\ &= 1 - P(S_1^c) P(S_2^c | S_1^c) P(S_3^c | S_1^c S_2^c) \\ &= 1 - (0.70)(0.50)(0.35) = 0.8775 \end{aligned}$$ ■

Tree diagrams. Tree diagrams are useful tools for computing probabilities. They often arise when events can be ordered sequentially (first one thing happens, then the next). They are also great visual aids that decompose a problem into smaller logical units. Probabilities are written on the branches of the tree, and outcomes are written at the end of each branch.

Figure 2.2 illustrates the random experiment of picking two balls from a bag containing two red and three blue balls. The outcome of picking two red balls is described by the top branch of the tree. First we select a red ball (with probability 2/5), and then we select a second red ball given that the first ball was red (with probability 1/4). The probability of the final outcome is obtained by multiplying along the branch ($1/10 = 2/5 \times 1/4$). Observe that the branches of the tree are labeled with conditional probabilities.

Example 2.8 lends itself naturally to a tree diagram analysis because of the sequential nature of the random experiment. The subject is given three tries to complete a task (See Fig. 2.3). The event that the subject eventually completes their task

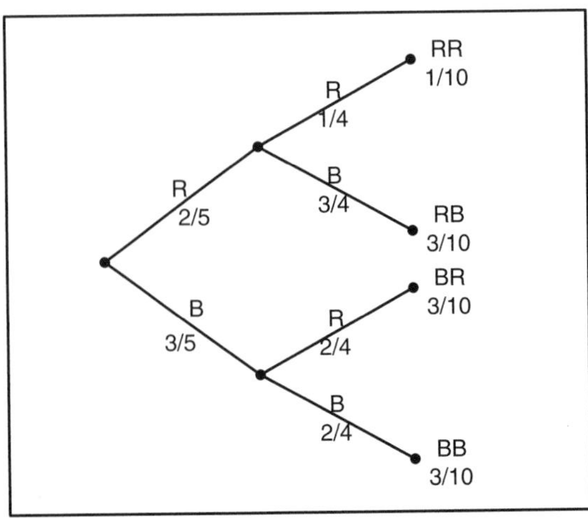

FIGURE 2.2: Tree diagram for picking two balls from a bag of two red and three blue balls.

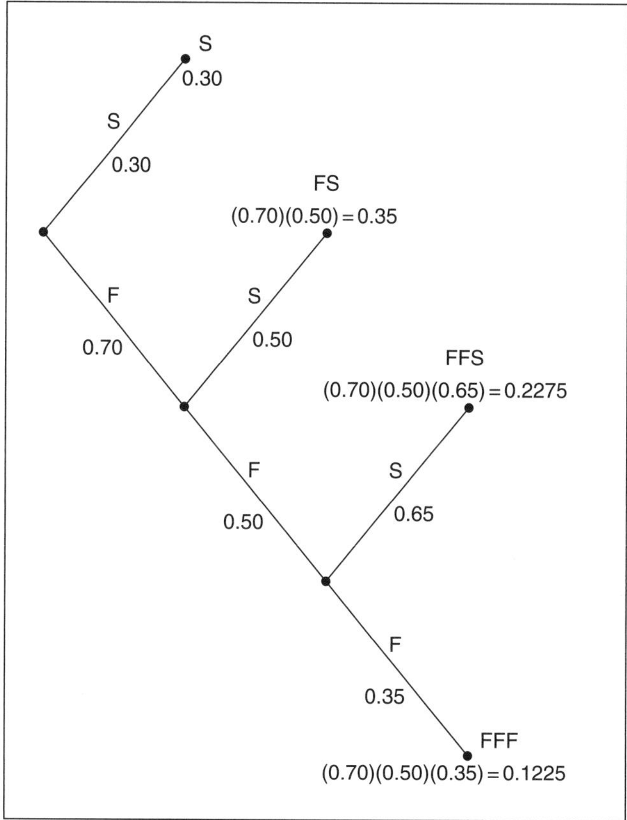

FIGURE 2.3: Tree diagram for Example 2.8.

is the disjoint union of the events that the subject completes their task on the first, second, or third try, respectively. Thus, the probability that the subject eventually completes their task is the sum

$$P(S) + P(FS) + P(FFS) = 0.30 + 0.35 + 0.2275 = 0.8775.$$

Example 2.9 Blackjack, or twenty-one, is a popular casino game. The player is dealt two cards. A *blackjack* is an ace and a ten card (10, jack, queen, or king). What is the probability of being dealt a blackjack?

We illustrate the solution with the tree diagram in Figure 2.4. Blackjack is obtained by either getting an ace on the first card and then a ten card, or a ten card first and then an ace. We use "Not" to denote any of the 32 cards that are neither an ace nor a ten card.

There are two outcomes that correspond to a blackjack—being dealt an ace on the first card and then a 10 on the second, or vice versa. The probability of each outcome is shown to be 16/663. The probability of blackjack is the sum of these outcomes,

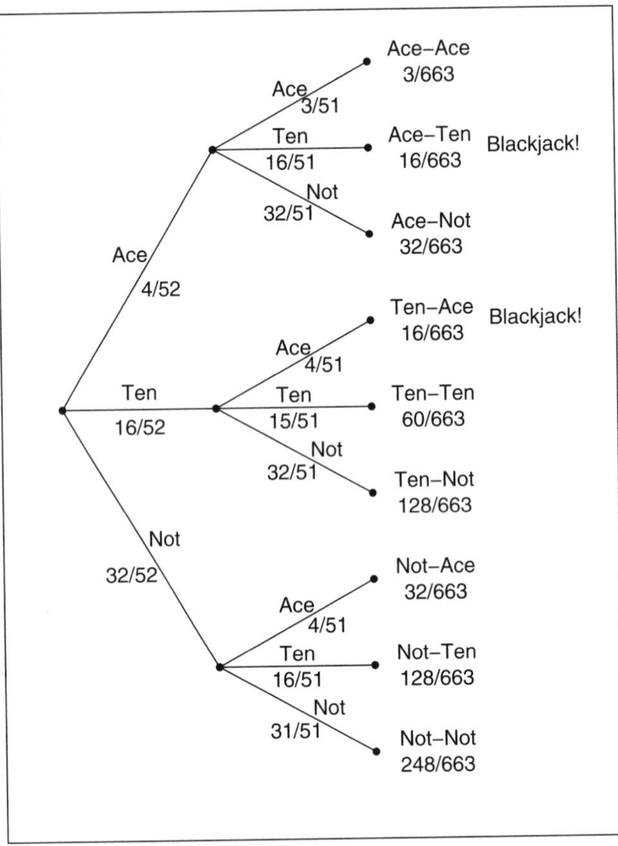

FIGURE 2.4: Tree diagram for blackjack.

which gives

$$P(\text{Blackjack}) = \frac{16}{663} + \frac{16}{663} = \frac{32}{663} = 0.048.$$

For a more formal approach, let A_1 and A_2 denote the events of getting an ace on the first and second cards, respectively. Similarly define T_1 and T_2 for getting a ten card on the first and second cards. Then

$$\begin{aligned} P(\text{Blackjack}) &= P(A_1T_2 \text{ or } T_1A_2) \\ &= P(A_1T_2) + P(T_1A_2) \\ &= P(T_2|A_1)P(A_1) + P(A_2|T_1)P(T_1) \\ &= \left(\frac{16}{52}\right)\left(\frac{4}{51}\right) + \left(\frac{4}{52}\right)\left(\frac{16}{51}\right) = \frac{2 \times 4 \times 16}{52 \times 51} = 0.048. \end{aligned}$$

■

R: SIMULATING BLACKJACK

The script **Blackjack.R** simulates the blackjack probability. The numbers 1–52 represent a deck of cards. We assign the four aces to the numbers 1, 2, 3, 4, and the 16 ten cards to the numbers 37–52. The command

```
> trial <- sample(1:52, 2, replace=FALSE)
```

chooses two cards from 1 to 52, sampling without replacement, and assigns them to the variable `trial`. The command

```
> success <- if (trial[1] <= 4 && trial[2] >= 37
|| trial[1] >= 37 && trial[2] <= 4) 1 else 0
```

uses the logical operators || (or) and && (and) to determine blackjack, returning a 1 if yes, and 0, otherwise.

```
# Blackjack.R
> n <- 50000
> simlist <- numeric(n)
> for (i in 1:n) {
   trial <- sample(1:52, 2, replace=FALSE)
   success <- if (trial[1] <= 4 && trial[2] >= 37 ||
   trial[1] >= 37 && trial[2] <= 4) 1 else 0
   simlist[i] <- success  }
> mean(simlist)
[1] 0.04644
```

2.3.1 Birthday Problem

The birthday problem is a classic probability delight first introduced by the mathematician Richard Von Mises in 1939. Von Mises asked, "How many people must be in a room before the probability that some share a birthday, ignoring the year and ignoring leap days, becomes at least 50%?"

For a group of k people, let B be the event that at least two people have the same birthday. We find $P(B)$. Remember the problem-solving strategy of taking complements for "at least" probabilities. The complement B^c is the event that none of the k people have the same birthday. We compute that probability with a tree diagram.

Consider asking people one by one their birthday and checking whether their birthday is different from the birthdays of those previously asked. The first person's birthday is fixed. The second person's birthday either matches the first birthday, which occurs with probability 1/365, or does not, with probability 364/365. Two branches grow out of the first node labeled with these probabilities as in Figure 2.5.

The full tree will have a lot of branches. But we are only interested in one path of the tree, where everyone's birthday is different. So it is not necessary to draw the entire tree, just that relevant path.

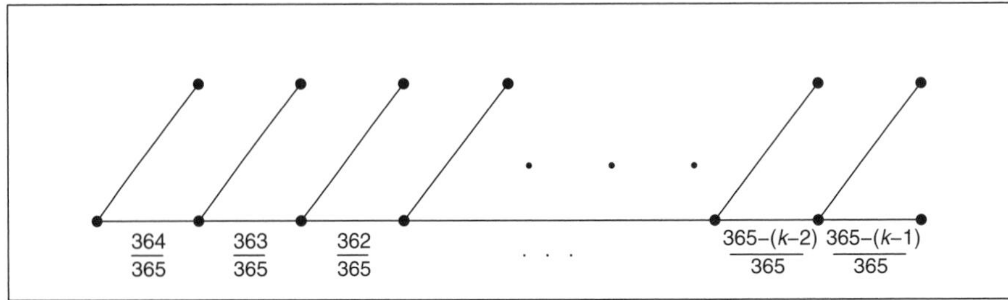

FIGURE 2.5: Solving the birthday problem with a tree diagram.

From the second node, the probability that the third person has a birthday different from the previous two, given that the previous two birthdays are different, is 363/365 (since two birthdays have been picked and there are 363 available ones left). Continuing in this way, we see that the ith branch of the tree gives the probability that the $(i+1)$st person's birthday is different from the previous i birthdays, given that the previous i birthdays are all different, which occurs with probability $(365-i)/365$. This gives

$$P(B^c) = \left(\frac{364}{365}\right)\left(\frac{363}{365}\right)\cdots\left(\frac{365-(k-1)}{365}\right) = \prod_{i=1}^{k-1}\left(1-\frac{i}{365}\right). \tag{2.5}$$

And thus the birthday probability that at least two people have the same birthday is

$$P(B) = 1 - P(B^c) = 1 - \prod_{i=1}^{k-1}\left(1-\frac{i}{365}\right). \tag{2.6}$$

At $k = 22$, $P(B) = 0.476$, and at $k = 23$, $P(B) = 0.507$. So the answer to Von Mises' question is, remarkably, 23 people. The number is much smaller than most people think. Table 2.5 gives birthday probabilities for different group sizes. With just $k = 15$ people there is a 25% chance of a birthday match. And with $k = 50$ people the likelihood of a match is virtually certain with $P(B) = 0.970$.

To explain the seemingly paradoxical result, intuitively observe that in a group of 23 people there are actually 253 ways for people to be paired. And we just need one of those pairs to have a common birthday for the desired event to occur.

There are many fun ways to illustrate the birthday problem. Consider the birthdays of the last 44 U.S. presidents. With $k = 44$, the probability of a match is 0.933. And we find that Warren G. Harding and James K. Polk were both born on November 2.

The 25-player active roster of a major league baseball team is a nice vehicle for checking the birthday problem. Each team's roster gives a slightly greater than 50%

chance that two players on the team have the same birthday. Among the 30 teams of major league baseball, we estimate that about 15 teams will have two players on the active roster with the same birthday. And we invite a baseball fan to check our conjecture.

The birthday problem can be cast in a very general framework. Suppose we distribute k balls into m boxes in such a way so that each of the m boxes is equally likely to receive any ball. Think of the boxes as birthdays and the balls as people. The probability that some box contains two or more balls is equivalent to the birthday probability that among k people there are two birthdays in common. This probability is equal to

$$P(\text{Some box contains at least two balls}) = 1 - \prod_{i=1}^{k-1}\left(1 - \frac{i}{m}\right). \tag{2.7}$$

If k is large, this probability does not lend itself to easy calculation or interpretation. A simpler closed form expression can be gotten with the help of calculus, in particular Taylor series. We give the derivation thinking of the birthday problem with $m = 365$.

Let $p = \prod_{i=1}^{k-1}(1 - i/365)$. Then

$$\ln p = \ln \prod_{i=1}^{k-1}\left(1 - \frac{i}{365}\right) = \sum_{i=1}^{k-1} \ln\left(1 - \frac{i}{365}\right).$$

The Taylor series expansion for $\ln(1 - x)$ is

$$\ln(1 - x) = -x - \frac{x^2}{2} - \frac{x^3}{3} - \cdots,$$

which converges for $-1 < x < 1$. Truncate off all but the first term of the series to obtain the approximation $\ln(1 - x) \approx -x$. The approximation is good for small values of x close to 0.

For $1 \le i < k < 365$, $i/365$ will be "small," justifying the use of the approximation. This gives

$$\ln p = \sum_{i=1}^{k-1} \ln\left(1 - \frac{i}{365}\right) \approx \sum_{i=1}^{k-1} -\frac{i}{365} = -\frac{k(k-1)}{2 \times 365},$$

using the fact that the sum of the first $k - 1$ integers is $k(k - 1)/2$. Exponentiating both sides gives $p \approx e^{-k(k-1)/(2\times 365)}$. And the birthday probability is

$$P(B) \approx 1 - e^{-k(k-1)/(2\times 365)} \approx 1 - e^{-k^2/(2\times 365)}. \tag{2.8}$$

TABLE 2.1: Birthday probabilities.

k	Exact	Approximate
15	0.253	0.250
23	0.507	0.500
30	0.706	0.696
40	0.891	0.882
50	0.970	0.965
60	0.994	0.992

In Table 2.1, we compare the approximation with the exact probabilities for select values of k.

More generally, in the balls-and-boxes setting with k balls and m boxes, the probability that some box contains two or more balls is approximately equal to $1 - e^{-k^2/(2m)}$.

There are many applied settings that fit the balls-into-boxes framework of the birthday problem.

Example 2.10 In 2001, the Arizona Department of Public Safety reported, in response to a court order, that a search of its state offender database of 65,493 DNA profiles found a "nine-locus" DNA match (the DNA of two samples agreed at nine positions on the chromosome). See Troyer et al., 2001. The estimated probability of such a match is about one in 754 million. At the time, the DNA match was said to be so unlikely as to call into question the reliability of the state's database and even of the use of DNA evidence in court. What is the probability of finding two such matching profiles in the database?

If we assume that all 754 million DNA outcomes are equally likely, then this gives an application of the birthday problem with $k = 65,493$ balls and $m = 7.54 \times 10^8$ boxes. The probability of a DNA match is

$$1 - \prod_{i=1}^{65,492} \left(1 - \frac{i}{7.54 \times 10^8}\right) \approx 1 - e^{-65,493^2/(2\times 7.54\times 10^8)} = 0.942.$$

What was originally thought to be an extremely rare coincidence actually has a very high probability of occurrence.

An intuitive explanation for this high probability is that in a database of 65,493 DNA profiles there are about two billion different *pairs* of profiles. An event which has a one in a billion chance of occurring, if it is repeated two billion times, will likely occur twice! ∎

Coincidences, like having the same birthday as your roommate, or having gone to the same high school as the person sitting next to you on the plane, or always seeing license plates that seem to start with the same three letters as your own, seem to defy logic. Yet they can often be explained by the laws of probability.

If an event has a one-in-a-million chance of occurring, then in a country of 300 million you would expect about 300 occurrences.

In "Methods for Studying Coincidences" Diaconis and Mosteller (1989), assert the *Law of Truly Large Numbers*: "With a large enough sample, any outrageous thing is likely to happen." Their highly readable and entertaining paper is part human psychology and gives a guide to the probabilistic and statistical techniques for studying coincidences.

Many variations and extensions of the birthday problem have been proposed and studied. In our discussion, we assume that the 365 days of nonleap years are equally likely. This is not the case in this country, as seen in Figure 2.6, which shows the distribution of birthday frequencies in the United States based on 20 years of census data. The birthday probabilities change very slightly for this distribution of birthday frequencies.

The birthday problem does *not* ask for the probability that among k people there will be a match of any one particular birthday, but rather that some pair of people will have the same birthday. The author's experience polling his classes of about 30 students is that roughly three-fourths of the time the birthday problem "works" and two students have the same birthday. But he has never, in about 20 classes, found a student who matches his own December 13 birthday.

2.4 CONDITIONING AND THE LAW OF TOTAL PROBABILITY

According to the Howard Hughes Medical Institute, about 7% of men and 0.4% of women are colorblind—either cannot distinguish red from green or see red and green differently from most people. In the United States, about 49% of the population is male and 51% female. A person is selected at random. What is the probability they are colorblind?

As you contemplate answering this question, you might find yourself saying, "Well, it depends—on whether you are male or female." The problem provides conditional information based on sex but the question asks for an *unconditional* probability.

In this section, we introduce a powerful technique for using conditional probability for solving "unconditional" problems.

The event $C = \{\text{Colorblind}\}$ can be decomposed into the disjoint union

$$\{\text{Colorblind}\} = \{\text{Colorblind and Male}\} \cup \{\text{Colorblind and Female}\}.$$

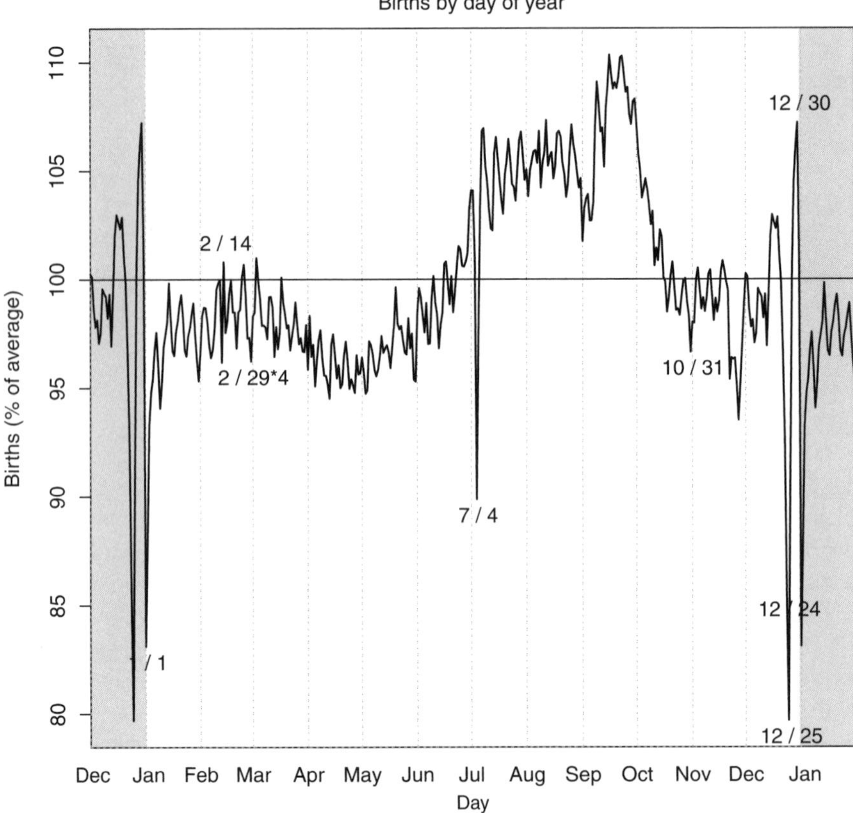

FIGURE 2.6: This graph of actual birthday frequencies in the United States is based on 20 years of census data. It was generated in R by Chris Mulligan (http://chmullig.com/2012/06/births-by-day-of-year/). Observe the seasonal trends and how unlikely it is for someone to be born on New Year's Day, July 4, or Christmas.
Source: National Vital Statistics System natality data 1969-1988, as provided by Google BigQuery. Graph by Chris Mulligan (chmullig.com)

We then obtain

$$\begin{aligned} P(C) &= P(CM \cup CF) = P(CM) + P(CF) \\ &= P(C|M)P(M) + P(C|F)P(F) \\ &= (0.07)(0.49) + (0.004)(0.51) = 0.03634. \end{aligned}$$

The approach to solving this problem is known as *conditioning*. In this case, we are conditioning on sex since the conditional probabilities $P(C|M)$ and $P(C|F)$ are easier and more "natural" to solve than the unconditional probability $P(C)$.

More generally, say that a collection of events $\{B_1, \ldots, B_k\}$ is a *partition* of the sample space Ω if (i) the events have no outcomes in common and (ii) their union is equal to Ω. Given an event A, the *law of total probability* shows how to find the unconditional probability $P(A)$ by conditioning on the B_is.

LAW OF TOTAL PROBABILITY

Suppose $B_1, \ldots, B_k$ is a partition of the sample space. Then

$$P(A) = \sum_{i=1}^{k} P(A|B_i)P(B_i). \tag{2.9}$$

Observe that we can decompose A into the disjoint union

$$A = AB_1 \cup \ldots \cup AB_k$$

as illustrated in Figure 2.7. The law of total probability follows by taking probabilities and applying the conditional probability formula to each term of the resulting sum since

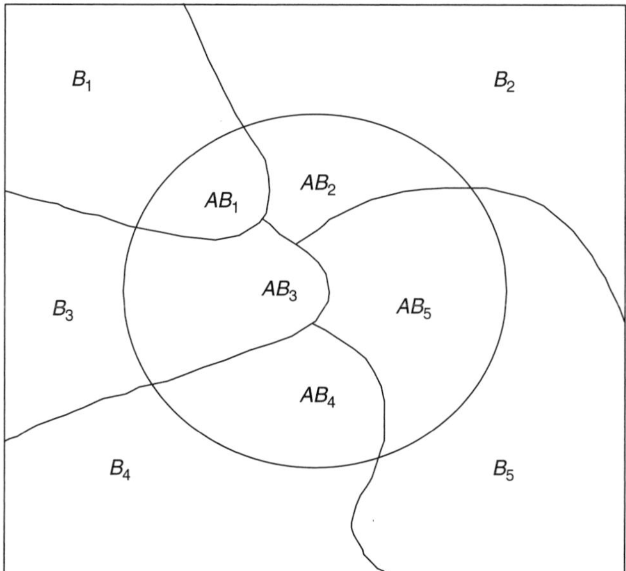

FIGURE 2.7: The events $B_1, \ldots, B_5$ partition the sample space. The circle represents event A, which is decomposed into the disjoint union $A = AB_1 \cup \cdots \cup AB_5$.

$$P(A) = P\left(\bigcup_{i=1}^{k} AB_i\right) = \sum_{i=1}^{k} P(AB_i) = \sum_{i=1}^{k} P(A|B_i)P(B_i).$$

A common special case of the law of total probability occurs when $k = 2$. For any event B, the sets B and B^c partition the sample space. This gives

$$P(A) = P(A|B)P(B) + P(A|B^c)P(B^c). \tag{2.10}$$

Solving a general probability $P(A)$ by conditioning refers to introducing "supplemental" disjoint events $B_1, \ldots, B_k$. We find the "partial" probabilities $P(A|B_1)$, ..., $P(A|B_k)$ in order to find the "total" probability $P(A)$.

Example 2.11 Table 2.2 gives an insurance company's prediction for the likelihood that a person in a particular age group will have an auto accident during the next year. The company's policyholders are 20% under the age of 25, 30% between 25 and 39, and 50% over the age of 40. What is the probability that a random policyholder will have an auto accident next year?

Denote the three age groups by 1, 2, and 3, respectively. Let A be the event of having an auto accident. Conditioning on age group, the law of total probability gives

$$\begin{aligned} P(A) &= P(A|1)P(1) + P(A|2)P(2) + P(A|3)P(3) \\ &= (0.11)(0.20) + (0.03)(0.30) + (0.02)(0.50) \\ &= 0.041. \end{aligned}$$

■

TABLE 2.2: Insurance predictions for probability of auto accident.

Under 25	25–39	Over 40
0.11	0.03	0.02

Example 2.12 How to ask a sensitive question? Statisticians are sometimes confronted with how to obtain information on sensitive issues. What proportion of people use illegal drugs? How many students ever cheated on an exam? Surveying people directly and asking these types of sensitive questions is not likely to get honest responses and useful data.

Using probabilistic methods, statisticians have developed interesting ways to ask sensitive questions that protect confidentiality. Here is one example.

Respondents are given a coin and told to flip it in private, not letting anyone see the outcome. If it lands heads, they answer the sensitive question of interest (e.g., "Have you ever taken illegal drugs?"). If tails, they answer an innocuous question such as "Were you born in the first half of the year—January through June?")

The respondent reports a yes or no, but does not say which question they actually answered. And from a sample of such yes–no responses, statisticians can estimate the parameter of interest, such as the proportion of people who have ever taken illegal drugs. How can this be done?

Let Y and N denote responses of yes and no, respectively. Let Q_S denote the sensitive question and Q_I the innocuous question. The unknown parameter that surveyors want to estimate is $p = P(Y|Q_S)$, the probability that someone answers yes given that they were asked the sensitive question. We assume that the innocuous question is (i) easy to answer and (ii) has a known probability of yes and no, in this case 50% each.

Consider the unconditional probability $P(Y)$. By the law of total probability,

$$\begin{aligned} P(Y) &= P(Y|Q_S)P(Q_S) + P(Y|Q_I)P(Q_I) \\ &= p\left(\frac{1}{2}\right) + \left(\frac{1}{2}\right)\left(\frac{1}{2}\right) \\ &= \frac{p}{2} + \frac{1}{4}. \end{aligned}$$

When this survey is given to n people, the final data will consist of n yes's and no's. The proportion of yes's is a simulated estimate of the unknown $P(Y)$. And thus

$$\frac{p}{2} + \frac{1}{4} = P(Y) \approx \frac{\text{Number of yes's in the sample}}{n}.$$

Solving for p gives

$$p \approx 2\left(\frac{\text{Number of yes's in the sample}}{n} - \frac{1}{4}\right),$$

which is the final estimate of the parameter of interest.

Thirty students participated in such a classroom experiment where the "sensitive" question was "Are you wearing running shoes?" The instructor gave each student a coin with instructions for doing the anonymous survey. Sixteen students responded yes and 14 students responded no. The instructor estimated that $p = P(Y|Q_S) \approx 2((16/30) - 1/4) = 0.567$. Since there were 30 students in the class, he guessed that $30 \times (0.567) = 17$ students were wearing running shoes. In fact, there were 15. ■

Example 2.13 Finding the largest number. The following problem, known originally as the secretary problem, was first introduced by Martin Gardner in his *Mathematical Games* column in *Scientific American*. It was originally cast in terms of a manager trying to hire the most qualified secretary from a group of n applicants. Versions include finding the best lottery prize among n prizes, and the following game to find the highest number in a list of n numbers.

On n pieces of paper are written n distinct numbers. Let z be the largest number in the group. You will be shown the pieces of paper one at a time and you must decide after each number whether to choose that number—and the game stops—or reject it and move on to the next number. Your goal is to find z. When shown the kth number, the only information you are given is the relative rank of that number compared to the previous $k-1$ numbers.

What should your strategy be for deciding which number to choose? And using such a strategy what is the probability of getting the top number?

It seems at first reading that any strategy will produce a very small probability of getting the top number. Remarkably, one can do reasonably well.

Consider the following strategy: For some r between 0 and n, reject the first r numbers and then choose the first number that is better than the first r. We find the probability of choosing the top number z for a fixed r. We will then find the choice of r which does best.

Let A be the event of choosing z. Let R be the relative rank of z. Assume all arrangements of the numbers are equally likely. (Also, we make no assumptions about the size of the numbers.) By conditioning on R,

$$P(A) = \sum_{i=1}^{n} P(A|R=i)P(R=i) = \frac{1}{n}\sum_{i=1}^{n} P(A|R=i).$$

Consider $P(A|R=i)$, the probability of choosing the top number z given that it is in position i.

Suppose z is at position i where $i \leq r$. Then z will be rejected, and you lose the game.

Suppose z is at position i where $i > r$. Then z will be chosen if and only if the largest of the first $i-1$ numbers is among the first r numbers. (Otherwise, the largest of the first $i-1$ numbers will be chosen.) The largest of the first $i-1$ numbers can be in one of $i-1$ equally likely positions. The probability that it is among the first r numbers is thus $r/(i-1)$. This gives

$$P(A) = \frac{1}{n}\sum_{i=1}^{n} P(A|R=i) = \frac{1}{n}\sum_{i=r+1}^{n} \frac{r}{i-1} = \frac{r}{n}\sum_{i=r+1}^{n} \frac{1}{i-1} = \frac{r}{n}\sum_{i=r}^{n-1} \frac{1}{i}.$$

The harmonic series $\sum_{i=1}^{n} 1/i$ diverges as $n \to \infty$. For large n, the sum of the series is approximately equal to $\log n$. This gives

$$P(A) = \frac{r}{n}\sum_{i=r}^{n-1} \frac{1}{i} \approx \frac{r}{n}(\log n - \log r) = \frac{r}{n}\log\left(\frac{n}{r}\right).$$

To find the r that does best, maximize the function $f(x) = (r/x)\log(x/r)$. Taking the derivative with respect to x and setting it equal to 0 finds that the maximum is achieved at $x = n/e$. For $r = n/e$, the probability that you will choose z is

$$P(A) \approx \frac{n/e}{n} \log\left(\frac{n}{n/e}\right) = \frac{1}{e} = 0.368.$$

In the following simulation, we choose from a list of $n = 100$ numbers, rejecting the first $n/e \approx 37$. The simulation is repeated 10,000 times.

R: FINDING THE LARGEST NUMBER

```
# TopNumber.R
# Numbers are 1, ... , n so x = n
> ntrials <- 10000
> n <- 100
> x <- n     # Top number
> r <- round(n/exp(1))    # r = n/e = 37
> simlist <- vector(length=ntrials)
> for (j in 1:ntrials) {
   numbers <- sample(1:n, n)
   best <- which(numbers==x) # position of x
   prob <- 0
   firstr <- max(numbers[1:r]) # largest of first r
   for (i in (r+1):n)  { # look after r-th number
      if (numbers[i] > firstmax)
         {if (numbers[i] == x) prob <- 1
        break}
    else {prob<-0}  }
     simlist[j] <- prob  }
 >  mean(simlist)
[1] 0.3694
```

■

Example 2.14 Random permutations. There are many settings where one wants to generate a uniformly random permutation. Random permutations are used in many computer algorithms. A common application is shuffling a deck of cards, which can be considered a permutation of $\{1, \ldots, 52\}$.

The following extremely fast method of generating a uniformly random permutation is known as the *Knuth shuffle,* named after the computer scientist Donald Knuth. Start with the list $(1, 2, \ldots, n)$. Move down the list from the first to the $(n-1)$st position. At each position i, swap the element in that position with a randomly chosen element from positions i to n. After $n - 1$ such swaps, the resulting list will have the desired distribution.

We show that a permutation produced by the Knuth shuffle has the desired probability distribution. Let $(R_1, R_2, \ldots, R_n)$ denote the final output of the Knuth algorithm. Suppose $(r_1, r_2, \ldots, r_n)$ is a permutation of $\{1, 2, \ldots, n\}$. We need to show that $P(R_1 = r_1, R_2 = r_2, \ldots, R_n = r_n) = 1/n!$.

Using the general formula for the intersection of n events Equation 2.4,

$$\begin{aligned} &P(R_1 = r_1, R_2 = r_2, \ldots, R_n = r_n) \\ &= P(R_1 = r_1)P(R_2 = r_2|R_1 = r_1) \cdots P(R_n = r_n|R_1 = r_1, \ldots, R_{n-1} = r_{n-1}). \end{aligned}$$

We have that $P(R_1 = r_1) = 1/n$, since R_1 can take any of n values, all of which are equally likely. Observe that $P(R_2 = r_2|R_1 = r_1) = 1/(n-1)$, since if $R_1 = r_1$, then R_2 can taken any value except r_1, all of which are equally likely. Similarly, for each $i = 2, \ldots, n-1$,

$$P(R_i = r_i|R_1 = r_1, \ldots, R_{i-1} = r_{i-1}) = \frac{1}{n-(i-1)}.$$

Finally, $P(R_n = r_n|R_1 = r_1, \ldots, R_{n-1} = r_{n-1}) = 1$, since if $n-1$ values have been assigned to the first $n-1$ positions of the list, the last remaining value must be assigned to the last position of the list. We have that

$$P(R_1 = r_1, R_2 = r_2, \ldots, R_n = r_n) = \prod_{i=1}^{n-1} \frac{1}{n-(i-1)} = \frac{1}{n!},$$

giving the result.

R: SIMULATING RANDOM PERMUTATIONS

The following code implements the Knuth shuffle to generate a uniformly random permutation. A permutation of size $n = 12$ is output.

```
> n <- 12
> perm <- 1:n
> for (i in 1:(n-1)) {
  x <- sample(i:n,1)
  old <- perm[i]
  perm[i] <- perm[x]
  perm[x] <- old }
> perm
 [1]  4  6 10  9  7  1  3 12  8  2 11  5
```

■

2.5 BAYES FORMULA AND INVERTING A CONDITIONAL PROBABILITY

It should be clear from many previous examples that in general $P(A|B) \neq P(B|A)$. The probability that someone uses hard drugs given that they smoke marijuana (fairly low) is not equal to the probability that they smoke marijuana given that they use hard drugs (fairly high—no pun intended).

When conditional probabilities arise in real-world problems, they can be confusing and subject to misinterpretation. Data may often be given in the form $P(A|B)$, but what is really desired is the "inverse probability" $P(B|A)$.

Bayes formula, also known as Bayes theorem, is a simple but remarkably powerful result for treating such conditional probability problems.

BAYES FORMULA

For events A and B,

$$P(B|A) = \frac{P(A|B)P(B)}{P(A|B)P(B) + P(A|B^c)P(B^c)}.$$

The result is a consequence of two applications of the basic conditional probability formula and the law of total probability, as

$$\begin{aligned} P(B|A) &= \frac{P(BA)}{P(A)} = \frac{P(AB)}{P(A)} \\ &= \frac{P(A|B)P(B)}{P(A)} \\ &= \frac{P(A|B)P(B)}{P(A|B)P(B) + P(A|B^c)P(B^c)}. \end{aligned}$$

Here is a more general form of the formula: Given event A and a sequence of events $B_1, \ldots, B_k$ that partition the sample space, then for each $j = 1, \ldots, k$,

$$P(B_j|A) = \frac{P(A|B_j)P(B_j)}{\sum_{i=1}^{k} P(A|B_i)P(B_i)}.$$

Example 2.15 Diagnostic tests. Diagnostic tests are commonly used to determine the likelihood of disease. Results are never certain with the possibility of false positives and false negatives. Confusion about conditional probability can lead to erroneous conclusions about the efficacy of a particular test.

Suppose a rare disease affects 1% of the population. A hypothetical blood test to detect the disease seems to be relatively accurate. On the one hand, the test has a 99% *sensitivity*, which means that if someone has the disease the chance that the test result is "positive" is 0.99. This also means that there is a 1% chance of error, called the false-negative rate. On the other hand, the test has a 90% *specificity*, which means that if someone does not have the disease the test will be "negative" 9 times out of 10. That is, there is a 10% false-positive rate.

The terms "sensitivity" and "specificity" are used by epidemiologists, public health workers who study the distribution patterns of disease and health events. A major tool in their arsenal is probability.

Suppose a random person gets tested, and the test comes back positive. What is the probability that they actually have the disease?

Before proceeding, you might want to test your intuition and guess the answer without doing any computations. Is the probability of having the disease close to 10, 50, or 90%?

Many people, even experienced doctors, when asked this question assume that the test is fairly accurate and give a high estimate for the probability of disease.

Let D be the event that a person has the disease. Let S be the event that the test comes back positive. The problem is asking for $P(D|S)$.

The data for this problem, however, are of the form $P(S|D)$ and $P(S|D^c)$. The 99% sensitivity rate means $P(S|D) = 0.99$. And the false-negative rate gives $P(S^c|D) = 0.01$. The 90% specificity rate means that $P(S|D^c) = 0.90$. And the false-positive rate gives $P(S^c|D^c) = 0.10$. We are also told that $P(D) = 0.01$.

The data in the problem are probabilities that are conditional on having or not having the disease. But the problem is asking for a conditional probability given the outcome of the test. In order to solve the problem, we need to *invert* the conditional probability $P(D|S)$ to use the available information. By Bayes formula,

$$P(D|S) = \frac{P(S|D)P(D)}{P(S|D)P(D) + P(S|D^c)P(D^c)}$$
$$= \frac{0.99(0.01)}{0.99(0.01) + 0.10(0.99)} = 0.091.$$

The chance of actually having the disease after testing positive is less than 10%. What looked like a fairly accurate diagnostic test is virtually worthless for deciding if someone has the disease.

While the final result might be perplexing, even paradoxical, the key to understanding it is the very low 1% probability of having the disease. Most people do not have the disease. Even though the diagnostic test has a low false-positive rate, the low rate applied to a large population of people who do not have the disease results in a lot of people with false positives.

Imagine a hypothetical town of 10,000. About 100 people (1%) will have the disease. If everyone takes the diagnostic test, about 99 people would test positive and one person would test negative. On the other hand, about 9900 people do not have the disease (99%). And if everyone takes the test, 8910 of them (90%) will test negative. But 990 (10%) will test positive.

This means that about $99 + 990 = 1089$ people test positive. And of those, $99/1089 = 0.091$ have the disease. ■

Example 2.16 Color blindness continued. Given the color blind rates for males and females presented at the beginning of Section 2.4, we found the probability that a random person is color blind. Even though color blindness is fairly unusual, it is much more common among men than women. Suppose a person is color blind. What is the probability they are male?

The problem asks for $P(M|C)$, and again we must invert the conditional probability in order to use the given data, which is conditional on sex, not color blindness. By Bayes formula,

$$P(M|C) = \frac{P(C|M)P(M)}{P(C|M)P(M) + P(C|F)P(F)}$$
$$= \frac{(0.07)(0.49)}{0.041} = 0.947.$$ ■

Example 2.17 Auto accidents continued. Based on insurance company data in Example 2.11, we found the probability that a random policyholder will have a car accident next year. The data show that adults under 25 years old are more likely to have an accident than older people. Suppose a policyholder has an accident. What is the probability they are under 25?

By Bayes formula,

$$P(1|A) = \frac{P(A|1)P(1)}{P(A|1)P(1) + P(A|2)P(2) + P(A|3)P(3)}$$
$$= \frac{(0.11)(0.20)}{(0.11)(0.20) + (0.03)(0.30) + (0.02)(0.50)} = 0.537.$$ ■

Bayesian statistics. Bayes formula is intimately connected to the field of Bayesian statistics. Statistical inference uses data to infer knowledge about an unknown parameter in a population. For instance, 100 fish are caught and measured to estimate the mean length of all the fish in a lake. The 100 fish measurements are the sampled data, and the mean length of all the fish in the lake is the unknown parameter.

In Bayesian statistics, the unknown population parameter is considered random and the tools of probability are used to make probabilistic estimates of the parameter. One conditions on the data in order to compute $P(\text{Parameter}|\text{Data})$.

For example, suppose your friend has three coins: one is fair, one is two-headed, and one is two-tailed. A coin is picked uniformly at random. It is tossed and comes up heads. Which coin is it?

In a Bayesian context, the type of coin is the unknown parameter. The outcome of the coin toss—heads in this case—is the data.

Let $C = 1, 2,$ or 3, depending upon whether the coin is fair, two-headed, or two-tailed, respectively. Let H denote heads. For $c = 1, 2, 3$, Bayes formula gives

$$P(\text{Parameter}|\text{Data}) = P(C = c|\text{H}) = \frac{P(\text{H}|C = c)P(C = c)}{P(\text{H})} = \frac{P(\text{H}|C = c)}{3P(\text{H})}.$$

By the law of total probability,

$$\begin{aligned} P(\text{H}) &= \\ &P(\text{H}|C = 1)P(C = 1) + P(\text{H}|C = 2)P(C = 2) + P(\text{H}|C = 3)P(C = 3) \\ &= \left(\frac{1}{2}\right)\left(\frac{1}{3}\right) + (1)\left(\frac{1}{3}\right) + (0)\left(\frac{1}{3}\right) = \frac{1}{2}. \end{aligned}$$

This gives

$$P(C = c|\text{H}) = 2P(\text{H}|C = c)/3 = \begin{cases} 1/3, & \text{if the coin is fair } (c = 1) \\ 2/3, & \text{if the coin is two-headed } (c = 2) \\ 0, & \text{if the coin is two-tailed } (c = 3). \end{cases}$$

In Bayesian statistics, this probability distribution is called the *posterior distribution* of the parameter (coin) given the data. A "best guess" of your friend's coin is that it is two-headed. It is twice as likely to be two-headed than it is to be fair. See the simulation in **Bayes.R**.

■ **Example 2.18 Bertrand's box paradox.** The French mathematician Joseph Louis François Bertrand posed the following problem in 1889. There are three boxes. One box contains two gold coins; one box contains two silver coins; and one box contains one gold and one silver coin. A box is picked uniformly at random. A coin is picked from the box and it is gold. What is the probability that the other coin in the box is also gold?

The correct answer is 2/3. Many people feel the answer should be 1/2, according to the following logic: The gold coin must have come from one of two boxes that are equally likely, either the gold–gold box or the gold–silver box. Thus, the gold–gold box is chosen half the time.

The fallacy is that once we know the coin is gold, the two boxes are not equally likely. There are *three* gold coins. Two of them come from the gold–gold box, and one from the gold–silver box. If the second coin is gold, it must have come from the gold–gold box and the resulting probability is two out of three.

Here is a conditional probability analysis. Let G_1 and G_2 denote that the first coin and second coin chosen are gold, respectively. Then, $P(G_2|G_1) = P(G_2G_1)/$

$P(G_1)$. The numerator is equal to the probability of picking the gold–gold box, which is 1/3. By conditioning on which box was chosen,

$$\begin{aligned} P(G_1) &= \frac{1}{3}\left(P(G_1|\text{gold–gold}) + P(G_1|\text{silver–silver}) + P(G_1|\text{gold–silver})\right) \\ &= \frac{1}{3}\left(1 + 0 + \frac{1}{2}\right) = \frac{1}{2}. \end{aligned}$$

The desired probability is $P(G_2|G_1) = (1/3)/(1/2) = 2/3$. ■

Perhaps the most well-known probability paradox of the past two decades is the infamous Monty Hall problem. This was popularized in Marilyn vos Savant's *Ask Marilyn* column in Parade magazine in 1990. She wrote,

> Suppose you're on a game show, and you're given the choice of three doors: Behind one door is a car; behind the others, goats. You pick a door, say No. 1, and the host, who knows what is behind the doors, opens another door, say No. 3, which has a goat. He then says to you, "Do you want to pick door No. 2?" Is it to your advantage to switch your choice of doors?

Savant answered—correctly—that it is always beneficial to switch. Without switching, the chance of picking the right door is 1/3. If you switch, the probability increases to 2/3.

One way to see that 2/3 is correct is to observe that with the switching strategy you always win if you initially pick a goat, which happens with probability 2/3.

Some 10,000 people wrote to Parade magazine, including many with PhDs and even some mathematicians, insisting that Savant was wrong. Savant's website (http:/marilynvossavant.com/game-show-problem/) contains many "testimonials" from people, especially high school teachers, who were convinced of her solution after they actually *simulated* the problem, either on a computer or by "playing the game show" in class. We invite the reader to search the web for the many articles, applets, simulations, and discussion of this intriguing problem.

> This branch of mathematics [probability] is the only one, I believe, in which good writers frequently get results which are entirely erroneous.
>
> —Charles S. Pierce

2.6 SUMMARY

Conditional probability is introduced. Conditional probability can often be useful for finding $P(A \text{ and } B)$. Tree diagrams are useful devices for finding probabilities, especially when a sequence of events occur in succession. The classic birthday problem is discussed. The law of total probability is a powerful tool for computing probabilities by *conditioning*. Sometimes a problem asks for $P(A|B)$, but the information we are given is of the form $P(B|A)$. This is a natural setting for Bayes formula, which can be thought of as a way to "invert" a conditional probability.

- **Conditional probability formula:** $P(A|B) = P(AB)/P(B)$.
- **Conditional probability as a probability function:** For fixed B, the conditional probability $P(A|B)$ as a function of its first argument *is* a probability function that satisfies the three defining properties.
- $P(A \text{ and } B) = P(A|B)P(B) = P(B|A)P(A)$.
- **Birthday problem** gives an excellent example of using tree diagrams and conditional probability.
- **Law of total probability:** If $B_1, \ldots, B_k$ partition Ω, then

  $$P(A) = P(A|B_1)P(B_1) + \cdots + P(A|B_k)P(B_k)$$

  For $k = 2$, this gives

  $$P(A) = P(A|B)P(B) + P(A|B^c)P(B^c).$$

  We say that we are *conditioning* on B.
- **Bayes formula:**

  $$P(B|A) = \frac{P(A|B)P(B)}{P(A|B)P(B) + P(A|B^c)P(B^c)}.$$

- **Problem-solving strategies**
  1. **Tree diagrams:** Tree diagrams are intuitive and useful tools for finding probabilities of events that can be ordered sequentially.
  2. **Conditioning:** Given an event A for which we want to find $P(A)$, introducing disjoint events $B_1, \ldots B_k$ and applying the law of total probability, whereby the conditional probabilities $P(A|B_i)$ are easier and more natural to solve than $P(A)$.

EXERCISES

Basics of conditional probability

2.1 Your friend missed probability class today. Explain to your friend, in simple language, the meaning of *conditioning*.

2.2 Suppose $P(A) = P(B) = 0.3$ and $P(A|B) = 0.5$. Find $P(A \cup B)$.

2.3 Suppose $P(A) = P(B) = p_1$ and $P(A \cup B) = p_2$. Find $P(A|B)$.

2.4 **A paradox?** John flips three pennies.

(a) Amy peeks and sees that the first coin lands heads. What is the probability of getting all heads?

(b) Zach peeks and sees that one of the coins lands heads. What is the probability of getting all heads? (The two probabilities are different.)

2.5 Find a simple expression for $P(A|B)$ under the following conditions:

(a) A and B are disjoint.

(b) $A = B$.

(c) A implies B.

(d) B implies A.

2.6 **Nontransitive dice**: Consider three nonstandard dice. Instead of the numbers 1 through 6, die A has two 3's, two 5's, and two 7's; die B has two 2's, two 4's, and two 9's; and die C has two 1's, two 6's, and two 8's, as in Figure 2.8.

Suppose dice A and B are rolled. Show that A is more likely to get the higher number. That is, $P(A > B) > 0.50$, where $\{A > B\}$ denotes the event that A beats B. Hint: Condition on the outcome of die A.

Now show that if B and C are rolled, B is more likely to get the higher number. And, remarkably, if C and A are rolled, C is more likely to get the higher number.

Many relationships in life are *transitive.* For instance, if Amy is taller than Ben and Ben is taller than Charlie, then Amy is taller than Charlie. But these dice show that the relation "more likely to roll a higher number" is not transitive.

The dice are the basis of a magic trick. You pick any die. Then I can always pick a die that is more likely to beat yours. If you pick A, I pick C. If you pick B, I pick A. And if you pick C, I pick B.

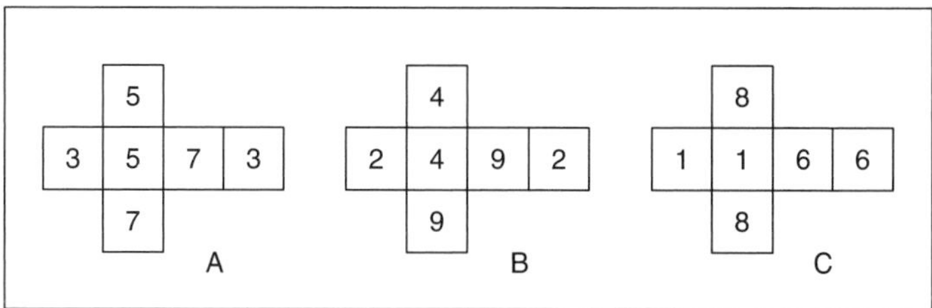

FIGURE 2.8: Nontransitive dice.

2.7

(a) True or false: $P(A|B) + P(A|B^c) = 1$. Either show it true for any event A and B or exhibit a counter-example.

(b) True or false: $P(A|B) + P(A^c|B) = 1$. Either show it true for any event A and B or exhibit a counter-example.

2.8 A bag of 15 Scrabble tiles contains three each of the letters A, C, E, H, and N. If you pick six letters one at a time, what is the chance that you spell C-H-A-N-C-E?

2.9 In the game of Poker, a *flush* is five cards of the same suit. Use conditional probability to find the probability of being dealt a flush.

2.10 Box A contains one white ball and two red balls. Box B contains one white ball and three red balls. A ball is picked at random from box A and put into box B. A ball is then picked at random from box B. Draw a tree diagram for this problem and use it to find the probability that the final ball picked is white.

2.11 Bob is taking a test. There are two questions he is stumped on and he decides to guess. Let A be the event that he gets the first question right; let B be the event he gets the second question right (adapted from Blom et al., 1991).

(a) Obtain an expression for p_1, the probability that he gets both questions right conditional on getting the first question right.

(b) Obtain an expression for p_2, the probability that he gets both questions right conditional on getting either of the two questions right (A or B).

(c) Show that $p_2 \leq p_1$. This may seem paradoxical. Knowledge that A or B has taken place makes the conditional probability that A and B happens smaller than when we know that A has happened. Can you untangle the paradox?

2.12 Suppose $P(A) = 1/2$, $P(B^c|AC) = 1/3$ and $P(C|A) = 1/4$. Find $P(ABC)$.

2.13 Prove the addition rule for conditional probabilities. That is, show that for events A, B, and C,

$$P(A \cup B|C) = P(A|C) + P(B|C) - P(AB|C).$$

Conditioning, law of total probability, and Bayes formula

2.14 The planet Mars revolves around the sun in 687 days. Answer Von Mises' birthday question for Martians. That is, how many Martians must be in a room before the probability that some share a birthday becomes at least 50%.

2.15 Jimi has 5000 songs on his iPod shuffle, which picks songs uniformly at random. Jimi plans to listen to 100 songs today. What is the chance he will hear at least one song more than once?

2.16 A standard deck of cards has one card missing. A card is then picked from the deck. What is the chance that it is a heart? Solve this problem in two ways:

(a) Condition on the missing card.

(b) Appeal to symmetry. That is, make a *qualitative* argument for why the answer should not depend on the heart suit.

2.17 Amy has two bags of candy. The first bag contains two packs of M&Ms and three packs of Gummi Bears. The second bag contains four packs of M&Ms

and two packs of Gummi Bears. Amy chooses a bag uniformly at random and then picks a pack of candy. What is the probability that the pack chosen is Gummi Bears? Solve (i) by using a tree diagram and (ii) by another method.

2.18 In a roll of two tetrahedron dice, each labeled one to four, let X be the sum of the dice. Let $A = \{X \text{ is prime}\}$ and

$$B_1 = \{X = 2\},\ B_2 = \{3 \leq X \leq 5\},\ B_3 = \{6 \leq X \leq 7\},\ \text{and}\ B_4 = \{X = 8\}.$$

Observe that the B_i's partition the sample space.

(a) Draw a diagram as in Figure 2.7. Label all 16 outcomes in the diagram.

(b) Illustrate the law of total probability by writing out formula 2.9 and finding the probabilities for each term in the equation.

2.19 Give a formula for $P(A|B^c)$ in terms of $P(A)$, $P(B)$, and $P(AB)$ only.

2.20 **Lewis Carroll's pillow problem #5.** Lewis Carroll, author of Alice's Adventures in Wonderland, is the pen name of Charles Lutwidge Dodgson, who was an Oxford mathematician and logician. Lewis Carroll's *Pillow Problems* (1958), is a collection of 72 challenging, and sometimes amusing, mathematical problems, several of which involve probability. Here is Problem #5.

A bag contains one counter, known to be either white or black. A white counter is put in, the bag shaken, and a counter drawn out, which proves to be white. What is now the chance of drawing a white counter?

2.21 **Pillow problem #72.** Here is Lewis Carroll's last pillow problem (1958).

A bag contains two counters, as to which nothing is known except that each is either black or white. Ascertain their colors without taking them out of the bag.

Carroll's answer is

One is black, and the other is white.

What do you think of his "solution"?

We know that, if a bag contained three counters, two being black and one white, the chance of drawing a black one would be 2/3; and that any *other* state of things would *not* give this chance.

Now the chances, that the given bag contains (a) BB, (b) BW, (c) WW, are respectively, 1/4, 1/2, 1/4. Add a black counter. Then the chances that it contains (a) BBB, (b) BWB, (c) WWB are, as before, 1/4, 1/2, 1/4.

Hence the chance of now drawing the black one is

$$\frac{1}{4}(1) + \frac{1}{2}\left(\frac{2}{3}\right) + \frac{1}{4}\left(\frac{1}{3}\right) = \frac{2}{3}.$$

Hence the bag now contains BBW (since any *other* state of things would *not* give this chance). Hence, before the black counter was added, it contained BW, i.e., one black counter and one white.

2.22 Consider flipping coins until either two heads HH or heads then tails HT first occurs. By conditioning on the first coin toss, find the probability that HT occurs before HH.

2.23 In a certain population of youth, the probability of being a smoker is 20%. The probability that at least one parent is a smoker is 30%. And if at least one parent is a smoker, the probability of being a smoker is 35%. Find the probability of being a smoker if neither parent is a smoker.

2.24 According to the National Cancer Institute, for women between 50 and 59, there is a 2.38% chance of being diagnosed with breast cancer. Screening mammography has a sensitivity of about 85% for women over 50 and a specificity of about 95%. That is, the false-negative rate is 15% and the false-positive rate is 5%. If a woman over 50 has a mammogram and it comes back positive for breast cancer, what is the probability that she has the disease?

2.25 A polygraph (lie detector) is said to be *90% reliable* in the following sense: There is a 90% chance that a person who is telling the truth will pass the polygraph test; and there is a 90% chance that a person telling a lie will fail the polygraph test.

(a) Suppose a population consists of 5% liars. A random person takes a polygraph test, which concludes that they are lying. What is the probability that they are actually lying?

(b) Consider the probability that a person is actually lying given that the polygraph says that they are. Using the definition of reliability, how reliable must the polygraph test be in order that this probability is at least 80%?

2.26 An eyewitness observes a hit-and-run accident in New York City, where 95% of the cabs are yellow and 5% are blue. A witness asserts the cab was blue. A police expert believes the witness is 80% reliable. That is, the witness will correctly identify the color of a cab 80% of the time. What is the probability that the cab actually was blue?

2.27 Your friend has three dice. One die is fair. One die has fives on all six sides. One die has fives on three sides and fours on three sides. A die is chosen at random. It comes up five. Find the probability that the chosen die is the fair one.

Simulation and R

2.28 The R command

```
> sample(1:365,23,replace=T)
```

simulates birthdays from a group of 23 people. The expression

```
> 2 %in% table(sample(1:365,23,replace=T))
```

can be used to simulate the birthday problem. It creates a frequency table showing how many people have each birthday, and then determines if two is in that table; that is, whether two or more people have the same birthday. Use and suitably modify the expression for the following problems.

(a) Simulate the probability that two people have the same birthday in a room of 23 people.

(b) Estimate the number of people needed so that the probability of a match is 95%.

(c) Find the approximate probability that three people have the same birthday in a room of 50 people.

(d) Estimate the number of people needed so that the probability that three people have the same birthday is 50%.

2.29 Simulate the nontransitive dice probabilities in Exercise 2.6.

2.30 The following problem appeared in the news column "Ask Marilyn" on September 19, 2010.

> Four identical sealed envelopes are on a table. One of them contains a \$100 bill. You select an envelope at random and hold it in your hand without opening it. Two of the three remaining envelopes are then removed and set aside, unopened. You are told that they are empty. You are given the choice of keeping the envelope you chose or exchanging it for the one on the table.
>
> What should you do? a) Keep your envelope. b) Switch it. c) It doesn't matter.

Write a simulation to find the probability of selecting the \$100 bill when you switch. Confirm the results of your simulation with an exact analysis.

2.31 Modify the **Blackjack.R** script to simulate the probability of being dealt two cards of the same suit. Compare with the exact answer.

2.32 See Example 2.14 for generating a random permutation. Implement the algorithm in R for shuffling a standard deck of cards. Use it to simulate the probability that in a randomly shuffled deck the top and bottom cards are the same suit.

2.33 Simulate Bertrand's box paradox (Example 2.18).

2.34 Make up your own random experiment involving conditional probability. Write an R script to simulate your problem and compare the simulation to your exact solution.

3

INDEPENDENCE AND INDEPENDENT TRIALS

> Independence? That's middle class blasphemy. We are all dependent on one another, every soul of us on earth.
>
> —George Bernard Shaw

3.1 INDEPENDENCE AND DEPENDENCE

The probability that you get an A in your math class is probably dependent on how much you study. But it probably is not dependent on the color of your roommate's hair. Intuitively, your grade and your roommate's hair color are independent events. On the other hand, you are more likely to get an A in math class if you study hard. Most likely,

$$P(\text{A in math class}|\text{Roommate is a red head}) = P(\text{A in math class})$$

while

$$P(\text{A in math class}|\text{Study hard}) > P(\text{A in math class}).$$

This suggests the definition of independent events.

Probability: With Applications and R, First Edition. Robert P. Dobrow.

INDEPENDENT EVENTS

Definition 3.1. *Events A and B are independent if*

$$P(A|B) = P(A). \tag{3.1}$$

Events that are not independent are said to be *dependent.*

Example 3.1 A card is drawn from a standard deck. Let A be the event that it is a spade. Let B be the event that it is an ace. Then

$$P(A|B) = \frac{P(AB)}{P(B)} = \frac{P(\text{Ace of Spades})}{P(\text{Ace})} = \frac{1/52}{1/13} = \frac{1}{4} = P(A).$$

The two events are independent. ■

This example illustrates that independence is *not* the same as mutually exclusive. Beginning students sometimes confuse the two. The events A and B are not mutually exclusive since $AB = \{\text{Ace of Spades}\} \neq \emptyset$. One can tell if two events are mutually exclusive by looking at the Venn diagram. Independence is more subtle. A Venn diagram alone will not identify independence. Knowledge of the probabilities $P(A)$, $P(B)$, and $P(AB)$ is required.

If A and B are independent, then B and A are independent. And thus $P(B|A) = P(B)$. This follows from the defining formula Equation 3.1 since

$$P(B|A) = \frac{P(AB)}{P(A)} = \frac{P(A|B)P(B)}{P(A)} = \frac{P(A)P(B)}{P(A)} = P(B).$$

If A and B are independent events, then rearranging the conditional probability formula Equation 2.1 gives

$$P(AB) = P(A|B)P(B) = P(A)P(B).$$

The equation

$$P(AB) = P(A)P(B) \tag{3.2}$$

is sometimes used as the primary definition of independent events as it is referred to as the *multiplication rule for independent events.*

The advantage of Definition 3.1 is that it is intuitive—events are independent if knowledge of whether or not one event occurs does not affect the probability of the other. On the other hand, Equation 3.2 highlights an important computational advantage of working with independent events—the probability that two independent events both occur is the *product* of their individual probabilities. This is sometimes called the simple multiplication rule for independent events.

From a practical modeling perspective, we often decide *a priori* that events are independent based on assumptions that seem plausible in a real-world context. For instance, successive coin flips are modeled as independent events. The model may be useful and a reasonable approximation of reality, but no model is 100% true.

COIN TOSSING IN THE REAL WORLD

In a fascinating investigation of coin tossing, Diaconis (2007) try to analyze the natural process of flipping a coin that is caught in the hand. They use high-speed slow motion cameras to record data of actual coin tosses. Their paper, which contains a lot of physics, shows that vigorously flipped coins are slightly biased to come up the same way they started. They show that

$$P(\text{Heads}|\text{Start with Heads}) = 0.508 \neq 0.50.$$

Example 3.2 What is the probability of getting "snake-eyes"—two ones—when rolling two dice?

In Chapter 1, we enumerated the sample space for two dice rolls. By the multiplication principle there are 36 outcomes. Assuming each outcome is equally likely, $P(\text{Snakeyes}) = 1/36$.

Now an alternate derivation can be given using independence. Let A_1 and A_2 denote getting a one on the first and second rolls, respectively. Then

$$P(\text{Snake-eyes}) = P(A_1 A_2) = P(A_1)P(A_2) = \left(\frac{1}{6}\right)\left(\frac{1}{6}\right) = \frac{1}{36}. \qquad \blacksquare$$

Independence of two events means that knowledge of whether or not one event occurs does not affect the probability of the other event occurring. Thus if A and B are independent events, then intuitively the pairs (A, B^c), (A^c, B), and (A^c, B^c) are also independent events.

To show independence of A and B^c, write $A = AB \cup AB^c$. Then $P(A) = P(AB) + P(AB^c)$. Hence,

$$\begin{aligned} P(AB^c) &= P(A) - P(AB) \\ &= P(A) - P(A)P(B) \\ &= P(A)[1 - P(B)] = P(A)P(B^c). \end{aligned}$$

Switching A and B shows independence of A^c and B. See Exercise 3.1 for showing that A^c and B^c are independent.

Mutual and pairwise independence. How should independence be defined for more than two events? Surely if three events A, B, C are independent, then each pair (A, B), (A, C), and (B, C) should also be independent. Generalizing the simple multiplication rule Equation 3.2, you might also guess that independence is equivalent to the identity $P(ABC) = P(A)P(B)P(C)$. This, however, is not enough. Examples can be found where this equation holds for A, B, and C, but no two of the three events are independent. (See Exercise 3.4.)

For three events A, B, C, independence requires the multiplication rule to hold for the collection of three events *and* for all subgroups of two events. That is, we need

$$P(ABC) = P(A)P(B)P(C)$$

and

$$P(AB) = P(A)P(B), \quad P(AC) = P(A)P(C), \quad \text{and} \quad P(BC) = P(B)P(C).$$

For larger collections of events—including infinite collections—independence requires the multiplication rule to hold for *all* finite subgroups. This gives the general definition.

INDEPENDENCE FOR COLLECTION OF EVENTS

Definition 3.2. *A collection of events is independent if for every finite subgroup* $A_1, \ldots, A_k$,

$$P(A_1 \cdots A_k) = P(A_1) \cdots P(A_k). \tag{3.3}$$

This definition of independence is also called *mutual independence.* That is, mutual independence is a synonym for independence.

If we restrict to the case $k = 2$ and only require pairs of events to satisfy Equation 3.3, we say that the collection is *pairwise independent.* That is, a collection of events is pairwise independent if the multiplication rule holds for every pair of events. Clearly mutual independence implies pairwise independence. However, the converse is not always true.

Example 3.3 Flip two coins. Let A be the event that the first coin comes up heads; B the event that the second comes up heads; and C the event that both coins come up the same, either heads or tails. Check that $P(A) = P(B) = P(C) = 1/2$. Also,

$$\begin{aligned} P(AB) = P(A)P(B) &= \frac{1}{4} \\ &= P(AC) = P(A)P(C) = P(BC) = P(B)P(C). \end{aligned}$$

The three events are pairwise independent. But $P(ABC) = P(\text{Two heads}) = 1/4$ and

$$P(A)P(B)P(C) = (1/2)(1/2)(1/2) = 1/8 \neq 1/4.$$

The three events are not mutually independent. The fact that the three events are not independent can be seen without any calculation since if A and B both occur, then so does C. ■

Example 3.4 Data from the Red Cross on the distribution of blood type in the United States are given in Table 3.1.

In the case of needed blood plasma, people with blood group O (O+ and O−) can receive plasma from anyone. Suppose three people are selected at random. What is the probability they are all blood group O?

Let O_1, O_2, O_3 be the events that the first, second, and third persons selected are from blood group O, respectively. By independence,

$$\begin{aligned} P(O_1O_2O_3) &= P(O_1)P(O_2)P(O_3) \\ &= (0.374 + 0.066)^3 = (0.44)^3 = 0.085. \end{aligned}$$ ■

TABLE 3.1: Distribution of blood type in the United States.

O+	A+	B+	AB+	O−	A−	B−	AB−
0.374	0.357	0.085	0.034	0.066	0.063	0.015	0.006

How might independence be violated in this last example? If we chose three people from the same family, or the same nationality or ethnicity, this would violate independence, as people from such groups might be more likely to have similar blood types. Note that without the property of independence it would not be possible to answer the probability question without additional information.

Sampling with and without replacement. Independence is often associated with sampling with replacement. For instance, a bowl contains 10 balls of different colors, including red and green. Pick two balls. Let R_1 be the event that the first ball is red. Let G_2 be the event that the second ball is green. If we sample with replacement, then $P(G_2|R_1) = 1/10 = P(G_2)$, and the events are independent. After the first ball is picked, it is returned to the bowl and the second selection is made as if nothing changed.

On the other hand, sampling without replacement gives $P(G_2|R_1) = 1/9$, since once the first red ball is picked there are nine balls remaining. For $P(G_2)$ appeal to symmetry. The second ball is equally likely to be any of the 10 colors. Thus, $P(G_2) = 1/10$. The events are not independent.

Note that for finding $P(G_2)$ one can also condition on whether or not the first ball selected is green, giving

$$P(G_2) = P(G_2|G_1)P(G_1) + P(G_2|G_1^c)P(G_1^c) = 0\left(\frac{1}{10}\right) + \frac{1}{9}\left(\frac{9}{10}\right) = \frac{1}{10}.$$

Typically, when sampling with replacement, successive outcomes are independent events. When sampling without replacement, they are not independent. However, when the population size is very big (when the number of balls in the bowl is large), the actual numerical probabilities resulting from the two sampling schemes are practically the same. As a mind stretch, you can think of sampling without replacement from a bowl of size n, and then let $n \to \infty$. Sampling without replacement from an "infinite bowl" gives sampling with replacement!

In statistical surveying, while many practical sampling schemes from large populations are done without replacement, the analysis is often done with replacement to exploit the computational advantages of working with independence.

Example 3.5 Coincidences and the birthday problem. It is first week on campus. Six new students are sitting together in the cafeteria. They start asking about each other. What dorm floor are you on? (There are 30 possibilities.) What frosh seminar are you in (with 40 to choose from)? What is your Zodiac sign?

Remarkably two students have a match. What a coincidence! But is it really that remarkable?

Generalizing the birthday problem (see Section 2.3.1), the probability that none of the six students are on the same dorm floor is $(29 \cdot 28 \cdot 27 \cdot 26 \cdot 25)/30^5$. Similar calculations are done for frosh seminar and Zodiac sign. If we assume that the three categories dorm floor, seminar, and Zodiac sign are independent, then the probability that there is no match for any category among the six students at the table is the product of the probabilities of no match in each category, which is

$$\left(\frac{29}{30} \cdots \frac{25}{30}\right)\left(\frac{39}{40} \cdots \frac{35}{40}\right)\left(\frac{11}{12} \cdots \frac{7}{12}\right) = 0.0882.$$

Therefore, the probability of a least one match among six people is

$$P(\text{Match}) = 1 - 0.0882 = 0.9118.$$

Not such a coincidence after all. ∎

Example 3.6 *A* before *B*. The following scenario is very general. A random experiment is performed repeatedly. Events A and B are mutually exclusive. What is the probability that A occurs before B?

We assume that $B \neq A^c$. Otherwise, the solution is not very interesting. If $B = A^c$, then either A or B occurs and the solution is just $P(A)$.

Let $p = P(A)$, $q = P(B)$, with $p + q < 1$. We give two solutions to this problem—one analytical, the other probabilistic.

Solution 1: Let E be the event that A occurs before B. For E to happen, either A occurs right away or neither A nor B occurs for one or more trials and then A occurs. That is, for some $k \geq 0$, neither A nor B occur for k trials, and then A occurs on trial $k + 1$. For each k, let E_k denote the event that (i) A first occurs on trial $k + 1$

and (ii) neither A nor B occur on the first k trials. Observe that the E_k's are mutually exclusive and $E = \bigcup_{k=0}^{\infty} E_k$.

The probability that on a particular trial neither A nor B occur is $1 - p - q$. By independence, $P(E_k) = (1 - p - q)^k p$. This gives,

$$\begin{aligned} P(E) &= P\left(\bigcup_{k=0}^{\infty} E_k\right) = \sum_{k=0}^{\infty} P(E_k) \\ &= \sum_{k=0}^{\infty} (1-p-q)^k p \\ &= p\left(\frac{1}{1-(1-p-q)}\right) = \frac{p}{p+q}. \end{aligned}$$

The geometric series converges since $0 < p + q < 1$ and thus $0 < 1 - p - q < 1$.

Solution 2: Condition on the first trial. There are three possibilities:

(i) If A occurs on the first trial, then E occurs.

(ii) If B occurs on the first trial, then E does not occur.

(iii) If neither A nor B occur on the first trial, then we "start over again" to determine whether or not A occurs first. It is as if the first trial did not happen, and the problem begins anew at the second trial. This is a consequence of independence. The event that neither A nor B occur on the first trial is independent of the event that A occurs before B. Letting C be the event that neither A nor B occurs on the first trial, this gives that $P(E|C) = P(E)$.

By the law of total probability,

$$\begin{aligned} P(E) &= P(E|A)P(A) + P(E|B)P(B) + P(E|C)P(C) \\ &= 1(p) + 0(q) + P(E)(1-p-q) \\ &= p + P(E)(1-p-q). \end{aligned}$$

The equation contains $P(E)$ on both sides. Solving for $P(E)$ gives $P(E) = p/(p+q)$. We restate this useful result.

A BEFORE B

In repeated independent trials, if A and B are mutually exclusive events, the probability that A occurs before B is

$$\frac{P(A)}{P(A)+P(B)} \tag{3.4}$$

■

For example, when repeatedly rolling pairs of dice, what is the probability that a sum of nine appears before a sum of 7?

The probability of getting nine is $P(\{(3,6),(4,5),(5,4),(6,3)\}) = 4/36$. The probability of getting 7 is 6/36. The desired probability is

$$P(9 \text{ before } 7) = \frac{4/36}{4/36 + 6/36} = \frac{2}{5}.$$

We make use of these ideas in the analysis of a popular casino game.

Example 3.7 Craps. The dice game craps is fast-paced, exciting, and typically offers the best odds at the casino. The player rolls two dice. If you get a 7 or 11, you win immediately. If you get a 2, 3, or 12, you lose. For any other outcome (4, 5, 6, 8, 9, 10), the number rolled is your *point.* You now roll again and keep rolling until you either roll your point again or roll a 7. If the 7 comes before the point, you lose. If the point comes before the 7, you win. What is the probability of winning at craps?

Let W denote winning. Conditioning on the first outcome,

$$\begin{aligned} P(W) &= \sum_{k=2}^{12} P(W|k)P(k) \\ &= (0)[P(2)+P(3)+P(12)] + (1)[P(7)+P(11)] \\ &\quad + \sum_{k\in\{4,5,6,8,9,10\}} P(W|k)P(k) \\ &= P(7) + P(11) + \sum_{k\in\{4,5,6,8,9,10\}} P(k \text{ before } 7)P(k), \end{aligned}$$

where $\{k \text{ before } 7\}$ denotes the event that the point k comes up before 7 in repeated rolls. The probability of winning at craps is thus

$$\begin{aligned} P(W) &= P(7) + P(11) + \sum_{k\in\{4,5,6,8,9,10\}} P(k \text{ before } 7)P(k) \\ &= \frac{6}{36} + \frac{2}{36} + \left(\frac{3/36}{3/36+6/36}\right)\frac{3}{36} + \left(\frac{4/36}{4/36+6/36}\right)\frac{4}{36} \\ &\quad + \left(\frac{5/36}{5/36+6/36}\right)\frac{5}{36} + \left(\frac{5/36}{5/36+6/36}\right)\frac{5}{36} \\ &\quad + \left(\frac{4/36}{4/36+6/36}\right)\frac{4}{36} + \left(\frac{3/36}{3/36+6/36}\right)\frac{3}{36} \\ &= 0.4929. \end{aligned}$$

See the script **Craps.R** to simulate the game. ■

3.2 INDEPENDENT RANDOM VARIABLES

The intuitive notion of independence extends to random variables. We say that random variables X and Y are independent to mean that knowledge of the outcome of X does not affect the probability of the outcome of Y.

Independence for random variables means that for all x and y, the events $\{X = x\}$ and $\{Y = y\}$ are independent events.

INDEPENDENCE OF RANDOM VARIABLES

Definition 3.3. *Random variables X and Y are said to be independent if*

$$P(X = x|Y = y) = P(X = x), \quad \textit{for all } x, y. \tag{3.5}$$

Equivalently,

$$P(X = x, Y = y) = P(X = x)P(Y = y), \quad \textit{for all } x, y. \tag{3.6}$$

A collection of random variables, such as an infinite sequence, is independent if for all finite subgroups $\mathrm{X}_1, \ldots, \mathrm{X}_\mathrm{k}$ *of the collection,*

$$P(X_1 = x_1, \ldots, X_k = x_k) = P(X_1 = x_1) \cdots P(X_k = x_k),$$

for all $x_1, \ldots, x_k$.

Example 3.8 The Current Population Survey of 2010 provides data on the proportion of family households in the United States by number of children under 18 years old. (See Table 3.2).

In a sample of four households, what is the probability that no household has children?

Although the actual sampling was done without replacement, since the population is so large, we assume sampling with replacement and independence, as the difference is negligible.

For $k = 1, 2, 3, 4$, let X_k be the number of children for the kth family household in the sample. Then the desired probability is

$$P(X_1 = 0, X_2 = 0, X_3 = 0, X_4 = 0) =$$
$$P(X_1 = 0)P(X_2 = 0)P(X_3 = 0)P(X_4 = 0) = (0.5533)^4 = 0.0938. \quad \blacksquare$$

TABLE 3.2: Distribution of number of children in U.S. households.

0	1	2	3 or more
0.5533	0.1922	0.1642	0.0903

For a subset A of real numbers, the event $\{X \in A\}$ is the event that X takes values in A. For instance, if $A = (1, 3)$, then $P(X \in A) = P(1 < X < 3)$. If A is the set of positive even numbers, then

$$P(X \in A) = P(X \text{ is even}) = P\left(\bigcup_{k=0}^{\infty} \{X = 2k\}\right).$$

With this notation, independence of random variables can be expressed more generally as follows.

INDEPENDENT RANDOM VARIABLES

Definition 3.4. *Random variables X and Y are independent if for all* $A, B \subseteq \mathbb{R}$,

$$P(X \in A, Y \in B) = P(X \in A)P(Y \in B).$$

It should be clear that this definition implies Equation 3.6. Simply take $A = \{x\}$ and $B = \{y\}$. Conversely, suppose Definition 3.3 holds. Then

$$\begin{aligned} P(X \in A, Y \in B) &= \sum_{x \in A} \sum_{y \in B} P(X = x, Y = y) \\ &= \sum_{x \in A} \sum_{y \in B} P(X = x)P(Y = y) \\ &= \left(\sum_{x \in A} P(X = x)\right)\left(\sum_{y \in B} P(Y = y)\right) \\ &= P(X \in A)P(Y \in B). \end{aligned}$$

3.3 BERNOULLI SEQUENCES

A random variable that takes only two values 0 and 1 is called a *Bernoulli random variable.*

BERNOULLI DISTRIBUTION

A random variable X has a *Bernoulli distribution with parameter* p if

$$P(X = 1) = p \quad \text{and} \quad P(X = 0) = 1 - p,$$

for $0 < p < 1$. We write $X \sim \text{Ber}(p)$.

There are a vast range of applications that can be modeled as sequences of Bernoulli random variables, starting with coin flips. We often refer to the states of a Bernoulli variable as "success" and "failure," with the parameter p identified as the *success parameter* or *success probability*.

Example 3.9 A manufacturing process produces electronic components that occasionally are defective. There is is a one-in-a-thousand chance that a component is defective. Furthermore, whether or not a component is defective is independent of any other component. If n components are produced in a day, find the probability that at least one is defective.

Let

$$X_k = \begin{cases} 1, & \text{if the } k\text{th component is defective} \\ 0, & \text{otherwise.} \end{cases}$$

for $k = 1, \ldots, n$. Each X_k has a Bernoulli distribution with parameter $p = 0.001$. The probability that at least one component is defective is

$$\begin{aligned} &P(\text{At least one component defective}) \\ &= 1 - P(\text{No defective components}) \\ &= 1 - P(X_1 = 0, \ldots, X_n = 0) \\ &= 1 - P(X_1 = 0) \cdots P(X_n = 0) \\ &= 1 - (1 - 0.001) \cdots (1 - 0.001) = 1 - (0.999)^n. \end{aligned}$$

If the manufacturer produces 500 components per day, then the probability that at least one component will be defective is $1 - 0.999^{500} = 39.36\%$

In this last example, the random variables $X_1, \ldots, X_n$ were all independent and had the same Bernoulli distribution. We say that $X_1, \ldots, X_n$ is an *independent and identically distributed (i.i.d.) sequence.* ■

INDEPENDENT AND IDENTICALLY DISTRIBUTED (i.i.d.) SEQUENCES

A sequence of random variables is said to be independent and identically distributed (i.i.d.) if the random variables are independent and have the same probability distribution.

An i.i.d. sequence consisting of Bernoulli random variables is called a *Bernoulli sequence.* Random samples in statistics are often modeled as i.i.d. sequences.

Example 3.10 A national pollster wants to estimate the president's approval rating. Let p be the unknown proportion of U.S. adults who approve the president's handling of his job. A random sample of 1000 adults is taken. Each person is asked, "Do you agree or disagree with the president's handling of his job?" The responses are modeled as an i.i.d. Bernoulli sequence $X_1, \ldots, X_{1000}$, where

$$X_k = \begin{cases} 1, & \text{if the } k\text{th person in the sample approves} \\ 0, & \text{if the } k\text{th person in the sample does not approve.} \end{cases}$$

The goal is to use these data to estimate the unknown proportion p. The proportion of people in the sample who approve the president's handling of his job is $(X_1 + \cdots + X_{1000})/1000$. This *sample proportion* is an estimate of the *population proportion* p. ∎

Example 3.11 A four-sided tetrahedron die labeled 1, 2, 3, 4 is thrown four times. What is the probability of rolling a three exactly twice?

The four die rolls are modeled by four independent Bernoulli trials with success probability $p = 1/4$. Let 3 denote rolling a three and N denote not rolling a 3. A not-so-elegant solution is to enumerate all the possible ways to get a 3 exactly twice: $33NN$, $3N3N$, $3NN3$, $N33N$, $N3N3$, $NN33$. There are six possibilities. By independence, each outcome occurs with probability $(1/4)^2(3/4)^2 = 9/256$. Thus, the probability of rolling a 3 exactly twice is $6(9/256) = 27/128 = 0.211$.

We will generalize this problem and simplify its solution when we introduce the binomial distribution next. ∎

A *binary sequence* is a list, each of whose elements can take one of two values, which we generically take to be zeros and ones (as opposed to N's and 3's). In the last example, we counted that there are six binary sequences of length four made up of two ones. The general counting problem suggested by the example is: How many binary sequences of length n contain exactly k ones?

3.4 COUNTING II

In Chapter 2, you learned how to count ordered lists and permutations. Here we count unordered sets and subsets.

We first show a simple yet powerful correspondence between subsets of a set and binary sequences, or lists. This correspondence will allow us to relate counting results for sets to those for lists, and vice versa.

To illustrate, consider a group of n people lined up in a row and numbered 1 to n. Each person holds a card. On one side of the card is a 0; on the other side is a 1. Initially, all the cards are turned to 0. The n cards from left to right form a binary list.

Select a subset of the n people. Each person in the subset turns over his/her card, from 0 to 1. The cards taken from left to right form a new binary list. For instance, if $n = 6$ and the first and third persons are selected, the corresponding list is $(1, 0, 1, 0, 0, 0)$.

Conversely, given a list of zeros and ones, we select those people corresponding to the ones in the list. That is, if a one is in the kth position in the list, then

TABLE 3.3: Correspondence between subsets and binary lists.

Subset	List
$\emptyset$	$(0,0,0)$
$\{1\}$	$(1,0,0)$
$\{2\}$	$(0,1,0)$
$\{3\}$	$(0,0,1)$
$\{1,2\}$	$(1,1,0)$
$\{1,3\}$	$(1,0,1)$
$\{2,3\}$	$(0,1,1)$
$\{1,2,3\}$	$(1,1,1)$

person k is selected. If the list is $(1,0,1,1,1,1)$, then all but the second person are selected.

This establishes a one-to-one correspondence between subsets of $\{1,\ldots,n\}$ and binary lists of length n. Table 3.3 shows the correspondence for the case $n = 3$.

A one-to-one correspondence between two finite sets means that both sets have the same number of elements. Our one-to-one correspondence shows that the number of subsets of an n-element set is equal to the number of binary lists of length n. The number of binary lists of length n is easily counted by the multiplication principle. Since there are two choices for each element of the list, there are 2^n binary lists. The number of subsets of an n-element set immediately follows.

COUNTING SUBSETS

There are 2^n subsets of an n-element set.

There are 2^n binary lists of length n.

Counting k-element subsets. In the subset-list correspondence, observe that every k-element subset of $\{1,\ldots,n\}$ corresponds to a binary list with k ones. And conversely, every binary list with exactly k ones corresponds to a k-element subset. This is true for each $k = 0,1,\ldots,n$. For instance, in the case $n = 4$ and $k = 2$, the subsets are

$$\{1,2\},\ \{1,3\},\ \{1,4\},\ \{2,3\},\ \{2,4\},\ \{3,4\}$$

with corresponding lists

$$(1,1,0,0),\ (1,0,1,0),\ (1,0,0,1),\ (0,1,1,0),\ (0,1,0,1),\ (0,0,1,1).$$

Our goal is to count the number of binary lists of length n with exactly k ones. We will do so by first counting the number of k-element subsets of an n-element set.

Given a k-element subset, there are $k!$ ordered lists that can be formed by permuting the elements of that subset. For instance, the three-element subset $\{1, 3, 4\}$ yields the $3! = 6$ lists (1,3,4), (1,4,3), (3,1,4), (3,4,1), (4,1,3), and (4,3,1).

It follows that the number of k-element subsets of $\{1, \ldots, n\}$ is equal to $k!$ times the number of lists of length k made up of the elements $\{1, \ldots, n\}$. By the multiplication principle, there are $n \times (n-1) \times \cdots \times (n-k+1)$ such lists. Thus, the number of k-element subsets of $\{1, \ldots, n\}$ is equal to

$$\frac{n \times (n-1) \times \cdots \times (n-k+1)}{k!} = \frac{n!}{k!(n-k)!}.$$

This quantity is so important it gets its own name. It is known as a *binomial coefficient*, written

$$\binom{n}{k} = \frac{n!}{k!(n-k)!},$$

and read as "n choose k."

By the one-to-one correspondence between k-element subsets and binary lists with exactly k ones, we have the following.

COUNTING k-ELEMENT SUBSETS AND LISTS WITH k ONES

There are $\binom{n}{k}$ k-element subsets of $\{1, \ldots, n\}$.

There are $\binom{n}{k}$ binary lists of length n with exactly k ones.

Example 3.12 A classroom of 10 students has 6 females and 4 males. (i) What are the number of ways to pick five students for a project? (ii) How many ways can we pick a group of two females and three males?

(i) There are $\binom{10}{5} = 252$ ways to pick five students.

(ii) There are $\binom{6}{2} = 15$ ways to pick the females, and $\binom{4}{3} = 4$ ways to pick the males. By the multiplication principle, there are $15 \times 4 = 60$ ways to pick the group. ■

Binomial coefficients are defined for nonnegative integers n and k, where $0 \leq k \leq n$. For $k < 0$ or $k > n$, set $\binom{n}{k} = 0$. Common values of binomial coefficients are given in Table 3.4.

TABLE 3.4: Common values of binomial coefficients.

Binomial coefficients
$\binom{n}{0} = 1 = \binom{n}{n}$
$\binom{n}{1} = n$
$\binom{n}{2} = n(n-1)/2$
$\binom{n}{k} = \binom{n}{n-k}$

Example 3.13 The best hand in the game of poker is a royal straight flush consisting of 10-Jack-Queen-King-Ace, all of the same suit. What is the probability of getting dealt a royal straight flush?

There are four possible royal straight flushes, one for each suit. A five-card hand in poker is a five-element subset of a 52-element set. Thus,

$$P(\text{Royal straight flush}) = \frac{4}{\binom{52}{5}} = 1.539 \times 10^{-6},$$

or about 1.5 in a million.

See Exercise 3.16 for a challenging counting exercise to find the probabilities for each of the hands in poker. ■

Example 3.14 **Texas hold 'em.** In Texas hold 'em poker, players are initially dealt two cards. What is the probability of being dealt at least one ace?

Consider the complementary event of being dealt no ace. There are $\binom{48}{2}$ ways of being dealt two cards neither of which are aces. The desired probability is

$$1 - \binom{48}{2} \Big/ \binom{52}{2} = 1 - \frac{188}{221} = \frac{33}{221} = 0.149321.$$ ■

Example 3.15 Twenty-five people will participate in a clinical trial, where 15 people receive the treatment and 10 people receive the placebo. In a group of six people who participated in the trial, what is the probability that four received the treatment and two received the placebo?

There are $\binom{15}{4}$ ways to pick the four who received the treatment, and $\binom{10}{2}$ ways to pick the two who received the placebo. There are $\binom{25}{6}$ possible subgroups of six people. The desired probability is

$$\binom{15}{4}\binom{10}{2} \Big/ \binom{25}{6} = \frac{351}{1012} = 0.347.$$ ■

Example 3.16 **Lottery.** In the Powerball lottery, the player picks five numbers between 1 and 59 and then a single "powerball" number between 1 and 35. To win the jackpot, you need to match all six numbers. What is the probability of winning the jackpot?

There are $35\binom{59}{5}$ possible plays. Of these, one is the jackpot winner. The desired probability is

$$P(\text{Jackpot}) = 1 \Big/ 35\binom{59}{5} = 5.707 \times 10^{-9},$$

almost 1 out of 200 million.

You can win \$10,000 in the Powerball lottery if you match the powerball and exactly four of the other five numbers. The number of ways to make such a match is $\binom{5}{4}\binom{54}{1}$, selecting the powerball winner, four of the five nonpowerball winners, and one loser. The desired probability is

$$P(\$10,000) = \binom{5}{4}\binom{54}{1} \Big/ 35\binom{59}{5} = 1.54089 \times 10^{-6},$$

about the same as the probability of being dealt a royal straight flush in poker. ■

Example 3.17 Bridge. In the game of bridge, all 52 cards are dealt out to four players. Each player gets 13 cards. A *perfect* bridge hand is getting all cards of the same suit. (i) What is the probability of being dealt a perfect hand? (ii) What is the probability that all four players will be dealt perfect hands?

(i) There are $\binom{52}{13}$ possible bridge hands. Of those, four contain all the same suit. Thus, the probability of being dealt a perfect hand is

$$\frac{4}{\binom{52}{13}} = \frac{1}{158,753,389,900} = 6.29908 \times 10^{-12}.$$

(ii) There are $\binom{52}{13}$ ways for the first player to be dealt 13 cards. Then $\binom{39}{13}$ ways to deal the remaining 39 cards to the second player. And so on. There are

$$\binom{52}{13}\binom{39}{13}\binom{26}{13}\binom{13}{13} = \frac{52!}{13!13!13!13!}$$

possible ways to deal out the deck. A perfect hand is determined by the suit. And there are $4! = 24$ ways to permute the four suits among the four players. The desired probability is

$$\frac{4!13!13!13!13!}{52!} = 4.47388 \times 10^{-28}.$$ ■

2,235,197,406,895,366,368,301,599,999 to one

According to the BBC News (http: //news. bbc.co.uk/2/Li/uk_news/50977.stw), on January 27, 1998, in a whist club in Bucklesham, England, four card players at a table were dealt perfect bridge hands.

"Witnesses in the village hall where the game was being played," reported the BBC, "have confirmed what happened. Eighty-seven-year-old Hilda Golding was the first to pick up her hand. She was dealt 13 clubs in the pack. . . . Hazel Ruffles had all the diamonds. Alison Chivers held the hearts. . . ."

"Chivers insists that the cards were shuffled properly. 'It was an ordinary pack of cards,' she said."

Another report of a perfect bridge hand appeared in *The Herald* of Sharon, Pa., on August 16, 2010.

Do you think these reports are plausible?

Example 3.18 In 20 coin tosses, what is the probability of getting exactly 10 heads?

Here is a purely counting approach. There are 2^{20} possible coin tosses. Of those, there are $\binom{20}{10}$ sequences of H's and F's with exactly 10 H's. The desired probability is

$$\frac{\binom{20}{10}}{2^{20}} = 0.176197.$$

A slightly different approach first counts the number of possible outcomes and then computes the probability of each. There are $\binom{20}{10}$ possible outcomes. By independence, each outcome occurs with probability $(1/2)^{20}$. This gives the same result. ■

Example 3.19 A DNA strand can be considered a sequence of A's, C's, G's, and T's. Positions on the DNA sequence are called *sites*. Assume that the letters at each site on a DNA strand are equally likely and independent of other sites (a simplifying assumption that is not true with actual DNA).

1. What is the probability that a DNA strand of length 20 is made up of four A's and 16 G's?

   There are $\binom{20}{4}$ binary sequences of length 20 with four A's and 16 G's. By independence, each sequence occurs with probability $1/4^{20}$. The desired probability is

$$\frac{\binom{20}{16}}{4^{20}} = 4.4065 \times 10^{-9}.$$

2. What is the probability that a DNA strand of length 20 is made up of four A's, five G, three T's, and eight C's?

   Consider the *positions* of the letters. There are $\binom{20}{4}$ choices for the positions of the A's. This leaves 16 positions for the G's. There are $\binom{16}{5}$ choices for those. Of the remaining 11 positions, there are $\binom{11}{3}$ positions for the T's. And the last eight positions are fixed for the C's. This gives

$$\binom{20}{4}\binom{16}{5}\binom{11}{3}\binom{8}{8}\Big/4^{20} = \frac{20!}{4!5!3!8!}\Big/4^{20} = 0.00317.$$

   This last expression is an example of a *multinomial* probability, discussed in Chapter 6.

■

The classic binomial theorem describes the algebraic expansion of powers of a polynomial with two terms. The algebraic proof uses induction and is somewhat technical. Here is a *combinatorial* proof.

Theorem 3.1. Binomial theorem. *For nonnegative integer* n, *and real numbers* x *and* y,

$$(x+y)^n = \sum_{k=0}^{n} \binom{n}{k} x^k y^{n-k}.$$

Proof. It will help the reader in following the proof to choose a small n, say $n = 3$, and expand $(x+y)^3$ by hand.

Observe that all the terms of the expansion have the form $x^k y^{n-k}$, for $k = 0, 1, \ldots, n$. Fix k and consider the coefficient of $x^k y^{n-k}$ in the expansion. The product $(x+y)^n = (x+y)\cdots(x+y)$ consists of n factors. There are k of these factors to choose an x from, and the remaining $n-k$ factors to choose a y from. There are $\binom{n}{k}$ ways to do this. Thus, the coefficient of $x^k y^{n-k}$ is $\binom{n}{k}$, which gives the result.

Let $x = y = 1$ in the binomial theorem. This gives

$$2^n = \sum_{k=0}^{n} \binom{n}{k} 1^k 1^{n-k} = \sum_{k=0}^{n} \binom{n}{k}.$$

There is a combinatorial interpretation of this identity. The left-hand side counts the number of sets of size n. The right-hand side counts the number of such sets by summing the number of subsets of size 0, size 1, ..., and size n.

Binomial coefficients appear in the famous Pascal's triangle.

1
1 1
1 2 1
1 3 3 1
1 4 6 4 1
1 5 10 10 5 1
1 6 15 20 55 6 1
1 7 21 35 35 21 7 1
1 8 28 56 70 56 28 8 1
1 9 36 84 126 126 84 36 9 1
1 10 45 120 210 252 210 120 45 10 1

Each entry of the triangle is the sum of the two numbers above it. The entries are all binomial coefficients. Enumerate the rows starting at $n = 0$ at the top. The entries of each row are numbered from the left starting at $k = 0$. The kth number on the nth row is $\binom{n}{k}$. The fact that each entry is the sum of the two entries above it gives the identity

$$\binom{n}{k} = \binom{n-1}{k-1} + \binom{n-1}{k}. \tag{3.7}$$

The algebraic proof of this identity is an exercise in working with factorials.

$$\begin{aligned}\binom{n-1}{k-1} + \binom{n-1}{k} &= \frac{(n-1)!}{(k-1)!(n-k)!} + \frac{(n-1)!}{k!(n-1-k)!}\\ &= \frac{(n-1)!k}{k!(n-k)!} + \frac{(n-k)(n-1)!}{k!(n-k)!}\\ &= \frac{(n-1)!(k+(n-k)}{k!(n-k)!}\\ &= \frac{(n-1)!n}{k!(n-k)!}\\ &= \frac{n!}{k!(n-k)!} = \binom{n}{k}.\end{aligned}$$

Here is a combinatorial proof:

Question: There are n students in the room, including Angel. How many ways are there to pick a group of k students?

Answer #1: Choose k students from the set of n students in $\binom{n}{k}$ ways.

Answer #2: Pick k students that include Angel. Then pick k students that do not include Angel. If Angel is included, there are $\binom{n-1}{k-1}$ ways to pick the remaining $k-1$ students in the group. If Angel is not included, there are $\binom{n-1}{k}$ ways to pick the group (choosing from everyone except Angel). Thus, there are $\binom{n-1}{k-1} + \binom{n-1}{k}$ ways to pick a group of k students.

The two solutions answer the same question, proving the desired identity.

Example 3.20 Ballot problem. This classic problem introduced by Joseph Louis François Bertrand in 1887 asks, "In an election where candidate A receives p votes and candidate B receives q votes with $p > q$, what is the probability that A will be strictly ahead of B throughout the count?" The problem assumes that votes for A and B are equally likely.

For instance, if A receives $p = 3$ votes and B receives $q = 2$ votes, the possible vote counts are given in Table 3.5. Of the 10 possible voting outcomes, only the first two show A always ahead throughout the count. The desired probability is $2/10 = 1/5$.

We show that the solution to the ballot problem is $(p-q)/(p+q)$.

A voting outcome can be thought of as a list of length $p+q$ with p A's and q B's. Thus, there are $\binom{p+q}{p}$ possible voting outcomes.

Consider the number of outcomes in which A is always ahead. Clearly such an outcome must begin with a vote for A. The number of outcomes that begin with A is $\binom{p+q-1}{p-1}$, since the first element of the list is fixed and there are $p+q-1$ positions to fill with $p-1$ A's. Some of these outcomes are "good" (A stays ahead throughout) and some are "bad." We need to subtract off the number of bad lists.

To count such lists, we give a one-to-one correspondence between bad lists that start with A and general lists that start with B. To do so we represent a voting outcome by a path where the vertical axis represents the number of votes for A. Thus, when a path crosses the horizontal axis, it represents a tie.

TABLE 3.5: Voting outcomes for the ballot problem. A receives three votes and B receives two votes.

Voting pattern	Total votes for A throughout the count
AAABB	1,2,3,2,1
AABAB	1,2,1,2,1
AABBA	1,2,1,0,1
ABABA	1,0,1,0,1
ABAAB	1,0,1,2,1
ABBAA	1,0,−1,0,1
BAAAB	−1,0,1,2,1
BAABA	−1,0,1,0,1
BABAA	−1,0,−1,0,1
BBAAA	−1,−2,0,1,2

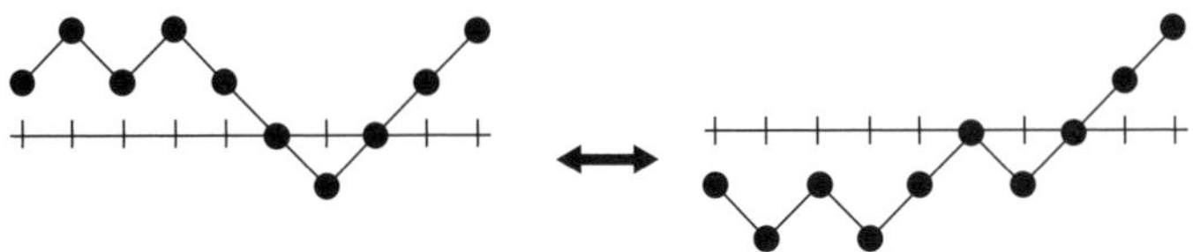

FIGURE 3.1: Illustrating the correspondence between "bad" lists that start with A and lists that start with B.

See the example in Figure 3.1. The left diagram corresponds to the voting outcome AABABBBAAA. The outcome is "bad" in that there is eventually a tie and the path crosses the horizontal axis. For such a path, "reflect" the portion of the path up until the tie across the x axis, giving the outcome in the right diagram. The reflection results in a path that starts with B.

Conversely, consider a path that starts with B. Since there are more A's than B's, at some point in the count there must be a tie and the path crosses the x-axis. Reflecting the portion of the path up until the tie across the x-axis gives a bad path that starts with A.

Having established a one-to-one correspondence we see that the number of bad lists that start with A is equal to the number of lists that start with B. There are $\binom{p+q-1}{q-1}$ lists that start with B. This gives the desired probability

$$
\begin{aligned}
&P(A \text{ is ahead throughout the count}) \\
&= \frac{\text{Number of good lists that start with } A}{\text{Number of voting outcomes}} \\
&= \frac{\text{Number of lists that start with } A - \text{Number of bad lists that start with } A}{\text{Number of voting outcomes}} \\
&= \frac{\left[\binom{p+q-1}{p-1} - \binom{p+q-1}{q-1}\right]}{\binom{p+q}{p}}.
\end{aligned}
$$

We leave it to the reader to check that this last expression simplifies to $(p-q)/(p+q)$. ∎

3.5 BINOMIAL DISTRIBUTION

Let $X_1, \ldots, X_n$ be an i.i.d. 0–1 Bernoulli sequence with success parameter p. Let 1 denote "success," and 0 "failure." The sum $X = X_1 + \cdots + X_n$ counts the number of successes in n trials.

For $k = 0, \ldots, n$, consider $P(X = k)$, the probability of exactly k successes in n trials. Each outcome of k successes and $n - k$ failures can be represented by a binary list of length n with exactly k ones. There are $\binom{n}{k}$ such lists. The probability of each such outcome, by independence, is $p^k(1-p)^{n-k}$, which gives the probability function of the binomial distribution.

BINOMIAL DISTRIBUTION

A random variable X is said to have a *binomial distribution with parameters* n *and* p if

$$P(X = k) = \binom{n}{k} p^k (1-p)^{n-k}, \text{ for } k = 0, 1, \ldots, n. \tag{3.8}$$

We write $X \sim \text{Binom}(n, p)$.

The most common setting in which the binomial distribution arises is modeling the number of successes in n independent Bernoulli trials. Examples include

- The number of heads in n coin tosses has a binomial distribution with parameters n and $p = 1/2$.
- Suppose 500 bits of data are sent through a digital transmission such that there is a 1% chance that any bit is received in error. If bit errors are independent of each other, then the number of errors has a binomial distribution with $n = 500$ and $p = 0.001$.
- Mutations on a DNA strand can be modeled as a sequence of independent Bernoulli trials. The total number of mutations has a binomial distribution. The parameters are the length of the strand n and the mutation rate p.

VISUALIZING THE BINOMIAL DISTRIBUTION

To visualize the binomial distribution using R, type

```
> n <- 8
> p <- 0.15
> barplot(dbinom(0:n,n,p), names.arg=0:n)
```

with your own choices of n and p. See Figure 3.2 for four examples.

The probability function for the binomial distribution is indeed a valid probability function as the terms are nonnegative and sum to 1. By the binomial theorem,

$$\sum_{k=0}^{n} P(X = k) = \sum_{k=0}^{n} \binom{n}{k} p^k (1-p)^{n-k} = (p + (1-p))^n = 1^n = 1.$$

Example 3.21 A multiple choice exam has 10 questions with four choices for each question. If a student guesses on each question from the available choices, what is the chance they will get exactly two questions right?

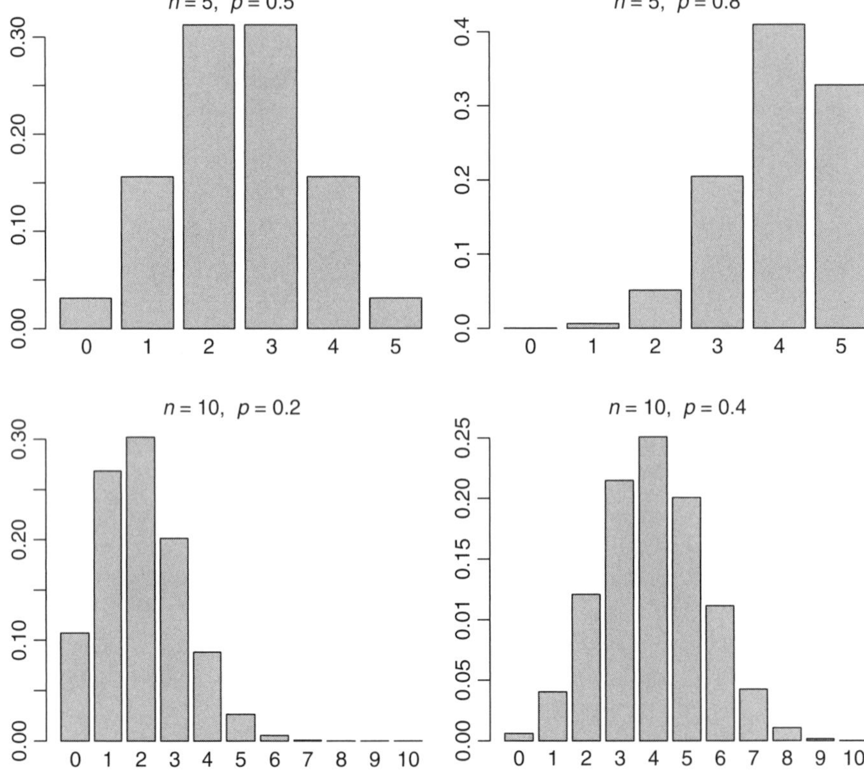

FIGURE 3.2: Four examples of the binomial distribution.

Let X be the number of questions the student gets correct. Then X has a binomial distribution with parameters $n = 10$ and $p = 1/4$. This gives

$$P(X = 2) = \binom{10}{2}\left(\frac{1}{4}\right)^2\left(\frac{3}{4}\right)^8 = 0.281.$$

If the question for this problem had asked, "What is the chance that the student will get two questions right?" without the word "exactly," the question could be considered vague. Does it mean *exactly* two questions right or *at least* two questions right? In the latter case,

$$\begin{aligned} P(X \geq 2) &= 1 - P(X \leq 1) = 1 - P(X = 0) - P(X = 1) \\ &= 1 - \binom{10}{0}\left(\frac{1}{4}\right)^0\left(\frac{3}{4}\right)^{10} - \binom{10}{1}\left(\frac{1}{4}\right)^1\left(\frac{3}{4}\right)^9 = 0.756. \end{aligned}$$

In this book, we will be careful about such vagaries. Be careful to clarify such issues if they arise in a real-world problem. ∎

Example 3.22 In a field of 100 trees, each tree has a 10% chance of being infected by a root disease independently of other trees. What is the probability that more than five trees are infected?

Let X be the number of infected trees. Then X has a binomial distribution with $n = 100$ and $p = 0.10$. This gives

$$P(X > 5) = 1 - P(X \leq 5) = 1 - \sum_{k=0}^{5} P(X = k)$$
$$= 1 - \sum_{k=0}^{5} \binom{100}{k} (0.1)^k (0.9)^{100-k} = 0.942. \quad \blacksquare$$

R: WORKING WITH PROBABILITY DISTRIBUTIONS

R has several commands for working with probability distributions like the binomial distribution. These commands are prefixed with `d`, `p`, and `r`. They take a suffix that describes the distribution. For the binomial distribution, these commands are shown.

R command	What it does
`dbinom(k,n,p)`	$P(X = k)$
`pbinom(k,n,p)`	$P(X \leq k)$
`rbinom(k,n,p)`	Simulates k random variables

To find the exact probability $P(X > 5)$ in the last Example 3.22, type

```
>  1-pbinom(5,100,0.10)
[1] 0.94242
```

To simulate the probability $P(X > 5)$ based on 10,000 repetitions, type

```
> n <- 10000
> simlist <- rbinom(n,100,0.10)
> sum(simlist>5)/n
[1] 0.9433
```

Example 3.23 According to Leder et al. (2002), many airlines consistently report that about 12% of all booked passengers do not show up to the gate due to cancellations and no-shows. If an airline sells 110 tickets for a flight that seats 100 passengers, what is the probability that the airline overbooked (sold more tickets than seats)?

Let X be the number of ticket holders who arrive at the gate. If we assume that passengers' gate arrivals are independent, then X has a binomial distribution with $n = 110$ and $p = 1 - 0.12 = 0.88$. The desired probability is

$$\begin{aligned} P(X > 100) &= \sum_{k=101}^{110} P(X = k) \\ &= \sum_{k=101}^{110} \binom{110}{k} (0.88)^k (0.12)^{110-k} \\ &= 0.137. \end{aligned}$$

In R, since $P(X > 100) = 1 - P(X \leq 100)$, type

```
> 1-pbinom(100,110,0.88)
[1] 0.1366599
```

For a 100-seat flight, suppose the airline would like to sell the maximum number of tickets such that the chance of overbooking is less than 5%. The airlines call this a "5% bump threshold" overbooking strategy. How many tickets should the airline sell?

Let n be the number of tickets sold. Find n so that $P(X > 100) \leq 0.05$, where $X \sim \text{Binom}(n, 0.88)$. Trial and error using R shows that for $n = 108$, $P(X > 100) = 0.0449$. And for $n = 109$, $P(X > 100) = 0.0823$. Therefore, sell $n = 108$ tickets.

```
> 1-pbinom(100,108,0.88)
[1] 0.04492587
> 1-pbinom(100,109,0.88)
[1] 0.08231748
```

R: SIMULATING THE OVERBOOKING PROBABILITY

To simulate the probability of overbooking when 108 tickets are sold, generate 108 Bernoulli random variables with $p = 0.88$. A Bernoulli variable is a binomial variable with $n = 1$. Type

```
> rbinom(108,1,0.88)
```

To check if more than 100 tickets are sold and the airline overbooks, type

```
> if (sum(rbinom(108,1,0.88))>100) 1 else 0
```

which returns a 1 if too many tickets are sold, and 0 otherwise.

Here we simulate the probability of overbooking based on 10,000 trials.

```
> simlist <- replicate(10000,
    if (sum(rbinom(108,1,0.88))>100) 1 else 0)
> mean(simlist)
[1] 0.0496
```

■

Probabilities that are modeled with the binomial distribution assume an underlying sequence of Bernoulli trials, which are independent and identically distributed. Without independence, or a constant probability p for each trial, the binomial model is not valid.

The following situations are *not* appropriate for a binomial model:

1. *Ask 50 people their height.* In a binomial setting, each trial has *two* possible outcomes. Heights have many values.

   However, if people are asked whether or not they are at least six feet, then there are two outcomes, and the binomial model could be appropriate, assuming independence.
2. *The chance that a manuscript has a typo on any page is 1%. We will read the manuscript until we find 10 typos.* In a binomial setting, the number of trials n is fixed. In this case, the number of pages read is not fixed.

   However, if we read 100 pages and consider the probability of getting 10 typos, then the binomial setting is appropriate.
3. *Five cards are dealt from a standard deck of cards. How many aces were dealt?* In dealing cards, we are sampling without replacement. Successive draws are neither independent nor have the same probability of success. That is, successive draws are not Bernoulli trials.

   However, if we replaced the card after each draw so that the sampling was with replacement, then the binomial setting would work.

Example 3.24 The approximate frequencies of the four nucleotides in human DNA are given in Table 3.6.

In two DNA strands of length 10, what is the probability that nucleotides will match in exactly seven positions?

We make the assumption that the two strands are independent of each other. The chance that at any site there will be a match is the probability that both sites are A, C, G, or T. The probability of a match is thus $(0.292)^2 + (0.207)^2 + (0.207)^2 + (0.292)^2 = 0.256$. The number of matches is a binomial random variable with

TABLE 3.6: Nucleotide frequencies in human DNA.

A	C	G	T
0.292	0.207	0.207	0.292

$n = 10$ and $p = 0.256$. The desired probability is $P(X = 7) = 0.0356$, where $X \sim \text{Binom}(10, 0.256)$.

```
> dbinom(7,10,0.256)
[1] 0.00356
```

■

Example 3.25 Tom and Danny independently each toss four coins. What is the probability that they get the same number of heads?

Let X be the number of heads Danny gets, and let Y be the number of heads Tom gets. The event that they get the same number of heads is

$$\{X = Y\} = \bigcup_{k=0}^{4} \{X = k, Y = k\}.$$

This gives

$$\begin{aligned} P(X = Y) &= P\left(\bigcup_{k=0}^{4} \{X = k, Y = k\}\right) \\ &= \sum_{k=0}^{4} P(X = k, Y = k) = \sum_{k=0}^{4} P(X = k)P(Y = k) \\ &= \sum_{k=0}^{4} \left[\binom{4}{k}\left(\frac{1}{2}\right)^4\right]^2 = \left(\frac{1}{2}\right)^8 \sum_{k=0}^{4} \binom{4}{k}^2 \\ &= \frac{1}{256}\left(1^2 + 4^2 + 6^2 + 4^2 + 1^2\right) = \frac{70}{256} = 0.273, \end{aligned}$$

where the third equality is because of independence. ■

Example 3.26 Random graphs. A *graph* is a set of vertices (or nodes), with edges joining them. In mathematics, the study of graphs is called graph theory. Here we describe a model for a *random graph.*

Start with a graph on n vertices and no edges. For each pair of vertices, flip a coin with heads probability p. If the coin lands heads, place an edge between that pair of vertices. If tails, do not place an edge. Note the two extreme cases: if $p = 0$, there are no edges in the graph; if $p = 1$, every pair of vertices gets an edge in what is called the complete graph. The parameter p is called the *edge probability.* Properties of random graphs are studied for large n as the edge probability varies from 0 to 1 (See Fig. 3.3).

Random graphs have been used as models for Internet traffic, social networks, and the spread of infectious diseases.

Let X be the number of edges in a random graph of n vertices. There are $\binom{n}{2}$ ways to pick two vertices. Thus, there are $\binom{n}{2}$ possible edges in a graph on n vertices. The number of edges X thus has a binomial distribution with parameters $\binom{n}{2}$ and p.

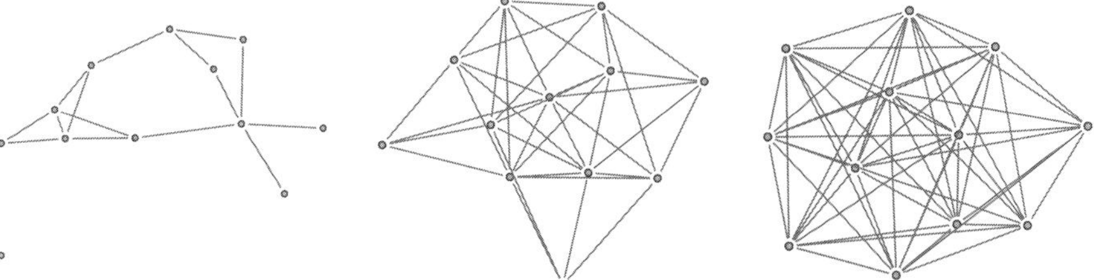

FIGURE 3.3: Three random graphs on $n = 12$ vertices generated, respectively, with $p = 0.2, 0.5$, and 0.9.

In a graph, the *degree* of a vertex is the number of edges that contain that vertex, also described as the number of edges incident to that vertex. Let $\deg(v)$ be the degree of vertex v in a random graph. There are $n - 1$ possible edges incident to v, since there are $n-1$ vertices left in the graph, other than v. Each of those edges occurs with probability p. Thus for any vertex v, the degree $\deg(v)$ is a random variable that has a binomial distribution with parameters $n - 1$ and p. That is,

$$P(\deg(v) = k) = \binom{n-1}{k} p^k (1-p)^{n-1-k}, \quad \text{for } k = 0, \ldots, n-1. \qquad \blacksquare$$

3.6 STIRLING'S APPROXIMATION

The factorial function $n!$ is pervasive in discrete probability. A good approximation when n is large is given by Stirling's approximation

$$n! \approx n^n e^{-n} \sqrt{2\pi n}. \tag{3.9}$$

More precisely,

$$\lim_{n \to \infty} \frac{n!}{n^n e^{-n} \sqrt{2\pi n}} = 1.$$

We say that $n!$ "is asymptotic to" the function $n^n e^{-n} \sqrt{2\pi n}$.

For first impressions it looks like the right-hand side of Equation 3.9 is more complicated than the left. However, the right-hand side is made up of relatively simple, elementary functions, which makes it possible to obtain useful approximations of factorials and binomial coefficients.

Example 3.27 Suppose we toss $2n$ fair coins. We expect to get about n heads. What is the probability of getting exactly n heads?

TABLE 3.7: Accuracy of Stirling's approximation.

n	Exact probability	Stirling's approximation
1	0.5	0.5642
5	0.246	0.252
10	0.1762	0.1784
50	0.0796	0.0798
10^2	0.0563	0.0564
10^3	0.017939	0.017841
10^4	5.6418×10^{-3}	5.6419×10^{-3}
10^6	5.6419×10^{-4}	5.6419×10^{-4}

Let X be the number of heads. Then X has a binomial distribution with parameters $2n$ and $p = 1/2$. This gives

$$P(X = n) = \binom{2n}{n}\left(\frac{1}{2}\right)^{2n}.$$

Stirling's approximation can be used on the binomial coefficient to obtain an expression that is easier to interpret. We have

$$\binom{2n}{n} = \frac{(2n)!}{n!n!} = \frac{(2n)^{2n}e^{-2n}\sqrt{2\pi 2n}}{(n^n e^{-n}\sqrt{2\pi n})(n^n e^{-n}\sqrt{2\pi n})} = \frac{2^{2n}}{\sqrt{\pi n}}.$$

This gives

$$P(X = n) = \binom{2n}{n}\left(\frac{1}{2}\right)^{2n} \approx \frac{2^{2n}}{\sqrt{\pi n}}\left(\frac{1}{2}\right)^{2n} = \frac{1}{\sqrt{\pi n}} \approx \frac{0.564}{\sqrt{n}}.$$

Table 3.7 compares exact probabilities, obtained in R with the command `dbinom(n,2*n,1/2)`, and Stirling's approximation for some values of n from one to a million. ∎

3.7 POISSON DISTRIBUTION

The binomial setting requires a fixed number n of independent trials. However, in many applications, we model counts of independent outcomes where there is no prior constraints on the number of trials. Examples include

- The number of alpha particles emitted by a radioactive substance in 1 minute.
- The number of wrong numbers you receive on your cell phone over a month's time.

- The number of babies born on a maternity ward in one day.
- The number of blood cells recorded on a hemocytometer (a device used to count cells).
- The number of chocolate chips in a cookie.
- The number of accidents on a mile-long stretch of highway.
- The number of soldiers killed by horse kick each year in each corps in the Prussian cavalry.

The last example may seem far-fetched, but it was actually one of the first uses of the distribution that we introduce in this section, called the Poisson distribution.

POISSON DISTRIBUTION

A random variable X has a *Poisson distribution with parameter* $\lambda > 0$ if

$$P(X = k) = \frac{e^{-\lambda}\lambda^k}{k!}, \quad \text{for } k = 0, 1, \ldots \tag{3.10}$$

We write $X \sim \text{Pois}(\lambda)$.

The probability function is nonnegative and sums to 1, as

$$\sum_{k=0}^{\infty} P(X = k) = \sum_{k=0}^{\infty} \frac{e^{-\lambda}\lambda^k}{k!} = e^{-\lambda} \sum_{k=0}^{\infty} \frac{\lambda^k}{k!} = e^{-\lambda}e^{\lambda} = 1.$$

See graphs of the Poisson distribution for four choices of λ in Figure 3.4.

The Poisson distribution is intimately connected to the binomial distribution introduced in the last section. To see how it arises in a real-world context, we elaborate on a specific example.

Consider developing a probability model for the number of babies born on a busy maternity ward in one day. Let X be the number of births. Note that this is not a binomial setting since there is neither a fixed number of trials nor at this stage even a probability to speak of. We give a heuristic argument to motivate the Poisson distribution.

Suppose that babies are born at some average rate of λ births per day. We break up the day into n subintervals of length $1/n$. Babies are born at the rate of λ/n births on each subinterval. For instance, if $n = 24$, babies are born at the rate of $\lambda/24$ births per hour. Let n get large. The lengths of the subintervals get small and *eventually* on each subinterval it is very unlikely that two babies will be born during that small time period. In each small subinterval, either a baby is born or not born with probability λ/n. If we assume that births on different subintervals are independent of each other,

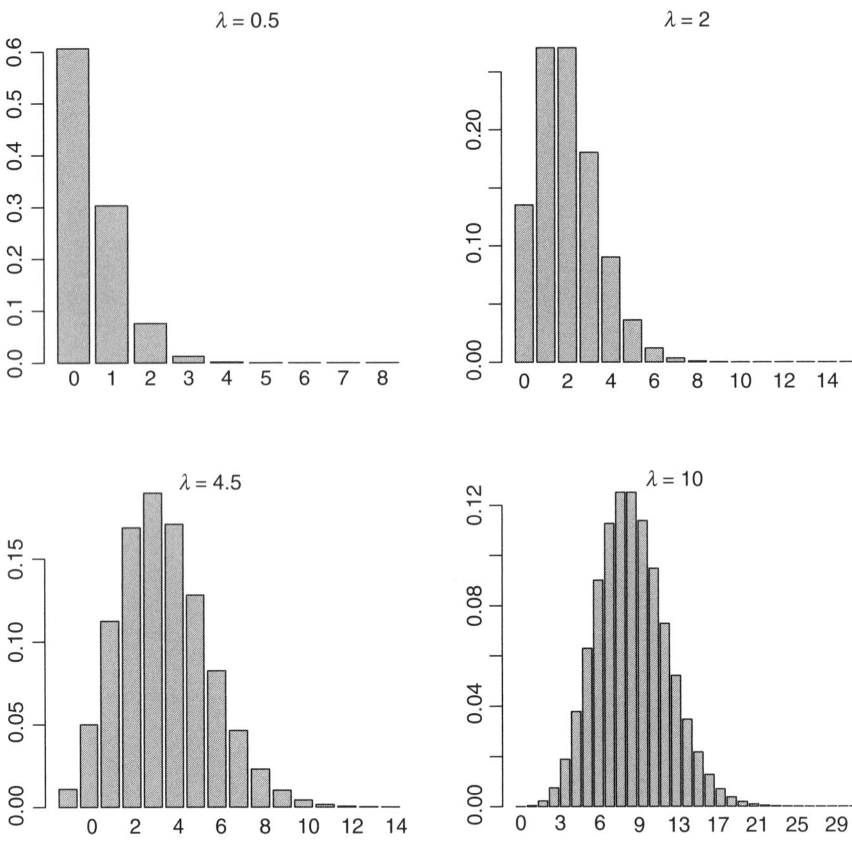

FIGURE 3.4: Four Poisson distributions.

then we can regard the occurrence of births on the subintervals as a sequence of n i.i.d. Bernoulli trials with success probability λ/n.

The total number of births in one day is the sum of the births on the n subintervals and has a binomial distribution with parameters n and λ/n. That is,

$$P(X = k) = \binom{n}{k} \left(\frac{\lambda}{n}\right)^k \left(1 - \frac{\lambda}{n}\right)^{n-k}, \quad \text{for } k = 0, \ldots, n.$$

The word "eventually" in the previous paragraph suggests a limiting process. Let n tend to infinity. We show in Section 3.7.2 that

$$\lim_{n\to\infty} \binom{n}{k} \left(\frac{\lambda}{n}\right)^k \left(1 - \frac{\lambda}{n}\right)^{n-k} = \frac{e^{-\lambda}\lambda^k}{k!}. \tag{3.11}$$

In other words, the number of babies born in one day has a Poisson distribution with parameter λ, where λ is the average rate of births per day.

Example 3.28 Data from a hospital maternity ward suggest that about 4.5 babies are born every day. What is the probability that there will be six births on the ward tomorrow?

Let X be the number of births on the ward tomorrow. Model X with a Poisson distribution with $\lambda = 4.5$. Then

$$P(X = 6) = \frac{e^{-4.5}(4.5)^6}{6!} = 0.128. \qquad \blacksquare$$

The Poisson distribution is sometimes called the law of rare events. The "rarity" of the events does not refer to the number of events that occur, but rather to whether or not the event occurs in some small interval of time or space. What the examples at the beginning of this section have in common is that some event (e.g., births, traffic accidents, wrong numbers, and deaths) occur in some fixed region of time or space at a constant rate such that occurrences in disjoint subregions are independent. The parameter λ has the interpretation of the average number of occurrences per unit of time or space.

Example 3.29 According to the United States Geological Survey, between 1970 and 2012 there have been 33 *major earthquakes* (7.0 or greater on the Richter scale) in the United States Assuming successive major earthquakes are independent what is the probability there will be at least three major earthquakes in the United States next year?

Let X be the number of major earthquakes next year. The data are based on 43 years. The rate of earthquake occurrence is 33 per 43 years, or $33/43$ per year. We model X with a Poisson distribution with $\lambda = 33/43$. Then

$$\begin{aligned} P(X \geq 3) &= 1 - P(X < 3) = 1 - P(X \leq 2) \\ &= 1 - P(X = 0) - P(X = 1) - P(X = 2) \\ &= 1 - e^{-33/43} - e^{-33/43}(33/43) - \frac{e^{33/43}(33/43)^2}{2} \\ &= 0.0428. \end{aligned}$$

Geologists might argue whether successive earthquakes are in fact independent. After a big earthquake, it is very common for there to be aftershocks. The dependence of these residual quakes on the initial earthquake would violate the assumption of independence. More sophisticated models that incorporate some dependency structure are typically used to model seismic events. ■

Example 3.30 **Death by horse kicks.** In 1898, Ladislaus Bortkiewicz, a Polish statistician, studied the distribution of soldier deaths by horse kicks in the Prussian cavalry. Over 20 years there were 122 deaths in 10 Prussian army corps. He divided the data into $20 \times 10 = 200$ corps-years. The average number of deaths per corps-

TABLE 3.8: Deaths by horse kicks in the Prussian cavalry.

Number of deaths	Observed	Poisson probability	Expected
0	109	0.543	108.7
1	65	0.331	66.3
2	22	0.101	20.2
3	3	0.021	4.1
4	1	0.003	0.6
5+	0	0.000	0.0

year was $122/200 = 0.61$. Bortkiewicz modeled the number of deaths with a Poisson distribution with parameter $\lambda = 0.61$. Table 3.8 contains the data Borkiewicz worked with, including observed number of deaths and expected numbers predicted by the Poisson model. The expected number is the Poisson probability times 200. Observe how close the model fits the data.

Bortkiewicz authored a book about the Poisson distribution in 1898 titled *The Law of Small Numbers*. In addition to analyzing deaths by horse kicks, he also considered data on child suicides in Prussia. ■

R: POISSON DISTRIBUTION

Commands for working with the Poisson distribution are

```
> dpois(x,lambda)
> ppois(x,lambda)
> rpois(n,lambda)
```

To obtain the expected counts for the number of deaths by horse kicks in the last column of Table 3.8, type

```
> probs <- dpois(0:4,0.61)
> probs <- c(probs,1-ppois(4,0.061))
> expected <- 200*probs
> expected
[1] 108.67  66.29  20.22   4.11   0.63   0.00
```

■ **Example 3.31** The number of accidents per month at a busy intersection has a Poisson distribution with parameter $\lambda = 7.5$. Conditions at the intersection have not changed much over time. Suppose each accident costs local government about $25,000 for clean-up. How much do accidents cost, on average, over a year's time?

With the interpretation that λ represents the average number of accidents per month, it is not hard to see that the average cost is about $7.5 \times 25,000 \times 12 =$

\$2,250,000. We approach the problem using simulation. The following commands simulate 12 months of accidents and the associated annual cost.

R: SIMULATING ANNUAL ACCIDENT COST

```
> accidents <- rpois(12,7.5)
> cost <- sum(25000*accidents)
> cost
[1] 2050000
> simlist <- replicate(1000,sum(25000*rpois(12,7.5)))
> mean(simlist)
[1] 2251975
```

Local government can expect to pay about \$2.25 million in costs.

■

Example 3.32 The Poisson distribution is a common model in genetics. The distribution is used to describe occurrences of mutations and chromosome crossovers. *Crossovers* occur when two chromosomes break and then reconnect at different end pieces resulting in an exchange of genes. This process is known as genetic recombination.

Suppose a genetics lab has a means to count the number of crossovers between two genes on a chromosome. In 100 samples, 50 cells have no crossovers, 25 cells have one crossover, 20 cells have two crossovers, and 5 cells have three crossovers. Find the probability that a new sample will show at least one crossover.

The average number of crossovers from the sample is

$$50(0) + 25(1) + 20(2) + 5(3) = 80 \text{ crossovers per 100 cells,}$$

or 0.80 crossovers per cell. Model the number of crossovers on the chromosome with a Poisson distribution with $\lambda = 0.80$. This gives

$$P(\text{At least one crossover}) = 1 - P(\text{No crossovers}) = 1 - e^{-0.80} = 0.55.$$ ■

3.7.1 Poisson Approximation of Binomial Distribution

We have seen the close connection between the binomial and Poisson distributions. The Poisson distribution arises as a limiting expression for the binomial distribution letting $n \to \infty$ and $p = \lambda/n \to 0$. For binomial problems with large n and small p, a Poisson approximation works well with $\lambda = np$.

Example 3.33 We have not determined the accuracy of the following story, which has appeared in several blogs and websites, including Krantz (2005):

> On Oct. 9, 1972, the mathematician Dr. Jeffrey Hamilton from Warwick University wanted to show his students the effect of chance by tossing a coin. Taking a two pence coin out of his pocket, he tossed it. The class joined him in watching the coin flip over and over and then land on the floor—on its edge! Dozens of students witnessed the amazing event, and after a stunned silence they all broke into wild applause. Hamilton later calculated that the chances of this happening are one in one billion.

Assume that the probability that a coin lands on its edge is one in one billion. Suppose everyone in the world flips a coin. With a population of about seven billion you would expect about seven coins to land on their edge. What is the probability that between six and eight coins land on their edge?

Let X be the number of coins that land on their edge. Then X has a binomial distribution with n = 7,000,000,000 and $p = 10^{-9} = 0.000000001$. The desired probability is

$$P(6 \leq X \leq 8) = \sum_{k=6}^{8} \binom{7 \times 10^9}{k} \left(10^{-9}\right)^k \left(1 - 10^{-9}\right)^{10^9 - k}.$$

Before calculators or computers it would have been extremely difficult to compute such a probability. In fact, the Poisson approximation was discovered almost 200 years ago in order to solve problems like this.

Let $\lambda = np = \left(7 \times 10^9\right) 10^{-9} = 7$. Then X has an approximate Poisson distribution with parameter $\lambda = 7$, and

$$P(6 \leq X \leq 8) \approx \frac{e^{-7}7^6}{6!} + \frac{e^{-7}7^7}{7!} + \frac{e^{-7}7^8}{8!} = 0.428.$$ ■

In this example, the random variable X has an *exact* binomial distribution and an *approximate* Poisson distribution.

ON THE EDGE

Computing the exact probability of a coin landing on its edge is extremely hard, especially for thin coins. In Murray and Teare (1993), a model is presented and supported by numerical simulations. Extrapolations based on the model suggest that the probability of a U.S. nickel landing on its edge is approximately 1 in 6,000.

Example 3.34 Mutations in DNA sequences occur from environmental factors, such as ultraviolet light and radiation, and mistakes that can happen when a cell copies its DNA in preparation for cell division. Nachman and Crowell (2000) estimate the mutation rate per nucleotide of human DNA as about 2.5×10^{-8}. There are about 3.3×10^9 nucleotide bases in the human DNA genome. Assume that whether or not a mutation occurs at a nucleotide site is independent of what occurs at other sites. We expect about $(3.3 \times 10^9)(2.5 \times 10^{-8}) = 82.5$ mutations. What is the probability that exactly 80 nucleotides will mutate in a person's DNA?

Let X be the number of mutations. Then X has an exact binomial distribution with $n = 3.3 \times 10^9$ and $p = 2.5 \times 10^{-8}$. Approximate X with a Poisson distribution with $\lambda = np = 82.5$. The approximate probability $P(X = 80)$ is

```
> dpois(80,82.5)
[1] 0.04288381408
```

The exact probability can be obtained in R.

```
> dbinom(80,3.3*10^9,2.5*10^(-8))
[1] 0.04288381456
```

The approximation is good to eight significant digits. The approximation works well because n is large and p is small.

But what if p is not small? Consider modeling "good" nucleotides, instead of mutations. The probability of a nucleotide not mutating is $1 - p$. What is the probability that $n - 80$ nucleotides do not mutate?

The number of "good" nucleotides has an exact binomial distribution with parameters n and $1 - p$. The exact probability is

```
> n <- 3.3*10^9
> p <- 2.5* 10^(-8)
> dbinom(n-80,n,1-p)
[1] 0.04288381
```

This is the same number gotten earlier, which makes sense since $n - 80$ "good" nucleotides is equivalent to 80 mutations. However, the Poisson approximation gives

```
> dpois(n-80,n*(1-p))
[1] 6.944694e-06
```

which is way off. The approximation is not good since $1 - p$ is large. ■

Example 3.35 Balls, bowls, and bombs. The following setting is very general. Suppose n balls are thrown into n/λ bowls so that each ball has an equal chance of landing in any bowl. If a ball lands in a bowl, call it a "hit." The chance that a ball hits a particular bowl is $1/(n/\lambda) = \lambda/n$. Keeping track of whether or not each ball hits that bowl, the successive hits form a Bernoulli sequence, and the number of hits has a binomial distribution with parameters n and λ/n. If n is large, the number of balls in each bowl is approximated by a Poisson distribution with parameter $n(\lambda/n) = \lambda$.

TABLE 3.9: Bomb hits over London during World War II.

Hits	0	1	2	3	4	≥ 5
Data	229	211	93	35	7	1
Expected	226.7	211.4	98.6	30.67	7.1	1.6

Many diverse applications can be fit into this ball and bowl setting. In his classic analysis of Nazi bombing raids on London during World War II, William Feller (1968) modeled bomb hits (balls) using a Poisson distribution. The city was divided into 576 small areas (bowls) of $1/4\,\text{km}^2$. The number of areas hit exactly k times was counted. There were a total of 537 hits, so the average number of hits per area was $537/576 = 0.9323$. Feller gives the data in Table 3.9.

See the script **Balls.R** to simulate the London bombing example. Change the numbers in the script file to simulate your own balls and bowls experiment. ■

3.7.2 Poisson Limit*

The Poisson probability function arises as the limit of binomial probabilities. Here we show the limit result

$$\lim_{n\to\infty} \binom{n}{k}\left(\frac{\lambda}{n}\right)^k\left(1-\frac{\lambda}{n}\right)^{n-k} = \frac{e^{-\lambda}\lambda^k}{k!}$$

presented at the beginning of this section. Consider

$$\begin{aligned}
&\binom{n}{k}\left(\frac{\lambda}{n}\right)^k\left(1-\frac{\lambda}{n}\right)^{n-k}\\
&\quad= \frac{n(n-1)\cdots(n-k+1)}{k!}\left(\frac{\lambda}{n}\right)^k\left(1-\frac{\lambda}{n}\right)^{n-k}\\
&\quad= \frac{n^k\left(1-\frac{1}{n}\right)\cdots\left(1-\frac{k-1}{n}\right)}{k!}\left(\frac{\lambda^k}{n^k}\right)\left(1-\frac{\lambda}{n}\right)^{-k}\left(1-\frac{\lambda}{n}\right)^{n}\\
&\quad= \frac{\lambda^k}{k!}\left[\left(1-\frac{1}{n}\right)\cdots\left(1-\frac{k-1}{n}\right)\right]\left(1-\frac{\lambda}{n}\right)^{-k}\left(1-\frac{\lambda}{n}\right)^{n}. \qquad (3.12)
\end{aligned}$$

Taking limits as $n \to \infty$, consider the four factors in this expression.

(i) Since λ and k are constants, $\lambda^k/k!$ stays unchanged in the limit.

(ii) For fixed k,

$$\left(1-\frac{1}{n}\right)\cdots\left(1-\frac{k-1}{n}\right) \to 1^{k-1} = 1;$$

and

(iii) $\left(1-\frac{\lambda}{n}\right)^{-k} \to 1^{-k} = 1.$

(iv) For the last factor $(1-\lambda/n)^n$, recall from calculus that the constant $e = 2.71828\ldots$ is defined as the limit

$$\lim_{x\to\infty}\left(1+\frac{1}{x}\right)^x = e.$$

Make the substitution $1/x = -\lambda/n$ so that $n = -\lambda x$. This gives

$$\lim_{n\to\infty}\left(1-\frac{\lambda}{n}\right)^n = \lim_{x\to\infty}\left(1+\frac{1}{x}\right)^{-\lambda x} = \left[\lim_{x\to\infty}\left(1+\frac{1}{x}\right)^x\right]^{-\lambda} = e^{-\lambda}.$$

Substituting in the four limits in Equation 3.12 gives

$$\binom{n}{k}\left(\frac{\lambda}{n}\right)^k\left(1-\frac{\lambda}{n}\right)^{n-k}$$
$$= \frac{\lambda^k}{k!}\left[\left(1-\frac{1}{n}\right)\cdots\left(1-\frac{k-1}{n}\right)\right]\left(1-\frac{\lambda}{n}\right)^{-k}\left(1-\frac{\lambda}{n}\right)^n \to \frac{e^{-\lambda}\lambda^k}{k!},$$

as $n \to \infty$.

3.8 PRODUCT SPACES

In this section, we treat some technical issues associated with modeling independent events.

When we repeat a random experiment several times, or combine the results of two or more different random experiments, we are in effect creating a new sample space.

For instance, in rolling one die, the sample space is $\{1,2,3,4,5,6\}$. But in rolling two dice, a larger sample space is used of all $6 \times 6 = 36$ pairs of dice rolls

$$\{(1,1),(1,2),\ldots(6,5),(6,6)\}.$$

If the two dice rolls are independent, then for each ordered pair (x,y) the probability of rolling an x on the first die and a y on the second die $P((x,y))$ is equal to $P(x)P(y)$, the product of the individual probabilities. The larger sample space makes it possible to consider probabilities involving two dice. To consider three dice rolls, the underlying sample space would be lists of length three

$$\{(1,1,1),(1,1,2),\ldots,(6,6,6)\}.$$

And so on.

Consider two random experiments with respective sample spaces Ω and Ω'. (The sample spaces may be the same, as in the dice example, or different.) The *product space* $\Omega \times \Omega'$ is the set of all ordered pairs (ω, ω') such that $\omega \in \Omega$ and $\omega' \in \Omega'$.

To define a probability function on that product space, let

$$P((\omega,\omega')) = P(\omega)P(\omega')$$

for all $\omega \in \Omega$ and $\omega' \in \Omega'$. This gives a valid probability function since

$$\begin{aligned}\sum_{(\omega,\omega')\in\Omega\times\Omega'} P((\omega,\omega')) &= \sum_{\omega\in\Omega}\sum_{\omega'\in\Omega'} P((\omega,\omega'))\\ &= \sum_{\omega\in\Omega}\sum_{\omega'\in\Omega'} P(\omega)P(\omega')\\ &= \sum_{\omega\in\Omega} P(\omega)\sum_{\omega'\in\Omega'} P(\omega')\\ &= (1)(1) = 1.\end{aligned}$$

With this construction the outcomes of the Ω random experiment are independent of the outcomes of the Ω' random experiment.

If $A \subseteq \Omega$ and $B \subseteq \Omega'$, the event $A \cap B$ is not necessarily defined on Ω or Ω' but it is defined on the product space $\Omega \times \Omega'$. And with the probability function just defined,

$$\begin{aligned}P(AB) &= \sum_{(\omega,\omega')\in AB} P((\omega,\omega'))\\ &= \sum_{\omega\in A}\sum_{\omega'\in B} P((\omega,\omega'))\\ &= \sum_{\omega\in A}\sum_{\omega'\in B} P(\omega)P(\omega')\\ &= \sum_{\omega\in A} P(\omega)\sum_{\omega'\in B} P(\omega')\\ &= P(A)P(B).\end{aligned}$$

(We use the same capital letter P for what is in effect three different probability functions—one on Ω, one on Ω', and one on $\Omega \times \Omega'$—corresponding to three different random experiments.)

Here is a worked out example. Roll a four-sided tetrahedron die. The sample space is $\Omega = \{1, 2, 3, 4\}$ with each outcome equally likely. Draw a letter at random from a bag that contains letters a, b, and c. The sample space is $\Omega' = \{a, b, c\}$, with each outcome equally likely. To model the random experiment of both rolling the die and drawing a letter so that outcomes of the first experiment are independent of the second, consider the 12-element product space

$$\Omega \times \Omega' = \{(1, a), (1, b), \ldots, (4, b), (4, c)\}$$

and define

$$P(\omega, \omega') = P(\omega)P(\omega') = \left(\frac{1}{4}\right)\left(\frac{1}{3}\right) = \frac{1}{12},$$

for all $\omega \in \Omega$ and $\omega' \in \Omega'$.

Suppose A is the event of rolling a 1 or 2 on the die. And B is the event of drawing the letter b from the bag. Then $P(A) = 1/2$ and $P(B) = 1/3$. And $P(AB) = P(\{(1, b), (2, b)\}) = P(1, b) + P(2, b) = 2/12 = 1/6$. The events A and B are independent.

Product spaces are a natural construction for working with repeated random experiments. They extend in an obvious way to more than two sample spaces. We can even define product spaces with infinite products for modeling, say, an infinite sequence of coin flips, although there are technical issues to be resolved when working with infinite sequences and infinite products that require more advanced analytical tools than what is presented in this first course.

3.9 SUMMARY

The concept of independence plays a central role in probability. In this chapter, we introduce independence and sequences of independent trials. Independent and identically distributed (i.i.d.) sequences are first introduced in the context of Bernoulli trials and coin-flipping. The binomial distribution arises naturally as the distribution of the sum of n i.i.d. Bernoulli trials. To derive the binomial distribution, we learn more counting techniques. Most importantly, binomial coefficients count: (i) the number of subsets of an n-element set and (ii) the number of n-element binary sequences with a fixed number of ones. Along with the binomial distribution, the Poisson distribution is one of the most important discrete probability distributions. It arises from the binomial setting when n, the number of fixed trials, is large and p, the success probability, is small.

- **Independent events:** Events A and B are independent if $P(A|B) = P(A)$. Equivalently, $P(AB) = P(A)P(B)$.
- **Mutual independence:** For general collections of events, independence means that for every finite subcollection $A_1, \ldots, A_k$,

$$P(A_1 \cdots A_k) = P(A_1) \cdots P(A_k).$$

  Mutual independence is a synonym for independence.
- **Pairwise independence:** A collection of events is pairwise independent if $P(A_i A_j) = P(A_i)P(A_j)$ for all pairs of events.
- **Independence note:** (i) A collection of independent events is pairwise independent. But events that are pairwise independent are not necessarily independent. (ii) Do not confuse independence with mutually exclusive.
- **Independent random variables:** Random variables X and Y are independent if $P(X = i, Y = j) = P(X = i)P(X = j)$ for all i, j. Equivalently, $P(X \in A, Y \in B) = P(X \in A)P(Y \in B)$, for all $A, B \subseteq \mathbb{R}$.
- **i.i.d. sequences:** A sequence of random variables is an independent and identically distributed sequence if the random variables are independent and all have the same distribution.
- **Bernoulli distribution:** A random variable X has a Bernoulli distribution with parameter $0 < p < 1$, if $P(X = 1) = p = 1 - P(X = 0)$.
- **Binomial coefficient:** The binomial coefficient $\binom{n}{k} = n!/(k!(n-k)!)$ "n choose k" counts: (i) the number of k-element subsets of $\{1, \ldots, n\}$ and (ii) the number of n element $0 - 1$ sequences with exactly k ones.
- **Binomial theorem:** For all x and y and nonnegative integer n,

$$(x + y)^n = \sum_{k=0}^{n} \binom{n}{k} x^k y^{n-k}.$$

- **Binomial distribution:** A random variable X has a binomial distribution with parameters n and p if

$$P(X = k) = \binom{n}{k} p^k (1-p)^{n-k}, \quad \text{for } k = 0, 1, \ldots, n.$$

- **Binomial setting:** The binomial distribution arises as the number of successes in n i.i.d. Bernoulli trials. The binomial setting requires: (i) a fixed number n of independent trials; (ii) trials take one of two possible values; and (iii) each trial has a constant probability p of success.
- **Stirling's approximation:** For large n,

$$n! \approx n^n e^{-n} \sqrt{2\pi n}.$$

- **Poisson distribution:** A random variable X has a Poisson distribution with parameter $\lambda > 0$, if

$$P(X = k) = \frac{e^{-\lambda}\lambda^k}{k!}, \quad \text{for } k = 0, 1, \ldots.$$

- **Poisson setting:** The Poisson setting arises in the context of discrete counts of "events" that occur over space or time with small probability and where successive events are independent.
- **Poisson approximation of binomial distribution:** Suppose $X \sim \text{Binom}(n, p)$ and $Y \sim \text{Pois}(\lambda)$. If $n \to \infty$ and $p \to 0$ in such a way so that $np \to \lambda > 0$, then for all k, $P(X = k) \to P(Y = k)$. The Poisson distribution with parameter $\lambda = np$ serves as a good approximation for the binomial distribution when n is large and p is small.

EXERCISES

Independence

3.1 Suppose A and B are independent events. Show that A^c and B^c are independent events.

3.2 Suppose A, B, and C are independent events with respective probabilities $1/3$, $1/4$, and $1/5$. Find

(a) $P(ABC)$

(b) $P(A \text{ or } B \text{ or } C)$

(c) $P(AB|C)$

(d) $P(B|AC)$

(e) $P(\text{At most one of the three events occurs})$

3.3 There is a 70% chance that a tree is infected with either root rot or bark disease. The chance that it does not have bark disease is 0.4. Whether or not a tree has root rot is independent of whether it has bark disease. Find the probability that a tree has root rot.

3.4 Toss two dice. Let A be the event that the first die rolls 1, 2, or 3. Let B be the event that the first die rolls 3, 4, or 5. Let C be the event that the sum of the dice is 9. Show that $P(ABC) = P(A)P(B)P(C)$, but no pair of events is independent.

3.5 **The first probability problem** A gambler's dispute in 1654 is said to have led to the creation of mathematical probability. Two French mathematicians, Blaise Pascal and Pierre de Fermat, considered the probability that in 24 throws of a pair of dice at least one "double six" occurs. It was commonly believed by gamblers at the time that betting on double sixes in 24 throws would be a profitable bet (i.e., greater than 50% chance of occurring). But Pascal and Fermat showed otherwise. Find this probability.

3.6 A lottery will be held. From 1000 numbers, one will be chosen as the winner. A lottery ticket is a number between 1 and 1000. How many tickets do you need to buy in order for the probability of winning to be at least 50%?

3.7 The original slot machine had 3 reels with 10 symbols on each reel. On each play of the slot machine, the reels spin and stop at a random position. Suppose each reel has one cherry on it. Let X be the number of cherries that show up from one play of the slot machine. Find $P(X = k)$, for $k = 0, 1, 2, 3$. Slot machines are also known as "one-armed bandits."

3.8 There is a 50-50 chance that the queen carries the gene for hemophilia. If she is a carrier, then each prince has a 50-50 chance of having hemophilia.

(a) If the queen has had three princes without the disease, what is the probability the queen is a carrier.

(b) If there is a fourth prince, what is the probability that he will have hemophilia?

3.9 Let X be a random variable such that $P(X = k) = k/10$, for $k = 1, 2, 3, 4$. Let Y be a random variable with the same distribution as X. Suppose X and Y are independent. Find $P(X + Y = k)$, for $k = 2, \ldots, 8$.

3.10 **Concidences** (Diaconis and Mosteller, 1989). See Section 2.3.1 on the birthday problem. Some categories (like birthdays) are equally likely to occur, with c possible values.

(a) Let k be the number of people needed so that the probability of at least one match is 95%. Show $k \approx 2.45\sqrt{c}$. (Hint: Use Equation (2.5).)

(b) Suppose there are m categories, all of which are independent and take c possible values. Let k be the number of people needed so that the probability of at least one match in any category is 95%. Show $k \approx 2.45\sqrt{c/m}$.

(c) A group of k people is comparing (i) their birthdays, (ii) the last two digits on their social security card, and (iii) the two-digit ticket number on their movie stubs. How big should k be so that there is a 50% chance of at least one match? A 95% chance?

3.11 Suppose X_1, X_2, X_3 are i.i.d. random variables, each uniformly distributed on $\{1, 2, 3\}$. Find the probability function for $X_1 + X_2 + X_3$. That is, find $P(X_1 + X_2 + X_3 = k)$, for $k = 3, \ldots, 9$.

Counting

3.12 There are 40 pairs of shoes in Bill's closet. They are all mixed up.

(a) If 20 shoes are picked, what is the chance that Bill's favorite sneakers will be in the group?

(b) If 20 shoes are picked, what is the chance that one shoe from each pair will be represented? (Remember, a left shoe is different than a right shoe.)

3.13 Many bridge players believe that the most likely distribution of the four suits (spades, hearts, diamonds, and clubs) in a bridge hand is 4-3-3-3 (four cards in one suit, and three cards of the other three).

(a) Show that the suit distribution 4-4-3-2 is more likely than 4-3-3-3.

(b) In fact, besides the 4-4-3-2 distribution, there are three other patterns of suit distributions that are more likely than 4-3-3-3. Can you find them?

3.14 Find the probability that a bridge hand contains a nine-card suit. That is, the number of cards of the longest suit is nine.

3.15 A chessboard is an eight-by-eight arrangement of 64 squares. Suppose eight chess pieces are placed on a chessboard at random so that each square can receive at most one piece. What is the probability that there will be exactly one piece in each row and in each column?

3.16

> There are few things that are so unpardonably neglected in our country as poker. The upper class knows very little about it. Now and then you find ambassadors who have sort of a general knowledge of the game, but the ignorance of the people is fearful. Why, I have known clergymen, good men, kind-hearted, liberal, sincere, and all that, who did not know the meaning of a "flush." It is enough to make one ashamed of one's species.
>
> —Mark Twain

Find the probabilities for the following poker hands. They are arranged in decreasing order of probability.

(a) Straight flush. (Five cards in a sequence and of the same suit.)

(b) Four of a kind. (Four cards of one face value and one other card.)

(c) Full house. (Three cards of one face value and two of another face value.)

(d) Flush. (Five cards of the same suit. Does not include a straight flush.)

(e) Straight. (Five cards in a sequence. Does not include a straight flush. Ace can be high or low.)

(f) Three of a kind. (Three cards of one face value. Does not include four of a kind or full house.)

(g) Two pair. (Does not include four of a kind or full house.)

(h) One pair. (Does not include any of the aforementioned conditions.)

3.17 A walk in the positive quadrant of the plane consists of a sequence of moves, each one from a point (a, b) to either $(a + 1, b)$ or $(a, b + 1)$.

(a) Show that the number of walks from the origin $(0, 0)$ to (x, y) is $\binom{x+y}{x}$.

(b) Suppose a walker starts at the origin $(0, 0)$ and at each discrete unit of time moves either up one unit or to the right one unit each with probability 1/2.

If $x > y$, find the probability that a walk from (0,0) to (x,y) always stays above the main diagonal.

3.18 See Example 3.16 for a description of the Powerball lottery. A \$100 prize is gotten by either (i) matching exactly three of the five balls and the powerball or (ii) matching exactly four of the five balls and not the powerball. Find the probability of winning \$100.

3.19 Give a combinatorial argument (not an algebraic one) for why

$$\binom{n}{k} = \binom{n}{n-k}.$$

3.20 Give a combinatorial proof that

$$\sum_{i=0}^{k} \binom{m}{i}\binom{n}{k-i} = \binom{m+n}{k}. \tag{3.13}$$

Hint: How many ways can you choose k people from a group of m men and n women? From Equation 3.13 show that

$$\sum_{k=0}^{n} \binom{n}{k}^2 = \binom{2n}{n}. \tag{3.14}$$

Binomial distribution

3.21 Every person in a group of 1000 people has a 1% chance of being infected by a virus. The process of being infected is independent from person to person. Using random variables, write expressions for the following probabilities and solve them with R.

(a) The probability that exactly 10 people are infected.

(b) That probability that at least 16 people are infected.

(c) The probability that between 12 and 14 people are infected.

(d) The probability that someone is infected.

3.22 **Newton–Pepys problem (Stigler, 2006).** In 1693, Samuel Pepys wrote a letter to Isaac Newton posing the following question.

Which of the following three occurrences has the greatest chance of success?

1. Six fair dice are tossed and at least one 6 appears.
2. Twelve fair dice are tossed and at least two 6's appear.
3. Eighteen fair dice are tossed and at least three 6's appear.

Answer Mr. Pepys' question.

3.23 For the following situations, identify whether or not X has a binomial distribution. If it does, give n and p; if not, explain why.

- **(a)** Every day Amy goes out for lunch there is a 25% chance she will choose pizza. Let X be the number of times she chose pizza last week.
- **(b)** Brenda plays basketball, and there is a 60% she makes a free throw. Let X be the number of successful baskets she makes in a game.
- **(c)** A bowl contains 100 red candies and 150 blue candies. Carl reaches and takes out a sample of 10 candies. Let X be the number of red candies in his sample.
- **(d)** Evan is reading a 600-page book. On even-numbered pages, there is a 1% chance of a typo. On odd-numbered pages, there is a 2% chance of a typo. Let X be the number of typos in the book.
- **(e)** Fran is reading a 600-page book. The number of typos on each page has a Bernoulli distribution with $p = 0.01$. Let X be the number of typos in the book.

3.24 See Example 3.26. Consider a random graph on $n = 8$ vertices with edge probability $p = 0.25$.

- **(a)** Find the probability that the graph has at least six edges.
- **(b)** A vertex of a graph is said to be *isolated* if its degree is 0. Find the probability that a particular vertex is isolated.

3.25 A bag of 16 balls contains one red, three yellow, five green, and seven blue balls. Suppose four balls are picked, sampling with replacement.

- **(a)** Find the probability that the sample contains at least two green balls.
- **(b)** Find the probability that each of the balls in the sample is a different color.
- **(c)** Repeat these two problems for the case when the sampling is without replacement.

3.26 Ecologists use *occupancy models* to study animal populations. Ecologists at the Department of Natural Resources use helicopter surveying methods to look for otter tracks in the snow along the Mississippi River to study which parts of the river are occupied by otter. The *occupancy rate* is the probability that an animal species is present at a particular site. The *detection rate* is the probability that the animal will be detected. (In this case, whether tracks will be seen from a helicopter.) If the animal is not detected, this might be due to the site not being occupied or because the site is occupied and the tracks were not detected.

A common model used by ecologists is a *zero-inflated binomial model.* If the region is occupied, then the number of detections is binomial with n the number of sites and p the detection rate. If the region is unoccupied, the number of detections is 0.

Let α be the occupancy rate, p the detection rate, and n the number of sites.

(a) Find the probability of zero detections. (Hint: Condition on occupancy.)

(b) DNR ecologists searched five sites along the Mississippi River for the presence of otter. Suppose $\alpha = 0.75$ and $p = 0.50$. Let Z be the number of observed detections. Give the probability function for Z.

3.27 **Sign test:** A new experimental drug is given to patients suffering from severe migraine headaches. Patients report their pain experience on a scale of 1 – 10 before and after the drug. The difference of their pain measurement is recorded. If their pain decreases, the difference will be positive $(+)$; if the pain increases, the difference will be negative $(-)$. The data are ignored if the difference is 0.

Under the *null hypothesis* that the drug is ineffective and there is no difference in pain experience before and after the drug, the number of $+$'s will have a binomial distribution with n equal to the number of $+$' s and $-$'s; and $p = 1/2$. This is the basis of a statistical test called the sign test.

Suppose a random sample of 20 patients are given the new drug. Of 16 nonzero differences, 12 report an improvement $(+)$. If one assumes that the drug is ineffective, what is the probability of obtaining 12 or more +'s, as observed in these data? Based on these data do you think the drug is ineffective?

3.28 I have two dice, one is a standard die. The other has three ones and three fours. I flip a coin. If heads, I will roll the standard die five times. If tails, I will roll the other die five times. Let X be the number of fours that appear. Find $P(X = 3)$. Does X have a binomial distribution?

Poisson distribution

3.29 Suppose X has a Poisson distribution and $P(X = 2) = 2P(X = 1)$. Find $P(X = 3)$.

3.30 Computer scientists have modeled the length of search queries on the web using a Poisson distribution. Suppose the average search query contains about three words. Let X be the number of words in a search query. Since one cannot have a query consisting of zero words, we model X as a "restricted" Poisson distribution that does not take values of 0. That is, let $P(X = k) = P(Y = k|Y \neq 0)$, where $Y \sim \text{Pois}(3)$.

(a) Find the probability function of X.

(b) What is the probability of obtaining search queries longer than 10 words?

(c) Arampatzis and Kamps (2008) observe that many data sets contain very large queries that are not predicted by a Poisson model, such as queries of 10 words or more. They propose a restricted Poisson model for short queries, for instance queries of six words or less. Find $P(Y = k|1 \leq Y \leq 6)$ for $k = 0, \ldots, 6$.

TABLE 3.10: No-hitter baseball games.

Number of games	0	1	2	3	4	5	6	7
Seasons	18	30	21	21	6	3	3	2

3.31 The number of eggs a chicken hatches is a Poisson random variable. The probability that the chicken hatches no eggs is 0.10. What is the probability that she hatches at least two eggs?

3.32 Hemocytometer slides are used to count cells and other microscopic particles. They consist of glass engraved with a laser-etched grid. The size of the grid is known making it possible to count the number of particles in a specific volume of fluid.

(a) A hemocytometer slide used to count red blood cells has 160 squares. The average number of blood cells per square is 4.375. What is the probability that one square contains between 3 and 6 cells?

(b) Modify the **Balls.R** script to simulate the distribution of 700 red blood cells on a hemocytometer slide. Use your simulation to estimate the probability that a square contains between 3 and 6 cells.

3.33 Cars pass a busy intersection at a rate of approximately 16 cars per minute. What is the probability that at least 1000 cars will cross the intersection in the next hour? (Hint: What is the rate per hour?)

3.34 Table 3.10 from Huber and Gleu (2007) shows the number of no hitter baseball games that were pitched in the 104 ball seasons between 1901 and 2004. For instance, the following data on the number of no-hitter baseball games that were pitched in the 104 baseball seasons between 1901 and 2004 are given in Huber and Glen (2007).

For instance, 18 seasons saw no no-hit games pitched; 30 seasons saw one no-hit game, etc. Use these data to model the number of no-hit games for a baseball season. Create a table that compares the observed counts with the expected number of no-hit games under your model (Table 3.10).

3.35 Suppose $X \sim \text{Pois}(\lambda)$. Find the probability that X is odd. (Hint: Consider Taylor expansions of e^{λ} and $e^{-\lambda}$.)

3.36 If you take the red pill, the number of colds you get next winter will have a Poisson distribution with $\lambda = 1$. If you take the blue pill, the number of colds will have a Poisson distribution with $\lambda = 4$. Each pill is equally likely. Suppose you get three colds next winter. What is the probability you took the blue pill?

3.37 A physicist estimated that the probability of a U.S. nickel landing on its edge is one in 6000. Suppose a nickel is flipped 10,000 times. Let X be the number of times it lands on its edge. Find the probability that X is between one and three using

(a) The exact distribution of X

(b) An approximate distribution of X

3.38 A chessboard is put on the wall and used as a dart board. Suppose 100 darts are thrown at the board and each of the 64 squares is equally likely to be hit.

(a) Find the exact probability that the left-top corner of the chessboard is hit by exactly two darts.

(b) Find an approximation of this probability using an appropriate distribution.

3.39 Give a probabilistic interpretation of the series

$$\frac{1}{e} + \frac{1}{2!e} + \frac{1}{4!e} + \frac{1}{6!e} + \cdots.$$

That is, pose a probability question for which the sum of the series is the answer.

3.40 Suppose that the number of eggs that an insect lays is a Poisson random variable with parameter λ. Further, the probability that an egg hatches and develops is p. Egg hatchings are independent of each other. Show that the total number of eggs that develop has a Poisson distribution with parameter λp. Hint: Condition on the number of eggs that are laid.

Simulation and R

3.41 Which is more likely: 5 heads in 10 coin flips, 50 heads in 100 coin flips, or 500 heads in 1000 coin flips? Use R's `dbinom` command to find out.

3.42 Simulate the 1654 gambler's dispute in Exercise 3.5.

3.43 Choose your favorite value of λ and let $X \sim \text{Pois}(\lambda)$. Simulate the probability that X is odd. See Exercise 3.35. Compare with the exact solution.

3.44 Write an R function `before(a,b)` to simulate the probability, in repeated independent throws of a pair of dice, that a appears before b, for $a, b = 2, \ldots, 12$.

3.45 Simulate a Prussian soldier's death by horse kick as in Example 3.8. Create a histogram based on 10,000 repetitions. Compare to a histogram of the Poisson distribution using the command `rpois(10000,0.61)`.

3.46 Poisson approximation of the binomial: Suppose $X \sim \text{Binom}(n, p)$. Write an R function `compare(n,p,k)` that computes (exactly) $P(X = k) - P(Y = k)$, where $Y \sim \text{Pois}(np)$. Try your function on numbers where you expect the Poisson probability to be a good approximation of the binomial. Also try it on numbers where you expect the approximation to be poor.

4

RANDOM VARIABLES

Iacta alea est. (*The die is cast.*)

—Julius Caesar, upon crossing the Rubicon.

Having introduced the binomial and Poisson distributions, two of the most important probability distributions for discrete random variables, we now look at random variables in general in some depth.

The probability function $P(X = x)$ of a discrete random variable X is called the *probability mass function (pmf) of* X.

PROBABILITY MASS FUNCTION

Definition 4.1. For a random variable X that takes values in a set S, the *probability mass function of* X is the probability function

$$m(x) = P(X = x), \quad \text{for } x \in S.$$

Probability mass functions are the central objects in discrete probability that allow one to compute probabilities. If we know the pmf of a random variable, in a sense we have "complete knowledge"—in a probabilistic sense—of the behavior of that random variable.

Table 4.1 summarizes the distributions that you have seen so far, together with their probability mass functions. See Appendix C for a more complete list.

Probability: With Applications and R, First Edition. Robert P. Dobrow.

TABLE 4.1: Discrete probability distributions.

Distribution	Parameters	Probability mass function
Bernoulli	p	$P(X=k)=\begin{cases}p, & \text{if } k=1\\ 1-p, & \text{if } k=0\end{cases}$
Binomial	n, p	$P(X=k)=\binom{n}{k}p^k(1-p)^{n-k},\ k=0,\ldots,n$
Poisson	λ	$P(X=k)=\frac{e^{-\lambda}\lambda^k}{k!},\ k=0,1,\ldots$
Uniform on $\{x_1,\ldots,x_n\}$		$P(X=x_k)=\frac{1}{n},\ k=1,\ldots,n$

A NOTE ON NOTATION

Although probability was studied for hundreds of years, many of the familiar symbols that we use today have their origin in the mid-twentieth century. In particular, their use was made popular in William Feller's remarkable probability textbook *An Introduction to Probability Theory and Its Applications* (Feller, 1968), first published in 1950 and considered by many to be the greatest mathematics book of the twentieth century.

According to Prof. John Aldrich of the University of Southhampton (http://jeff560.fripod.com/stat.html), some of the notation and symbols made popular by their use in Feller's book include

1. $P(A)$ the notation for the probability of an event.
2. $P(X=x)$ the use of capital and lowercase letters to denote a random variable and its value.
3. $P(A|B)$ the use of the vertical bar to denote conditional probability.

4.1 EXPECTATION

The expectation is a numerical measure that summarizes the typical, or average, behavior of a random variable.

EXPECTATION

If X is a discrete random variable that takes values in a set S, the *expectation* $E[X]$ is defined as

$$E[X]=\sum_{x\in S} xP(X=x).$$

The sum in the definition is over all values of X. If the sum is a divergent infinite series, we say that the expectation of X *does not exist.*

Expectation is a *weighted average* of the values of X, where the weights are the corresponding probabilities of those values. The expectation places more weight on values that have greater probability.

In the case when X is uniformly distributed on a finite set $\{x_1, \ldots, x_n\}$, that is, all outcomes are equally likely,

$$E[X] = \sum_{i=1}^{n} x_i P(X = x_i) = \sum_{i=1}^{n} x_i \left(\frac{1}{n}\right) = \frac{x_1 + \cdots + x_n}{n}.$$

With equally likely outcomes the expectation is just the regular average of the values.

Other names for expectation are *mean* and *expected value*. In the context of games and random experiments involving money, expected value is often used.

Example 4.1 Scrabble. In the game of Scrabble, there are 100 letter tiles with the distribution of point values given in Table 4.2. Let X be the point value of a random Scrabble tile. What is the expectation of X?

Convert the entries of Table 4.2 to probabilities by dividing number of tiles by 100. The expected point value of a Scrabble tile is

$$E[X] = 0(0.02) + 1(0.68) + 2(0.07) + 3(0.08) + 4(0.10) + 5(0.01)$$
$$+ 8(0.02) + 10(0.02) = 1.87.$$

■

TABLE 4.2: Tile values in Scrabble.

Point value	0	1	2	3	4	5	8	10
Number of tiles	2	68	7	8	10	1	2	2

Example 4.2 Roulette. In the game of roulette, a ball rolls around a roulette wheel landing on one of 38 numbers. Eighteen numbers are red; 18 are black; and two—0 and 00—are green. A bet of "red" costs \$1 to play and pays off even money if the ball lands on red.

Let X be a player's winnings at roulette after one bet of red. What is the distribution of X? What is the expected value $E[X]$?

The player either wins or loses \$1. So $X = 1$ or -1, with $P(X = 1) = 18/38$ and $P(X = -1) = 20/38$. Once the distribution of X is found, the expected value can be computed:

$$E[X] = (1)P(X = 1) + (-1)P(X = -1)$$
$$= \frac{18}{38} - \frac{20}{38} = \frac{-2}{38} = -0.0526.$$

The expected value of the game is about -5 cents. That is, the expected loss is about a nickel.

■

What does $E[X] = -0.0526$ or, in the previous example, $E[X] = 1.87$, really mean since you cannot lose 5.26 cents at roulette or pick a Scrabble tile with a value of 1.87?

We interpret $E[X]$ as a long-run average. That is, if you pick Scrabble tiles repeatedly for a long time (with replacement), then the average of those tile values will be about 1.87. If you play roulette for a long time making many red bets, then the average of all your one dollar wins and losses will be about -5 cents. What that also means is that if you play, say, 10,000 times then your *total* loss will be about $5 \times 10{,}000 = 50{,}000$ cents, or about \$500.

More formally, let $X_1, X_2, \ldots$ be an i.i.d. sequence of outcomes of roulette bets, where X_k is the outcome of the kth bet. Then the interpretation of expectation is that

$$E[X] \approx \frac{X_1 + \cdots + X_n}{n}, \tag{4.1}$$

when n is large. This gives a prescription for simulating the expectation of a random variable X: choose n large, simulate n copies of X, and take the average as an approximation for $E[X]$.

R: PLAYING ROULETTE

The command

```
> sample(1:38,1)
```

simulates a uniform random integer between 1 and 38. Let the numbers 1 – 18 represent red. The command

```
> if (sample(1:38,1) <= 18) 1 else -1
```

returns a 1 if you roll red, and -1, otherwise.

Repeat 10,000 times and take the average to simulate the expected value.

```
> simlist <- replicate(10000,
    if (sample(1:38,1) <= 18) 1 else -1)
> mean(simlist)
[1] -0.051
```

Example 4.3 Expectation of uniform distribution. Let $X \sim \text{Unif}\{1, \ldots, n\}$. The expectation of X is

$$E[X] = \sum_{x=1}^{n} xP(X = x) = \sum_{x=1}^{n} \frac{x}{n} = \frac{1}{n}\left(\frac{(n+1)n}{2}\right) = \frac{n+1}{2}. \quad \blacksquare$$

Example 4.4 Expectation of Poisson Distribution. Let $X \sim \text{Pois}(\lambda)$. Before doing the calculation, make an educated guess at the expectation of X.

$$
\begin{aligned}
E[X] &= \sum_{k=0}^{\infty} kP(X=k) = \sum_{k=0}^{\infty} k\frac{e^{-\lambda}\lambda^k}{k!} \\
&= e^{-\lambda}\sum_{k=1}^{\infty} \frac{\lambda^k}{(k-1)!} = \lambda e^{-\lambda}\sum_{k=1}^{\infty} \frac{\lambda^{k-1}}{(k-1)!} \\
&= \lambda e^{-\lambda}\sum_{k=0}^{\infty} \frac{\lambda^k}{k!} = \lambda e^{-\lambda}e^{\lambda} = \lambda.
\end{aligned}
$$

The parameter of a Poisson distribution is the mean of the distribution. ■

4.2 FUNCTIONS OF RANDOM VARIABLES

Suppose X is a random variable and f is some function. Then $Y = f(X)$ is a random variable that is a function of X. The values of this new random variable are found as follows. If $X = x$, then $Y = f(x)$.

Functions of random variables, like X^2, e^X, and $1/X$, show up all the time in probability. Often we apply some function to the outcomes of a random experiment, as we will see in many examples later. In statistics, it is common to transform data using an elementary function such as the log or exponential function.

Suppose X is uniformly distributed on $\{-2,-1,0,1,2\}$. That is,

$$P(X=k) = \frac{1}{5}, \quad \text{for } k = -2,-1,0,1,2.$$

Take $f(x) = x^2$, and let $Y = f(X) = X^2$. The possible outcomes of Y are $(-2)^2, (-1)^2, 0^2, 1^2, 2^2$, that is, 0, 1, and 4. Since $Y = 0$ if and only if $X = 0$, $Y = 1$ if and only if $X = \pm 1$, and $Y = 4$ if and only if $X = \pm 2$, the probability mass function of Y is

$$
\begin{aligned}
&P(Y=0) = P(X^2=0) = P(X=0) = \frac{1}{5}, \\
&P(Y=1) = P(X^2=1) = P(X=\pm 1) = P(X=-1) + P(X=1) = \frac{2}{5}, \\
&P(Y=4) = P(X^2=2) = P(X \pm 2) = P(X=-2) + P(X=2) = \frac{2}{5}.
\end{aligned}
$$

Here is one way to think of functions of random variables. Suppose there are two rooms labeled X and Y. In the first X room, a random experiment is performed.

The outcome of the experiment is x. A messenger takes x and delivers it to the Y room. But before he gets there he applies the f function to x and delivers $f(x)$ to the Y room. As the random experiment is repeated, an observer looking into the X room sees the x outcomes. An observer looking into the Y room sees the $f(x)$ outcomes.

Example 4.5 Timmy spends \$2 in supplies to set up his lemonade stand. He charges 25 cents a cup. Suppose the number of cups he sells in a day has a Poisson distribution with $\lambda = 10$. Describe his profit as a function of a random variable and find the probability that the lemonade stand makes a positive profit.

Let X be the number of cups Timmy sells in a day. Then $X \sim \text{Pois}(10)$. If he sells x cups then his profit is $25x - 200$ cents. The random variable $Y = 25X - 200$ defines his profit as a function of X, the number of cups sold.

Since X takes values $0, 1, 2, \ldots$, Y takes values $-200, -175, -150, \ldots$ The probability that Timmy makes a positive profit is

$$\begin{aligned} P(Y > 0) &= P(25X - 200 > 0) \\ &= P(X > 8) = 1 - P(X \le 8) \\ &= 1 - \sum_{k=0}^{8} \frac{e^{-10} 10^k}{k!} = 0.667. \end{aligned}$$

```
> 1-ppois(8,10)
[1] 0.6671803
```

Observe carefully the use of algebraic operations with random variables. We have $\{25X - 200 > 0\}$ if and only if $\{25X > 200\}$ if and only if $\{X > 8\}$. We can add, subtract, multiply, and do any allowable algebraic operation on both sides of the expression. You are beginning to see the power of working with random variables.

R: LEMONADE PROFITS

To simulate one day's profit in cents for the lemonade stand, type

```
> rpois(1,10)*25 - 200
```

To simulate the probability of making a positive profit, type

```
> simlist <- rpois(10000,10)*25 - 200
> sum(simlist>0)/10000
[1] 0.6592
```

■

The next result is most important for finding the expectation of a function of a random variable.

EXPECTATION OF FUNCTION OF A RANDOM VARIABLE

Let X be a random variable that takes values in a set S. Let f be a function. Then,

$$E[f(X)] = \sum_{x \in S} f(x)P(X = x). \tag{4.2}$$

The result seems straightforward and perhaps even "obvious." But its proof is somewhat technical, and we leave the details for the interested reader at the end of the chapter. It is sometimes called "the law of the unconscious statistician."

Example 4.6 A number X is picked uniformly at random from 1 to 100. What is the expected value of X^2?

The random variable X^2 is equal to $f(X)$, where $f(x) = x^2$. This gives

$$\begin{aligned} E[X^2] &= \sum_{x=1}^{100} x^2 P(X = x) = \sum_{x=1}^{100} \frac{x^2}{100} \\ &= \left(\frac{1}{100}\right) \frac{100(101)(201)}{6} \\ &= \frac{(101)(201)}{6} = 3383.5, \end{aligned}$$

using the fact that the sum of the first n squares is $n(n+1)(2n+1)/6$.

You might have first thought that since $E[X] = 101/2 = 50.5$, then $E[X^2] = (101/2)^2 = 2550.25$. We see that this is not correct. It is not true that $E[X^2] = (E[X])^2$. Consider a simulation. We take one million replications to get a good estimate. ■

```
> mean(sample(1:100,1000000,replace=T)^2)
[1] 3382.928
```

The simulation should reinforce the fact that $E[X^2] \neq E[X]^2$. More generally, it is not true that $E[f(X)] = f(E[X])$. The operations of expectation and function evaluation cannot be interchanged. It is very easy, and common, to make this kind of a mistake. To be forewarned is to be forearmed!

Example 4.7 Create a "random sphere" whose radius R is determined by the roll of a die. Let V be the volume of the sphere. Find $E[V]$.

The formula for the volume of a sphere as a function of radius is $v(r) = (4\pi/3)r^3$. The expected volume of the sphere is

$$
\begin{aligned}
E[V] = E\left[\frac{4\pi}{3}R^3\right] &= \sum_{r=1}^{6}\left(\frac{4\pi}{3}r^3\right)P(R=r) \\
&= \frac{4\pi}{3}\left(\frac{1}{6}\right)(1^3+2^3+3^3+4^3+5^3+6^3) \\
&= 98\pi.
\end{aligned}
$$

■

Example 4.8 Suppose X has a Poisson distribution with parameter λ. Find $E[1/(X+1)]$.

$$
\begin{aligned}
E\left[\frac{1}{X+1}\right] &= \sum_{k=0}^{\infty}\frac{1}{k+1}P(X=k) \\
&= \sum_{k=0}^{\infty}\frac{1}{k+1}\left(\frac{e^{-\lambda}\lambda^k}{k!}\right) \\
&= \frac{e^{-\lambda}}{\lambda}\sum_{k=0}^{\infty}\frac{\lambda^{k+1}}{(k+1)!} \\
&= \frac{e^{-\lambda}}{\lambda}\sum_{k=1}^{\infty}\frac{\lambda^k}{k!} = \frac{e^{-\lambda}}{\lambda}\left(\sum_{k=0}^{\infty}\frac{\lambda^k}{k!}-1\right) \\
&= \frac{e^{-\lambda}}{\lambda}(e^{\lambda}-1) = \frac{1-e^{-\lambda}}{\lambda}.
\end{aligned}
$$

■

We gave a dire warning a few paragraphs back that in general it is not true that $E[f(X)] = f(E[X])$. However, there is one notable exception. That is the case when f is a linear function.

EXPECTATION OF A LINEAR FUNCTION OF X

For constants a and b,

$$E[aX+b] = aE[X]+b.$$

Let $f(x) = ax+b$. By the law of the unconscious statistician,

$$
\begin{aligned}
E[aX+b] &= \sum_x (ax+b)P(X=x) \\
&= a\sum_x xP(X=x) + b\sum_x P(X=x) \\
&= aE[X]+b.
\end{aligned}
$$

In words, the expectation of a linear function of X is that function evaluated at the expectation of X.

Example 4.9 A lab in France takes temperature measurements of data that are randomly generated. The mean temperature for their data is 5°C. The data are transferred to another lab in the United States, where temperatures are recorded in Fahrenheit. When the data are sent, they are first transformed using the Celsius to Fahrenheit conversion formula $f = 32 + (9/5)c$. Find the mean temperature of the Fahrenheit data.

Let F and C denote the random temperature measurements in Fahrenheit and Celsius, respectively. Then

$$\begin{aligned} E[F] &= E\left[32 + \left(\frac{9}{5}\right) C\right] = 32 + \left(\frac{9}{5}\right) E[C] \\ &= 32 + \left(\frac{9}{5}\right) 5 = 41°\mathrm{F}. \end{aligned}$$

■

4.3 JOINT DISTRIBUTIONS

In the case of two random variables X and Y, a *joint distribution* specifies the values and probabilities for all pairs of outcomes. The *joint probability mass function of X and Y* is the function of two variables $P(X = x, Y = y)$.

Since the joint pmf is a probability function, it sums to 1. If X takes values in a set S and Y takes values in a set T, then

$$\sum_{x \in S} \sum_{y \in T} P(X = x, Y = y) = 1.$$

As in the one variable case, probabilities of events are obtained by summing over the individual outcomes contained in the event. For instance, for constants $a < b$ and $c < d$,

$$P(a \le X \le b, c \le Y \le d) = \sum_{x=a}^{b} \sum_{y=c}^{d} P(X = x, Y = y).$$

A joint probability mass function can be defined for any finite collection of discrete random variables $X_1, \ldots, X_n$ defined on a common sample space. The joint pmf is the function of n variables $P(X_1 = x_1, \ldots, X_n = x_n)$.

Example 4.10 The joint probability mass function of X and Y is

$$P(X = x, Y = y) = cxy, \quad \text{for } x, y = 0, 1, 2.$$

(i) Find the constant c. (ii) Find $P(X \leq 1, Y \geq 1)$.

(i) To find c,

$$1 = \sum_{x=0}^{2} \sum_{y=0}^{2} cxy = c(0+0+0+0+0+1+2+2+4) = 9c,$$

and thus $c = 1/9$.

(ii) The desired probability is

$$P(X \leq 1, Y \geq 1) = \sum_{x=0}^{1} \sum_{y=1}^{2} \frac{xy}{9} = \frac{1}{9}(1+2) = \frac{1}{3}.$$ ■

Example 4.11 Red ball, blue ball. A bag contains four red, three white, and two blue balls. A sample of two balls is picked without replacement. Let R and B be the number of red and blue balls, respectively, in the sample. (i) Find the joint probability mass function of R and B. (ii) Use the joint pmf to find the probability that the sample contains at most one red and one blue ball.

Consider the event $\{R = r, B = b\}$. The number of red and blue balls in the sample must be between 0 and 2. For $0 \leq r + b \leq 2$, if r red balls and b blue balls are picked, then $2-r-b$ white balls must also be picked. Selecting r red, b blue, and $2-r-b$ white balls can be done in $\binom{4}{r}\binom{2}{b}\binom{3}{2-r-b}$ ways. There are $\binom{9}{2} = 36$ ways to select two balls from the bag. Thus, the joint probability mass function of (R, B) is

$$P(R = r, B = b) = \binom{4}{r}\binom{3}{2-r-b}\binom{2}{b} \Big/ 36\,, \quad \text{for} \quad 0 \leq r + b \leq 2.$$

The joint pmf of R and B is described by the joint probability table

		B 0	1	2
	0	3/36	6/36	1/36
R	1	12/36	8/36	0
	2	6/36	0	0.

The desired probability is

$$P(R \leq 1, B \leq 1) = \sum_{r=0}^{1} \sum_{b=0}^{1} P(R = r, B = b)$$
$$= \frac{3}{36} + \frac{12}{36} + \frac{6}{36} + \frac{8}{36} = \frac{5}{6}.$$ ■

From the joint distribution of X and Y, one can obtain the univariate, or *marginal distribution* of each variable. Since

$$\{X = x\} = \bigcup_{y \in T} \{X = x, Y = y\},$$

the probability mass function of X is

$$P(X = x) = P\left(\bigcup_{y \in T} \{X = x, Y = y\}\right) = \sum_{y \in T} P(X = x, Y = y).$$

The marginal distribution of X is obtained from the joint distribution of X and Y by summing over the values of y. Similarly the probability mass function of Y is obtained by summing the joint pmf over the values of x.

MARGINAL DISTRIBUTIONS

$$P(X = x) = \sum_{y \in T} P(X = x, Y = y)$$

and

$$P(Y = y) = \sum_{x \in S} P(X = x, Y = y).$$

Example 4.12 Red ball, blue ball, continued. For the last Example 4.11, find the marginal distributions of the number of red and blue balls, respectively. Use these distributions to find the expected number of red balls and the expected number of blue balls in the sample.

Given a joint probability table, the marginal distributions are obtained by summing over the rows and columns of the table.

		B 0	1	2	
	0	3/36	6/36	1/36	10/36
R	1	12/36	8/36	0	20/36
	2	6/36	0	0	6/36
		21/36	14/36	1/36	

That is,

$$P(R=r)=\begin{cases}10/36, & \text{if } r=0\\ 20/36, & \text{if } r=1\\ 6/36, & \text{if } r=2,\end{cases}$$

and

$$P(B=b)=\begin{cases}21/36, & \text{if } b=0\\ 14/36, & \text{if } b=1\\ 1/36, & \text{if } b=2.\end{cases}$$

For the expectations,

$$E[R]=0\left(\frac{10}{36}\right)+1\left(\frac{20}{36}\right)+2\left(\frac{6}{36}\right)=\frac{8}{9}$$

and

$$E[B]=0\left(\frac{21}{36}\right)+1\left(\frac{14}{36}\right)+2\left(\frac{1}{36}\right)=\frac{4}{9}. \quad \blacksquare$$

Example 4.13 A computer store has modeled the number of computers C it sells per day, together with the number of extended warranties W. The joint probability mass function is

$$P(C=c,W=w)=\left(\frac{5}{2}\right)^c\frac{e^{-5}}{w!(c-w)!},$$

for $c=0,1,\ldots,$ and $w=0,\ldots,c$. We explore the marginal distributions of C and W.

To find the marginal distribution of C sum over all w. Since $0\le w\le c$,

$$\begin{aligned}P(C=c)&=\sum_w P(C=c,W=w)\\&=\sum_{w=0}^{c}\left(\frac{5}{2}\right)^c\frac{e^{-5}}{w!(c-w)!}\\&=\left(\frac{5}{2}\right)^c e^{-5}\sum_{w=0}^{c}\frac{1}{w!(c-w)!}\\&=\left(\frac{5}{2}\right)^c\frac{e^{-5}}{c!}\sum_{w=0}^{c}\binom{c}{w}\\&=\left(\frac{5}{2}\right)^c\frac{e^{-5}}{c!}2^c=\frac{e^{-5}5^c}{c!},\end{aligned}$$

for $c = 0, 1, \ldots$. The final expression is the probability mass function of a Poisson random variable with $\lambda = 5$. Thus, the marginal distribution of C, the number of computers sold, is a Poisson distribution with $\lambda = 5$.

For the marginal distribution of W, sum over c values. Since $0 \leq w \leq c$, the sum is over all $c \geq w$, which gives

$$\begin{aligned}
P(W = w) &= \sum_c P(W = w, C = c) \\
&= \sum_{c=w}^{\infty} \left(\frac{5}{2}\right)^c \frac{e^{-5}}{w!(c-w)!} \\
&= \frac{e^{-5}}{w!} \sum_{c=w}^{\infty} \left(\frac{5}{2}\right)^c \frac{1}{(c-w)!} \\
&= \frac{e^{-5}}{w!} \left(\frac{5}{2}\right)^w \sum_{c=w}^{\infty} \left(\frac{5}{2}\right)^{c-w} \frac{1}{(c-w)!} \\
&= \frac{e^{-5}}{w!} \left(\frac{5}{2}\right)^w \sum_{c=0}^{\infty} \left(\frac{5}{2}\right)^c \frac{1}{c!} \\
&= \frac{e^{-5}}{w!} \left(\frac{5}{2}\right)^w e^{5/2} = \frac{e^{-5/2}(5/2)^w}{w!},
\end{aligned}$$

for $w = 0, 1, \ldots$. This gives the pmf of a Poisson random variable with $\lambda = 5/2$. That is, the marginal distribution of W, the number of extended warranties sold, is a Poisson distribution with $\lambda = 5/2$.

Computers, according to the model, sell roughly at the rate of five per day. And warranties sell at half that rate.

Suppose the store sells c computers on a particular day, what is the probability they will sell w extended warranties?

We have posed the conditional probability

$$\begin{aligned}
P(W = w | C = c) &= \frac{P(W = w, C = c)}{P(C = c)} \\
&= \frac{(5/2)^c e^{-5}/(w!(c-w)!)}{e^{-5}5^c/c!} \\
&= \binom{c}{w} \left(\frac{1}{2}\right)^c.
\end{aligned}$$

This is a binomial probability. We discovered that conditional on selling c computers the number of warranties sold has a binomial distribution with parameters c and $p = 1/2$. This is the same distribution as the number of heads in c fair coin flips.

Here is a probabilistic description of computer and warranty sales at the computer store. About five computers, on average, are sold per day according to a Poisson distribution. For each computer sold, there is a 50–50 chance that the extended warranty will be purchased. And thus about 2.5 warranties are sold, per day, on average, according to a Poisson distribution.

Random variables can arise as functions of two or more random variables. Suppose $f(x, y)$ is a real-valued function of two variables. Then $Z = f(X, Y)$ is a random variable, which is a function of two random variables. For expectations of such random variables, there is a multivariate version of Equation 4.2. ■

EXPECTATION OF FUNCTION OF TWO RANDOM VARIABLES

$$E[f(X,Y)] = \sum_{x \in S} \sum_{y \in T} f(x,y) P(X = x, Y = y). \tag{4.3}$$

4.4 INDEPENDENT RANDOM VARIABLES

If X and Y are independent, then the joint probability mass function of X and Y has a particularly simple form. In the case of independence,

$$P(X = x, Y = y) = P(X = x)P(Y = y), \quad \text{for all } x \text{ and } y.$$

The joint distribution is the product of the marginal distributions.

Example 4.14 Angel rolls a die four times and flips a coin twice. Let X be the number of ones she gets on the die. Let Y be the number of heads she gets on the coin. Assume die rolls are independent of coin flips. (i) Find the joint probability mass function. (ii) Find the probability that Angel gets the same number of ones on the die as heads on the coin.

(i) The joint probability mass function of X and Y is

$$\begin{aligned} P(X = x, Y = y) &= P(X = x)P(Y = y) \\ &= \binom{4}{x} \left(\frac{1}{6}\right)^x \left(\frac{5}{6}\right)^{4-x} \binom{2}{y} \left(\frac{1}{2}\right)^2, \end{aligned}$$

for $x = 0, 1, 2, 3, 4$ and $y = 0, 1, 2$. The joint probability is the product of two binomial probabilities.

(ii) The desired probability is

$$\begin{aligned} P(X=Y) &= P(X=0,Y=0)+P(X=1,Y=1)+P(X=2,Y=2) \\ &= \sum_{k=0}^{2} \binom{4}{k}\left(\frac{1}{6}\right)^{k}\left(\frac{5}{6}\right)^{4-k}\binom{2}{k}\left(\frac{1}{2}\right)^{2} \\ &= 0.1206+0.1929+0.0289=0.3424 \end{aligned}$$

R: DICE AND COINS

Here is a quick "one-liner" simulation of the probability $P(X=Y)$.

```
> n = 1000000; sum(rbinom(n,4,1/6)==rbinom(n,2,1/2))/n
[1] 0.342423
```

■

If X and Y are independent random variables, then knowledge of whether or not X occurs gives no information about whether or not Y occurs. It follows that if f and g are functions, then $f(X)$ gives no information about whether or not $g(Y)$ occurs and hence $f(X)$ and $g(Y)$ are independent random variables.

FUNCTIONS OF INDEPENDENT RANDOM VARIABLES ARE INDEPENDENT

Suppose X and Y are independent random variables, and f and g are functions, Then the random variables $f(X)$ and $g(Y)$ are independent.

The following result for independent random variables has wide application.

EXPECTATION OF A PRODUCT OF INDEPENDENT RANDOM VARIABLES

Let X and Y be independent random variables. Then for any functions f and g,

$$E[f(X)g(Y)] = E[f(X)]E[g(Y)]. \tag{4.4}$$

Letting f and g be the identity function gives

$$E[XY] = E[X]E[Y] \tag{4.5}$$

The expectation of a product of independent random variables is the product of their expectations. If X and Y are independent,

$$\begin{aligned}
E[XY] &= \sum_{x \in S} \sum_{y \in T} xyP(X = x, Y = y) \\
&= \sum_{x \in S} \sum_{y \in T} xyP(X = x)P(Y = y) \\
&= \sum_{x \in S} xP(X = x) \sum_{y \in T} yP(Y = y) \\
&= E[X]E[Y].
\end{aligned}$$

The general result $E[f(X)g(Y)] = E[f(X)]E[g(Y)]$ follows similarly.

Example 4.15 Random cone. Suppose the radius R and height H of a cone are independent and each uniformly distributed on $\{1, \ldots, 10\}$. Find the expected volume of the cone.

The volume of a cone is given by the formula $v(r,h) = \pi r^2 h/3$. Let V be the volume of the cone. Then $V = \pi R^2 H/3$ and

$$\begin{aligned}
E[V] &= E\left[\frac{\pi}{3} R^2 H\right] \\
&= \frac{\pi}{3} E[R^2 H] = \frac{\pi}{3} E[R^2]E[H] \\
&= \frac{\pi}{3} \left(\sum_{r=1}^{10} \frac{r^2}{10}\right) \left(\sum_{h=1}^{10} \frac{h}{10}\right) \\
&= \frac{\pi}{3} \left(\frac{77}{2}\right) \left(\frac{11}{2}\right) \\
&= \frac{847\pi}{12} = 221.744,
\end{aligned}$$

where the third equality uses the independence of R^2 and H, which is a consequence of the independence of R and H.

R: EXPECTED VOLUME

The expectation is simulated with the commands

```
> simlist <- replicate(100000,
     (pi/3)*sample(1:10,1)^2*sample(1:10,1))
> mean(simlist)
[1] 220.9826
```

■

4.4.1 Sums of Independent Random Variables

Sums of random variables figure prominently in probability and statistics. To find probabilities of the form $P(X + Y = k)$, observe that $X + Y = k$ if and only if $X = i$ and $Y = k - i$ for some i. This gives

$$P(X+Y=k) = P\left(\bigcup_i \{X=i, Y=k-i\}\right) = \sum_i P(X=i, Y=k-i). \tag{4.6}$$

If X and Y are independent, then

$$P(X+Y=k) = \sum_i P(X=i, Y=k-i) = \sum_i P(X=i)P(Y=k-i). \tag{4.7}$$

The limits of the sum $\sum_i$ will depend on the possible values of X and Y. For instance, if X and Y are nonnegative integers, then

$$P(X+Y=k) = \sum_{i=0}^{k} P(X=i, Y=k-i), \quad \text{for } k \geq 0.$$

Example 4.16 A nationwide survey collected data on TV usage in the United States The distribution of U.S. households by number of TVs per household is given in Table 4.3. If two households are selected at random, find the probability that there are two TVs in both houses combined.

Let T_1 and T_2 be the number of TVs in the two households, respectively. Then,

$$\begin{aligned}
&P(T_1 + T_2 = 2) \\
&= P(T_1 = 0, T_2 = 2) + P(T_1 = 1, T_2 = 1) + P(T_1 = 2, T_2 = 0) \\
&= P(T_1 = 0)P(T_2 = 2) + P(T_1 = 1)P(T_2 = 1) + P(T_1 = 0)P(T_2 = 2) \\
&= (0.01)(0.33) + (0.21)(0.21) + (0.33)(0.01) = 0.051.
\end{aligned}$$

■

Example 4.17 During rush hour, the number of minivans M on a fixed stretch of highway has a Poisson distribution with parameter λ_M. The number of sports cars S on the same stretch has a Poisson distribution with parameter λ_S. If the number

TABLE 4.3: Distribution of U.S. households by number of TVs.

TVs	0	1	2	3	4	5
Proportion of households	0.01	0.21	0.33	0.23	0.13	0.09

of minivans and sports cars is independent, find the probability mass function of the total number of vehicles $M + S$.

For $k \geq 0$,

$$\begin{aligned}
P(M + S = k) &= \sum_{i=0}^{k} P(M = i, S = k - i) \\
&= \sum_{i=0}^{k} P(M = i)P(S = k - i) \\
&= \sum_{i=0}^{k} \left(\frac{e^{-\lambda_M}\lambda_M^i}{i!}\right)\left(\frac{e^{-\lambda_S}\lambda_S^{k-i}}{(k-i)!}\right) \\
&= e^{-(\lambda_M+\lambda_S)} \sum_{i=0}^{k} \frac{\lambda_M^i \lambda_S^{k-i}}{i!(k-i)!} \\
&= \frac{e^{-(\lambda_M+\lambda_S)}}{k!} \sum_{i=0}^{k} \binom{k}{i} \lambda_M^i \lambda_S^{k-i} \\
&= \frac{e^{-(\lambda_M+\lambda_S)}}{k!} (\lambda_M + \lambda_S)^k.
\end{aligned}$$

The last equality follows from the binomial theorem. We see from the final form of the pmf that $M + S$ has a Poisson distribution with parameter $\lambda_M + \lambda_S$. ■

The last example illustrates a general result.

THE SUM OF INDEPENDENT POISSON RANDOM VARIABLES IS POISSON

Let $X_1, \ldots, X_k$ be a sequence of independent Poisson random variables with respective parameters $\lambda_1, \ldots, \lambda_k$. Then

$$X_1 + \cdots + X_k \sim \text{Pois}(\lambda_1 + \cdots + \lambda_k).$$

Example 4.18 Sum of uniforms. Let X and Y be independent random variables both uniformly distributed on $\{1, \ldots, n\}$. Find the probability mass function of $X + Y$.

The sum $X + Y$ takes values between 2 and $2n$. We have that

$$\begin{aligned}
P(X + Y = k) &= \sum_{i} P(X = i, Y = k - i) \\
&= \sum_{i} P(X = i)P(Y = k - i), \text{ for } k = 2, \ldots, 2n.
\end{aligned}$$

We need to take care with the limits of the sum $\sum_i$. The limits of i in the sum will depend on k. For $2 \le k \le n$,

$$P(X+Y=k) = \sum_{i=1}^{k} P(X=i)P(Y=k-i) = \sum_{i=1}^{k} \left(\frac{1}{n}\right)\left(\frac{1}{n}\right) = \frac{k}{n^2}.$$

For $n+1 \le k \le 2n$,

$$\begin{aligned} P(X+Y=k) &= \sum_{i=k-n}^{n} P(X=i)P(Y=k-i) \\ &= \sum_{i=k-n}^{n} \left(\frac{1}{n}\right)\left(\frac{1}{n}\right) = \frac{2n-k+1}{n^2}. \end{aligned}$$

Summarizing,

$$P(X+Y=k) = \begin{cases} k/n^2, & \text{for } k = 2,\ldots,n \\ (2n-k+1)/n^2, & \text{for } k = n+1,\ldots,2n. \end{cases}$$

Observe the case $n = 6$ gives the probability mass function of the sum of two independent dice rolls. ■

4.5 LINEARITY OF EXPECTATION

Most important for computing expectations is the *linearity* property of expectation.

> **LINEARITY OF EXPECTATION**
>
> For random variables X and Y,
>
> $$E[X+Y] = E[X] + E[Y].$$

To prove this result, we apply formula Equation 4.3 for the expectation of a function of two random variables with the function $f(x,y) = x + y$. This gives

$$
\begin{aligned}
E[X+Y] &= \sum_{x\in S}\sum_{y\in T}(x+y)P(X=x,Y=y)\\
&= \sum_{x\in S}\sum_{y\in T} xP(X=x,Y=y) + \sum_{x\in S}\sum_{y\in T} yP(X=x,Y=y)\\
&= \sum_{x\in S} x\left(\sum_{y\in T} P(X=x,Y=y)\right) + \sum_{y\in T} y\left(\sum_{x\in S} P(X=x,Y=y)\right)\\
&= \sum_{x\in S} xP(X=x) + \sum_{y\in T} yP(Y=y)\\
&= E[X]+E[Y].
\end{aligned}
$$

The expectation of a sum is equal to the sum of the expectations.

Linearity of expectation is an enormously useful result. Note carefully that it makes no assumptions about the distribution of X and Y. In particular, it does not assume independence. It applies to *all* random variables regardless of their joint distribution.

Linearity of expectation extends to finite sums. That is,

$$E[X_1+\cdots+X_n] = E[X_1]+\cdots+E[X_n]. \tag{4.8}$$

It does not, however, extend to infinite sums in general. That is, it is not always true that $E[\sum_{x=1}^{\infty} X_i] = \sum_{x=1}^{\infty} E[X_i]$. However, if all the X_i's are nonnegative and if the infinite sum $\sum_{i=1}^{\infty} E[X_i]$ converges, then it is also true that

$$E\left[\sum_{i=1}^{\infty} X_i\right] = \sum_{i=1}^{\infty} E[X_i].$$

4.5.1 Indicator Random Variables

Given an event A, define a random variable I_A such that

$$I_A = \begin{cases} 1, & \text{if } A \text{ occurs}\\ 0, & \text{if } A \text{ does not occur.}\end{cases}$$

Therefore, I_A equals 1, with probability $P(A)$, and 0, with probability $P(A^c)$. Such a random variable is called an *indicator variable.* An indicator is a Bernoulli random variable with $p = P(A)$.

The expectation of an indicator variable is important enough to highlight.

EXPECTATION OF INDICATOR VARIABLE

$$E[I_A] = (1)P(A) + (0)P(A^c) = P(A).$$

This is fairly simple, but nevertheless extremely useful and interesting, because it means that probabilities of events can be thought of as expectations of indicator random variables.

Often random variables involving *counts* can be analyzed by expressing the count as a sum of indicator variables. We illustrate this powerful technique in the next examples.

Example 4.19 Expectation of binomial distribution. Let $I_1, \ldots, I_n$ be a sequence of i.i.d. Bernoulli (indicator) random variables with success probability p. Let $X = I_1 + \cdots + I_n$. Then X has a binomial distribution with parameters n and p. By linearity of expectation,

$$E[X] = E\left[\sum_{k=1}^{n} I_k\right] = \sum_{k=1}^{n} E[I_k] = \sum_{k=1}^{n} p = np.$$

This result should be intuitive. For instance, if you roll 600 dice, you would expect 100 ones. The number of ones has a binomial distribution with $n = 600$ and $p = 1/6$. And $np = 100$.

We emphasize the simplicity and elegance of the last derivation, a result of thinking probabilistically about the problem. Contrast this with the algebraic approach, using the definition of expectation. If X has a binomial distribution with parameters n and p, then

$$\begin{aligned}
E[X] &= \sum_{k=0}^{n} kP(X = k) = \sum_{k=0}^{n} k\binom{n}{k}p^k(1-p)^{n-k} \\
&= \sum_{k=1}^{n} \frac{n!}{(k-1)!(n-k)!}p^k(1-p)^{n-k} \\
&= np\sum_{k=1}^{n} \frac{(n-1)!}{(k-1)!(n-k)!}p^{k-1}(1-p)^{n-k} \\
&= np\sum_{k=1}^{n} \binom{n-1}{k-1}p^{k-1}(1-p)^{n-k} \\
&= np\sum_{k=0}^{n-1} \binom{n-1}{k}p^k(1-p)^{n-k-1} \\
&= np(p + (1-p))^{n-1} = np,
\end{aligned}$$

where the next-to-last equality follows from the binomial theorem. ■

For the binomial example, the sequence of Bernoulli indicator variables are independent. But they need not be to use linearity of expectation.

Example 4.20 Problem of coincidences. The problem of coincidences, also called the matching problem, was introduced by Pierre Rémond de Montmort in 1703. In the French card game Recontres (*Coincidences*), two persons, each having a full deck of cards, draw from their deck at the same time one card after the other, until they both draw the same card. Montmort asked for the probability that a match occurs. We consider the expected number of matches, with a modern twist.

At graduation ceremony, a class of n seniors, upon hearing that they have graduated, throw their caps up into the air in celebration. Their caps fall back to the ground uniformly at random and each student picks up a cap. What is the expected number of students who get their original cap back (a "match")?

Let X be the number of matches. Define

$$I_k = \begin{cases} 1, & \text{if the } k\text{th student gets their cap back} \\ 0, & \text{if the } k\text{th student does not get their cap back,} \end{cases}$$

for $k = 1, \ldots, n$. Then $X = I_1 + \cdots + I_n$. The expected number of matches is

$$\begin{aligned} E[X] = E\left[\sum_{k=1}^{n} I_k\right] &= \sum_{k=1}^{n} E\left[I_k\right] \\ &= \sum_{k=1}^{n} P(k\text{th student gets their cap back}) \\ &= \sum_{k=1}^{n} \frac{1}{n} = n\left(\frac{1}{n}\right) = 1. \end{aligned}$$

The probability that the kth student gets their cap back is $1/n$ since there are n caps to choose from and only one belongs to the kth student.

Remarkably, the expected number of matches is one, independent of the number of people n. If everyone in China throws their hat up in the air, on average about one person will get their hat back. ■

The indicator random variables $I_1, \ldots, I_n$ in the matching problem are not independent. In particular, if $I_1 = \cdots = I_{n-1} = 1$, that is, if the first $n-1$ people get their hats back, then necessarily $I_n = 1$, the last person must also get their hat back.

One can cast the matching problem in terms of permutations. Given a permutation of $\{1, \ldots, n\}$, a *fixed point* is a number k such that the number k is in position k in the permutation. For instance, the permutation $(2, 4, 3, 1, 5)$ has two fixed points—3 and 5. The permutation $(3, 4, 5, 1, 2)$ has no fixed points. Table 4.4 gives the permutations and number of fixed points for the case $n = 3$.

TABLE 4.4: Fixed points of permutations for $n = 3$.

Permutation	Number of fixed points
(1,2,3)	3
(1,3,2)	1
(2,1,3)	1
(2,3,1)	0
(3,1,2)	0
(3,2,1)	1

The number of fixed points in a permutation is equal to the number of matches in the problem of coincidences. If we pick a permutation above uniformly at random, then the probability mass function of F, the number of fixed points, is

$$P(F = k) = \begin{cases} 2/6, & \text{if } k = 0 \\ 3/6, & \text{if } k = 1 \\ 1/6, & \text{if } k = 3, \end{cases}$$

with expectation $E[F] = 0(2/6) + 1(3/6) + 3(1/6) = 1$.

R: SIMULATING THE MATCHING PROBLEM

In order to simulate the matching problem, we simulate random permutations with the command

```
> sample(n,n)
```

It is remarkable how fast this is done for even large n. Even though there are $n! \approx n^n$ permutations of an n-element set, the algorithm for generating a random permutation, takes on the order of n steps, not n^n steps, and is thus extremely efficient.

To count the number of fixed points in a uniformly random permutation, type

```
> sum(sample(n,n)==1:n)
```

Here we simulate the expected number of matchings in the original game of Recontres with a standard deck of 52 cards.

```
> n <- 52
> mean(replicate(10000,sum(sample(n,n)==(1:n))))
[1] 0.9917
```

Now pause to reflect upon what you have actually done. You have simulated a random element from a sample space that contains $52! \approx 8 \times 10^{67}$ elements, checked the number of fixed points, and then repeated the operation 10,000 times, finally computing the average number of fixed points—all in a second or two on your computer. It would not be physically possible to write down a table of all 52! permutations, their number of fixed points, or their corresponding probabilities. And yet by generating random elements and taking simulations the problem becomes computationally feasible. Of course, in this case we already know the exact answer and do not need simulation to find the expectation. However, that is not the case with many complex, real-life problems. In many cases, simulation is the only way to go.

To learn more about the problem of coincidences, see Takacs (1980).

Example 4.21 St. Petersburg paradox. I offer you the following game. Flip a coin until heads appears. If it takes n tosses, I will pay you $\$2^n$. Thus if heads comes up on the first toss, I pay you \$2. If it first comes up on the 10th toss, I pay you \$1024.

How much would you pay to play this game? Would you pay \$5, \$50, \$500?

Let X be the payout. Your expected payout is

$$E[X] = \sum_{n=1}^{\infty} 2^n \frac{1}{2^n} = \sum_{n=1}^{\infty} 1 = +\infty.$$

The expected value is infinite. The expectation does not exist.

This problem, discovered by the eighteenth-century Swiss mathematician Daniel Bernoulli, is the St. Petersburg paradox. The "paradox" is that most people would not pay very much to play this game. And yet the expected payout is infinite. ∎

4.6 VARIANCE AND STANDARD DEVIATION

Expectation is a measure of the *average* behavior of a random variable. Variance and standard deviation are measures of *variability*. They describe how near or far typical outcomes are to the mean.

VARIANCE

Let X be a random variable with mean $E[X] = \mu < \infty$. The *variance of* X is

$$V[X] = E[(X - \mu)^2] = \sum_x (x - \mu)^2 P(X = x). \tag{4.9}$$

In the variance formula, $(x - \mu)$ is the difference or "deviation" of an outcome from the mean. Thus, the variance is a weighted average of the squared deviations from the mean.

The *standard deviation of* X is the square root of the variance.

STANDARD DEVIATION

$$\text{SD[X]} = \sqrt{V[X]}.$$

Variance and standard deviation are always nonnegative. The greater the variability of outcomes, the larger the deviations from the mean, and the greater the two measures. If X is a constant, and hence has *no* variability, then $X = E[X] = \mu$ and we see from the variance formula that $V[X] = 0$. The converse is also true. That is, if $V[X] = 0$, then X is almost certainly a constant. We leave the proof to Exercise 4.39.

The graphs in Figure 4.1 show four probability distributions all with expectations equal to 4, but with different variances.

Example 4.22 We find the variances for each of the distributions in Figure 4.1. Let W, X, Y, Z be the corresponding random variables, respectively. We have that $E[W] = E[X] = E[Y] = E[Z] = 4$.

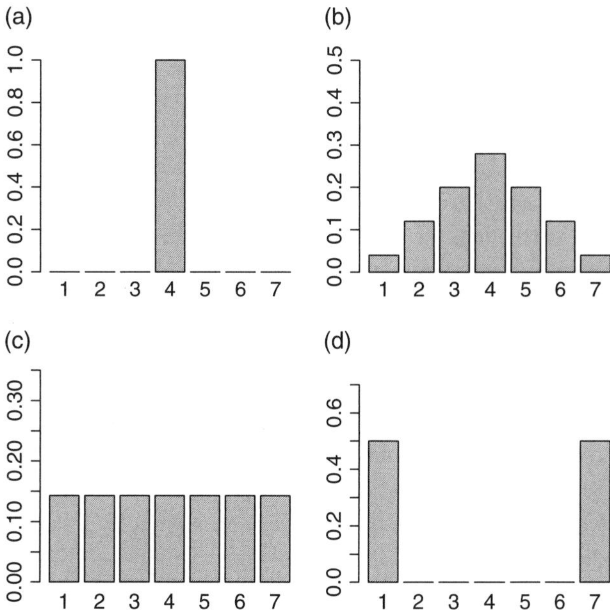

FIGURE 4.1: Four distributions with $\mu = 4$. Variances are (a) 0, (b) 2.08, (c) 4, and (d) 9.

(a) W is a constant equal to 4. The variance is 0.

(b) The probability mass function for X is

$$P(X = k) = \begin{cases} 1/25, & \text{if } k = 1, 7 \\ 3/25, & \text{if } k = 2, 6 \\ 5/25, & \text{if } k = 3, 5 \\ 7/25, & \text{if } k = 4, \end{cases}$$

with variance

$$V[X] = 2(1-4)^2\frac{1}{25} + 2(2-4)^2\frac{3}{25} + 2(3-4)^2\frac{5}{25} = 2.08.$$

(c) Outcomes are equally likely. This gives

$$V[Y] = \sum_{k-1}^{7}(k-4)^2\frac{1}{7} = 4.$$

(d) We have $P(Z = 1) = P(Z = 7) = 1/2$. This gives

$$V[Z] = (1-4)^2\frac{1}{2} + (7-4)^2\frac{1}{2} = 9.$$

■

The variance is a "cleaner" mathematical formula than the standard deviation because it does not include the square root. For simplicity and for connections with other areas of mathematics, mathematicians and probabilists often prefer variance when working with random variables.

However, the standard deviation can be easier to interpret particularly when working with data. In statistics, random variables are often used to model data, which have some associated units attached to their measurements. For instance, we might take a random person's height and assign it to a random variable H. The units are inches. The expected height $E[H]$ is also expressed in inches. However, because of the square in the variance formula, the units of the variance $V[H]$ are square inches. The square root in the standard deviation brings the units back to the units of the data.

In statistics, there are analogous definitions of mean, variance, and standard deviation for a collection of data. For a list of measurements $x_1, \ldots, x_n$, the *sample mean* is the average $\bar{x} = (x_1 + \cdots + x_n)/n$. The *sample variance* is defined as $\sum_{i=1}^{n}(x_i - \bar{x})^2/(n-1)$. (The reason the denominator is $n - 1$ rather than n is a topic for a statistics class.) The *sample standard deviation* is the square root of the sample variance.

The R commands `mean(vec)`, `var(vec)`, and `sd(vec)` compute these quantities for a vector `vec`.

When we simulate a random variable X with repeated trials, the sample mean of the replications is a Monte Carlo approximation of $E[X]$. Similarly, the sample variance of the replications is a Monte Carlo approximation of $V[X]$.

Within two standard deviations of the mean. Many probability distributions are fairly concentrated near their expectation in the sense that the probability that an outcome is within a few standard deviations from the mean is high. This is particularly true for symmetric, "bell-shaped" distributions such as the Poisson distribution when λ is large and the binomial distribution when p is close to 1/2. The same is true for many datasets. For instance, the average height of women in this country is roughly 64 inches with standard deviation 3 inches. Roughly 95% of all adult women's heights are within two standard deviations of the mean, which is within $64 \pm 2(3)$, between 58 and 70 inches.

When we discuss the normal distribution (the "bell-shaped curve") in Chapter 7, this will be made more precise. In the meantime, a rough "rule of thumb" is that for many symmetric and near-symmetric probability distributions, the mean and variance (or standard deviation) are good summary measures for describing the behavior of "typical" outcomes of the underlying random experiment. Typically most outcomes from such distributions fall within two standard deviations of the mean.

From the definition of the variance a little manipulation goes a long way. A useful computation formula for the variance is

$$V[X] = E[X^2] - E[X]^2. \tag{4.10}$$

This is derived using properties of expectation. Remember that $\mu = E[X]$ is a constant. Then

$$\begin{aligned} V[X] &= E[(X-\mu)^2] = E[X^2 - 2\mu X + \mu^2] \\ &= E[X^2] - 2\mu E[X] + \mu^2 \\ &= E[X^2] - 2\mu^2 + \mu^2 = E[X^2] - \mu^2 \\ &= E[X^2] - E[X]^2. \end{aligned}$$

Example 4.23 Variance of uniform distribution. Suppose X is uniformly distributed on $\{1, \ldots, n\}$. Find the variance of X.

First find

$$\begin{aligned} E[X^2] &= \sum_{k=1}^{n} k^2 P(X=k) = \frac{1}{n}\sum_{k=1}^{n} k^2 \\ &= \left(\frac{1}{n}\right)\frac{n(n+1)(2n+1)}{6} = \frac{(n+1)(2n+1)}{6}. \end{aligned}$$

This gives

$$V[X] = E[X^2] - E[X]^2 = \frac{(n+1)(2n+1)}{6} - \left(\frac{n+1}{2}\right)^2 = \frac{n^2-1}{12}.$$

For large n, the mean of the uniform distribution on $\{1, \ldots, n\}$ is $(n+1)/2 \approx n/2$, and the standard deviation is $\sqrt{(n^2-1)/12} \approx n/\sqrt{12} \approx n/3.5$. ■

Example 4.24 Variance of an indicator. For an event A, let I_A be the corresponding indicator random variable. Since I_A only takes values 0 and 1, it follows that $(I_A)^2 = I_A$, which gives

$$\begin{aligned} V[I_A] &= E[I_A^2] - E[I_A]^2 \\ &= E[I_A] - E[I_A]^2 \\ &= P(A) - P(A)^2 \\ &= P(A)(1 - P(A)) = P(A)P(A^c). \end{aligned}$$

Results for indicators are worthwhile to remember. ■

EXPECTATION AND VARIANCE OF INDICATOR VARIABLE

$$E[I_A] = P(A) \qquad \text{and} \qquad V[I_A] = P(A)P(A^c).$$

Properties of variance. The linearity properties of expectation do *not* extend to the variance. It is not true that $V[aX + b] = aV[X] + b$.

If X is a random variable with expectation μ, and a and b are constants, then $aX + b$ has expectation $a\mu + b$ and

$$\begin{aligned} V[aX + b] &= E[(aX + b - (a\mu + b))^2] = E[(aX - a\mu)^2] \\ &= E[a^2(X - \mu)^2] = a^2E[(X - \mu)^2] = a^2V[X]. \end{aligned}$$

We summarize the important properties.

PROPERTIES OF VARIANCE, STANDARD DEVIATION, AND EXPECTATION

For constants a and b,

$$\begin{aligned} V[aX + b] &= a^2V[X] \\ \text{SD}[aX + b] &= |a|\text{SD}[X] \\ E[aX + b] &= aE[X] + b. \end{aligned}$$

Example 4.25 See Example 4.9. Suppose the French lab's temperature measurements have a variance of 2°C. Then upon conversion to Fahrenheit, the variance of the U.S. lab's temperature measurements is

$$V[F] = V\left[32 + \left(\frac{9}{5}\right)C\right] = \left(\frac{9}{5}\right)^2 V[C] = \frac{162}{25},$$

with standard deviation $\text{SD}[F] = \sqrt{162/25} = 2.55°\text{F}$. ■

Variance of Poisson distribution. In the end of chapter Exercise 4.30, we invite you to show that the variance of a Poisson random variable X with parameter λ is equal to λ. Hence,

$$E[X] = V[X] = \lambda.$$

This is a special property of the Poisson distribution. Using the heuristic that *most* observations are within two standard deviations of the mean, it would follow that most outcomes of a Poisson random variable are contained in the interval $\lambda \pm 2\sqrt{\lambda}$. This is generally true, at least for large λ.

R: SIMULATION OF POISSON DISTRIBUTION

You can observe this phenomenon for a specific choice of λ by typing

```
> hist(rpois(100000,lambda), prob=T)
> abline(v=lambda-2*sqrt(lambda))
> abline(v=lambda+2*sqrt(lambda))
```

The output of the commands, with $\lambda = 25$, is shown in Figure 4.2. The graph is a histogram. Think of the rectangles as bins. For each x-value, the height of the rectangle corresponds to the number of outcomes of x. The vertical axis is scaled so that the areas of the rectangles sum to 1. The histogram is generated from 100,000 simulations of a Poisson(25) random variable. It thus simulates the probability distribution of X. Vertical lines are drawn at two standard deviations from the mean, that is, at $x = 15$ and $x = 35$.

Variances of sums and sums of variances. For random variables X and Y, consider the variance of the sum $V(X + Y)$. We have

$$V(X+Y) = E\left[(X+Y)^2\right] - (E[X+Y])^2.$$

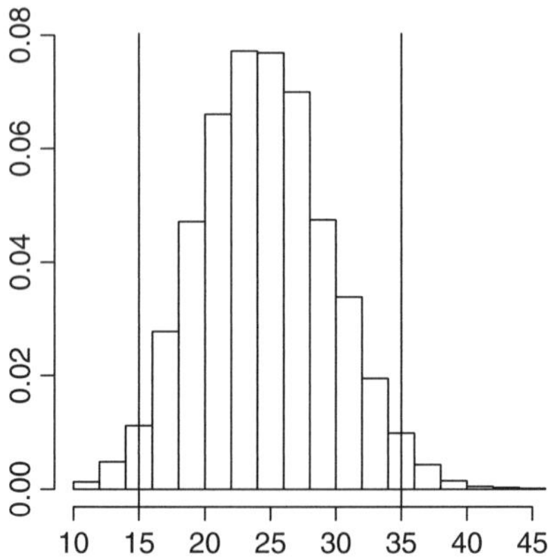

FIGURE 4.2: Simulation of Poisson(25) distribution. Vertical lines are drawn at $x = 15$ and $x = 35$, two standard deviations from the mean.

Let $\mu_X = E[X]$ and $\mu_Y = E[Y]$. Then

$$E[(X+Y)^2] = E[X^2 + 2XY + Y^2] = E[X^2] + 2E[XY] + E[Y^2]$$

and

$$(E[X+Y])^2 = (\mu_X + \mu_Y)^2 = \mu_X^2 + 2\mu_X\mu_Y + \mu_Y^2.$$

Hence,

$$\begin{aligned} V[X+Y] &= E[(X+Y)^2] - (E[X+Y])^2 \\ &= E[X^2] + 2E[XY] + E[Y^2] - (\mu_X^2 + 2\mu_X\mu_Y + \mu_Y^2) \\ &= (E[X^2] - \mu_X^2) + (E[Y^2 - \mu_Y^2) + 2(E[XY] - \mu_X\mu_Y) \\ &= V[X] + V[Y] + 2(E[XY] - E[X]E[Y]). \end{aligned} \tag{4.11}$$

If X and Y are independent, then $E[XY] = E[X]E[Y]$, and the variance of $X + Y$ has a simple form.

VARIANCE OF THE SUM OF INDEPENDENT VARIABLES

If X and Y are independent, then

$$V[X+Y] = V[X] + V[Y]. \tag{4.12}$$

For the difference of independent random variables, it might be tempting to conclude that $V[X - Y] = V[X] - V[Y]$. But this is not true. Rather,

$$\begin{aligned} V[X - Y] &= V[X + (-1)Y] = V[X] + V[(-1)Y] \\ &= V[X] + (-1)^2 V[Y] \\ &= V[X] + V[Y]. \end{aligned}$$

Variances always add. (If you are not completely convinced and have some nagging doubts about $V]X - Y]$, suppose that X and Y are independent with $V[X] = 1$ and $V[X] = 2$. If $V[X - Y] = V[X] - V[Y] = 1 - 2 = -1$, we have a problem, since variances can never be negative.)

Example 4.26 Variance of binomial distribution. Recall how indicator variables are used to find the expectation of a binomial distribution. Use them again to find the variance.

Suppose $X = I_1 + \cdots + I_n$ is the sum of n independent indicator variables with success probability p. Then X has a binomial distribution with parameters n and p. Since the I_k's are independent, the variance of the sum of indicators is equal to the sum of the variances and thus

$$V[X] = V\left[\sum_{k=1}^{n} I_k\right] = \sum_{k=1}^{n} V[I_k] = \sum_{k=1}^{n} p(1-p) = np(1-p). \qquad \blacksquare$$

Example 4.27 Danny said he flipped 100 pennies and got 70 heads. Is this believable?

The number of heads has a binomial distribution with parameters $n = 100$ and $p = 1/2$. The mean number of heads is $np = 50$. The standard deviation is $\sqrt{np(1-p)} = \sqrt{25} = 5$. We expect most outcomes from tossing 100 coins to fall with two standard deviations of the mean, that is between 40 and 60 heads. As 70 heads represents an outcome that is four standard deviations from the mean, we are a little suspicious of Danny's claim. ∎

Example 4.28 Roulette continued—how the casino makes money. In Example 4.2, we found the expected value of a bet of "red" in roulette. We will shift gears and look at things from the casino's perspective. Your loss is their gain. Let G be the casino's gain after a player makes one red bet. Then

$$P(G = 1) = \frac{20}{38} \quad \text{and} \quad P(G = -1) = \frac{18}{38}$$

with $E[G] = 2/38$. The casino's expected gain from one red bet is about five cents. For the variance, $E[G^2] = (1)(20/38) + (1)(18/38) = 1$, and

$$V[G] = E[G^2] - E[G]^2 = 1 - (2/38)^2 = 0.99723,$$

with standard deviation 0.998614, almost one dollar.

Suppose in one month the casino expects customers to make n red bets. What is the expected value and standard deviation of the casino's total gain?

Let G_k be the casino's gain from the kth red bet of the month, for $k = 1, \ldots, n$. Let T be the casino's total gain. Write $T = G_1 + \cdots + G_n$. Unless someone is cheating, we can assume that the G_k's are independent.

By linearity of expectation,

$$E[T] = E[G_1 + \cdots + G_n] = E[G_1] + \cdots + E[G_n] = \frac{2n}{38},$$

or about n nickels. If one million bets are placed, that is an expected gain of \$52,631.58.

Of interest to the casino's accountants is the variability of total gain. If the variance is large, the casino might see big swings from month to month where some months they make money and some months they do not.

By independence of the G_k's,

$$V[T] = V[G_1 + \cdots + G_n] = V[G_1] + \cdots + V[G_n] = (0.99723)n,$$

with standard deviation

$$\text{SD}[T] = (0.998614)\sqrt{n}.$$

For one million bets, the standard deviation is \$998.61. This is about one-fiftieth the size of the mean.

By the heuristic that says that most observations are within about two standard deviations from the mean, it is virtually certain that every month the casino will see a hefty positive gain of about \$52,000 give or take about \$2,000.

Which is why for the customers it is risky entertainment, but for the casino it is a business.

R: A MILLION RED BETS

Here we simulate the casino's gain from three different months of play, assuming one million red bets are placed each month.

```
> n=1000000
> sum(sample(c(-1,1),n,prob=c(18/38,20/38), replace=T))
```

```
[1] 52972
> sum(sample(c(-1,1),n,prob=c(18/38,20/38), replace=T))
[1] 52604
> sum(sample(c(-1,1),n,prob=c(18/38,20/38), replace=T))
[1] 53392
```

With the sample command we can simulate from any finite distribution. The syntax is

```
> sample(val,n,prob=,replace=),
```

where val is a vector of values, n is the number of trials, prob is a vector of probabilities, and replace is either T(rue) or F(alse).

■

4.7 COVARIANCE AND CORRELATION

Having looked at measures of variability for individual and independent random variables, we now consider measures of variability between dependent random variables. The *covariance* is a measure of the association between two random variables.

COVARIANCE

Definition 4.2. For random variables X and Y, with respective means μ_X and μ_Y, the *covariance between X and Y* is

$$\text{Cov}(X, Y) = E[(X - \mu_X)(Y - \mu_Y)]. \tag{4.13}$$

Equivalently,

$$\text{Cov}(X, Y) = E[XY] - \mu_X\mu_Y = E[XY] - E[X]E[Y]. \tag{4.14}$$

We leave it as an exercise to show that Equation 4.14 follows from the definition Equation 4.13

For independent random variables, $E[XY] = E[X]E[Y]$ and thus $\text{Cov}(X, Y) = 0$. The covariance will be positive when large values of X are associated with large values of Y and small values of X are associated with small values of Y. In particular, for outcomes x and y products of the form $(x - \mu_X)(y - \mu_Y)$ in the covariance formula will tend to either both be positive or both be negative, both cases resulting in positive values.

On the other hand, if X and Y are inversely related, most terms $(x - \mu_X)$ $(y - \mu_Y)$ will be negative, since when X takes values above the mean, Y will tend

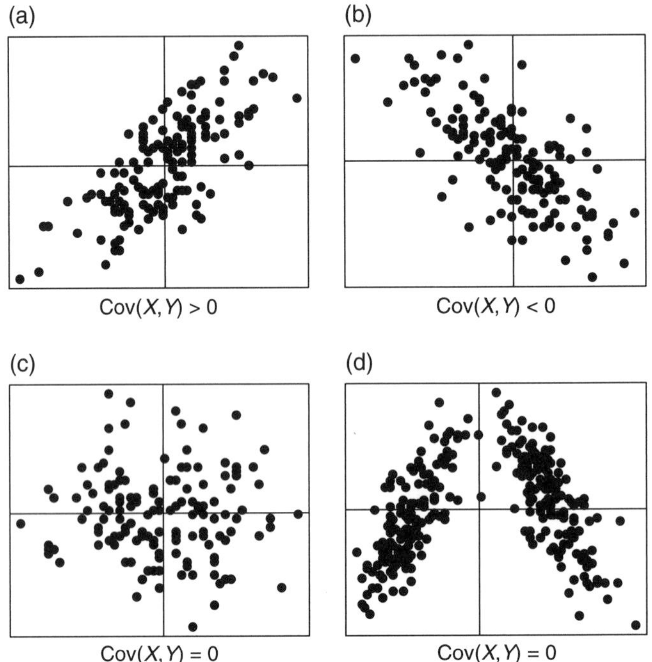

FIGURE 4.3: Covariance is a measure of linear association between two random variables. Vertical and horizontal lines are drawn at the mean of the marginal distributions.

to fall below the mean, and vice versa. In this case, the covariance between X and Y will be negative.

See Figure 4.3. Points on the graphs are simulations of the points (X, Y) from four joint distributions. Vertical and horizontal lines are drawn at the mean of the marginal distributions. In the graph in (a), the main contribution to the covariance is from points in the first and third quadrants, where products are positive. In (b), the main contribution is in the second and fourth quadrants where products are negative. In (c), the random variables are independent and each of the four quadrants are near equally representated so positive terms cancel negative terms and the covariance is 0.

Covariance is a measure of *linear association* between two variables. In a sense, the "less linear" the relationship, the closer the covariance is to 0. In the fourth graph (d), the covariance will be close to 0 as positive products tend to cancel out negative products. But here the random variables are not independent. There is a strong relationship between them, although it is not linear. Both large and small values of X are associated with small values of Y.

The sign of the covariance indicates whether two random variables are positively or negative associated. But the magnitude of the covariance can be difficult to interpret. The *correlation* is an alternative measure.

CORRELATION

The correlation between X and Y is

$$\text{Corr}(X,Y) = \frac{\text{Cov}(X,Y)}{\text{SD}[X]\text{SD}[Y]}.$$

Properties of correlation

1. $-1 \leq \text{Corr}(X,Y) \leq 1$.
2. If $Y = aX + b$ is a linear function of X for constants a and b, then $\text{Corr}(X,Y) = \pm 1$, depending on the sign of a.

Correlation is a common summary measure in statistics. Dividing the covariance by the standard deviations creates a "standardized" covariance, which is a unitless measure that takes values between -1 and 1. The correlation is exactly equal to ± 1 if Y is a linear function of X. We prove these results in Section 4.9.

Random variables that have correlation, and covariance, equal to 0 are called *uncorrelated.*

UNCORRELATED RANDOM VARIABLES

Definition 4.3. *We say random variables* X *and* Y *are uncorrelated if*

$$E[XY] = E[X]E[Y],$$

that is, if $\text{Cov}(X, Y) = 0$.

If random variables X and Y are independent, then they are uncorrelated. However, the converse is not necessarily true.

Example 4.29 Let X be uniformly distributed on $\{-1, 0, 1\}$. Let $Y = X^2$. The two random variables are not independent as Y is a function of X. However,

$$\text{Cov}(X,Y) = \text{Cov}(X, X^2) = E[X^3] - E[X]E[X^2] = 0 - 0 = 0.$$

The random variables are uncorrelated. ■

Example 4.30 The number of defective parts in a manufacturing process is modeled as a binomial random variable with parameters n and p. Let X be the number of defective parts, and let Y be the number of nondefective parts. Find the covariance between X and Y.

Observe that $Y = n - X$ is a linear function of X. Thus $\text{Corr}(X,Y) = -1$. Since "failures" for X are "successes" for Y, Y has a binomial distribution with parameters n and $1-p$. To find the covariance, rearrange the correlation formula

$$\begin{aligned}\text{Cov}(X,Y) &= \text{Corr}(X,Y)\text{SD}[X]\text{SD}[Y]\\ &= (-1)\sqrt{np(1-p)}\sqrt{n(1-p)p}\\ &= -np(1-p).\end{aligned}$$

■

Example 4.31 Red ball, blue ball continued. For Example 4.11, find $\text{Cov}(R,B)$, the covariance between the number of red and blue balls.

The joint distribution table is

		B			
		0	1	2	
	0	3/36	6/36	1/36	10/36
R	1	12/36	8/36	0	20/36
	2	6/36	0	0	6/36
		21/36	14/36	1/36.	

From the joint table,

$$E[RB] = \sum_r \sum_b rbP(R=r, B=b) = \frac{8}{36} = \frac{2}{9}.$$

(Eight of the nine terms in the sum are equal to 0.) Also,

$$E[R] = (1)\frac{5}{9} + (2)\frac{1}{6} = \frac{8}{9}$$

and

$$E[B] = (1)\frac{7}{18} + (2)\frac{1}{36} = \frac{4}{9}.$$

This gives

$$\begin{aligned}\text{Cov}(R,B) &= E[RB] - E[R]E[B]\\ &= \frac{2}{9} - \left(\frac{8}{9}\right)\left(\frac{4}{9}\right)\\ &= -\frac{14}{81} = -0.17284.\end{aligned}$$

The negative result should not be surprising. There is an inverse association between R and B since the more balls there are of one color in the sample the less balls there will be of another color. ■

One place where the covariance regularly appears is in the variance formula for sums of random variables. We summarize results for the variance of a sum.

GENERAL FORMULA FOR VARIANCE OF A SUM

For random variables X and Y with finite variance,

$$V[X+Y] = V[X] + V[Y] + 2\text{Cov}(X,Y),$$

and

$$V[X-Y] = V[X] + V[Y] - 2\text{Cov}(X,Y).$$

If X and Y are uncorrelated,

$$V[X \pm Y] = V[X] + V[Y].$$

Example 4.32 After a severe storm, the number of claims received by an insurance company for hail H and tornado T damages are each modeled with a Poisson distribution with respective parameters 400 and 100. The correlation between H and T is 0.75. Let Z be the total number of claims from hail and tornado. Find the variance and standard deviation of Z.

We have

$$E[Z] = E[H+T] = E[H] + E[T] = 400 + 100 = 500,$$

and

$$\begin{aligned} V[Z] = V[H+T] &= V[H] + V[T] + 2\text{Cov}(H,T) \\ &= 400 + 100 + 2\text{Corr}(H,T)\text{SD}[H]\text{SD}[T] \\ &= 500 + 2(0.75)\sqrt{400}\sqrt{100} \\ &= 500 + 600 = 1100, \end{aligned}$$

with standard deviation $\text{SD}[Z] = \sqrt{1100} = 33.17$. ■

The covariance of a random variable X with itself $\text{Cov}(X,X)$ is just the variance $V[X]$, as $E[(X-\mu_X)(X-\mu_X)] = E[(X-\mu_X)^2]$. Also, the covariance is symmetric in its terms. That is, $\text{Cov}(X,Y) = \text{Cov}(Y,X)$. So another way of writing the variance of a sum of two random variables is

$$V[X+Y] = \text{Cov}(X,X) + \text{Cov}(Y,Y) + \text{Cov}(X,Y) + \text{Cov}(Y,X),$$

that is as a sum over all possible pairings of X and Y.

For the variance of a sum of more than two random variables, take covariances of all possible pairs.

VARIANCE OF SUM OF N RANDOM VARIABLES

$$\begin{aligned} V[X_1+\cdots+X_n] &= \sum_{i=1}^{n}\sum_{j=1}^{n}\text{Cov}(X_i,X_j) \\ &= V[X_1]+\cdots+V[X_n]+\sum_{i\neq j}\text{Cov}(X_i,X_j) \\ &= V[X_1]+\cdots+V[X_n]+2\sum_{i<j}\text{Cov}(X_i,X_j). \end{aligned}$$

The penultimate sum is over all pairs of indices $i \neq j$. For the last equality, since $\text{Cov}(X_i,X_j) = \text{Cov}(X_j,X_i)$, we can just consider indices such that $i < j$.

Example 4.33 Let X, Y, Z be random variables with variances equal to 1, 2, and 3, respectively. Also, $\text{Cov}(X,Y) = -1$, $\text{Cov}(X,Z) = 0$, and $\text{Cov}(Y,Z) = 3$. The variance of $X+Y+Z$ is

$$\begin{aligned} &V[X+Y+Z] \\ &\quad = V[X]+V[Y]+V[Z]+2\left(\text{Cov}(X,Y)+\text{Cov}(X,Z)+\text{Cov}(Y,Z)\right) \\ &\quad = 1+2+3+2(-1+0+3) = 10. \end{aligned}$$

■

Example 4.34 Coincidences continued. In the matching problem, Example 4.20, the expectation of the number of students who get their cap back was shown to be one using the method of indicators. What is the variance?

Let X be the number of students who get their own cap back. Write $X = I_1+\cdots+I_n$, where I_k is equal to 1, if the kth student gets their cap, and 0, otherwise, for $k = 1,\ldots,n$. Since the I_k's are not independent, use the general sum formula

$$V[X] = V\left[\sum_{k=1}^{n} I_k\right] = \sum_{k=1}^{n} V[I_k] + \sum_{i \neq j} \text{Cov}(I_i, I_j).$$

We have

$$E[I_k] = P(k\text{th student gets their cap}) = \frac{1}{n},$$

and

$$\begin{aligned} V[I_k] &= P(k\text{th student gets their cap})P(k\text{th student does not get their cap}) \\ &= \left(\frac{1}{n}\right)\left(\frac{n-1}{n}\right) = \frac{n-1}{n^2}. \end{aligned}$$

For the covariance terms, consider $E[I_iI_j]$ where $i \neq j$. Since I_i and I_j are 0–1 variables, the product I_iI_j is equal to 1 if and only if both I_i and I_j are equal to 1, which happens if the ith and jth students both get their caps back. That is, I_iI_j is the indicator of the event that the ith and jth students get their caps back. Thus,

$$E[I_iI_j] = P(i\text{th and } j\text{th students get their caps}) = P(I_i = 1, I_j = 1).$$

Conditional on the jth student getting their cap back, the probability that the ith student gets their cap is $1/(n-1)$ since there are $n-1$ caps to choose from. Thus,

$$P(I_1 = 1, I_j = 1) = P(I_i = 1|I_j = 1)P(I_j = 1) = \left(\frac{1}{n-1}\right)\left(\frac{1}{n}\right).$$

For $i \neq j$, this gives

$$\text{Cov}(I_i, I_j) = E[I_iI_j] - E[I_i]E[I_j] = \frac{1}{n(n-1)} - \frac{1}{n^2} = \frac{1}{n^2(n-1)}.$$

Putting all the pieces together, we have

$$V[X] = \sum_{i=1}^{n} \frac{n-1}{n^2} + \sum_{i \neq j} \frac{1}{n^2(n-1)} = \frac{n-1}{n} + \sum_{i \neq j} \frac{1}{n^2(n-1).}$$

Finally, we need to count the number of terms in the last sum. There are n^2 terms in the full double sum $\sum_{i=1}^{n}\sum_{j=1}^{n}$. And there are n terms for which $i = j$. This leaves $n^2 - n$ terms for which $i \neq j$. Hence,

$$V[X] = \frac{n-1}{n} + \sum_{i \neq j} \frac{1}{n^2(n-1)}$$
$$= \frac{n-1}{n} + (n^2 - n)\left(\frac{1}{n^2(n-1)}\right)$$
$$= \frac{n-1}{n} + \frac{1}{n} = 1.$$

The variance, and standard deviation, of the number of matchings is also equal to 1! Thus, if everyone in China throws their hat up in the air, we expect about one person to get their hat back give or take one or two. It would be extremely unlikely if four or more people get their hat back. ■

4.8 CONDITIONAL DISTRIBUTION

Chapter 2 introduced conditional probability. For jointly distributed random variables, we have the more general notion of *conditional distribution.* In the discrete setting, we define the conditional probability mass function.

CONDITIONAL PROBABILITY MASS FUNCTION

If X and Y are jointly distributed discrete random variables, then the *conditional probability mass function* of Y given $X = x$ is

$$P(Y = y|X = x) = \frac{P(X = x, Y = y)}{P(X = x)},$$

given $P(X = x) > 0$.

The conditional probability mass function *is* a probability mass function, as was shown in Section 2.2. It is nonnegative and sums to 1.

If X and Y are independent, then the conditional probability mass function of Y given $X = x$ reduces to the regular pmf Y, as $P(Y = y|X = x) = P(Y = y)$.

Example 4.35 In a study of geriatric health and the risk of hip fractures, Schechner et al. (2010) model the occurrence of falls and hip fractures for elderly individuals. They assume that the number of times a person falls during one year has a Poisson distribution with parameter λ. Each fall may independently result in a hip fracture with probability p.

Let X be the number of falls. If $X = x$, then conditional on the number of falls x, the number of hip fractures Y has a binomial distribution with parameters x and p. That is,

$$P(Y = y|X = x) = \binom{x}{y} p^y (1-p)^{x-y}, \text{ for } y = 0, \ldots x.$$

That is, the conditional distribution of Y given $X = x$ is binomial with parameters x and p. ∎

Example 4.36 Red ball, blue ball continued. For Example 4.11, find the conditional probability mass function of the number of red balls given that there are no blue balls in the sample.

The joint distribution table is

		B 0	1	2	
	0	3/36	6/36	1/36	10/36
R	1	12/36	8/36	0	20/36
	2	6/36	0	0	6/36
		21/36	14/36	1/36.	

We have

$$P(R = r|B = 0) = \frac{P(R = r, B = 0)}{P(B = 0)} = \frac{P(R = r, B = 0)}{21/36}.$$

This gives

$$P(R = r|B = 0) = \begin{cases} (3/36)/(21/36) = 1/7, & \text{if } r = 0 \\ (12/36)/(21/36) = 4/7, & \text{if } r = 1 \\ (6/36)/(21/36) = 2/7, & \text{if } r = 2. \end{cases}$$ ∎

Example 4.37 Hip fractures continued. As in Example 4.35, let Y be the number of hip fractures and X the number of falls, where $X \sim \text{Pois}(\lambda)$. We found that the conditional distribution of Y given $X = x$ is binomial with parameters x and p. (i) Find the joint pmf of X and Y. (ii) Find the marginal distribution of the number of hip fractures Y.

(i) Rearranging the conditional probability formula gives

$$\begin{aligned} P(X = x, Y = y) &= P(Y = y|X = x)P(X = x) \\ &= \binom{x}{y} p^y (1-p)^{x-y} \frac{e^{-\lambda}\lambda^x}{x!} \\ &= \frac{p^y (1-p)^{x-y} e^{-\lambda}\lambda^x}{y!(x-y)!}, \end{aligned}$$

for $x = 0, 1, \ldots$, and $y = 0, \ldots, x$.

(ii) The marginal distribution of Y is found by summing the joint pmf over values of x. Since $0 \le y \le x$, we sum over $x \ge y$. This gives

$$\begin{aligned}
P(Y = y) &= \sum_{x=y}^{\infty} P(X = x, Y = y) \\
&= \sum_{x=y}^{\infty} \frac{p^y (1-p)^{x-y} e^{-\lambda} \lambda^x}{y!(x-y)!} \\
&= \frac{e^{-\lambda} p^y}{y!} \sum_{x=y}^{\infty} \frac{(1-p)^{x-y} \lambda^x}{(x-y)!} \\
&= \frac{e^{-\lambda} (\lambda p)^y}{y!} \sum_{x=y}^{\infty} \frac{(\lambda(1-p))^{x-y}}{(x-y)!} \\
&= \frac{e^{-\lambda} (\lambda p)^y}{y!} \sum_{x=0}^{\infty} \frac{(\lambda(1-p))^{x}}{x!} \\
&= \frac{e^{-\lambda} (\lambda p)^y}{y!} e^{\lambda(1-p)} \\
&= \frac{e^{-\lambda p} (\lambda p)^y}{y!},
\end{aligned}$$

for $y = 0, 1, \ldots$. Thus, the marginal distribution of the number of hip fractures is Poisson with parameter λp. We can interpret this distribution as a "thinned" version of the distribution of the number of falls X. Both are Poisson distributions. The mean parameter of Y is p times the mean parameter of X. ■

Example 4.38 During the day, Sam receives text message and phone calls. The numbers of each are independent Poisson random variables with parameters λ and μ, respectively. If Sam receives n texts and phone calls during the day, find the conditional distribution of the number of texts he receives.

Let T be the number of texts Sam receives. Let C be the number of phone calls. We show in Section 4.4.1 that the sum of independent Poisson variables has a Poisson distribution, and thus $T + C \sim \text{Pois}(\lambda + \mu)$. For $0 \le t \le n$, the conditional probability mass function of T given $T + C = n$ is

$$\begin{aligned}
P(T = t | T + C = n) &= \frac{P(T = t, T + C = n)}{P(T + C = n)} \\
&= \frac{P(T = t, C = n - t)}{P(T + C = n)} \\
&= \frac{P(T = t) P(C = n - t)}{P(T + C = n)}
\end{aligned}$$

$$= \left(\frac{e^{-\lambda}\lambda^t}{t!}\right)\left(\frac{e^{-\mu}\mu^{n-t}}{(n-t)!}\right)\left(\frac{e^{-(\lambda+\mu)}(\lambda+\mu)^n}{n!}\right)^{-1}$$
$$= \binom{n}{t}\frac{\lambda^t\mu^{n-t}}{(\lambda+\mu)^n}$$
$$= \binom{n}{t}\left(\frac{\lambda}{\lambda+\mu}\right)^t\left(\frac{\mu}{\lambda+\mu}\right)^{n-t}.$$

The conditional distribution of T given $T + C = n$ is binomial with parameters n and $p = \lambda/(\lambda + \mu)$. ■

Example 4.39 At the beginning of this section, we said that conditional distribution generalizes the idea of conditional probability. Let A and B be events. Define corresponding indicator random variables $X = I_A$ and $Y = I_B$. The conditional distribution of Y given the outcomes of X is straightforward:

$$P(Y = y|X = 1) = \begin{cases} P(A|B), & \text{if } y = 1 \\ P(A^c|B), & \text{if } y = 0 \end{cases}$$

and

$$P(Y = y|X = 0) = \begin{cases} P(A|B^c), & \text{if } y = 1 \\ P(A^c|B^c), & \text{if } y = 0. \end{cases}$$ ■

4.8.1 Introduction to Conditional Expectation

A *conditional expectation* is an expectation with respect to a conditional distribution.

CONDITIONAL EXPECTATION OF Y GIVEN $X = X$

For discrete random variables X and Y, the *conditional expectation of Y given* $X = x$ is

$$E[Y|X = x] = \sum_y yP(Y = y|X = x).$$

Similarly, a conditional variance is a variance with respect to a conditional distribution. Conditional expectation and conditional variance will be treated in depth in Chapter 8. We introduce this important concept in the discrete setting with several examples.

Example 4.40 In Example 4.38, we find that the conditional distribution of the number of Sam's texts T given $T + C = n$ is binomial with parameters n and $p = \lambda/(\lambda + \mu)$. From properties of the binomial distribution, it follows immediately that

$$E[T|T + C = n] = np = \frac{n\lambda}{\lambda + \mu}.$$

Similarly, the conditional variance is

$$V[T|T + C = n] = np(1 - p) = n\left(\frac{\lambda}{\lambda + \mu}\right)\left(1 - \frac{\lambda}{\lambda + \mu}\right) = \frac{n\lambda\mu}{(\lambda + \mu)^2}. \quad \blacksquare$$

Example 4.41 The joint probability mass function of X and Y is given in the following joint table, along with marginal probabilities. Find $E[Y|X = x]$.

		Y			
		1	2	6	
	−1	0.0	0.1	0.1	0.2
X	1	0.1	0.3	0.2	0.6
	4	0.1	0.1	0.0	0.2
		0.2	0.5	0.3	

For $x = -1$,

$$\begin{aligned} E[Y|X = -1] &= \sum_{y \in \{1,2,6\}} yP(Y = y|X = -1) \\ &= \sum_{y \in \{1,2,6\}} y\frac{P(Y = y, X = -1)}{P(X = -1)} \\ &= 1(0.0/0.2) + 2(0.1/0.2) + 6(0.1/0.2) = 4. \end{aligned}$$

Similarly, for $x = 1$,

$$\begin{aligned} E[Y|X = 1] &= 1(0.1/0.6) + 2(0.3/0.6) + 6(0.2/0.6) \\ &= 1.9/0.6 = 3.167. \end{aligned}$$

And for $x = 4$,

$$\begin{aligned} E[Y|X = 4] &= 1(0.1/0.2) + 2(0.1/0.2) + 6(0.0/0.2) \\ &= 0.3/0.2 = 1.5 \end{aligned}$$

This gives

$$E[Y|X=x] = \begin{cases} 4, & \text{if } x=-1 \\ 3.167, & \text{if } x=1 \\ 1.5, & \text{if } x=4. \end{cases}$$ ■

Example 4.42 Xavier picks a number X uniformly at random from $\{1,2,3,4\}$. Having picked $X=x$ he shows the number to Yolanda, who picks a number Y uniformly at random from $\{1,\ldots,x\}$. (i) Find the expectation of Yolanda's number if Xavier picks x. (ii) Yolanda picked the number one. What do you think Xavier's number was?

(i) The problem asks for $E[Y|X=x]$. The conditional distribution of Y given $X=x$ is uniform on $\{1,\ldots,x\}$. The expectation of the uniform distribution gives

$$E[Y|X=x] = \frac{x+1}{2}.$$

(ii) If Yolanda picked one, we will infer Xavier's number based on the conditional expectation of X given $Y=1$. We have

$$\begin{aligned} E[X|Y=1] &= \sum_{x=1}^{4} xP(X=x|Y=1) \\ &= \sum_{x=1}^{4} x\frac{P(X=x,Y=1)}{P(Y=1)}. \end{aligned}$$

In the numerator,

$$P(X=x,Y=1) = P(Y=1|X=x)P(X=x) = \frac{1}{x}\left(\frac{1}{4}\right) = \frac{1}{4x}.$$

For the denominator, condition on X and apply the law of total probability.

$$\begin{aligned} P(Y=1) &= \sum_{x=1}^{4} P(Y=1|X=x)P(X=x) \\ &= \sum_{x=1}^{4} \frac{1}{4x} = \frac{1}{4}+\frac{1}{8}+\frac{1}{12}+\frac{1}{16} = \frac{25}{48}. \end{aligned}$$

We thus have

$$E[X|Y=1] = \sum_{x=1}^{4} x\left(\frac{1}{4x}\right)\frac{48}{25} = \sum_{x=1}^{4}\frac{12}{25} = \frac{48}{25} = 1.92.$$

A good guess at Xavier's number is two!

R: SIMULATING A CONDITIONAL EXPECTATION

We simulate the conditional expectation $E[X|Y=1]$.

```
> trials <- 10000
> ctr <- 0
> simlist <- numeric(trials)
> while (ctr < trials) {
   xav <- sample(1:4,1)
   yol <- sample(1:x,1)
   if (yol==1) {
    ctr <- ctr + 1
    simlist[ctr] <- xav}
   }
> mean(simlist)
[1] 1.9121
```

■

4.9 PROPERTIES OF COVARIANCE AND CORRELATION*

The covariance $\text{Cov}(X, Y)$ takes two arguments. In each argument, the function is linear in the following sense.

COVARIANCE PROPERTY: LINEARITY

For random variables X, Y, and Z, and constants a, b, c,

$$\text{Cov}(aX + bY + c, Z) = a\text{Cov}(X, Z) + b\text{Cov}(Y, Z) \tag{4.15}$$

and

$$\text{Cov}(X, aY + bZ + c) = a\text{Cov}(X, Y) + b\text{Cov}(X, Z). \tag{4.16}$$

Showing these properties is a straightforward application of the definition of covariance, which we leave to the exercises.

Given a random variable X with mean μ and variance σ^2, the *standardized* variable X^* is defined as

$$\boxed{X^* = \frac{X-\mu}{\sigma}.}$$

Observe that

$$E[X^*] = E\left[\frac{X-\mu}{\sigma}\right] = \frac{1}{\sigma}(E[X]-\mu) = \frac{1}{\sigma}(\mu-\mu) = 0$$

and

$$V[X^*] = V\left[\frac{X-\mu}{\sigma}\right] = \frac{1}{\sigma^2}(V[X-\mu) = \frac{\sigma^2}{\sigma^2} = 1.$$

"Standardlizing" a random variable gives a new random variable with mean 0 and variance one.

The following theorem shows that for any random variables X and Y, the correlation $\text{Corr}(X, Y)$ is always between -1 and 1. Further, the correlation is equal to ± 1 if and only if one variable is a linear function of the other. Most proofs of this result use linear algebra. The following probabilistic treatment is based on Feller (1968).

Theorem 4.1. *For random variables* X *and* Y,

$$-1 \leq Corr(\mathrm{X}, \mathrm{Y}) \leq 1.$$

If Corr$(\mathrm{X}, \mathrm{Y}) = \pm 1$, *then there exists constants* a *and* b *such that* $\mathrm{Y} = \mathrm{aX} + \mathrm{b}$.

Proof. Given X and Y, let X^* and Y^* be the standardized variables. Observe that

$$\begin{aligned}
\text{Cov}(X^*, Y^*) &= \text{Cov}\left(\frac{X-\mu_X}{\sigma_X}, \frac{Y-\mu_Y}{\sigma_Y}\right) \\
&= \frac{1}{\sigma_X \sigma_Y}\text{Cov}(X, Y) \\
&= \text{Corr}(X, Y).
\end{aligned}$$

Consider the variance of $X^* \pm Y^*$:

$$\begin{aligned}
V(X^* + Y^*) &= V(X^*) + V(Y^*) + 2\text{Cov}(X^*, Y^*) \\
&= 2 + 2\text{Corr}(X, Y)
\end{aligned}$$

and

$$\begin{aligned}V(X^* - Y^*) &= V(X^*) + V(Y^*) - 2\text{Cov}(X^*, Y^*) \\ &= 2 - 2\text{Corr}(X, Y).\end{aligned}$$

This gives

$$\text{Corr}(X, Y) = \frac{V(X^* + Y^*)}{2} - 1 \geq -1$$

and

$$\text{Corr}(X, Y) = -\frac{V(X^* - Y^*)}{2} + 1 \leq 1$$

since the variance is nonnegative. That is, $-1 \leq \text{Corr}(X, Y) \leq 1$.

Suppose $Y = aX + b$ is a linear function of X. Show in Exercise 4.45 that $\text{Corr}(X, Y) = \pm 1$ depending upon the sign of a. Conversely, if $\text{Corr}(X, Y) = 1$, then $V(X^* - Y^*) = 0$ and thus

$$X^* - Y^* = \frac{X - \mu_X}{\sigma_X} - \frac{Y - \mu_Y}{\sigma_Y}$$

is a constant. Solving for X gives

$$X = \left(\frac{\sigma_X}{\sigma_Y}\right) Y + \text{constant}.$$

Similarly for $\text{Corr}(X, Y) = -1$.

4.10 EXPECTATION OF A FUNCTION OF A RANDOM VARIABLE*

Let X be a random variable. Let f be a function. Following is the proof that

$$E[f(X)] = \sum_x f(x)P(X = x),$$

the so-called law of the unconscious statistician.

Proof. Let $Y = f(X)$. Then

$$\begin{aligned}
E[f(X)] = E[Y] &= \sum_y yP(Y = y) \\
&= \sum_y yP(f(X) = y) \\
&= \sum_y P(X \in f^{-1}(y)) \\
&= \sum_y y \sum_{x:x \in f^{-1}(y)} P(X = x) \\
&= \sum_y y \sum_{x:f(x)=y} P(X = x) \\
&= \sum_x \left(\sum_{y:y=f(x)} y \right) P(X = x) \\
&= \sum_x f(x)P(X = x).
\end{aligned}$$

Note that if f is a one-to-one function and has an inverse f^{-1}, then $f^{-1}(y)$ is a single number. However, in general, $f^{-1}(y)$ is a set of numbers such that $x \in f^{-1}(y)$ if and only if $f(x) = y$.

4.11 SUMMARY

Random variables are the central objects of probability and this chapter introduces them in the discrete setting. Many important concepts are first introduced in this chapter. The probability mass function of a random variable is its probability function $P(X = k)$. The expectation of a random variable is a measure of its "average" or typical value. The variance is a measure of spread and variability. We often work with functions of random variables $f(X)$ and these are presented along with methods for finding their expectation. Numerous properties of expectation and variance are given. Most important is linearity of expectation: the expectation of a sum is equal to the sum of the expectations. The property holds without any condition on the distribution of the underlying random variables. If the random variables are also independent, then the variance of a sum is equal to the sum of the variances.

For two or more random variables defined on the same sample space, we have a joint distribution and joint pmf. The univariate marginal distributions are obtained from the joint distribution by summing over the other variable(s). Indicator random variables I_A are introduced and shown to be a useful method for representing counts. The covariance and correlation of two jointly distributed random variables is a measure of the linear association between them. The general formula for the variance of a sum requires a covariance term.

At the end of the chapter, conditional distributions are introduced. The conditional probability mass function is defined, along with conditional expectation.

- **Probability mass function:** For a random variable X, the pmf is $m(x) = P(X = x)$.
- **Expectation:** $E[X] = \sum_x xP(X = x)$.
  1. For $X \sim \text{Unif}(\{1, \ldots, n\})$, $E[X] = (n + 1)/2$.
  2. For $X \sim \text{Ber}(p)$, $E[X] = p$
  3. For $X \sim \text{Binom}(n, p)$, $E[X] = np$.
  4. For $X \sim \text{Pois}(\lambda)$, $E[X] = \lambda$.
- **Functions of random variables:** If f is a function, then $f(X)$ is a random variable that takes values $f(x)$ whenever $X = x$.
- **Expectation of function of a random variable and the "law of the unconscious statistician":** $E[f(X)] = \sum_x f(x)P(X = x)$. Warning: It is not true in general that $E[f(X)] = f(E[X])$. In particular, it is not true that $E[X^2] = E[X]^2$.
- **Expectation of linear function:** For constants a, b, $E[aX + b] = aE[X] + b$.
- **Joint probability mass function:** For jointly distributed random variables X and Y, the joint pmf is $P(X = x, Y = y)$.
- **Marginal distributions:** If X and Y are jointly distributed, the marginal distribution of X is found by summing over the Y variable. That is $P(X = x) = \sum_y P(X = x, Y = y)$. Similarly, $P(Y = y) = \sum_x P(X = x, Y = y)$.
- **Joint pmf for independent random variables:** If X and Y are independent, $P(X = x, Y = y) = P(X = x)P(Y = y)$.
- **Expectation of function of two random variables:**

$$E[f(X, Y)] = \sum_x \sum_y f(x, y)P(X = x, Y = y).$$

- **Independent random variables:** Functions of independent random variables are independent. That is, if f and g are functions of independent random variables X and Y, then $f(X)$ and $g(Y)$ are independent.
- **Expectation of product of independent random variables:** If X and Y are independent, then $E[XY] = E[X]E[Y]$. Similarly,

$$E[f(X)g(Y)] = E[f(X)]E[g(Y)].$$

- **Sum of independent random variables:** If X and Y are independent, then

$$P(X + Y = k) = \sum_i P(X = i)P(Y = k - i).$$

- **Linearity of expectation:** $E[X+Y] = E[X] + E[Y]$.
- **Variance:** $V[X] = E[(X - E[X])^2] = E[X^2] - (E[X])^2$.
  1. For $X \sim \text{Unif}(\{1, \dots, n\})$, $V[X] = (n^2 - 1)/12$.
  2. For $X \sim \text{Ber}(p)$, $V[X] = p(1-p)$.
  3. For $X \sim \text{Binom}(n, p)$, $V[X] = np(1-p)$.
  4. For $X \sim \text{Pois}(\lambda)$, $V[X] = \lambda$.
- **Standard deviation:** $\text{SD}[X] = \sqrt{V[X]}$.
- **Properties of variance and standard deviation:**
  1. $V[X] = 0$ if and only if X is a constant.
  2. $V[aX + b] = a^2 V[X]$.
  3. $\text{SD}[aX + b] = |a| \text{SD}[X]$.
- **Within two standard deviation heuristic:** For many near-symmetric probability distributions, "most" outcomes are (roughly) within two standard deviations from the mean.
- **Indicator random variables:** For an event A, the indicator random variable is defined such that $I_A = 1$, if A occurs, and 0, if A does not occur.
- **Properties of indicators:** $E[I_A] = P(A)$ and $V[I_A] = P(A)P(A^c)$.
- **Covariance:**

$$\text{Cov}(X, Y) = E[(X - E[X])(Y - E[Y])] = E[XY] - E[X]E[Y].$$

  Covariance is a measure of linear association. Independent random variables have covariance equal to 0. If X and Y are inversely associated, the covariance between them is negative.
- **Correlation:** $\text{Corr}(X, Y) = \text{Cov}(X, Y)/(\text{SD}[X]\text{SD}[Y])$. Correlation always takes values between -1 and 1. If $Y = aX + b$ is a linear function of X, then $\text{Corr}(X, Y) = \pm 1$, with the sign determined by a.
- **Variance of sums**:
  1. $V[X + Y] = V[X] + V[Y] + 2\text{Cov}(X, Y)$.
  2. If X and Y are independent, $V[X + Y] = V[X] + V[Y]$.
  3. $V[X_1 + \cdots + X_n] = \sum_{k=1}^{n} V[X_k] + 2 \sum_{i<j} \text{Cov}(X_i, X_j)$.
- **Uncorrelated:** Random variables X and Y are uncorrelated if $E[XY] = E[X]E[Y]$. Independent random variables are uncorrelated. But the converse is not necessarily true.
- **Conditional probability mass function:** For jointly distributed random variables X and Y, the conditional pmf of Y given $X = x$ is $P(Y = y|X = x) = P(X = x, Y = y)/P(X = x)$.
- **Conditional expectation:** $E[Y|X = x] = \sum_y yP(Y = y|X = x)$.
- **Problem-solving strategies**
  1. **Using indicator random variables for counts:** Indicators are often used to model counts. For instance, if X is the number of successes in n trials, then

write $X = I_1 + \cdots I_n$, where I_k is the indicator that success occurs on the kth trial. The representation of a count as a sum of indicators allows us to use linearity of expectation in finding the expectation of X, as

$$E[X] = E[I_1 + \cdots I_n] = E[I_1] + \cdots + E[I_n].$$

EXERCISES

Expectation

4.1 What is the average number of vehicles per household in the United States? Table 4.5 gives data from the 2010 U.S. Census on the distribution of available vehicles per household. "Available vehicles" refers to the number of cars, vans, and pickup trucks kept at home and available for use by household members. Let X be the number of available vehicles for a randomly chosen household. Find the expectation of X.

4.2 Find the expectation of a random variable uniformly distributed on $\{a, \ldots, b\}$.

4.3 The following dice game costs \$10 to play. If you roll 1, 2, or 3, you lose your money. If you roll 4 or 5, you get your money back. If you roll a 6, you win \$24.

(a) Find the distribution of your winnings W.

(b) Find the expected value of the game.

4.4 Let $X \sim \text{Unif}\{-2, -1, 0, 1, 2\}$.

(a) Find $E[X]$.

(b) Find $E[e^X]$.

(c) Find $E[1/(X + 3)]$.

4.5 See Example 4.21 on the St. Petersburg paradox. Modify the game so that you only receive $\$2^{10} = \$1024$ if any number greater than or equal to 10 tosses are required to obtain the first tail. Find the expected value of this game.

4.6 Suppose $E[X^2] = 1$, $E[Y^2] = 2$, and $E[XY] = 3$. Find $E[(X + Y)^2]$.

4.7 Suppose $P(X = 1) = p$ and $P(X = 2) = 1 - p$. Show that there is no value of $0 < p < 1$ such that $E[1/X] = 1/E[X]$.

4.8 You are dealt five cards from a standard deck. Let X be the number of aces in your hand. Find $E[X]$.

TABLE 4.5: Household size by vehicles available.

Available vehicles	0	1	2	3	4
Proportion of households	0.092	0.341	0. 376	0.135	0.056

Source: U.S. Census.

4.9 Let $X \sim \text{Pois}(\lambda)$. Find $E[X!]$. For what values of λ does the expectation not exist?

4.10 On November 28, 2012, the jackpot of the powerball lottery was \$587.5 million. The website `http://www.powerball.com/powerball/pb_prizes.asp` gives the payouts and corresponding probabilities for the *Powerball* lottery. Find the expected value of the November 28 game.

Joint distribution

4.11 The number of tornadoes T and earthquakes E over a month's time in a particular region is independent and has a Poisson distribution with parameters four and two, respectively.

(a) Find the joint pmf of T and E.

(b) What is the probability of no tornadoes and no earthquakes in that region next month?

(c) What is the probability of at least two tornadoes and at least two earthquakes?

(d) What is the probability of at least two tornadoes or at most two earthquakes?

4.12 The joint pmf of X and Y is

$$P(X = x, Y = y) = \frac{x+1}{12},$$

for $x = 0, 1$ and $y = 0, 1, 2, 3$. Find the marginal distributions of X and Y. Describe their distributions *qualitatively*. That is, identify their distributions as one of the known distributions you have worked with (e.g., Bernoulli, binomial, Poisson, or uniform).

4.13 Suppose X, Y, and Z have joint pmf

$$P(X = x, Y = y, Z = z) = c, \text{ for } x = 1, \ldots, n, \ y = 1, \ldots, x, \ z = 1, \ldots y.$$

(a) Find the constant c. Of use will be the formula for the sum of the first n squares

$$\sum_{k=1}^{n} k^2 = \frac{n(n+1)(2n+1)}{6}.$$

(b) For $n = 4$, find $P(X \leq 3, Y \leq 2, Z = 1)$.

4.14 Suppose X, Y, and Z are independent random variables that take values 1 and 2 with probability 1/2 each. Find the pmf of (X, Y, Z).

4.15 Suppose X, Y, and Z are independent Bernoulli random variables with respective parameters $1/2$, $1/3$, and $1/4$.

(a) Find $E[XYZ]$.

(b) Find $E[e^{X+Y+Z}]$.

4.16 Suppose X and Y are independent random variables. Does $E[X/Y] = E[X]/E[Y]$? Either prove it true or exhibit a counterexample.

4.17 The joint pmf of (X, Y) is

$$P(X = x, Y = y) = \frac{1}{nx}, \quad \text{for } x = 1, \ldots, n,\ y = 1, \ldots, x.$$

(a) Find the marginal distribution of X. Describe the distribution qualitatively.

(b) Find $E[Y/(X + 1)]$.

(c) For the case $n = 3$, write out explicitly the joint probability table and confirm your result in (b).

4.18 Let X and Y be the first and second numbers obtained in two draws from the set $\{1, 2, 3, 4\}$ sampling with replacement.

(a) Give the joint distribution table for X and Y.

(b) Find $P(X \leq Y)$.

(c) Repeat the previous two exercises if the sampling is done without replacement.

4.19 A joint probability mass function is given by

$$P(X = 1, Y = 1) = \frac{1}{8} \quad P(X = 1, Y = 2) = \frac{1}{4}$$

$$P(X = 2, Y = 1) = \frac{1}{8} \quad P(X = 2, Y = 2) = \frac{1}{2}.$$

(a) Find the marginal distributions of X and Y.

(b) Are X and Y independent?

(c) Compute $P(XY \leq 3)$.

(d) Compute $P(X + Y > 2)$.

4.20 (Elevator problem) An elevator containing p passengers is at the ground floor of a building with n floors. On its way to the top of the building, the elevator will stop if a passenger needs to get off. Passengers get off at a particular floor with probability $1/n$. Find the expected number of stops the elevator makes. (Hint: Use indicators, letting $I_k = 1$ if the kth floor is a stop. Be careful: more than one passenger can get off at a floor.)

FIGURE 4.4: Randomly colored six-by-six board. There are 19 one-by-one black squares, two two-by-two black boards, and no three-by-three or larger black boards.

4.21 A bag contains r red and g green candies. We draw n candies from the bag without replacement. Find the expected number of red candies drawn by using indicator variables.

4.22 Take an n-by-n board divided into one-by-one squares and color each square white or black uniformly at random. That is, for each square flip a coin and color it white, if heads, and black, if tails. Let X be the number of two-by-two square "subboards" of the chessboard that are all black (See Fig. 4.4). Use indicator variables to find the expected number of black two-by-two subboards.

4.23 In a class of 25 students, what is the expected number of months in which at least two students are born? Assume birth months are equally likely. Hint: Use indicators.

Variance, standard deviation, covariance, correlation

4.24 Find the variance for the outcome of a fair die roll two ways: (i) using the definition of variance and (ii) using Equation 4.10. Which method do you prefer?

4.25 Suppose X takes values -1, 0, and 3, with respective probabilities 0.1, 0.3, and 0.6. Find $V[X]$.

4.26 Suppose $E[X] = a$, $E[Y] = b$, $V[X] = c$, and $V[Y] = d$. If X and Y are independent, find $V[2X - 3Y + 4]$.

4.27 Find the variance of the sum of n independent tetrahedron dice rolls.

4.28 In a random experiment, let A and B be two independent events with $P(A) = P(B) = p$. In an outcome of the experiment, let X be the number of these events that occur (0, 1, or 2). Find $E[X]$ and $V[X]$.

4.29 Suppose $E[X] = 2$ and $V[X] = 3$. Find

(a) $E[(3 + 2X)^2]$.

(b) $V[4 - 5X]$.

4.30 Let X be a Poisson random variable with parameter λ. Find the variance of X. Hint: Write

$$X^2 = (X^2 - X) + X = X(X - 1) + X.$$

4.31 Suppose A and B are events with $P(A) = a$, $P(B) = b$ and $P(AB) = c$. Define indicator random variables I_A and I_B. Find $V[I_A + I_B]$.

4.32 Define a sequence $X_1, \ldots, X_n$ of independent random variables such that for each $k = 1, \ldots, n$, $X_k = \pm 1$, with probability 1/2 each. Let $S = X_1 + \cdots + X_n$. This model describes a *simple symmetric random walk* that starts at the origin and moves left or right at each unit of time. Find $E[S]$ and $V[S]$.

4.33 In the random walk model, suppose the distribution of the X_k's is given by $P(X_k = +1) = p$ and $P(X_k = -1) = 1 - p$. If $p > 1/2$, this describes a *random walk with positive drift*. Find $E[S]$ and $V[S]$.

4.34 A bag contains one red, two blue, three green, and four yellow balls. A sample of three balls is taken without replacement. Let B be the number of blue balls and Y the number of yellow balls in the sample.

(a) Find the joint probability table.

(b) Find $\text{Cov}(B, Y)$.

4.35 Random variables X and Y have joint distribution

$$P(X = i, Y = j) = c(i + j), \text{ for } i = 0, 1; j = 1, 2, 3$$

for some constant c.

(a) Find c.

(b) Find the marginal distributions of X and Y.

(c) Find $\text{Cov}(X, Y)$.

4.36 Suppose X and Y have the same distribution and $\text{Corr}(X, Y) = -0.5$. Find $V[X + Y]$.

4.37 Suppose X and Y are independent random variables with $V[X] = \sigma_X^2$, and $V[Y] = \sigma_Y^2$. Let $Z = wX + (1 - w)Y$, for $0 < w < 1$. Thus, Z is a weighted average of X and Y. Find the variance of Z. What value of w minimizes this variance?

4.38 Let D_1 and D_2 be the outcomes of two dice rolls. Let $X = D_1 + D_2$ be the sum of the two numbers rolled. Let $Y = D_1 - D_2$ be their difference. Show that X and Y are uncorrelated, but not independent.

4.39 It is apparent from the definition of variance that if X is a constant, then $V[X] = 0$. Here we show the converse.

(a) Suppose X is a nonnegative discrete random variable with $E[X] = 0$. Show that $P(X = 0) = 1$.

(b) Let X be a discrete random variable with $E[X] = \mu$ and $V[X] = 0$. Show that $P(X = \mu) = 1$.

4.40 Let $E[X] = 1$, $E[X^2] = 2$, $E[X^3] = 5$, and $E[X^4] = 15$. Also $E[Y] = 2$, $E[Y^2] = 6$, $E[Y^3] = 22$, and $E[Y^4] = 94$. Suppose X and Y are independent.

(a) Find $V[3X^2 - Y]$.

(b) Find $E[X^4 Y^4]$.

(c) Find $\text{Cov}(X, X^2)$.

(d) Find $V[X^2 Y^2]$. (See also Exercise 4.58.)

4.41 See Exercise 4.22. Using indicators, find the variance of the number of black two-by-two subboards. Note that for the natural choice of indicators, the random variables are not independent.

4.42 Show the linearity properties for covariance as given in Equations 4.15 and 4.16.

4.43 Show that $\text{Cov}(X, Y) = E[XY] - E[X]E[Y]$, as a consequence of Equation 4.13.

4.44 Suppose $E[X] = 1$, $V[X] = 2$, $E[Y] = 3$, $V[Y] = 4$, and $\text{Cov}(X, Y) = -1$. Find $\text{Cov}(3X + 1, 2Y - 8)$.

4.45 Suppose $Y = aX + b$ for constants a and b. Find $\text{Cov}(X, Y)$.

Conditional distribution, expectation, functions of random variables

4.46 The joint probability mass function of X and Y is

$$P(X = x, Y = y) = \frac{1}{e^2 y!(x - y)!}, \quad x = 0, 1, \ldots, \quad y = 0, 1, \ldots, x.$$

(a) Find the conditional distribution of Y given $X = x$.

(b) Describe the distribution in terms of distributions that you know.

(c) Without doing any calculations, find $E[Y|X = x]$ and $V[Y|X = x]$.

4.47 Leiter and Hamdan (1973) model traffic accidents and fatalities at a specific location in a given time interval. They suppose that the number of accidents X has a Poisson distribution with parameter λ. If $X = x$, then the number of fatalities Y has a binomial distribution with parameter p.

(a) Find the marginal distribution of the number of fatalities.

(b) Show that the conditional distribution of X given $Y = y$ is a Poisson distribution with parameter $\lambda(1-p)$, which has been "shifted" y units to the right.

4.48 Let X be the first of two fair die rolls. Let M be the maximum of the two rolls.

(a) Find the conditional probability mass function of M given $X = x$.

(b) Find $E[M|X = x]$.

(c) Find the joint probability mass function of X and M.

4.49 Given events A and B, let I_A and I_B be the corresponding indicator variables. Find a simple expression for the conditional expectation of I_A given $I_B = 1$.

4.50 Suppose X and Y are independent with the following binomial distributions: $X \sim \text{Binom}(n, p)$ and $Y \sim \text{Binom}(m, p)$. Show that $X + Y$ has a binomial distribution and give the parameters. (Hint: Use Equation 3.13.)

4.51 Let X and Y be independent and identically distributed random variables. For each of the following questions, either show that it is true or exhibit a counter-example.

(a) Does $X + Y$ have the same distribution as $2X$?

(b) Does $X + Y$ have the same expectation as $2X$?

(c) Does $X + Y$ have the same variance as $2X$?

4.52 Let $X_1, \ldots, X_n$ be an i.i.d. Bernoulli sequence with parameter p.

(a) Find the conditional distribution of X_1 given $X_1 + \cdots + X_n = k$.

(b) Find $E[X_1|X_1 + \cdots + X_n = k]$ and $V[X_1|X_1 + \cdots + X_n = k]$.

4.53 Suppose X and Y are independent random variables both uniformly distributed on $\{1, \ldots, n\}$. Find the probability mass function of $X + Y$.

4.54 In the original matching problem, Montmort asked for the probability of at least one match. Find this probability and show that for large n, the probability of a match is about $1 - e^{-1} = 0.632$. Hint: Use inclusion–exclusion.

Simulation and R

4.55 Simulate the probability of obtaining at least one match in the problem of coincidences (see Exercise 4.54).

4.56 Simulate the dice game in Example 4.3. Estimate the expectation and variance of your winnings.

4.57 See Exercise 4.34. Simulate the mean and variance of the total number of blue and yellow balls in the sample.

4.58 In Exercise 4.40, the random variables X and Y have Poisson distributions with respective parameters 1 and 2. Simulate the results in that exercise.

4.59 Simulate the variance of the matching problem for a large value of n.

4.60 Let X be a random variable that takes values $x = (x_1, \ldots, x_n)$ with respective probabilities $p = (p_1, \ldots, p_n)$. Write two R functions `mymean(x,p)`, and `myvariance(x,p)`, which find the mean and variance of X, respectively. Use your function to find the mean and variance of the point value of a random Scrabble tile, as in Example 4.1.

4.61 In Texas hold 'em poker, players are initially dealt two cards each. (i) In a game of six players, simulate the probability that at least one of the players will have a pair. (ii) A hand is said to be *suited* if both cards are the same suit. Simulate the probability that at least one of the six players' hands is suited.

4.62 See Exercise 4.35. Write an R function `joint(i,j)` for computing $P(X = i, Y = j)$, for $i = 0, 1; j = 1, 2, 3$. Use this function to compute the covariance and correlation of X and Y.

5

A BOUNTY OF DISCRETE DISTRIBUTIONS

Essentially, all models are wrong, but some are useful.

—George Box

5.1 GEOMETRIC DISTRIBUTION

There are many questions one can ask about an underlying sequence of Bernoulli trials. The binomial distribution describes the number of successes in n trials. The geometric distribution arises as the number of trials until the first success occurs.

John likes to plays the "pick-3" lottery game. In the "pick-3" game, you choose three single-digit numbers, each from 0 to 9, in order to match the winning number. There are 1000 possible choices so the probability of winning is $1/1000$.

If John buys a "pick-3" ticket every day, what is the probability he will win during the next year? And how many days can he expect to wait until he wins the lottery?

To model the number of days until John wins the lottery, consider successive outcomes as an i.i.d. Bernoulli sequence with success probability $p = 0.001$. Let X be the the number of trials until the first success occurs. That is, the number of days until John wins the lottery. To find the probability mass function of X observe that $X = k$ if success occurs on the kth trial and the first $k - 1$ trials are failures. This occurs with probability $(1 - p)^{k-1}p$.

Probability: With Applications and R, First Edition. Robert P. Dobrow.

GEOMETRIC DISTRIBUTION

The random variable X has a *geometric distribution with parameter* p if

$$P(X = k) = (1 - p)^{k-1}p, \text{ for } k = 1, 2, \ldots. \tag{5.1}$$

We write $X \sim \text{Geom}(p)$.

The parameter of the geometric distribution is sometimes called the *success parameter.* The geometric probability mass function is nonnegative and sums to 1, as

$$\sum_{k=1}^{\infty}(1-p)^{k-1}p = p\frac{1}{1-(1-p)} = 1.$$

Let X be the number of days until John wins the lottery. Then $X \sim \text{Geom}(0.001)$. The probability that John will win the lottery during the following year is

$$\begin{aligned} P(X \leq 365) &= \sum_{k=1}^{365} P(X = k) = \sum_{k=1}^{365} 0.999^{k-1}(0.001) \\ &= (0.001)\frac{1 - 0.999^{365}}{1 - 0.999} \\ &= 1 - 0.999^{365} = 0.3059, \end{aligned}$$

where we have used the partial sum of the geometric series formula

$$\sum_{k=1}^{n} r^{k-1} = \sum_{k=0}^{n-1} r^k = \frac{1 - r^n}{1 - r}, \text{ for } r \neq 1.$$

Another way to find the probability $P(X \leq 365)$ is to take complements and think probabilistically. Consider $P(X \leq 365) = 1 - P(X > 365)$. The event $\{X > 365\}$ is equal to the event that no lottery tickets are winners in the first 365 days. This occurs with probability $(1 - p)^{365} = 0.999^{365}$. The desired probability is $1 - 0.999^{365}$.

This derivation gives the general closed form expression for the "tail probability" of a geometric random variable with success parameter p.

TAIL PROBABILITY OF GEOMETRIC DISTRIBUTION

If $X \sim \text{Geom}(p)$, then for $k > 0$,

$$P(X > k) = (1 - p)^k.$$

■ **Example 5.1 Expectation of geometric distribution.** Let $X \sim \text{Geom}(p)$. Then

$$E[X] = \sum_{k=1}^{\infty} kP(X = k) = \sum_{k=1}^{\infty} k(1-p)^{k-1}p = p\sum_{k=1}^{\infty} k(1-p)^{k-1}. \tag{5.2}$$

To make progress on the last sum observe that the expression looks like a derivative. Consider the geometric series

$$\sum_{k=0}^{\infty} r^k = \frac{1}{1-r}, \quad \text{for } |r| < 1.$$

The series is absolutely convergent and can be differentiated termwise. That is, the derivative of the sum is equal to the sum of the derivatives. Differentiating with respect to r gives

$$\frac{d}{dr}\sum_{k=0}^{\infty} r^k = \sum_{k=1}^{\infty} kr^{k-1} = \frac{d}{dr}\left(\frac{1}{1-r}\right) = \frac{1}{(1-r)^2}.$$

Apply to Equation 5.2 with $r = 1 - p$ to get

$$E[X] = p\sum_{k=1}^{\infty} k(1-p)^{k-1} = p\frac{1}{(1-(1-p))^2} = \frac{p}{p^2} = \frac{1}{p}.$$ ■

The expected number of trials until the first success is the inverse of the success probability. You can expect to flip two coins, on average, until you see heads; and roll 36 pairs of dice until you see snake-eyes. And John can expect to wait $1/(0.001) = 1000$ days until he wins the lottery.

We leave for the reader the pleasure of finding the variance of a geometric distribution. You will need to take two derivatives (See Exercise 5.3.)

5.1.1 Memorylessness

The geometric distribution has a unique property among discrete distributions. It is what is called *memoryless*. We illustrate the concept with an example.

■ **Example 5.2** A traffic inspector is monitoring a busy intersection for moving violations. The inspector wants to know how many cars will pass by before the first moving violation occurs. In particular, she would like the find the probability that the first moving violation occurs after the 30th car. Let X be the number of cars which pass the intersection until the first moving violation. The desired probability is $P(X > 30)$.

After 20 cars go by without a moving violation, a second inspector arrives on the scene. He would also like to know the probability that the first moving violation occurs after the 30th car. But given that 20 cars have gone by without any violations, this probability is equal to the conditional probability that 50 cars go by without moving violations, given that there were no violations among the first 20 cars, that is, $P(X > 50|X > 20)$.

If the distribution of X is memoryless it means that these two probabilities are the same, that is, $P(X > 50|X > 20) = P(X > 30)$. The cars that go by once the second inspector arrives on the scene do not "remember" the 20 cars that went by before.

Assume that whether or not cars have moving violations is independent of each other, and that the probability that any particular car has a violation is p. Then the number of cars which pass the intersection until the first moving violation X has a geometric distribution with parameter p. In that case,

$$\begin{aligned} P(X > 50|X > 20) &= \frac{P(X > 50, X > 20)}{P(X > 20)} \\ &= \frac{P(X > 50)}{P(X > 20)} = \frac{(1-p)^{50}}{(1-p)^{20}} \\ &= (1-p)^{30} = P(X > 30), \end{aligned}$$

where we use the fact that if there are no violations among the first 50 cars then there are no violations among the first 20 cars. That is, $\{X > 50\}$ implies $\{X > 20\}$ and thus $\{X > 50, X > 20\} = \{X > 50\}$. ■

MEMORYLESS PROPERTY

A random variable X has the *memoryless property* if for all $0 < s < t$,

$$P(X > t|X > s) = P(X > t - s).$$

We see that a geometric random variable is memoryless. If $X \sim \text{Geom}(p)$, then for $0 < s < t$,

$$\begin{aligned} P(X > t|X > s) &= \frac{P(X > t, X > s)}{P(X > s)} = \frac{P(X > t)}{P(X > s)} \\ &= \frac{(1-p)^t}{(1-p)^s} = (1-p)^{t-s} \\ &= P(X > t - s). \end{aligned}$$

Example 5.3 A baseball player's batting average is 0.300. That is, when the player comes up to bat there is a 30% chance of getting a hit. After three times at bat, the player has not had a hit. What is the probability the player will not get a hit after seven times at bat?

The number of times at bat until getting a hit is a geometric random variable with $p = 0.3$. By the memoryless property, the desired probability is $(1-p)^{7-3} = (0.70)^4 = 0.24$. ∎

Example 5.4 A hard day's night at the casino. Bob has been watching the roulette table all night, and suddenly sees the roulette ball land on black 10 times in a row. It is time for red, he reasons, by the "law of averages." But his partner Bill, a student of probability, recognizes that the number of times until red comes up has the geometric distribution and is thus memoryless. It does not "remember" the last 10 outcomes. The probability that red will come up after the 11th bet given past history is the same as the probability of it coming up after the first.

Remarkably, the geometric distribution is the *only* discrete distribution that is memoryless. Here is a short derivation assuming X takes positive integer values.

Let $f(x) = P(X > x)$. The defining property of memorylessness gives that $f(t-s) = f(t)/f(s)$ or $f(t) = f(s)f(t-s)$ for all integers $0 < s < t$.

With $s = 1$ and $t = 2$, $f(2) = f(1)f(1) = f(1)^2$. With $t = 3$, $f(3) = f(1)f(2) = f(1)f(1)^2 = f(1)^3$. In general, we see that $f(n) = f(1)^n$, for all integer n. This gives

$$P(X = n) = P(X > n-1) - P(X > n)$$
$$= f(n-1) - f(n) = f(1)^{n-1} - f(1)^n = f(1)^{n-1}[1 - f(1)],$$

for $n = 1, 2, \ldots$, which is the probability mass function of a geometric distribution with parameter $p = 1 - f(1) = P(X = 1)$. ∎

5.1.2 Coupon Collecting and Tiger Counting

In a national park in India, an automatic camera photographs the number of tigers n passing by over a 12-month period. From the photographs it was determined that the number of *different* tigers observed was t. From t and n the total number of tigers in the park is to be estimated. (Cited in Finkelstein et al. (1998).)

The problem of estimating the number of tigers in a park will introduce the *coupon collector's problem,* with many applications from wildlife management to electrical engineering. The coupon collector's problem asks: If one repeatedly samples with replacement from a collection of n distinct items ("coupons"), what is the expected number of draws required so that each item is selected at least once?

When the author had small children, a certain fast food chain was frequented which sold takeout meals with a Star Wars character in each meal. There were 10 characters in all. The coupon collector's problem asks: How many meals need be bought to get a complete set of 10 characters?

Given n coupons, let X be the number of draws required to obtain a complete set sampling with replacement. We find $E[X]$.

As is often the case with random variables which represent counts, we will decompose the count into simpler units. Let $G_1 = 1$. For $k = 2, \ldots, n$, let G_k be the number of draws required to increase the set of distinct coupons from $k-1$ to k. Then $X = G_1 + \cdots + G_n$.

For instance, if the set of coupons is $\{a, b, c, d\}$ and the sequence of successive draws is

$$a, a, d, b, d, a, d, b, c,$$

then $G_1 = 1$, $G_2 = 2$, $G_3 = 1$, $G_4 = 5$, and $X = G_1 + G_2 + G_3 + G_4 = 9$.

We find the distribution of the G_k's. Consider the process of collecting coupons. Once the first coupon is picked, successive draws might result in it being picked again, but eventually a second coupon will get picked. There are $n-1$ possibilities for the second coupon. Successive draws are Bernoulli trials where "success" means picking a second coupon. The number of draws until the second new coupon is picked is a geometric random variable with parameter $p = (n-1)/n$. That is, $G_2 \sim \text{Geom}((n-1)/n)$.

Similarly, once the second coupon is picked, successive draws until the third coupon is picked form a Bernoulli sequence with $p = (n-2)/n$, and thus $G_3 \sim \text{Geom}((n-2)/n)$. In general, once the $(k-1)$st coupon is picked, successive draws until the kth new coupon is picked form a Bernoulli sequence with parameter $p = (n-(k-1))/n$, and $G_k \sim \text{Geom}((n-(k-1))/n)$.

Since the expectation of a geometric distribution is the inverse of the success probability,

$$\begin{aligned} E[X] &= E[G_1 + G_2 + \cdots + G_n] \\ &= E[G_1] + E[G_2] + \cdots + E[G_n] \\ &= \frac{n}{n} + \frac{n}{n-1} + \cdots + \frac{n}{1} = n\left(\frac{1}{n} + \frac{1}{n-1} + \cdots + \frac{1}{1}\right) && (5.3) \\ &= n\left(1 + \frac{1}{2} + \cdots + \frac{1}{n}\right). && (5.4) \end{aligned}$$

For example, the expected number of meals needed to get the full set of 10 Star Wars characters is

$$E[X] = 10\left(1 + \frac{1}{2} + \cdots + \frac{1}{10}\right) = 29.3 \text{ meals.}$$

Observe that $G_1, \ldots, G_n$ are independent random variables. Consider G_k. Once $k-1$ coupons have been picked, the number of additional trials needed to select the next coupon does not depend on the past history of selections. For each k, G_k is independent of the previous $G_1, \ldots, G_{k-1}$.

The last expression in Equation 5.4 is the partial harmonic series. For large n the series is approximated by $\ln n$, as

$$\sum_{i=1}^{n} \frac{1}{i} \approx \int_1^n \frac{1}{x}\, dx = \ln n,$$

which gives $E[X] \approx n \ln n$.

For the tiger estimation problem introduced at the beginning of this section, we consider a slightly different version of the coupon collector's problem, as given in Ross (2012). Suppose there are n coupons to choose from and t items have been selected after repeated selections with replacement. What is the expected number of distinct coupons in the group of t items?

For instance, for Star Wars characters, if 12 meals are gotten what is the expected number of different characters collected? Here $n = 10$ and $t = 12$.

Let X be the number of different items collected after drawing t items. By the method of indicators write $X = I_1 + \cdots + I_n$, where

$$I_k = \begin{cases} 1, & \text{if coupon } k \text{ is contained in the set of } t \text{ items} \\ 0, & \text{if coupon } k \text{ is not contained in the set of } t \text{ items,} \end{cases}$$

for $k = 1, \ldots, n$. Then

$$\begin{aligned} E[I_k] &= P(\text{Coupon } k \text{ is contained in the set of } t \text{ items}) \\ &= 1 - P(\text{Coupon } k \text{ is not contained in the set of } t \text{ items}) \\ &= 1 - \left(1 - \frac{1}{n}\right)^t, \end{aligned}$$

which gives

$$E[X] = E[I_1] + \cdots + E[I_n] = n\left[1 - \left(1 - \frac{1}{n}\right)^t\right]. \tag{5.5}$$

Continuing with the Star Wars story, if 12 meals are gotten, the expected number of different characters is

$$10\left[1 - \left(1 - \frac{1}{10}\right)^{12}\right] = 7.18.$$

What all this has to do with tigers we come to presently. The unknown quantity of interest is n the number of tigers in the park. A group of t tigers has been "collected" by the park's camera. The expectation in Equation 5.5 gives the expected number of distinct tigers in a group of t tigers seen on camera. Suppose the photographic analysis shows d distinct tigers. Then

$$d \approx E[X] = n\left[1 - \left(1 - \frac{1}{n}\right)^t\right].$$

The data consist of t and d. Solving for n gives an estimate of the total number of tigers in the park.

Suppose the park's camera observes 100 tigers throughout the year, and identifies 50 distinct tigers from the photographs. To estimate the number of tigers in the park, solve

$$50 = n\left[1 - \left(1 - \frac{1}{n}\right)^{100}\right]. \tag{5.6}$$

R: NUMERICAL SOLUTION TO TIGER PROBLEM

There is no closed form algebraic solution to the nonlinear equation Equation 5.6. But there are numerical ways to solve it. In R, the `uniroot` command finds the root of a general function. The script **tiger.R** finds the numerical solution.

```
# tiger.R
# Define function f(n) in order to solve f(n) = 0.
> f <- function(n) n*(1-(1-1/n)^100) - 50
  # Find root numerically, searching inside (50, 200)
> uniroot(f,c(50,200))$root
[1] 62.40844
```

We estimate about 62–63 tigers in the park.

There are many generalizations and extensions of the coupon collector's problem. To learn more, see Dawkins (1991).

5.1.3 How R Codes the Geometric Distribution

An alternate formulation of the geometric distribution counts the number of *failures* $\tilde{X}$ until the first success, rather than the number of trials required for the first success. Since the number of trials required for the first success is the number of such failures plus one, $X = \tilde{X} + 1$. The probability mass function of $\tilde{X}$ is

$$P(\tilde{X} = k) = P(X - 1 = k) = P(X = k + 1) = (1 - p)^{(k+1)-1}p = (1 - p)^k p,$$

for $k = 0, 1, \ldots$ The expected number of failures until the first success is

$$E[\tilde{X}] = E[X - 1] = E[X] - 1 = \frac{1}{p} - 1 = \frac{1-p}{p}.$$

And the variance is

$$V[\tilde{X}] = V[X - 1] = V[X].$$

This variation of the geometric distribution is the one which R uses in its related commands.

R: GEOMETRIC DISTRIBUTION

The respective R commands for the geometric distribution are

```
> dgeom(k,p)
> pgeom(k,p)
> rgeom(n,p)
```

where k represents the number of failures before the first success.

If you are working with the number of trials required for the first success, as we do in this section, then use the following modifications:

```
> dgeom(k-1,p)
> pgeom(k-1,p)
> rgeom(n,p)+1
```

For example, recall John's issues with the lottery. To find the probability $P(X \leq 365)$ that John will win the lottery next year, type

```
> pgeom(364,0.001)
[1] 0.3059301
```

5.2 NEGATIVE BINOMIAL—UP FROM THE GEOMETRIC

The geometric distribution counts the number of trials until the first success occurs in i.i.d. Bernoulli trials. The *negative binomial distribution* counts the number of trials until the rth success occurs.

NEGATIVE BINOMIAL DISTRIBUTION

A random variable X has the *negative binomial distribution with parameters* r *and* p if

$$P(X = k) = \binom{k-1}{r-1} p^r (1-p)^{k-r}, \quad r = 1, 2, \ldots, \quad k = r, r+1, \ldots \tag{5.7}$$

We write $X \sim \text{NegBin}(r, p)$.

The probability mass function is derived by observing that if k trials are required for the rth success, then (i) the kth trial is a success and (ii) the previous $k - 1$ trials have $r - 1$ successes and $k - r$ failures. The first event (i) occurs with probability p. For (ii), there are $\binom{k-1}{r-1}$ outcomes of $k - 1$ trials with $r - 1$ successes. By independence, each of these outcomes occurs with probability $p^{r-1}(1-p)^{k-r}$.

Observe that the negative binomial distribution reduces to the geometric distribution when $r = 1$.

Using R with the negative binomial distribution poses similar issues as for the geometric distribution. There are several alternate formulations of the negative binomial distribution, and the one that R uses is based on the number of *failures* $\tilde{X}$ until the rth success, where $\tilde{X} = X - r$.

R: NEGATIVE BINOMIAL DISTRIBUTION

The R commands are

```
> dnbinom(k,r,p)
> pnbinom(k,r,p)
> rnbinom(n,r,p)
```

where k represents the number of failures before r successes.

If you are working with the number of trials required for the first r successes, as we do in this section, then use the following modifications.

```
> dnbinom(k-r,r,p)
> pnbinom(k-r,r,p)
> rnbinom(n,r,p)+r
```

Example 5.5 Applicants for a new student internship are accepted with probability $p = 0.15$ independently from person to person. Several hundred people are expected to apply. Find the probability that it will take no more than 100 applicants to find 10 students for the program.

Let X be the number of people who apply for the internship before the 10th student is accepted. Then X has a negative binomial distribution with parameters $r = 10$ and $p = 0.15$. The desired probability is

$$\begin{aligned} P(X \leq 100) &= \sum_{k=10}^{100} P(X = k) \\ &= \sum_{k=10}^{100} \binom{k-1}{r-1} p^r (1-p)^{k-r} \\ &= \sum_{k=10}^{100} \binom{k-1}{9} (0.15)^{10} (0.85)^{k-10} \\ &= 0.945. \end{aligned}$$

In R, type

```
> pnbinom(100-10,10,0.15)
[1] 0.9449054
```

■

Example 5.6 World Series. Baseball's World Series is a best-of-seven playoff between the top teams in the American and National Leagues. The team that wins the series is the one which first wins four games. Thus a series can go for 4, 5, 6, or 7 games. If both teams are evenly matched, what is the expected length of a World Series?

Let X be the number of games played in a World Series. Let Z be a random variable with a negative binomial distribution with parameters $r = 4$ and $p = 1/2$. For each $k = 4, 5, 6, 7$, the probability that team A wins the series in k games is equal to the probability that it takes k games for four "successes," which is $P(Z = k)$. Since either team A or team B could win the series, for $k = 4, 5, 6, 7$,

$$\begin{aligned} P(X = k) &= 2P(Z = k) \\ &= 2\binom{k-1}{4-1}\left(\frac{1}{2}\right)^4\left(\frac{1}{2}\right)^{k-4} = 2\binom{k-1}{3}\left(\frac{1}{2}\right)^k, \end{aligned}$$

which gives

$$P(X = k) = \begin{cases} 0.125, & \text{if } k = 4 \\ 0.25, & \text{if } k = 5 \\ 0.3125, & \text{if } k = 6, 7. \end{cases}$$

The expected length of the World Series is

$$E[X] = 4(0.125) + 5(0.25) + 6(0.3125) + 7(0.3125) = 5.8125 \text{ games.}$$

The actual lengths of the 104 World Series held between 1903 and 2011, not counting the four series which went to eight games because ties were called, is given in Table 5.1.

The average length is

$$\frac{1}{104}\left(4(18) + 5(26) + 6(24) + 7(36)\right) = 5.75 \text{ games.}$$ ■

TABLE 5.1: Lengths of 104 World Series, 1903–2011.

4	5	6	7
18	26	24	36

Expectation and variance. By expressing a negative binomial random variable as a sum of independent geometric random variables we show how to find the expectation and variance of the negative binomial distribution.

Consider an i.i.d. sequence of Bernoulli trials $X_1, X_2, \ldots$ with parameter p. Let G_1 be the number of trials required for the first success. Then $G_1 \sim \text{Geom}(p)$. Suppose it takes $G_1 = g$ trials for the first success. Let G_2 be the number of *additional* trials required for the second success. Then G_2 also has a geometric distribution with parameter p. And G_2 is independent of G_1. The reason is that after trial g, the sequence of random variables $X_{g+1}, X_{g+2}, \ldots$ is also an i.i.d. Bernoulli sequence. The number of trials required for the first success for this sequence is exactly the number of trials required for the second success for the original sequence. We say that the Bernoulli sequence "restarts" itself anew after trial g.

Continuing in this way, for each $k = 1, \ldots, r$, let G_k be the number of additional trials required after the $(k-1)$st success for the kth success to occur. Then $G_1, \ldots, G_r$ is an i.i.d. sequence of geometric random variables with parameter p.

Let X be the number of trials required for the rth success. Then X has a negative binomial distribution with parameters r and p, and $X = G_1 + \cdots + G_r$.

The mean and variance of the geometric distribution are $1/p$ and $(1-p)/p^2$, respectively. It follows that for the negative binomial distribution,

$$E[X] = E[G_1 + \cdots + G_r] = E[G_1] + \cdots + E[G_r] = \frac{r}{p}$$

and

$$V[X] = V[G_1 + \cdots + G_r] = V[G_1] + \cdots + V[G_r] = \frac{r(1-p)}{p^2}.$$

Example 5.7 Sophia is making tomato sauce for dinner and needs 10 ripe tomatoes. In the produce department of the supermarket, there is a 70% chance that a tomato is ripe. (i) How many tomatoes can Sophia expect to sample until she gets what she needs? (ii) What is the standard deviation? (iii) What is the probability Sophia will need to sample more than 15 tomatoes before she gets 10 ripe tomatoes?

Let X be the number of tomatoes that Sophia needs to sample in order to get 10 ripe ones. Then $X \sim \text{NegBin}(10, 0.7)$, with expectation $E[X] = 10/(0.7) = 14.29$ and standard deviation $\text{SD}[X] = \sqrt{10(0.3)/(0.7)^2} = 2.47$.

The desired probability is

$$\begin{aligned} P(X > 15) &= 1 - P(X \leq 15) \\ &= 1 - \sum_{k=10}^{15} \binom{k-1}{9} (0.7)^{10}(0.3)^{k-10} \\ &= 0.278. \end{aligned}$$

In R, type

```
> 1-pnbinom(15-10,10,.7)
[1] 0.2783786
```

■

Why "negative binomial"? You might be wondering what "negative" and "binomial" have to do with the negative binomial distribution. The reason for the choice of words is that the distribution, in a sense, is inverse to the binomial distribution.

Consider an i.i.d. Bernoulli sequence with success probability p. The event (i) that there are r or fewer successes in the first n trials is equal to the event (ii) that the $(r+1)$st success occurs after the nth trial.

Let B be the number of successes in the first n trials. Then B has a binomial distribution with parameters n and p, and the probability of the first event (i) is $P(B \leq r)$. Let Y be the number of trials required for the $(r+1)$st success. Then Y has a negative binomial distribution with parameters $r+1$ and p. The probability of the second event (ii) is $P(Y > n)$. Hence,

$$P(X \leq r) = P(Y > n), \tag{5.8}$$

giving

$$\sum_{k=0}^{r} \binom{n}{k} p^k (1-p)^{n-k} = \sum_{k=n+1}^{\infty} \binom{k-1}{r} p^{r+1}(1-p)^{k-r-1},$$

which is an interesting identity in its own right.

Example 5.8 **Problem of points.** This problem is said to have motivated the beginnings of modern probability through a series of letters between French mathematicians Blaise Pascal and Pierre de Fermat in the 1600s.

Two players are repeatedly playing a game of chance, say tossing a coin. The stakes are 100 francs. If the coin comes up heads, player A gets a point. If it comes up tails, player B gets a point. The players agree that the first person to get a set number of points will win the pot. But the game is interrupted. Player A needs a more points to win; Player B needs b more points. How should the pot be divided?

Pascal and Fermat not only solved the problem, but in their correspondence developed concepts that are fundamental to probability to this day. Their insight, remarkable for its time, was that the division of the stakes should not depend on the history of the game, not on what already took place, but what might have happened if the game were allowed to continue.

By today's standards the solution is fairly simple, especially now that you know about the negative binomial distribution. We generalize the problem from its original formulation and allow the coin they are playing with to have probability p of coming up heads. Suppose the game were to continue. Let X be the number of coin tosses required for A to win a points. Then X has a negative binomial distribution with parameters a and p. Since A needs a points and B needs b points, the game will be decided in $a + b - 1$ plays. Hence A will win the game if and only if $X \leq a + b - 1$. The probability that A wins is

$$P(\text{A wins}) = P(X \leq a + b - 1) = \sum_{k=a}^{a+b-1} \binom{k-1}{a-1} p^a (1-p)^{k-a}.$$

The pot would be divided according to this winning probability. That is, $100 \times P(\text{A wins})$ francs goes to player A and the remainder $100 \times P(\text{B wins})$ goes to player B.

R: PROBLEM OF POINTS

The exact probability that A wins is gotten by typing

```
> pnbinom(b-1,a,p)
```

See the script **Points.R** to play the "problem of points" for your choices of a, b, and p.

For an enjoyable book-length treatment of the history and mathematics of the problem of points see Devlin (2008). ∎

5.3 HYPERGEOMETRIC—SAMPLING WITHOUT REPLACEMENT

Whereas the binomial distribution arises from sampling with replacement, the hypergeometric distribution often arises when sampling is without replacement.

A bag of N balls contains r red balls and $N - r$ blue balls. A sample of n balls is picked. If the sampling is with replacement then the number of red balls in the sample has a binomial distribution with parameters n and $p = r/N$. Here we consider when the sampling is without replacement. Let X be the number of red balls in the sample. Then X has a *hypergeometric distribution.*

The probability mass function of X is obtained by a straightforward counting argument. Consider the event $\{X = k\}$ that there are k red balls in the sample. The number of possible samples of size n with k red balls is $\binom{r}{k}\binom{N-r}{n-k}$. (Choose k reds from the n red balls in the bag; and choose the remaining $n - k$ blues from the $N - r$ blue balls in the bag.) There are $\binom{N}{n}$ possible samples of size n. This gives

$$P(X = k) = \frac{\binom{r}{k}\binom{N-r}{n-k}}{\binom{N}{n}},$$

for $\max(0, n - (N - r)) \le k \le \min(n, r)$. The values of k are restricted by the domain of the binomial coefficients as $0 \le k \le r$ and $0 \le n - k \le N - r$.

The origins of this distribution go back to the 1700s when the word "hypergeometric numbers" was used for what we now call factorials.

R: HYPERGEOMETRIC DISTRIBUTION

The R commands are

```
> dhyper(k,r,N-r, n)
> phyper(k,r,N-r, n)
> rhyper(repeats,k,r,N-r,n)
```

Example 5.9 Independents. Suppose there are 100 political independents in the student body of 1000. A sample of of 50 students is picked. What is the probability there will be six independents in the sample?

The number of independents in the sample I has a hypergeometric distribution with $n = 50$, $r = 100$, and $N = 1000$. The desired probability is

$$P(I = 6) = \frac{\binom{100}{6}\binom{900}{44}}{\binom{1000}{50}} = 0.158.$$

In R, type

```
> dhyper(6,100,900,50)
[1] 0.1579155
```

■

Example 5.10 Inferring independents and maximum likelihood. Continuing from the last example, suppose the number of independents on campus c is not known. A sample of 50 students yields six independents. How can we use these data to infer an estimate of c?

The probability of obtaining six independents in a sample of 50 students when c is unknown is

$$P(I=6) = \frac{\binom{c}{6}\binom{1000-c}{44}}{\binom{1000}{100}}. \tag{5.9}$$

The *maximum likelihood method* in statistics says to estimate c with the value that will maximize this probability. That is, estimate c with the number which gives the highest probability (or "likelihood") of obtaining six independents in a sample of 50 students.

R: MAXIMIZING THE HYPERGEOMETRIC PROBABILITY

The R expression `0:1000` creates a vector of values from 0 to 1000. The command

```
> dhyper(6,0:1000,1000-(0:1000),50)
```

creates a vector of probabilities for $P(I=6)$ at values of c from 0 to 1000. To find the maximum value in that vector use `max()`. To find the index of the vector for where that maximum is located use `which.max()`.

The following commands find the value of c which maximizes the probability Equation 5.9.

```
> clist <- dhyper(6,0:1000,1000-(0:1000),50)
> which.max(clist)-1
[1] 120
```

Note that one is subtracted in the second line because the vector `clist` starts at $c = 0$, rather than 1.

The maximum likelihood estimate is $c = 120$. We infer there are about 120 independents on campus. This estimate is intuitive since $6/50 = 12\%$ of the sample are independents. And 12% of the population size is 120.

As mentioned above, the hypergeometric distribution often arises when sampling is without replacement; the binomial distribution arises when sampling is with replacement. As discussed in Section 3.1, when the population size N is large there is not much difference between sampling with and without replacement. One can show analytically that the hypergeometric probability mass function converges to the binomial pmf as $N \to \infty$. And the binomial distribution serves as a good approximation of the hypergeometric distribution when N is large. See Exercise 5.23, which includes a more precise statement of the limiting process. ∎

Expectation and variance. The expectation of the hypergeometric distribution can be obtained by the method of indicator variables. Let N be the number of balls in the bag, r the number of red balls in the bag, and X the number of red balls in the sample. Define a sequence of 0–1 indicators $I_1, \ldots, I_n$, where

$$I_k = \begin{cases} 1, & \text{if the } k\text{th ball in the sample is red} \\ 0, & \text{if the } k\text{th ball in the sample is not red,} \end{cases}$$

for $k = 1, \ldots, n$. Then $X = I_1 + \cdots + I_n$.

The probability that the kth ball in the sample is red is r/N. This can be shown most simply by a symmetry argument as the kth ball can be any of the N balls with equal probability. (If you need more convincing, use the law of total probability and find the probability that the second ball in the sample is red by conditioning on the first ball.) This gives

$$\begin{aligned} E[X] &= E[I_1 + \cdots + I_n] = E[I_1] + \cdots + E[I_n] \\ &= \sum_{k=1}^{n} P(k\text{th ball in the sample is red}) \\ &= \sum_{k=1}^{n} \frac{r}{N} = \frac{nr}{N}. \end{aligned}$$

The variance of the hypergeometric distribution can also be found by the indicator method. However, the indicators are not independent. The derivation of the variance has similarities to the variance calculation in the matching problem. (See Example 4.34.) Since $V[I_k] = (r/N)(1 - r/N)$, we have

$$\begin{aligned} V[X] &= V[I_1 + \cdots + I_n] \\ &= \sum_{i=1}^{n} V[I_i] + \sum_{i \neq j} \text{Cov}(I_i, I_j) \\ &= n\left(\frac{r}{N}\right)\left(1 - \frac{r}{N}\right) + (n^2 - n)\left(E[I_iI_j] - E[I_i]E[I_j]\right) \\ &= \frac{nr(N-r)}{N^2} + (n^2 - n)\left(E[I_iI_j] - \frac{r^2}{N^2}\right). \end{aligned}$$

For $i \neq j$, the product $I_i I_j$ is the indicator of the event that both the ith and jth balls in the sample are red. Thus

$$E[I_i I_j] = P(I_i = 1, I_j = 1) = P(I_i = 1 | I_j = 1) P(I_j = 1) = \left(\frac{r-1}{N-1}\right)\frac{r}{N},$$

since the sampling is without replacement. This gives

$$\begin{aligned} V[X] &= \frac{nr(N-r)}{N^2} - (n^2 - n)\left(E[I_i I_j] - \frac{r^2}{N^2}\right) \\ &= \frac{nr(N-r)}{N^2} - (n^2 - n)\left(\frac{r(r-1)}{N(N-1)} - \frac{r^2}{N^2}\right). \\ &= \frac{n(N-n)r(N-r)}{N^2(N-1)}. \end{aligned}$$

Example 5.11 In a bridge hand (13 cards from a standard deck), what is the mean and variance of the number of aces?

Let X be the number of aces in a bridge hand. Then X has a hypergeometric distribution. In our balls-in-the-bag imagery, the "bag" is the deck of cards with $N = 52$. The "red balls" are the aces with $r = 4$. The "sample" is the bridge hand with $n = 13$. This gives

$$E[X] = \frac{13(4)}{52} = 1$$

and

$$V[X] = \frac{13(52-13)(4)(52-4)}{52^2(51)} = \frac{12}{17} = 0.706.$$

R: SIMULATING ACES IN A BRIDGE HAND

To simulate, let 1, 2, 3, and 4 represent the aces in a 52-card deck. Type

```
> aces <-replicate(10000,sum(sample(1:52,13)<=4))
> mean(aces)
[1] 1.0034
> var(aces)
[1] 0.7132598
```

■

Example 5.12 **Counting the homeless.** So-called *capture–recapture* methods have been used for many years in ecology, public health, and social science to count rare and elusive populations. In order to estimate the size N of a population, like fish in a lake, researchers "capture" a sample of size r. Subjects are "tagged" and returned to the general population. Researchers then "recapture" a second sample of size n and count the number K which are found tagged.

If the first sample is sufficiently mixed up in the general population then the proportion of those tagged in the second sample should be approximately equal to the proportion tagged in the population. That is,

$$\frac{K}{n} \approx \frac{r}{N}, \text{ so } \frac{nr}{K} \approx N$$

gives an estimate of the population size.

The number tagged in the sample K has a hypergeometric distribution and $E[K] = nr/N$. The approximation formula follows based on $K \approx E[K]$.

Williams (2010) describes such a methodology for estimating the homeless population in Plymouth, a city of a quarter million people in Britain. On a particular day at various social service agency locations, homeless people are identified using several variables including sex and date of birth. At a later date, a sample of homeless people is taken and the number of people originally identified are counted.

Suppose on the first day, 600 homeless persons are counted and identified. At a later date a sample of 800 homeless people is taken, which includes 100 of the original persons. Here $r = 600$, $K = 100$, and $n = 800$. This gives the estimate $N \approx (nr)/K = (800 \times 600)/100 = 4800$ for the size of the homeless population.

Williams points out many difficulties with trying to reliably count the homeless, but states that "For all its shortcomings it remains possibly the most rigorous method of estimation. Some researchers argue that the homeless are a population we cannot reliably count, but this is a doctrine of despair and though capture-recapture is far from a perfect method, like so many other methods its use will undoubtedly lead to technical improvements." ■

5.4 FROM BINOMIAL TO MULTINOMIAL

In a binomial setting, successive trials take one of two possible values (e.g., success or failure). The multinomial distribution is a generalization of the binomial distribution which arises when successive independent trials can take more than two values. The multinomial distribution is used to model such things as:

1. The number of ones, twos, threes, fours, fives, and sixes in 25 dice rolls.
2. The frequencies of r different alleles among n individuals.
3. The number of outcomes of an experiment that has m possible results when repeated n times.
4. The frequencies of six different colors in a sample of 10 candies.

Consider a random experiment repeated independently n times. At each trial, the experiment can assume one of r values. The probability of obtaining the kth value is p_k, for $k = 1, \ldots, r$, with $p_1 + \cdots + p_r = 1$. For each k, let X_k denote the number of times value k occurs. Then $(X_1, \ldots, X_r)$ has a multinomial distribution.

MULTINOMIAL DISTRIBUTION

Suppose $p_1, \ldots, p_r$ are nonnegative numbers such that $p_1 + \cdots + p_r = 1$. Random variables $X_1, \ldots, X_r$ have a *multinomial distribution with parameters* $n, p_1, \ldots, p_r$ if

$$P(X_1 = x_1, \ldots, X_r = x_r) = \frac{n!}{x_1! \cdots x_r!} p_1^{x_1} \cdots p_r^{x_r},$$

for nonnegative integers $x_1, \ldots, x_r$ such that $x_1 + \cdots + x_r = n$.

We write $(X_1, \ldots, X_r) \sim \text{Multi}(n, p_1, \ldots, p_r)$.

Example 5.13 In a bag of candies, colors are distributed according to the probability distribution in Table 5.2.

TABLE 5.2: Distribution of colors in a bag of candies.

Red	Orange	Yellow	Green	Blue	Purple
0.24	0.14	0.16	0.20	0.12	0.14

Denote colors with the letters R, O, Y, G, B, P. In a sample of 10 candies, let $X_R, X_O, X_Y, X_G, X_B, X_P$ denote the number of respective colors in the sample. Then $(X_R, X_O, X_Y, X_G, X_B, X_P)$ has a multinomial distribution with parameters $10, 0.24, 0.14, 0.16, 0.20, 0.12, 0.14$. ■

Deriving the joint probability mass function for the multinomial distribution is similar to the derivation of the binomial pmf. Consider the event

$$\{X_1 = x_1, X_2 = x_2, \ldots, X_r = x_r\}, \quad \text{for } x_1 + x_2 + \cdots + x_r = n. \tag{5.10}$$

Each outcome contained in this event can be represented as a sequence of length n with x_1 elements of type one, x_2 elements of type two, and so on. Count the number of such sequences. Of the n positions in the sequence, there are x_1 positions for the type one elements. Choose these positions in $\binom{n}{x_1}$ ways. Of the remaining $n - x_1$ positions, choose x_2 positions for the type two elements in $\binom{n-x_1}{x_2}$ ways. Continuing in this way gives the number of such sequences as

$$\begin{aligned}
&\binom{n}{x_1}\binom{n-x_1}{x_2}\cdots\binom{n-x_1-\cdots-x_{n-1}}{x_n}\\
&=\left(\frac{n!}{x_1!(n-x_1)!}\right)\left(\frac{(n-x_1)!}{x_2!(n-x_1-x_2)!}\right)\cdots\left(\frac{(n-x_1-\cdots-x_{n-1})!}{x_n!(n-x_1-\ldots-x_n)!}\right)\\
&=\frac{n!}{x_1!x_2!\cdots x_n!},
\end{aligned}$$

where the final simplification for the last equality happens because of the telescoping nature of the previous product.

Having counted the number of sequences in the event $\{X_1, X_2, \ldots, X_r\}$, by independence the probability of each sequence is $p_1^{x_1}p_2^{x_2}\cdots p_r^{x_r}$. This gives the joint probability mass function for the multinomial distribution.

Observe how the multinomial distribution generalizes the binomial distribution. Consider a sequence of n i.i.d. Bernoulli trials with success parameter p. Let X_1 denote the number of successes in n trials. Let X_2 denote the number of failures. Then $X_2 = n - X_1$ and

$$P(X_1 = k) = P(X_1 = k, X_2 = n-k) = \frac{n!}{k!(n-k)!}p^k(1-p)^{n-k}.$$

This shows that $X_1 \sim$ Binom(n, p) is equivalent to $(X_1, X_2) \sim$ Multi$(n, p, 1-p)$, where $X_2 = n - X_1$.

5.4.1 Multinomial Counts

The quantity $n!/(x_1!\cdots x_r!)$ is known as a *multinomial coefficient* and generalizes the binomial coefficient where $r = 2$. It is sometimes written

$$\binom{n}{x_1,\ldots,x_r} = \frac{n!}{x_1!\cdots x_r!}.$$

Multinomial coefficients enumerate sequences of length n in which (i) each element can take one of r possible values and (ii) exactly x_k elements of the sequence take the kth value, for $k = 1, \ldots, r$.

Example 5.14 The *answer* is

$$\binom{11}{4,2,1,4} = \frac{11!}{4!2!1!4!} = 34{,}650.$$

Some possible *questions* are:

1. How many ways can eleven students fill four committees, of respective sizes four, two, one, and four?

2. How many ways can 11 balls be put into four boxes such that the first box gets four balls, the second box gets two balls, the third box gets one ball, and the fourth box gets four balls?
3. How many distinct ways can the letters M-I-S-S-I-S-S-I-P-P-I be permuted?

■

Recall the binomial theorem. Here is the multinomial generalization.

Theorem 5.1. Multinomial theorem. *For any positive integer r and n, and real numbers $z_1, \ldots, z_r$,*

$$(z_1 + \cdots + z_r)^n = \sum_{a_1+\cdots+a_r=n} \frac{n!}{a_1!a_2!\cdots a_r!} z_1^{a_1} z_2^{a_2} \cdots z_r^{a_r},$$

where the sum is over all lists of r nonnegative integers $(a_1, \ldots, a_r)$ which sum to n.

Proof. The proof is analogous to that given for the Binomial Theorem.

It will be instructive for the reader to work through this identity for the case $r = 3$, $z_1 = z_2 = z_3 = 1$, and $n = 3$.

Verifying that the multinomial probability mass function sums to one is an application of the multinomial theorem as

$$\sum_{x_1+\cdots+x_r=n} P(X_1 = x_1, \ldots, X_r = x_r) = \sum_{x_1+\cdots+x_r=n} \frac{n!}{x_1!\cdots x_r!} p_1^{x_1} \cdots p_r^{x_r}$$
$$= (p_1 + \cdots + p_r)^n = 1^n = 1.$$

Example 5.15 In a parliamentary election, it is estimated that parties A, B, C, and D will receive 20, 25, 30, and 25% of the vote, respectively. In a sample of 10 voters, find the probability that there will be two supporters each of parties A and B, and three supporters each of parties C and D.

Let X_A, X_B, X_C, X_D denote the number of supporters in the sample of parties A, B, C, and D, respectively. Then

$$(X_A, X_B, X_C, X_D) \sim \text{Multi}(10, 0.20, 0.25, 0.30, 0.25).$$

The desired probability is

$$P(X_A = 2, X_B = 2, X_C = 3, X_D = 3)$$
$$= \frac{10!}{2!2!3!3!}(0.20)^2(0.25)^2(0.30)^3(0.25)^3 = 0.0266.$$

R: MULTINOMIAL CALCULATION

The R commands for working with the multinomial distribution are `rmultinom` and `dmultinom`. The desired probability above is found by typing

```
> dmultinom(c(2,2,3,3),prob=c(0.20,0.25,0.30,0.25))
[1] 0.02657813
```

■

Example 5.16 Genetics. The Hardy–Weinberg principle in genetics states that the long-term gene frequencies in a population remain constant. An allele is a form of a gene. Suppose an allele takes one of two forms A and a. For a particular biological trait you receive two alleles, one from each parent. The *genotypes* are the possible genetic makeups of a trait: AA, Aa, and aa.

For instance, fruit flies contain a gene for body color: A is the allele for black, and a for brown. Genotypes AA and Aa correspond to black body color; and aa corresponds to brown. We say black is dominant and brown is recessive.

Suppose an allele takes the form A with probability p, and a with probability $1 - p$. The Hardy–Weinberg principle asserts that the proportion of individuals with genotype AA, Aa, and aa should occur in a population according to the frequencies p^2, $2p(1 - p)$, and $(1 - p)^2$, respectively. The frequency for Aa, for instance, is because you can get an A allele from your mother (with probability p) and an a allele from your father (with probability $1 - p$), or vice versa.

In a sample of n fruit flies, let (X_1, X_2, X_3) be the number of flies with genotypes AA, Aa, and aa, respectively. Then

$$(X_1, X_2, X_3) \sim \text{Multi}\left(n, p^2, 2p(1-p), (1-p)^2\right).$$

Suppose that allele A occurs 60% of the time. What is the probability, in a sample of six fruit flies, that AA occurs twice, Aa occurs three times, and aa occurs once?

$$\begin{aligned} P(X_1 = 2, X_2 = 3, X_3 = 1) & \\ &= \frac{6!}{2!3!1!}\left((0.60)^2\right)^2 (2(0.60)(0.40))^3 \left((0.40)^2\right)^1 = 0.1376. \end{aligned}$$

A common problem in statistical genetics is to estimate the allele probability p from a sample of n individuals, where the data consist of observed genotype frequencies. For instance, suppose in a sample of 60 fruit flies we observe the gene distribution given in Table 5.3

TABLE 5.3: Genotype frequencies for a sample of 60 fruit flies.

AA	Aa	aa
35	17	8

If p is unknown,

$$\begin{aligned} P(X_1 = 35, X_2 = 17, X_3 = 8) \\ &= \frac{60!}{35!17!8!}(p^2)^{35}(2p(1-p))^{17}((1-p)^2)^8. \\ &= \text{constant} \times p^{87}(1-p)^{33}. \end{aligned} \tag{5.11}$$

The maximum likelihood principle, introduced in Example 5.10, says to estimate p with the value that maximizes the probability of obtaining the observed data, that is, the value of p that maximizes the probability given in Equation 5.11.

Differentiating this expression and setting it equal to 0 gives the equation

$$87p^{86}(1-p)^{33} = 33(1-p)^{32}p^{87}.$$

Solving for p gives the maximum likelihood estimate $p = 87/120 = 0.725$. ■

Marginal distribution, expectation, variance. Let $(X_1, \ldots, X_r)$ have a multinomial distribution with parameters $n, p_1, \ldots, p_r$. The X_i's are not independent as they are constrained by $X_1 + \cdots + X_r = n$.

For each $k = 1, \ldots, r$, the marginal distribution of X_k is binomial with parameters n and p_k. We will derive this result two ways: easy and hard.

The easy way is a consequence of a simple probabilistic argument. Think of each of the underlying n independent trials as either resulting in outcome k or not. Then X_k counts the number of "successes" in n trials, which gives the binomial result $X_k \sim \text{Binom}(n, p_k)$.

The hard way is an exercise in working sums. To find a marginal distribution sum the joint probability mass function. Consider the marginal distribution of X_1. Summing over the outcomes of $X_2, \ldots, X_r$ gives

$$\begin{aligned} P(X_1 = x_1) &= \sum_{x_2, \ldots, x_r} P(X_1 = x_1, X_2 = x_2, \ldots, X_r = x_r) \\ &= \sum_{x_2 + \cdots + x_r = n - x_1} \frac{n!}{x_1! x_2! \cdots x_r!} p_1^{x_1} p_2^{x_2} \cdots p_r^{x_r} \\ &= \frac{n!}{x_1!} p_1^{x_1} \sum_{x_2 + \cdots x_r = n - x_1} \frac{1}{x_2! \cdots x_r!} p_2^{x_2} \cdots p_r^{x_r} \end{aligned}$$

$$= \frac{n!}{x_1!(n-x_1)!} p_1^{x_1} \sum_{x_2+\cdots x_r = n-x_1} \frac{(n-x_1)!}{x_2!\cdots x_r!} p_2^{x_2}\cdots p_r^{x_r}$$

$$= \frac{n!}{x_1!(n-x_1)!} p_1^{x_1} (p_2+\cdots+p_r)^{n-x_1}$$

$$= \frac{n!}{x_1!(n-x_1)!} p_1^{x_1} (1-p_1)^{n-x_1},$$

for $0 \le x_1 \le n$. The sum is over all nonnegative integers $x_2, \ldots, x_n$ which sum to $n - x_1$. The penultimate equality is because of the multinomial theorem.

We see that X_1 has a binomial distribution with parameters n and p_1. Similarly, for each $k = 1, \ldots, r$, $X_k \sim \text{Binom}(n, p_k)$.

The marginal result gives

$$E[X_k] = np_k \quad \text{and} \quad V[X_k] = np_k(1-p_k).$$

Covariance. For $i \neq j$, consider $\text{Cov}(X_i, X_j)$. Use indicators and write $X_i = I_1 + \cdots + I_n$, where

$$I_k = \begin{cases} 1, & \text{if the } k\text{th trial results in outcome } i \\ 0, & \text{otherwise,} \end{cases}$$

for $k = 1, \ldots, n$. Similarly, write $X_j = J_1 + \cdots + J_n$, where

$$J_k = \begin{cases} 1, & \text{if the } k\text{th trial results in outcome } j \\ 0, & \text{otherwise,} \end{cases}$$

for $k = 1, \ldots, n$.

Because of the independence of the Bernoulli trials, pairs of indicator variables involving different trials are independent. That is, I_g and J_h are independent if $g \neq h$, and hence $\text{Cov}(I_g, J_h) = 0$. On the other hand, if $g = h$, then $I_g J_g = 0$, since the gth trial cannot result in both outcomes i and j. Thus

$$\text{Cov}(I_g, J_g) = E[I_g J_g] - E[I_g]E[J_g] = -p_i p_j.$$

Using the linearity properties of covariance (see Section 4.9),

$$\text{Cov}(X_i, X_j) = \text{Cov}\left(\sum_{g=1}^{n} I_g, \sum_{h=1}^{n} J_h\right) = \sum_{g=1}^{n}\sum_{h=1}^{n} \text{Cov}(I_g, J_h)$$

$$= \sum_{g=1}^{n} \text{Cov}(I_g, J_g) = \sum_{g=1}^{n}(-p_i p_j) = -np_i p_j.$$

5.5 BENFORD'S LAW

> It has been observed that the pages of a much used table of common logarithms show evidences of a selective use of the natural numbers. The pages containing the logarithms of the low numbers 1 and 2 are apt to be more stained and frayed by use than those of the higher numbers 8 and 9. Of course, no one could be expected to be greatly interested in the condition of a table of logarithms, but the matter may be considered more worthy of study when we recall that the table is used in the building up of our scientific, engineering, and general factual literature. There may be, in the relative cleanliness of the pages of a logarithm table, data on how we think and how we react when dealing with things that can be described by means of numbers.
>
> —Frank Benford, *The Law of Anomalous Numbers*

In this book's Introduction we describe Benford's law and suggest the following classroom activity: Pick a random book in your backpack. Open up to a random page. Let your eyes fall on a random number in the middle of the page. Write down the number and circle the first digit (ignore zeros).

We collect everybody's first digits and before looking at the data ask students to guess what the distribution of these numbers looks like. Many say that they will be roughly equally distributed between 1 and 9, following a discrete uniform distribution. Remarkably they are not. Ones are most likely, then twos, which are more likely than threes, etc. They tend to follow Benford's law, the distribution given in Table 5.4.

Benford's law is named after physicist Frank Benford who was curious about the wear and tear of the large books of logarithms which were widely used for scientific calculations before computers or calculators. Benford eventually looked at 20,229 data sets—everything from areas of rivers to death rates. In all cases, the first digit was one about 30% of the time, and the digits followed a remarkably similar pattern.

Benford eventually discovered the formula

$$P(d) = \log_{10}\left(\frac{d+1}{d}\right),$$

for the probability that the first digit is d. Observe that P is in fact a probability distribution on $\{1, \ldots, 9\}$, as

$$\begin{aligned}
\sum_{k=1}^{9} P(d) &= \sum_{k=1}^{9} \log_{10}\left(\frac{d+1}{d}\right)\\
&= \sum_{k=1}^{9} [\log_{10}(d+1) - \log_{10} d]\\
&= (\log_{10} 2 - \log_{10} 1) + (\log_{10} 3 - \log_{10} 2) + \cdots + (\log_{10} 10 - \log_{10} 9)\\
&= \log_{10} 10 - \log_{10} 1 = 1,
\end{aligned}$$

as the terms are telescoping.

TABLE 5.4: Benford's law.

1	2	3	4	5	6	7	8	9
0.301	0.176	0.125	0.097	0.079	0.067	0.058	0.051	0.046

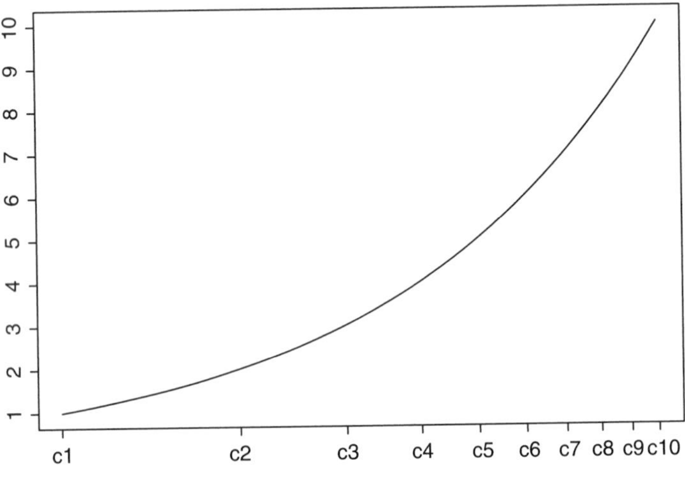

FIGURE 5.1: Graph of $P(n) = ar^n$.

It is not easy to explain why Benford's law works for so many empirical datasets, and much of the mathematics is outside the scope of this book.

Ross (2011) shows why phenomenon that exhibit exponential growth or decline—populations of cities, powers of 2, pollution levels, radioactive decay—exhibit Benford's law. As explained by Ross, suppose we have data of the form $P(n) = ar^n$ for constants a and r. For instance, $P(n)$ might be the population of a city at year n, or the half-life of a radioactive isotope after n minutes. Consider values of n such that $1 \leq P(n) < 10$. For each $d = 1, \ldots, 10$, let $[c_d, c_{d+1})$ be the interval of values for which the first digit of $P(n)$ is d (see Fig. 5.1).

For all values $c_1 \leq n < c_2$, the first digit of $P(n)$ is one. For $c_2 \leq n < c_3$, the first digit of $P(n)$ is two, and so on. For each $d = 1, \ldots, 10$, $P(c_d) = ar^{c_d} = d$. Taking logarithms (base 10) gives

$$\log a + c_d \log r = \log d.$$

Also $\log a + c_{d+1} \log r = \log(d+1)$. Subtracting equations gives

$$(c_{d+1} - c_d)(\log r) = \log(d+1) - \log d.$$

Thus the length of each interval $[c_d, c_{d+1})$ is

$$c_{d+1} - c_d = \frac{\log(d+1) - \log d}{\log r}.$$

The length of the entire interval $[c_1, c_{10})$ is

$$\sum_{d=1}^{9}(c_{d+1} - c_d) = \frac{\log 10 - \log 1}{\log r} = \frac{1}{\log r}.$$

Thus,

$$\frac{c_{d+1} - c_d}{c_{10} - c_1} = \log(d+1) - \log d = \log\left(\frac{d+1}{d}\right)$$

is the "fraction of the time" that the first digit of ar^n is d.

The argument here for values of $P(n)$ between 1 and 10 actually applies to all intervals of values between 10^k and 10^{k+1}, for $k = 0, 1, \ldots$

For a wonderful introduction to Benford's law, and its use in detecting fraud, hear the Radiolab podcast *From Benford to Erdős* at `www.radiolab.org`.

5.6 SUMMARY

This chapter explores several new discrete probability distributions. Many of them are generalizations or extensions of the ones we have studied. The geometric distribution arises as the number of i.i.d. Bernoulli trials required for the first success to occur. The negative binomial generalizes the geometric distribution and counts the number of trials required for the rth first success. The hypergeometric distribution arises when sampling is done without replacement, in contrast to the binomial distribution which arises when sampling is done with replacement. The multinomial distribution is a multivariate distribution which generalizes the binomial distribution. Also introduced are multinomial coefficients and the multinomial theorem. Finally, Benford's law is a fascinating distribution which governs the distribution of first digits for many empirical data sets.

- **Geometric distribution:** A random variable X has a geometric distribution with parameter p, if $P(X = k) = (1-p)^{k-1}p,\;$ for $k = 1, 2, \ldots$. The distribution arises as the number of trials required for the first success in repeated i.i.d. Bernoulli trials.
- **Memorylessness:** A random variable X has the memorylessness property if $P(X > s+t)|X > t) = P(X > s)$ for all $s, t > 0$. The geometric distribution is the only discrete distribution that is memoryless.
- **Properties of geometric distribution:**
  1. $E[X] = 1/p$.
  2. $V[X] = (1-p)/p^2$.
  3. $P(X > x) = (1-p)^x$, for $x > 0$.

- **Negative binomial distribution:** A random variable X has a negative binomial distribution with parameters r and p if

$$P(X = k) = \binom{k-1}{r-1} p^r (1-p)^{k-r}, \quad \text{for } r = 1, 2, \ldots, \; k = r, r+1, \ldots.$$

The distribution arises as the number of trials required for the rth success in repeated i.i.d. Bernoulli trials. If $r = 1$ we get the geometric distribution.

- **Properties of negative binomial distribution:**
  1. $E[X] = r/p$.
  2. $V[X] = r(1-p)/p^2$.

Hypergeometric distribution: Given a bag of N balls which contains r red balls and $N - r$ blue balls, let X be the number of red balls in a sample of size n taken without replacement. Then X has a hypergeometric distribution with pmf

$$P(X = k) = \frac{\binom{r}{k}\binom{N-r}{n-k}}{\binom{N}{n}}$$

for $\max(0, n - (N - r)) \leq k \leq \min(n, r)$.

- **Properties of hypergeometric distribution:**
  1. $E[X] = nr/N$.
  2. $V[X] = n(N-n)r(N-r)/(N^2(N-1))$.
- **Multinomial distribution:** Random variables $X_1, \ldots, X_r$ have a multinomial distribution with parameters $n, p_1, \ldots, p_r$, where $p_1 + \cdots + p_r = 1$, if

$$P(X_1 = x_1, \ldots, X_r = x_r) = \frac{n!}{x_1! \cdots x_r!} p_1^{x_1} \cdots p_r^{x_r},$$

for nonnegative integers $x_1, \ldots, x_r$ such that $x_1 + \cdots + x_r = n$. The distribution generalizes the binomial distribution (with $r = 2$) and arises when successive independent trials can take more than two values.

- **Multinomial counts:** The multinomial coefficient

$$\binom{n}{x_1, \ldots, x_r} = \frac{n!}{x_1! \cdots x_r!}$$

counts the number of n element sequences in which (i) each element can take one of r possible values and (ii) exactly x_k elements of the sequence take the kth value, for $k = 1, \ldots, r$.

- **Multinomial theorem:** For any positive integer r and n and real numbers $z_1, \ldots, z_r$,

$$(z_1 + \cdots + z_r)^n = \sum_{a_1+\cdots+a_r=n} \frac{n!}{a_1! \cdots a_r!} z_1^{a_1} \cdots z_r^{a_r},$$

where the sum is over all lists of r nonnegative integers $(a_1, \ldots, a_r)$ which sum to n.

- **Properties of multinomial distribution:**
  1. $X_k \sim \text{Binom}(n, p_k)$, for each $k = 1, \ldots, r$.
  2. $\text{Cov}(X_i, X_j) = -np_ip_j$.
- **Benford's law:** A random variable X has the Benford's law distribution if $P(X = d) = \log_{10}((d+1)/d)$, for $d = 1, \ldots 9$. The distribution arises as the distribution of the first digit for many datasets.

EXERCISES

Geometric distribution

5.1 Tom is playing poker with his friends. What is the expected number of hands it will take before he gets a full house?

5.2 What is the probability that it takes an even number of die rolls to get a four?

5.3 Find the variance of a geometric distribution. Hint: To find $E[X^2]$, write $k^2 = k^2 - k + k = k(k-1) + k$. You will need to take two derivatives.

5.4 A manufacturing process produces components which have a 1% chance of being defective. Successive components are independent.

(a) Find the probability that it takes exactly 110 components to be produced before a defective one occurs.

(b) Find the probability that it takes at least 110 components to be produced before a defective one occurs.

(c) What is the expected number of components that will be produced before a defective one occurs?

5.5 Danny is applying to college and sending out many applications. He estimates there is a 25% chance that an application will be successful. How many applications should he send out so that the probability of at least one acceptance is at least 95%?

5.6 There are 15 professors in the math department. Every time Tina takes a math class each professor is equally likely to be the instructor. What is the expected number of math classes which Tina needs to take in order to be taught by every math professor?

5.7 In the coupon collector's problem, let X be the number of draws of n coupons required to obtain a complete set. Find the variance of X.

5.8 A bag has r red and b blue balls. Balls are picked at random without replacement. Let X be the number of selections required for the first red ball to be picked.

(a) Explain why X does not have a geometric distribution.

(b) Show that the probability mass function of X is

$$P(X = k) = \frac{\binom{r+b-k}{r-1}}{\binom{r+b}{r}} \quad \text{for } k = 1, 2, \ldots, b+1.$$

5.9 Let $X \sim \text{Geom}(p)$. Find $E\left[2^X\right]$ for those values of p for which the expectation exists.

5.10 Make up your own example to show that the Poisson distribution is *not* memoryless. That is, pick values for λ, s, and t and show that $P(X > t | X > s) \neq P(X > t - s)$.

Negative binomial distribution

5.11 A fair coin is tossed until heads appears four times.

(a) Find the probability that it took exactly 10 flips.

(b) Find the probability that it took at least 10 flips.

(c) Let Y be the number of tails that occur. Find the pmf of Y.

5.12 Baseball teams A and B face each other in the World Series. For each game, the probability that A wins the game is p, independent of other games. Find the expected length of the series.

5.13 Let X and Y be independent geometric random variables with parameter p. Find the pmf of $X + Y$.

5.14 (i) In a bridge hand of 13 cards, what is the probability of being dealt exactly two red cards? (ii) In a bridge hand of 13 cards, what is the probability of being dealt four spades, four clubs, three hearts and two diamonds?

5.15 People whose blood type is O-negative are universal donors—anyone can receive a blood transfusion of O-negative blood. In the U.S., 7.2% of the people have O-negative blood. A blood donor clinic wants to find 10 O-negative individuals. In repeated screening, what is the chance of finding such individuals among the first 100 people screened?

5.16 Using the relationship in Equation 5.8 between the binomial and negative binomial distributions, recast the solution to the Problem of Points in terms of the binomial distribution. Give the exact solution in terms of a binomial probability.

5.17 Andy and Beth are playing a game worth \$100. They take turns flipping a penny. The first person to get 10 heads will win. But they just realized that they have to be in math class right away and are forced to stop the game. Andy had four heads and Beth had seven heads. How should they divide the pot?

5.18 **Banach's matchbox problem.** This famous problem was posed by mathematician Hugo Steinhaus as an affectionate honor to fellow mathematician Stefan Banach, who was a heavy pipe smoker. A smoker has two matchboxes, one in each pocket. Each box has n matches in it. Whenever the smoker needs a match he reaches into a pocket at random and takes a match from the box. Suppose he reaches into a pocket and finds that the matchbox is empty. Find the probability that the box in the other pocket has exactly k matches.

(a) Let X be the number of matches in the right box when the left box is found empty. Show that X has a negative binomial distribution with parameters $n+1$ and 1/2.

(b) Show that the desired probability is $2 \times P(X = 2n + 1 - k)$.

(c) Work out the problem in detail for the case $n = 1$.

5.19 Let $R \sim \text{Geom}(p)$. Conditional on $R = r$, suppose X has a negative binomial distribution with parameters r and p. Show that the marginal distribution of X is geometric. What is the parameter?

Hypergeometric distribution

5.20 There are 500 deer in a wildlife preserve. A sample of 50 deer are caught and tagged and returned to the population. Suppose that 20 deer are caught.

(a) Find the mean and standard deviation of the number of tagged deer in the sample.

(b) Find the probability that the sample contains at least three tagged deer.

5.21 **The Lady Tasting Tea.** This is one of the most famous experiments in the founding history of statistics. In his 1935 book *The Design of Experiments* (1935), Sir Ronald A. Fisher writes,

> A Lady declares that by tasting a cup of tea made with milk she can discriminate whether the milk or the tea infusion was first added to the cup. We will consider the problem of designing an experiment by means of which this assertion can be tested ... Our experiment consists in mixing eight cups of tea, four in one way and four in the other, and presenting them to the subject for judgment in a random order. ... Her task is to divide the 8 cups into two sets of 4, agreeing, if possible, with the treatments received.

Consider such an experiment. Four cups are poured milk first and four cups are poured tea first and presented to a friend for tasting. Let X be the number of milk-first cups that your friend correctly identifies as milk-first.

(a) Identify the distribution of X.

(b) Find $P(X = k)$ for $k = 0, 1, 2, 3, 4$.

(c) If in reality your friend had no ability to discriminate and actually guessed, what is the probability they would correctly identify all four cups correctly?

5.22 In a town of 20,000, there are 12,000 voters, of whom 6000 are registered democrats and 5000 are registered republicans. An exit poll is taken of 200 voters. Assume all registered voters actually voted. Use R to find: (i) the mean and standard deviation of the number of democrats in the sample; (ii) the probability that more than half the sample are republicans.

5.23 Consider the hypergeometric distribution with parameters r, n, and N. Suppose r depends on N in such as way that $r/N \to p$ as $N \to \infty$, where $0 < p < 1$. Show that the mean and variance of the hypergeometric distribution converges to the mean and variance, respectively, of a binomial distribution with parameters n and p, as $N \to \infty$.

Multinomial distribution

5.24 A Halloween bag contains three red, four green, and five blue candies. Tom reaches in the bag and takes out three candies. Let R, G, and B denote the number of red, green, and blue candies, respectively, that Tom got.

(a) Is the distribution of (R, G, B) multinomial? Explain.

(b) What is the probability that Tom gets one of each color?

5.25 In a city of 100,000 voters, 40% are Democrat, 30% Republican, 20% Green, and 10% Undecided. A sample of 1000 people is selected.

(a) What is the expectation and variance for the number of Greens in the sample?

(b) What is the expectation and variance for the number of Greens and Undecideds in the sample?

5.26 A random experiment takes r possible values, with respective probabilities $p_1, \ldots, p_r$. Suppose the experiment is repeated N times, where N has a Poisson distribution with parameter λ. For $k = 1, \ldots, r$, let N_k be the number of occurrences of outcome k. In other words, if $N = n$, then $(N_1, \ldots, N_r)$ has a multinomial distribution.

Show that $N_1, \ldots, N_r$ form a sequence of independent Poisson random variables and for each $k = 1, \ldots, r$, $N_k \sim \text{Pois}(\lambda p_k)$.

5.27 Suppose a gene allele takes two forms A and a, with $P(A) = 0.20 = 1 - P(a)$. Assume a population is in Hardy–Weinberg equilibrium.

(a) Find the probability that in a sample of eight individuals, there is one AA, two $Aa's$, and five aa's.

(b) Find the probability that there are at least seven aa's.

5.28 In 10 rolls of a fair die, let X be the number of fives rolled, let Y be the number of even numbers rolled, and let Z be the number of odd numbers rolled.

(a) Find $\text{Cov}(X, Y)$.

(b) Find $\text{Cov}(X, Z)$.

5.29 Prove the identity

$$4^n = \sum_{a_1+\cdots a_4=n} \frac{n!}{a_1!a_2!a_3!a_4!},$$

where the sum is over all nonnegative integers that sum to 4.

Benford's law

5.30 Find the expectation and variance of the Benford's law distribution.

5.31 Find a real-life dataset to test whether Benford's law applies.

Other

For the next five problems 5.32 to 5.34, identify a random variable and describe its distribution before doing any computations. (For instance, for Problem 5.32 start your solution with "Let X be the number of days when there is no homework. Then X has a binomial distribution with $n = 30$ and $p = \ldots$.")

5.32 A professor starts each class by picking a number from a hat that contains the numbers 1–30. If a prime number is chosen, there is no homework that day. There are 42 class periods in the semester. How many days can the students expect to have no homework?

5.33 Suppose eight cards are drawn from a standard deck *with replacement*. What is the probability of obtaining two cards from each suit?

5.34 Among 30 raffle tickets six are winners. Felicia buys 10 tickets. Find the probability that she got three winners.

5.35 A teacher writes an exam with 20 problems. There is a 5% chance that any problem has a mistake. The teacher tells the class that if the exam has three or more problems with mistakes he will give everyone an A. The teacher repeats this in 10 different classes. Find the probability that the teacher gave out all A's at least once.

Simulation and R

5.36 Conduct a study to determine how well the binomial distribution approximates the hypergeometric distribution. Consider a bag with n balls, 25% of which are red. A sample of size $(0.10)n$ is taken. Let X be the number of red balls in the sample. Find $P(X \leq (0.02)n)$ for increasing values of n when sampling is (i) with replacement and (ii) without replacement. Use R.

5.37 Write a function `coupon(n)` for simulating the coupon collector's problem. That is, let X be the number of draws required to obtain all n items when sampling with replacement. Use your function to simulate the mean and standard deviation of X for $n = 10$ and $n = 52$.

5.38 Let $p = (p_1, \ldots, p_n)$ be a list of probabilities with $p_1 + \cdots + p_n = 1$. Write a function `coupon(n, p)` which generalizes the function, above, and simulates the coupon collector's problem for unequal probabilities, where the probability of choosing item i is p_i. For $n = 52$, let p be a list of binomial probabilities with parameters 52 and 1/2. Use your function to simulate the mean and standard deviation of X.

5.39 Read about the World Series in Example 5.1. Suppose the World Series is played between two teams A and B such that for any matchup between A and B, the probability that A wins is $0 < p < 1$. For $p = 0.25$ and $p = 0.60$ simulate the expected length and standard deviation of the series.

6

CONTINUOUS PROBABILITY

Objective experience is discrete, but the world hangs together by the grace of continuity.

—A.W. Holt

Shooting an arrow at a target and picking a real number between 0 and 1 are examples of random experiments where the sample space is a continuum of values. Such sets have no gaps between elements; they are not discrete. The elements are uncountable and cannot be listed. We call such sample spaces *continuous*. The most common continuous sample spaces in one dimension are intervals such as (a, b), $(-\infty, c]$, and $(-\infty, \infty)$.

An archer shoots an arrow at a target. The target C is a circle of radius 1. The bullseye B is a smaller circle in the center of the target of radius $1/4$. What is the probability $P(B)$ of hitting the bullseye?

Our probability model will be *uniform* on the target—all points are equally likely. But how to make sense of that?

We start with some intuitive discussion. The technical details will come later. If this were a discrete finite problem, the solution would be to count the number of points in the bullseye and divide by the number of points in the target. But the sets are uncountable. The way to "count" points in continuous sets is by integration. Since all points on the target are equally likely we will integrate in such a way so that no point gets more "weight" than any other—all points are treated equal. That is, integrate a constant c over the bullseye region. This leads to

$$P(B) = \iint_B c \, dx \, dy = c\,[\text{Area}\,(B)] = \frac{c\pi}{16}. \tag{6.1}$$

Probability: With Applications and R, First Edition. Robert P. Dobrow.

What is c? Since this is a probability model, the points in the sample space should add up (oops, I mean integrate) to 1. The sample space is C, the target. This gives

$$1 = P(C) = \iint_C c\,dx\,dy = c\,[\text{Area}\,(C)] = c\pi$$

and thus $c = 1/\pi$. Plugging in c to Equation 6.1 gives

$$P(B) = \frac{1}{\text{Area}\,(C)} \iint_B dx\,dy = \frac{1}{16},$$

the proportion of the total area of the target taken up by the bullseye.

This gives the beginnings of a continuous uniform probability model. If Ω is a continuous set with all points equally likely, then for subsets $S \subseteq \Omega$,

$$P(S) = \frac{1}{\text{Area}\,(\Omega)} \iint_S dx\,dy = \frac{\text{Area}\,(S)}{\text{Area}\,(\Omega)}.$$

In one dimension, the double integral becomes a single integral and area becomes length. In three dimensions we have a triple integral and volume.

The following examples are all meant to be approached intuitively.

Example 6.1

- A real number is picked uniformly at random from the interval $(-5, 2)$. The probability the number is positive is

$$\frac{\text{Length}\,(0,2)}{\text{Length}\,(-5,2)} = \frac{2}{7}.$$

- A sphere of radius 1 is inscribed in a cube with side length 2. A point in the cube is picked uniformly at random. The probability that the point is contained in the sphere is

$$\frac{\text{Volume (Sphere)}}{\text{Volume (Cube)}} = \frac{4\pi/3}{8} = \frac{\pi}{6}.$$

- A student arrives to class at a uniformly random time between 8:45 and 9:05 a.m. Class starts at 9:00 a.m. The probability that the student arrives on time is

$$\frac{15 \text{ minutes}}{20 \text{ minutes}} = \frac{3}{4}.$$

■

All of these examples can be cast in terms of random variables. A variable X might be a point on a target or a time in a continuous interval. But to work with random variables in the continuous world will require some new mathematical tools.

6.1 PROBABILITY DENSITY FUNCTION

A continuous random variable X is a random variable that takes values in a continuous set. If S is a subset of the real numbers, then $\{X \in S\}$ is the event that X takes values in S. For instance, $\{X \in (a, b)\} = \{a < X < b\}$, and $\{X \in (-\infty, c]\} = \{X \leq c\}$.

In the discrete setting, to compute $P(X \in S)$ add up values of the probability mass function. That is, $P(X \in S) = \sum_{x \in S} P(X = x)$.

If X, however, is a continuous random variable, to compute $P(X \in S)$ we integrate the *probability density function (pdf)* over S. The probability density function plays the role of the pmf. It is the function used to compute probabilities. Observe the similarities between the pdf for continuous random variables and the probability mass function.

PROBABILITY DENSITY FUNCTION

Let X be a continuous random variable. A function f is a *probability density function* of X if

1. $f(x) \geq 0$, for all $-\infty < x < \infty$.

2. $$\int_{-\infty}^{\infty} f(x)\,dx = 1.$$

3. For $S \subseteq \mathbb{R}$,

$$P(X \in S) = \int_S f(x)\,dx. \tag{6.2}$$

When writing density functions we sometimes use subscripts, e.g., $f_X(x) = f(x)$, to identify the associated random variable. The integral $\int_S f(x)\,dx$ is taken over the set of values in S. For instance, if $S = (a, b)$, then

$$P(X \in S) = P(a < X < b) \text{ and } \int_S f(x)\,dx = \int_a^b f(x)\,dx.$$

If $S = (-\infty, c]$, then

$$P(X \in S) = P(X \leq c) = \int_{-\infty}^{c} f(x)\, dx.$$

For any real number a,

$$P(X = a) = P(X \in \{a\}) = \int_{a}^{a} f(x)\, dx = 0.$$

For a continuous random variable, the probability of any particular number occurring is 0. Non-zero probabilities are assigned to intervals, not to individual or discrete outcomes.

Since $[a, b) = (a, b) \cup \{a\}$ and $P(X = a) = 0$ it follows that

$$P(X \in [a, b)) = P(a \leq X < b) = P(a < X < b).$$

Similarly,

$$P(a < X < b) = P(a \leq X < b) = P(a < X \leq b) = P(a \leq X \leq b).$$

Example 6.2 A random variable X has density function $f(x) = 3x^2/16$, for $-2 < x < 2$, and 0, otherwise. Find $P(X > 1)$.

We have

$$P(X > 1) = \int_{1}^{2} f(x)\, dx = \int_{1}^{2} \frac{3x^2}{16}\, dx = \frac{1}{16} \left(x^3\right)\Big|_1^2 = \frac{7}{16}.$$

The density function is shown in Figure 6.1. ■

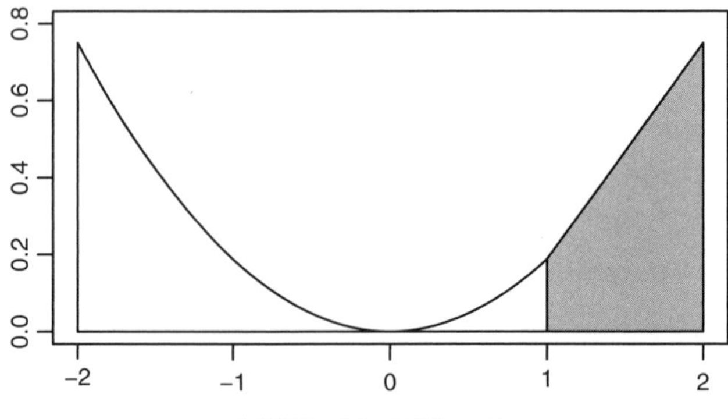

FIGURE 6.1: $P(X > 1)$.

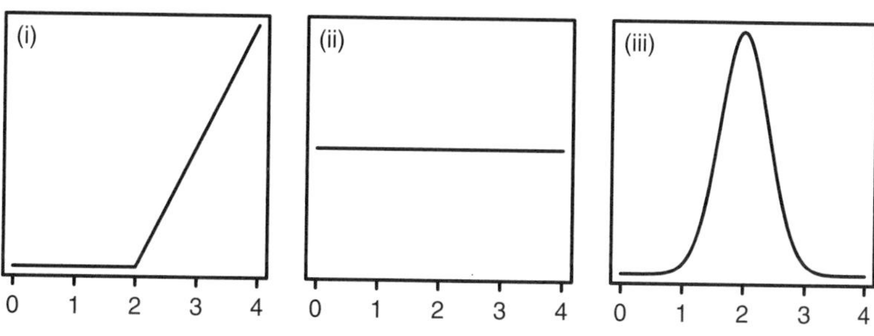

FIGURE 6.2: Three density shapes.

Example 6.3 A random variable X has density function of the form $f(x) = ce^{-|x|}$, for all x. (i) Find c. (ii) Find $P(0 < X < 1)$.

(i) Solve

$$\begin{aligned} 1 &= \int_{-\infty}^{\infty} ce^{-|x|}\,dx = \int_{-\infty}^{0} ce^{x}\,dx + \int_{0}^{\infty} ce^{-x}\,dx \\ &= 2c\int_{0}^{\infty} e^{-x}\,dx = 2c\left(-e^{-x}\right)\Big|_0^{\infty} = 2c, \end{aligned}$$

giving $c = 1/2$.

$$\text{(ii) } P(0 < X < 1) = \int_0^1 \frac{e^{-|x|}}{2}\,dx = \frac{1}{2}\left(-e^{-x}\right)\Big|_0^1 = \frac{1-e^{-1}}{2} = 0.316. \quad \blacksquare$$

Pdf and pmf—similarities and differences. It is important to understand the similarities and differences between probability density and mass functions. Both give measures of how likely or "probable" a particular value is. The graph of a density function is comparable to a "smooth" probability histogram for a pmf.

See the graphs of the three density functions in Figure 6.2. In (i) the density models a random variable which takes values between 2 and 4, with values near 4 most likely and those near two least likely. In (ii), values between 0 and 4 are equally likely. In (iii), outcomes near 2 are most likely, with probability decreasing quickly for outcomes far from 2.

Unlike the probability mass function, however, the density $f(x)$ is *not* the probability that X is equal to x. That probability is always 0 for continuous variables. The value of $f(x)$ is a unitless measure of probability "mass" with the property that the total probability mass, that is, the area under the density curve, is equal to 1.

For more insight into what $f(x)$ "measures," consider an interval $(x - \epsilon/2, x + \epsilon/2)$ centered at x of length ϵ, where ϵ is small. If f is continuous at x then the probability that X falls in the interval is

$$P\left(x - \frac{\epsilon}{2} < X < x + \frac{\epsilon}{2}\right) = \int_{x-\epsilon/2}^{x+\epsilon/2} f(t)\,dt \approx f(x)\epsilon,$$

where the integral is approximated by the area of the rectangle of height $f(x)$ and width ϵ. This gives

$$f(x) \approx \frac{1}{\epsilon} P\left(x - \frac{\epsilon}{2} < X < x + \frac{\epsilon}{2}\right). \tag{6.3}$$

In physics, "density" is a measure of mass per unit volume, area, or length. In probability, Equation 6.3 shows that the density function is a measure of "probability mass" per unit length.

A function f is said to be *proportional* to a function g if the ratio of the two functions is constant, that is, $f(x)/g(x) = c$ for some constant c. The constant c is called the *proportionality constant.* We write $f(x) \propto g(x)$. In probability models density functions are often specified up to a proportionality constant.

Example 6.4 Suppose the random time T of a radioactive emission is proportional to a decaying exponential function of the form $e^{-\lambda t}$ for $t > 0$, where $\lambda > 0$ is a constant. Find the probability density function of T.

Write $f(t) = ce^{-\lambda t}$, for $t > 0$. Solving for the proportionality constant c gives

$$1 = \int_0^\infty ce^{-\lambda t}\, dt = \frac{c}{\lambda}\left(-e^{-\lambda t}\right)\Big|_0^\infty = \frac{c}{\lambda}.$$

Thus $c = \lambda$ and $f(t) = \lambda e^{-\lambda t}$, for $t > 0$. ∎

Example 6.5 A random variable X has density function proportional to $x + 1$ on the interval $(2, 4)$. Find $P(2 < X < 3)$.

Write $f(x) = c(x + 1)$. To find c set

$$1 = \int_2^4 c(x+1)\, dx = c\left(\frac{x^2}{2} + x\right)\Big|_2^4 = 8c,$$

giving $c = 1/8$. Then,

$$P(2 < X < 3) = \int_2^3 \frac{1}{8}(x+1)\, dx = \frac{1}{8}\left(\frac{x^2}{2} + x\right)\Big|_2^3 = \frac{1}{8}\left(\frac{7}{2}\right) = \frac{7}{16}.$$ ∎

6.2 CUMULATIVE DISTRIBUTION FUNCTION

One way to connect and unify the treatment of discrete and continuous random variables is through the *cumulative distribution function (cdf)* which is defined for *all* random variables.

CUMULATIVE DISTRIBUTION FUNCTION

Definition 6.1. *Let* X *be a random variable. The cumulative distribution function of* X *is the function*

$$F(x) = P(X \leq x),$$

defined for all real numbers x.

The cdf plays an important role for continuous random variables in part because of its relationship to the density function. For a continuous random variable X

$$F(x) = P(X \leq x) = \int_{-\infty}^{x} f(t)\, dt.$$

If F is differentiable at x, then taking derivatives on both sides and invoking the fundamental theorem of calculus gives

$$\boxed{F'(x) = f(x).}$$

The probability density function is the derivative of the cumulative distribution function. Given a density for X we can, in principle, obtain the cdf, and vice versa. Thus either function can be used to specify the probability distribution of X.

If two random variables have the same density function then they have the same probability distribution. Similarly, if they have the same cdf they have the same distribution.

If we know the cdf of a random variable we can compute probabilities on intervals. For $a < b$, observe that $(-\infty, b] = (\infty, a] \cup (a, b]$. This gives

$$\begin{aligned} P(a < X \leq b) &= P(X \in (a, b]) \\ &= P(X \in (-\infty, b]) - P(X \in (-\infty, a]) \\ &= P(X \leq b) - P(X \leq a) \\ &= F(b) - F(a). \end{aligned}$$

For continuous variables,

$$\begin{aligned} &F(b) - F(a) \\ &\quad = P(a < X \leq b) = P(a < X < b) = P(a \leq X < b) = P(a \leq X \leq b). \end{aligned}$$

Example 6.6 The density for a random variable X is

$$f(x) = 2xe^{-x^2}, \quad \text{for } x > 0.$$

Find the cdf of X and use it to compute $P(1 < X < 2)$.

For $x > 0$,

$$F(x) = P(X \leq x) = \int_0^x 2te^{-t^2}\,dt = \int_0^{x^2} e^{-u}\,du = 1 - e^{-x^2},$$

where we use the u-substitution $u = t^2$. For $x \leq 0$, $F(x) = 0$.

We could find the probability $P(1 < X < 2)$ by integrating the density function over $(1,2)$. But since we have already found the cdf there is no need.

$$\begin{aligned} P(1 < X < 2) &= F(2) - F(1) \\ &= (1 - e^{-4}) - (1 - e^{-1}) = e^{-1} - e^{-4} = 0.350. \end{aligned}$$ ■

Example 6.7 A random variable X has density function

$$f(x) = \begin{cases} 2/5, & \text{if } 0 < x \leq 1 \\ 2x/5, & \text{if } 1 \leq x < 2 \\ 0, & \text{otherwise.} \end{cases}$$

(i) Find the cdf of X. (ii) Find $P(0.5 < X < 1.5)$.

(i) We need to take care since this density is defined differently on two intervals. If $0 < x < 1$,

$$F(x) = P(X \leq x) = \int_0^x \frac{2}{5}\,dt = \frac{2x}{5}.$$

If $1 < x < 2$,

$$\begin{aligned} F(x) = P(X \leq x) &= \int_0^1 \frac{2}{5}\,dt + \int_1^x \frac{2t}{5}\,dt \\ &= \frac{2}{5} + \frac{x^2 - 1}{5} = \frac{x^2 + 1}{5}. \end{aligned}$$

This gives

$$F(x) = \begin{cases} 0, & \text{if } x \leq 0 \\ 2x/5, & \text{if } 0 < x \leq 1 \\ (x^2 + 1)/5, & \text{if } 1 < x \leq 2 \\ 1, & \text{if } x > 2. \end{cases}$$

Observe that $F(x)$ is continuous at all points x. See the graphs of f and F in Figure 6.3.

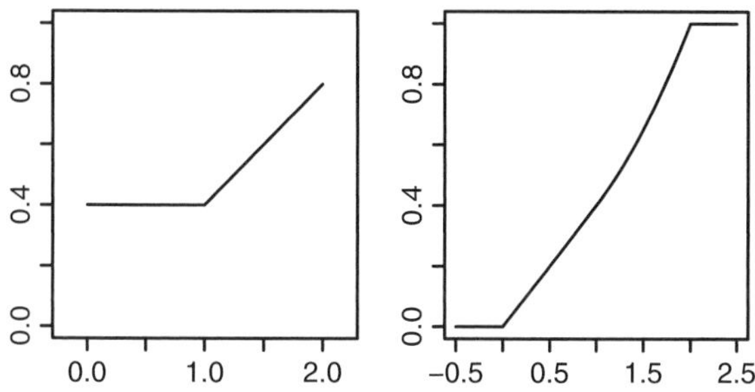

FIGURE 6.3: Density function and cdf.

(ii) The desired probability is

$$P(0.5 < X < 1.5) = F(1.5) - F(0.5) = \frac{(1.5)^2 + 1}{5} - \frac{2(0.5)}{5} = 0.85 \quad \blacksquare$$

Cumulative distribution functions for discrete random variables. Continuous random variables have continuous cdfs. However, the cdf of a discrete random variable has points of discontinuity at the discrete values of the variable.

For example, suppose $X \sim \text{Binom}(2, 1/2)$. Then the probability mass function of X is

$$P(X = x) = \begin{cases} 1/4, & \text{if } x = 0 \\ 1/2, & \text{if } x = 1 \\ 1/4, & \text{if } x = 2. \end{cases}$$

The cumulative distribution function of X, graphed in Figure 6.4, is

$$F(x) = P(X \leq x) = \begin{cases} 0, & \text{if } x < 0 \\ 1/4, & \text{if } 0 \leq x < 1 \\ 3/4, & \text{if } 1 \leq x < 2 \\ 1, & \text{if } x \geq 2, \end{cases}$$

with points of discontinuity at $x = 0, 1, 2$.

In general, a cdf need not be continuous. However, it is always right-continuous. The cdf is also an increasing function. That is, if $x \leq y$, then $F(x) \leq F(y)$. This holds since if $x \leq y$, the event $\{X \leq x\}$ implies $\{X \leq y\}$ and thus $P(X \leq x) \leq P(X \leq y)$. The cdf is a probability and takes values between 0 and 1. It has the property that $F(x) \to 0$, as $x \to -\infty$, and $F(x) \to 1$, as $x \to +\infty$.

We summarize the four defining properties of a cumulative distribution function.

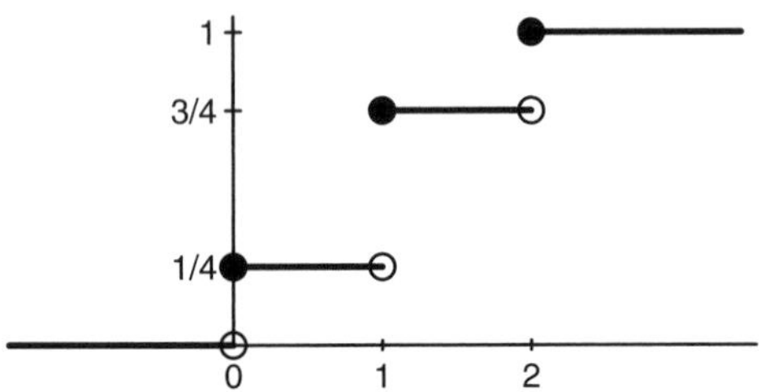

FIGURE 6.4: Cumulative distribution function $P(X \leq x)$ for $X \sim \text{Binom}(2, 1/2)$. The cdf has points of discontinuity at $x = 0, 1, 2$.

CUMULATIVE DISTRIBUTION FUNCTION

A function F is a cumulative distribution function, that is, there exists a random variable X whose cdf is F, if it satisfies the following properties.

1. $$\lim_{x \to -\infty} F(x) = 0.$$

2. $$\lim_{x \to +\infty} F(x) = 1.$$

3. If $x \leq y$, then $F(x) \leq F(y)$.
4. F is right-continuous. That is, for all real a,

$$\lim_{x \to a+} F(x) = F(a).$$

6.3 UNIFORM DISTRIBUTION

A uniform model on the interval (a, b) is described by a density function that is flat. That is, $f(x)$ is constant for all $a < x < b$. Since the density integrates to 1, to find that constant solve

$$1 = \int_a^b c\, dx = c(b - a),$$

giving $c = 1/(b - a)$, the reciprocal of the length of the interval.

UNIFORM DISTRIBUTION

A random variable X is *uniformly distributed on* (a, b) if the density function of X is

$$f(x) = \frac{1}{b-a}, \text{ for } a < x < b,$$

and 0, otherwise. Write $X \sim \text{Unif}(a, b)$.

If X is uniformly distributed on (a, b), the cdf of X is

$$F(x) = P(X \le x) = \begin{cases} 0, & \text{if } x \le a \\ (x-a)/(b-a), & \text{if } a < x \le b \\ 1, & \text{if } x > b. \end{cases}$$

For the case when X is uniformly distributed on $(0, 1)$, $F(x) = x$, for $0 < x < 1$.

Probabilities in the uniform model reduce to length. That is, if $X \sim \text{Unif}(a, b)$ and $a < c < d < b$, then

$$P(c < X < d) = F(d) - F(c) = \frac{d-a}{b-a} - \frac{c-a}{b-a} = \frac{d-c}{b-a} = \frac{\text{Length}\,(c,d)}{\text{Length}\,(a,b)}.$$

Example 6.8 Beth is taking a train from Minneapolis to Boston, a distance of 1400 miles. Her position is uniformly distributed between the two cities. What is the probability that she is past Chicago, which is 400 miles from Minneapolis?

Let X be Beth's location. Then $X \sim \text{Unif}(0, 1400)$. The probability is

$$P(X > 400) = \frac{1400 - 400}{1400} = \frac{5}{7} = 0.7143. \quad \blacksquare$$

R: UNIFORM DISTRIBUTION

The R commands for the continuous uniform distribution on (a, b) are

```
> punif(x,a,b)     # P(X ≤ x)
> dunif(x,a,b)     # f(x)
> runif(n,a,b)     # Simulates n random variables.
```

Default parameters are $a = 0$ and $b = 1$. To generate a uniform variable on $(0, 1)$, type

```
> runif(1)
[1] 0.973387
```

6.4 EXPECTATION AND VARIANCE

Formulas for expectation and variance for continuous random variables follow as expected from the discrete formulas: integrals replace sums, and densities replace probability mass functions.

EXPECTATION AND VARIANCE FOR CONTINUOUS RANDOM VARIABLES

For random variable X with density function f,

$$E[X] = \int_{-\infty}^{\infty} x f(x)\, dx$$

and

$$V[X] = \int_{-\infty}^{\infty} (x - E[X])^2 f(x)\, dx.$$

Properties of expectation and variance introduced in Chapter 4 for discrete random variables transfer to continuous random variables. We remind the reader of the most important.

PROPERTIES OF EXPECTATION AND VARIANCE

For constants a and b, and random variables X and Y,

- $E[aX + b] = aE[X] + b$
- $E[X + Y] = E[X] + E[Y]$
- $V[X] = E[X^2] - E[X]^2$
- $V[aX + b] = a^2 V[X]$

Example 6.9 Uniform distribution: expectation and variance. Let $X \sim$ Unif (a, b). Then

$$E[X] = \int_a^b x \left(\frac{1}{b-a}\right) dx = \left(\frac{1}{b-a}\right) \frac{x^2}{2}\bigg|_a^b = \frac{b+a}{2},$$

the midpoint of the interval (a, b). Also,

$$E[X^2] = \int_a^b x^2 \left(\frac{1}{b-a}\right) dx = \left(\frac{1}{b-a}\right) \frac{x^3}{3}\bigg|_a^b = \frac{b^2 + ab + a^2}{3},$$

and

$$V[X] = E[X^2] - E[X]^2 = \frac{b^2 + ab + a^2}{3} - \left(\frac{b+a}{2}\right)^2 = \frac{(b-a)^2}{12}.$$

It is worthwhile to remember these results as they occur frequently. The mean of a uniform distribution is the midpoint of the interval. The variance is one-twelfth the square of its length. ■

Example 6.10 Random variable X has density proportional to x^{-4}, for $x > 1$. Find $V[1 - 4X]$.

First find the constant of proportionality. Solve

$$1 = \int_1^\infty \frac{c}{x^4}\, dx = c\left(\frac{1}{-3x^3}\right)\Big|_1^\infty = \frac{1}{3}$$

giving $c = 3$. To find the variance first find the expectation

$$E[X] = \int_1^\infty x\left(\frac{3}{x^4}\right) dx = \int_1^\infty \frac{3}{x^3}\, dx = 3\left(\frac{1}{-2x^2}\right)\Big|_1^\infty = \frac{3}{2}.$$

Also,

$$E[X^2] = \int_1^\infty x^2\left(\frac{3}{x^4}\right) dx = \int_1^\infty \frac{3}{x^2}\, dx = 3\left(\frac{-1}{x}\right)\Big|_1^\infty = 3.$$

This gives

$$V[X] = E[X^2] - E[X]^2 = 3 - \left(\frac{3}{2}\right)^2 = \frac{3}{4}.$$

Therefore,

$$V[1 - 4X] = 16V[X] = 16\left(\frac{3}{4}\right) = 12.$$

■

The expectation of a function of a random variable is given similarly as in the discrete case.

EXPECTATION OF FUNCTION OF CONTINUOUS RANDOM VARIABLE

If X has density function f, and g is a function, then

$$E[g(X)] = \int_{-\infty}^\infty g(x)f(x)\, dx.$$

■ **Example 6.11** A balloon has radius uniformly distributed on (0,2). Find the expectation and standard deviation of the volume of the balloon.

Let V be the volume. Then $V = (4\pi/3)R^3$, where $R \sim$ Unif(0,2). The expected volume is

$$E[V] = E\left[\frac{4\pi}{3}R^3\right] = \frac{4\pi}{3}E[R^3] = \frac{4\pi}{3}\int_0^2 r^3\left(\frac{1}{2}\right)dr$$
$$= \frac{2\pi}{3}\left(\frac{r^4}{4}\right)\Big|_0^2 = \frac{8\pi}{3} = 8.378.$$

Also

$$E[V^2] = E\left[\left(\frac{4\pi}{3}R^3\right)^2\right] = \frac{16\pi^2}{9}E\left[R^6\right]$$
$$= \frac{16\pi^2}{9}\int_0^2 \frac{r^6}{2}\,dr = \frac{16\pi^2}{9}\left(\frac{128}{14}\right) = \frac{1024\pi^2}{63}$$

giving

$$\text{Var}[V] = E[V^2] - (E[V])^2 = \frac{1024\pi^2}{63} - \left(\frac{8\pi}{3}\right)^2 = \frac{64\pi^2}{7},$$

with standard deviation $\text{SD}[V] = 8\pi/\sqrt{7} \approx 9.50$.

R: SIMULATING BALLOON VOLUME

To simulate the balloon volume with one million trials is easy stuff for R.

```
> volume <-(4/3)*pi*runif(1000000,0,2)^3
> mean(volume)
[1] 8.386274
> sd(volume)
[1] 9.498317
```

■

6.5 EXPONENTIAL DISTRIBUTION

What is the size of a raindrop? When will your next text message arrive? How long does a bee spend gathering nectar at a flower? The applications of the exponential distribution are vast. The distribution is one of the most important in probability both for its practical and theoretical use.

EXPONENTIAL DISTRIBUTION

A random variable X has an *exponential distribution with parameter* $\lambda > 0$ if its density function has the form

$$f(x) = \lambda e^{-\lambda x}, \quad \text{for } x > 0,$$

and 0, otherwise. We write $X \sim \text{Exp}(\lambda)$.

The cumulative distribution function of the exponential distribution is

$$F(x) = \int_0^x \lambda e^{-\lambda t}\, dt = \lambda \left(\frac{-1}{\lambda} e^{-\lambda t} \right)\Big|_0^x = 1 - e^{-\lambda x},$$

for $x > 0$, and 0, otherwise. The "tail probability" formula

$$\boxed{P(X > x) = 1 - F(x) = e^{-\lambda x}, \quad \text{for } x > 0,}$$

is commonly used. We leave it to the exercises to show that

$$E[X] = \frac{1}{\lambda} \quad \text{and} \quad V[X] = \frac{1}{\lambda^2}.$$

For the exponential distribution the mean is equal to the standard deviation.

Example 6.12 A school's help desk receives calls throughout the day. The time T (in minutes) between calls is modeled with an exponential distribution with mean 4.5. A call just arrived. What is the probability no call will be received in the next 5 minutes?

The parameter of the exponential distribution is $\lambda = 1/4.5$. If no call is received in the next 5 minutes, then the time of the next call is greater than five. The desired probability is

$$P(T > 5) = e^{-5/4.5} = 0.329.$$ ■

R: EXPONENTIAL DISTRIBUTION

The R commands for the exponential distribution are

```
> dexp(x,λ)
> pexp(x,λ)
> rexp(n,λ)
```

6.5.1 Memorylessness

A most important property of the exponential distribution is memorylessness. You were introduced to this property in the discrete setting for the geometric distribution.

To illustrate in the continuous setting, suppose Amy and Zach are both waiting for a bus. Buses arrive about every 30 minutes according to an exponential distribution. Amy gets to the bus stop at time $t = 0$. The time until the next bus arrives has an exponential distribution with $\lambda = 1/30$.

Zach arrives at the bus stop 10 minutes later, at time $t = 10$. The memorylessness of the exponential distribution means that the time that Zach waits for the bus will also have an exponential distribution with $\lambda = 1/30$. They will *both* wait about the same amount of time.

Memorylessness means that if the bus does not arrive in the first 10 minutes then the probability that it will arrive after time $t = 10$ (for Zach) is the same as the probability that a bus arrives after time t (for Amy). After time $t = 10$, the next bus "does not remember" what happened in the first 10 minutes.

This may seem amazing, even paradoxical. Run the script file **Memory.R** to convince yourself it is true. See Figure 6.5 for the simulated distributions of Amy's and Zach's waiting times.

As long as the bus does not come in the first 10 minutes, Zach waits more than t minutes for the bus if and only if Amy waits more than $t + 10$ minutes. Letting A and Z denote Amy and Zach's waiting times, respectively, this gives,

$$P(Z > t) = P(A > t + 10 | A > 10) = \frac{P(A > t + 10)}{P(A > 10)}$$

$$= \frac{e^{-(t+10)/30}}{e^{-10/30}} = e^{-t/30} = P(A > t).$$

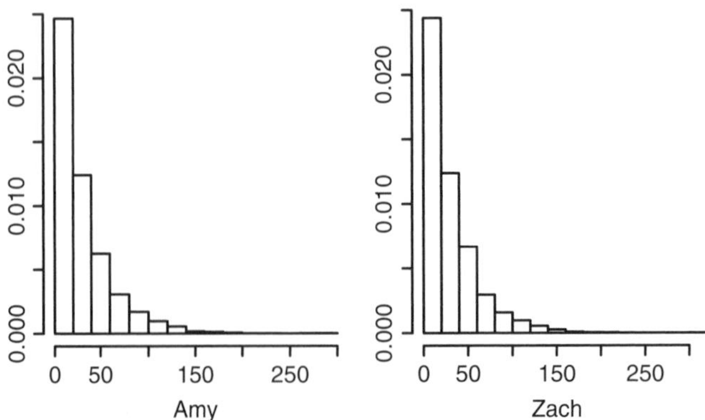

FIGURE 6.5: Zach arrives at the bus station 10 minutes after Amy. But the distribution of their waiting times is the same.

R: BUS WAITING TIME

```
# Memory.R
    # Amy arrives at time t=0
    # Zach arrives at time t=10
> n <- 10000
> Amy <- rexp(n,1/30)
> bus <- rexp(n,1/30)
> Zach <- bus[bus>10]-10
> mean(Amy)
[1] 29.8969
> mean(Zach)
[1] 29.58798
> par(mfrow=c(1,2))
> hist(Amy, prob=T)
> hist(Zach,prob=T)
```

The exponential distribution is the only continuous distribution that is memoryless. Here is the general property.

MEMORYLESSNESS FOR EXPONENTIAL DISTRIBUTION

Let $X \sim \text{Exp}(\lambda)$. For $0 < s < t$,

$$P(X > s + t | X > s) = \frac{P(X > s + t)}{P(X > s)} = \frac{e^{-\lambda(s+t)}}{e^{-\lambda s}} = e^{-\lambda t} = P(X > t).$$

Many random processes that evolve in time and/or space exhibit a lack of memory in the sense described here, which is key to the central role of the exponential distribution in applications. Novel uses of the exponential distribution include:

- The diameter of a raindrop plays a fundamental role in meteorology, and can be an important variable in predicting rain intensity in extreme weather events. In 1948, Marshall and Palmer (1948) proposed an exponential distribution for raindrop size. The parameter λ is a function of rainfall intensity (see Example 7.14). The empirically derived model has held up for many years and has found wide application in hydrology.
- Dorsch et al. (2008) use exponential distributions to model patterns in ocean storms off the southern coast of Australia as a way to study changes in ocean "storminess" as a result of global climate change. They fit storm duration and time between storms to exponential distributions with means 21.1 hours and 202.0 hours, respectively.

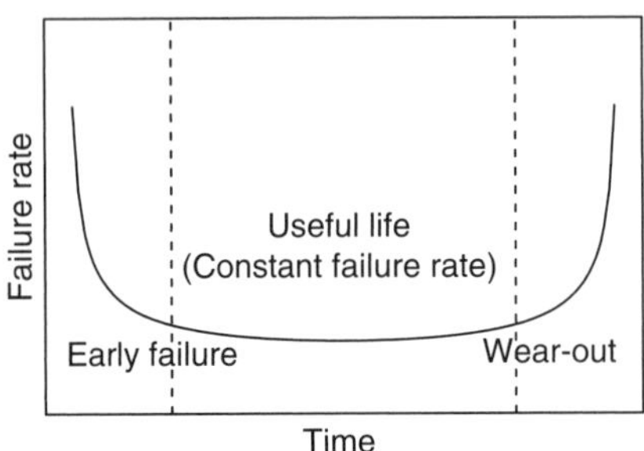

FIGURE 6.6: Bathtub curves are used in reliability engineering to model failure rates of a product or component.

- The "bathtub curve" is used in reliability engineering to represent the lifetime of a product. The curve models the failure rate, which is the frequency with which a product or component fails, often expressed as failures per time unit. See Figure 6.6. In the middle of the curve, the failure rate is constant and the exponential distribution often serves as a good model for the time until failure.

Exponential random variables often arise in applications as a sequence of successive times between *arrivals.* These times might represent arrivals of phone calls, buses, accidents, or component failures.

Let $X_1, X_2, \ldots$ be an independent sequence of Exp(λ) random variables where X_k is the time between the $(k-1)$st and kth arrival. Since the common expectation of the X_k's is $1/\lambda$ we expect arrival times to be about $1/\lambda$ time units apart, and there are about λ arrivals per one unit of time. Thus λ represents the rate of arrivals (number of arrivals per unit time). For this reason, the parameter λ is often called the *rate* of the exponential distribution.

Example 6.13 The time it takes for each customer to be served at a local restaurant has an exponential distribution. Serving times are independent of each other. Typically customers are served at the rate of 20 customers per hour. Arye is waiting to be served. What is the mean and standard deviation of his serving time? Find the probability that he will be served within 5 minutes?

Model Arye's serving time S with an exponential distribution with $\lambda = 20$. Then in hour units, $E[S] = \text{SD}[S] = 1/20$, or 3 minutes. Since the given units are hours, the desired probability is

$$P\left(S \leq \frac{5}{60}\right) = F_S\left(\frac{1}{12}\right) = 1 - e^{-20/12} = 0.811.$$

In R, the solution is obtained by typing

```
> pexp(1/12,20)
[1] 0.8111244
```

■

Sequences of i.i.d. exponential random variables as described above form the basis of an important class of random processes called the Poisson process, introduced in the next chapter.

6.6 FUNCTIONS OF RANDOM VARIABLES I

The diameter of a subatomic particle is modeled as a random variable. The volume of the particle is a function of that random variable. The rates of return for several stocks in a portfolio are modeled as random variables. The portfolio's maximum rate of return is a function of those random variables.

In this section you will learn how to find the distribution of a function of a random variable. We start with an example which captures the essence of the general approach.

Example 6.14 The radius R of a circle is uniformly distributed on $(0,2)$. Let A be the area of the circle. Find the probability density function of A.

The area of the circle is a function of the radius. Write $A = f(R)$, where $f(r) = \pi r^2$. Our approach to finding the density of A will be to (i) find the cdf of A in terms of the cdf of R, and (ii) take derivatives to find the desired density function.

Since R takes values between 0 and 2, A takes values between 0 and 4π. For $0 < a < 4\pi$,

$$F_A(a) = P(A \le a) = P\left(\pi R^2 \le a\right) = P\left(R \le \sqrt{a/\pi}\right) = F_R\left(\sqrt{a/\pi}\right).$$

Now take derivatives with respect to a. The lefthand side gives $f_A(a)$. The righthand side, using the chain rule, gives

$$\begin{aligned}\frac{d}{da}F_R\left(\sqrt{a/\pi}\right) &= f_R\left(\sqrt{a/\pi}\right)\left(\frac{d}{da}\sqrt{a/\pi}\right) = f_R\left(\sqrt{a/\pi}\right)\frac{1}{2\sqrt{a\pi}}\\ &= \left(\frac{1}{2}\right)\frac{1}{2\sqrt{a\pi}} = \frac{1}{4\sqrt{a\pi}}.\end{aligned}$$

That is,

$$f_A(a) = \frac{1}{4\sqrt{a\pi}}, \quad \text{for } 0 < a < 4\pi,$$

and 0, otherwise.

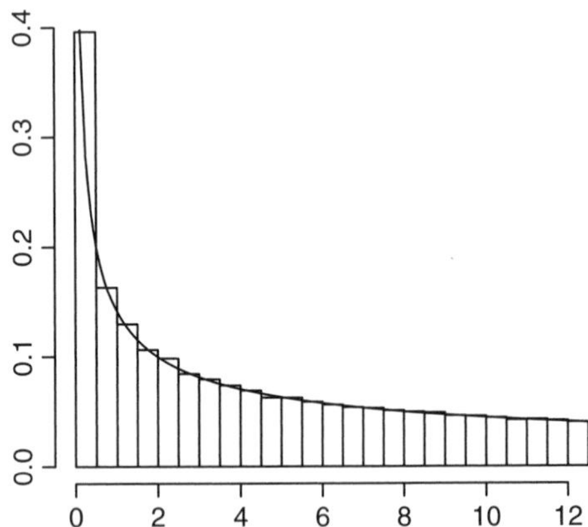

FIGURE 6.7: Simulated distribution of $A = \pi R^2$, where $R \sim$ Unif(0,2). The curve is density function $f(a) = 1/(4\sqrt{\pi a})$.

R: COMPARING THE EXACT DISTRIBUTION WITH A SIMULATION

We simulate $A = \pi R^2$. The theoretical density curve $f_A(a)$ is superimposed on the simulated histogram in Figure 6.7.

```
> simlist <- pi*runif(10000,0,2)^2
> hist(simlist,prob=T)
> curve(1/(4*sqrt(x*pi)),0,4*pi,add=TRUE)
```

■

This example illustrates the general approach for finding the density of a function of a random variable. If $Y = f(X)$, start with the cumulative distribution function of Y. Express the cdf of Y in terms of the cdf of X. Then take derivatives.

HOW TO FIND THE DENSITY OF $Y = F(X)$

1. Determine the possible values of Y based on the values of X and the function f.
2. Begin with the cdf $F_Y(y) = P(Y \leq y) = P(f(X) \leq y)$. Express the cdf in terms of the original random variable X.
3. From $P(f(X) \leq y)$ obtain an expression of the form $P(X \leq \ldots)$. The righthand side of the expression will be a function of y. If f is invertible then $P(f(X) \leq y) = P(X \leq f^{-1}(y))$.
4. Differentiate with respect to y to obtain the density $f_Y(y)$.

■ **Example 6.15 Linear function of a uniform random variable.** Suppose $X \sim$ Unif(0,1). For $a < b$, let $Y = (b - a)X + a$. Find the density function of Y.

Since X takes values between 0 and 1, $(b-a)X+a$ takes values between a and b. For $a < y < b$,

$$F_Y(y) = P(Y \leq y) = P((b-a)X + a \leq y)$$
$$= P\left(X \leq \frac{y-a}{b-a}\right) = F_X\left(\frac{y-a}{b-a}\right) = \frac{y-a}{b-a}.$$

To find the density of Y differentiate to get

$$f_Y(y) = \frac{1}{b-a}, \quad \text{for } a < y < b.$$

The distribution of Y is uniform on (a, b).

Observe that we could have made this conclusion earlier by noting that the cdf of Y is the cdf of a uniform distribution on (a, b).

The result shows how to simulate a uniform random variable on an interval (a, b) given a uniform random variable on (0,1). Given $a < b$, in R, type

```
> (b-a)*runif(1)+a
```

Of course, in R this is a little silly, since we could always simply type

```
> runif(1,a,b)
```

■

Example 6.16 Linear function of a random variable. Suppose X is a random variable with density f_X. Let $Y = rX + s$, where r and s are constants. Find the density of Y.

Suppose $r > 0$. The cdf of Y is

$$F_Y(y) = P(Y \leq y) = P(rX + s \leq y)$$
$$= P\left(X \leq \frac{y-s}{r}\right) = F_X\left(\frac{y-s}{r}\right).$$

Differentiating with respect to y gives

$$f_Y(y) = \frac{1}{r} f_X\left(\frac{y-s}{r}\right).$$

If $r < 0$, we have

$$F_y(y) = P\left(X \geq \frac{y-s}{r}\right) = 1 - F_X\left(\frac{y-s}{r}\right).$$

Differentiating gives

$$f_Y(y) = -\frac{1}{r} f_X\left(\frac{y-s}{r}\right).$$

In either case we have

$$f_Y(y) = \frac{1}{|r|} f_X\left(\frac{y-s}{r}\right).$$

Observe the relationship between the density functions of X and Y. The density of Y is obtained by translating the density of X s units to the right and stretching by a factor of r. Then compress by a factor of $|r|$ vertically. ∎

Example 6.17 A lighthouse is one mile off the coast from the nearest point O on a straight, infinite beach. The lighthouse sends out pulses of light at random angles Θ uniformly distributed from $-\pi/2$ to $\pi/2$. Find the distribution of the distance X between where the light hits the shore and O. Also find the expected distance.

See Figure 6.8. Let $\Theta \sim \text{Unif}(-\pi/2, \pi/2)$ be the angle of the light to the shore. Let X be the distance from where the light hits the shore to O. By the geometry of the problem, $X = \tan\Theta$. For all real x,

$$\begin{aligned} F_X(x) &= P(X \le x) = P(\tan\Theta \le x) \\ &= P(\Theta \le \tan^{-1} x) = \frac{\tan^{-1} x + \pi/2}{\pi}. \end{aligned}$$

Taking derivatives with respect to x gives,

$$f_X(x) = \frac{1}{\pi(1+x^2)}, \quad -\infty < x < \infty.$$

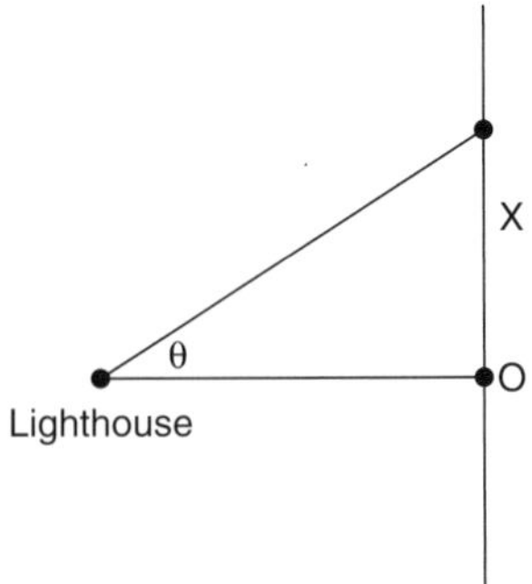

FIGURE 6.8: The geometry of the lighthouse problem.

This function is the density of the *Cauchy distribution*, also known as the Lorentz distribution in physics, where it is used to model energy states in quantum mechanics.

To find the expectation $E[X]$, we can set up the integral two ways.

1. Use the distribution of X.

$$E[X] = \int_{-\infty}^{\infty} x \frac{1}{\pi(1+x^2)}\, dx.$$

2. Use the distribution of Θ (and the law of the unconscious statistician).

$$E[X] = E[\tan \Theta] = \int_{-\pi/2}^{\pi/2} \frac{\tan \theta}{\pi}\, d\theta.$$

However, in either case, the integral does not converge. The Cauchy distribution has the property that its expectation does not exist.

What happens when one tries to simulate from a distribution with no expectation?

R: SIMULATING AN EXPECTATION THAT DOES NOT EXIST

The function `noexp()` simulates 100,000 trials of X, the distance from where the light hits the shore and O, and then computes the average.

```
> noexp <- function()
  mean(tan(runif(100000,-pi/2,pi/2)))}
```

Repeated simulations show no equilibrium. The averages are erratic; some are close to 0; others show large magnitudes.

```
> noexp()
[1] 0.1142573
> noexp()
[1] 1.031229
> noexp()
[1] 13.26526
> noexp()
[1] -1.303423
> noexp()
[1] -4.078775
> noexp()
[1] 27.48636
> noexp()
[1] -51.91099
> noexp()
[1] -2.806952
```

∎

6.6.1 Simulating a Continuous Random Variable

A random variable X has density $f(x) = 2x$, for $0 < x < 1$, and 0, otherwise. Suppose we want to simulate observations from X. In this section we present a simple and flexible method for simulating from a continuous distribution. It requires that the cdf of X is invertible.

INVERSE TRANSFORM METHOD

Suppose X is a continuous random variable with cumulative distribution function F, where F is invertible with inverse function F^{-1}. Let $U \sim \text{Unif}(0,1)$. Then the distribution of $F^{-1}(U)$ is equal to the distribution of X. To simulate X first simulate U and output $F^{-1}(U)$.

To illustrate with the initial example, the cdf of X is

$$F(x) = P(X \leq x) = \int_0^x 2t\,dt = x^2, \quad \text{for } 0 < x < 1.$$

On the interval $(0,1)$ the function $F(x) = x^2$ is invertible and $F^{-1}(x) = \sqrt{x}$. The inverse transform method says that if $U \sim \text{Unif}(0,1)$, then $F^{-1}(U) = \sqrt{U}$ has the same distribution as X. Thus to simulate X, generate $\sqrt{U}$.

R: IMPLEMENTING THE INVERSE TRANSFORM METHOD

The R commands

```
> simlist <- sqrt(runif(10000))
> hist(simlist,prob=T,main="",xlab="")
> curve(2*x,0,1,add=T)
```

generate the histogram in Figure 6.9, along with the super-imposed curve of the theoretical density $f(x) = 2x$.

The proof of the inverse transform method is quick and easy. We need to show that $F^{-1}(U)$ has the same distribution as X. For x in the range of X,

$$P(F^{-1}(U) \leq x) = P(U \leq F(x)) = F(x) = P(X \leq x),$$

using the fact that the cdf of the uniform distribution on (0,1) is the identity function.

■ **Example 6.18** We show how to simulate an exponential random variable using the inverse transform method. The cdf of the exponential distribution is $F(x) = 1 - e^{-\lambda x}$

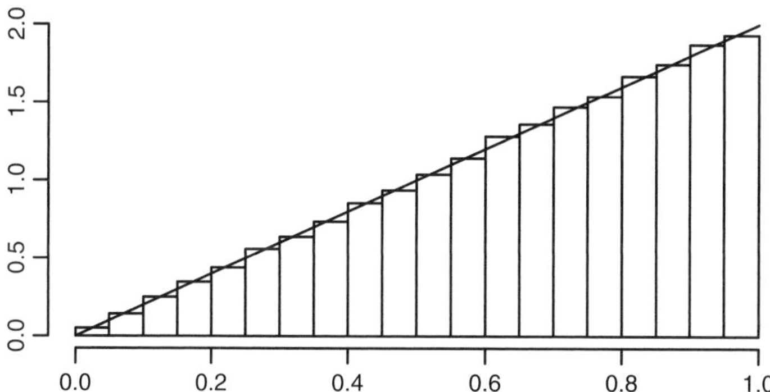

FIGURE 6.9: Simulating from density $f(x) = 2x$, for $0 < x < 1$ using the inverse transform method.

with inverse function

$$F^{-1}(x) = \frac{\ln(1-x)}{-\lambda}.$$

To simulate an exponential random variable with parameter λ, simulate a uniform (0,1) variable U and generate $-\ln(1-U)/\lambda$.

Observe that if U is uniform on (0,1), then so is $1 - U$. Thus an even simpler method for simulating an exponential random variable is to output $-\ln U/\lambda$. ∎

6.7 JOINT DISTRIBUTIONS

For two or more random variables, the *joint density function* plays the role of the joint pmf for discrete variables. Single integrals become multiple integrals.

JOINT DENSITY FUNCTION

For continuous random variables X and Y defined on a common sample space, the *joint density function* $f(x, y)$ of X and Y has the following properties.

1. $f(x, y) \geq 0$ for all real numbers x and y
2. $$\int_{-\infty}^{\infty}\int_{-\infty}^{\infty} f(x, y)\, dx\, dy = 1$$
3. For $S \subseteq \mathbb{R}^2$,
$$P((X, Y) \in S) = \iint_S f(x, y)\, dx\, dy. \tag{6.4}$$

Typically in computations the integral in Equation 6.4 will be an iterated integral whose limits of integration are determined by S. Since the joint density is a function of two variables, its graph is a *surface* over a two-dimensional domain.

For continuous random variables $X_1, \ldots, X_n$ defined on a common sample space, the joint density function $f(x_1, \ldots, x_n)$ is defined similarly.

Example 6.19 Suppose the joint density of X and Y is

$$f(x, y) = cxy, \text{ for } 1 < x < 4 \text{ and } 0 < y < 1.$$

Find c and compute $P(2 < X < 3, Y > 1/4)$.

First solve

$$\begin{aligned} 1 &= \int_0^1 \int_1^4 cxy \, dx \, dy = c \int_0^1 y \left(\int_1^4 x \, dx \right) dy \\ &= c \int_0^1 y \left(\frac{15}{2} \right) dy = c \left(\frac{15}{2} \right) \left(\frac{1}{2} \right) = \frac{15c}{4}, \end{aligned}$$

giving $c = 4/15$. Then,

$$P(2 < X < 3, Y > 1/4) = \int_{1/4}^1 \int_2^3 \frac{4xy}{15} \, dx \, dy = \left(\frac{4}{15} \right) \left(\frac{5}{2} \right) \left(\frac{15}{32} \right) = \frac{5}{16}.$$

The joint density function is graphed in Figure 6.10. The density is a surface over the domain $1 < x < 4$ and $0 < y < 1$. ■

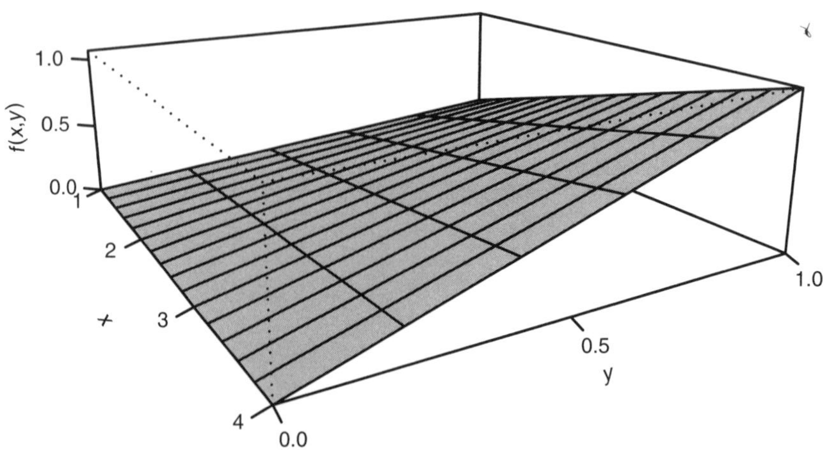

FIGURE 6.10: Joint density $f(x, y) = 4xy/15, \ 1 < x < 4, \ 0 < y < 1.$

JOINT CUMULATIVE DISTRIBUTION FUNCTION

If X and Y have joint density function f, the *joint cumulative distribution function* of X and Y is

$$F(x,y) = P(X \leq x, Y \leq y) = \int_{-\infty}^{x} \int_{-\infty}^{y} f(s,t)\, dt\, ds$$

defined for all x and y. Differentiating with respect to both x and y gives

$$\frac{\partial^2}{\partial x \partial y} F(x,y) = f(x,y). \tag{6.5}$$

Uniform model. Let S be a bounded region in the plane. By analogy with the development of the one-dimensional uniform model, if (X,Y) is distributed uniformly on S, the joint density will be constant on S. That is, $f(x,y) = c$ for all $(x,y) \in S$. Solving

$$1 = \iint_S f(x,y)\, dx\, dy = \iint_S c\, dx\, dy = c[\text{Area}\,(S)]$$

gives $c = 1/\text{Area}\,(S)$ and the uniform model on S.

UNIFORM DISTRIBUTION IN TWO DIMENSIONS

Let S be a bounded region in the plane. Then random variables (X,Y) are uniformly distributed on S if the joint density function of X and Y is

$$f(x,y) = \frac{1}{\text{Area}\,(S)}, \quad \text{for } (x,y) \in S,$$

and 0, otherwise. We write $(X,Y) \sim \text{Unif}(S)$.

Example 6.20 Suppose X and Y have joint density $f(x,y)$. (i) Find a general expression for $P(X < Y)$. (ii) Solve for the case when (X,Y) is uniformly distributed on the circle centered at the origin of radius one.

(i) The region determined by the event $\{X < Y\}$ is the set of all points in the plane (x,y) such that $x < y$. Setting up the double integral gives

$$P(X < Y) = \iint_{\{(x,y):x<y\}} f(s,t)\, ds\, dt = \int_{-\infty}^{\infty} \int_{-\infty}^{y} f(s,t)\, ds\, dt.$$

(ii) For the special case when (X,Y) is uniformly distributed on the circle, first draw the picture. See Figure 6.11. The joint density is $f(x,y) = 1/\pi$, if (x,y) is in the circle, and 0, otherwise. This gives

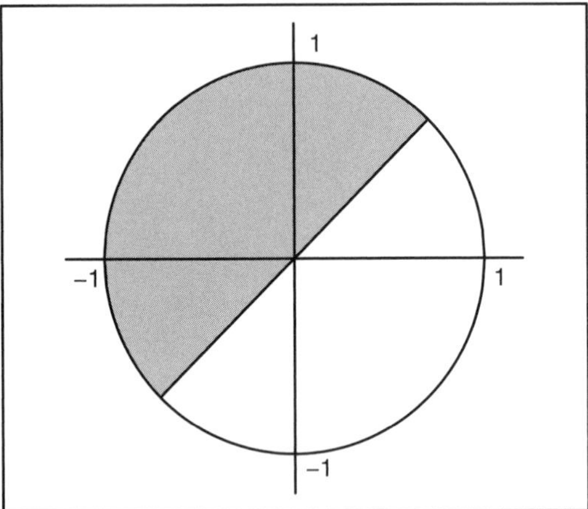

FIGURE 6.11: Shaded region defined by $P(X < Y)$.

$$P(X < Y) = \int_{-1}^{1} \int_{s}^{\sqrt{1-s^2}} \frac{1}{\pi} \, dt \, ds = \frac{1}{\pi} \int_{-1}^{1} \left(\sqrt{1-s^2} - s \right) \, ds.$$

Solving the integral requires trigonometric substitution. But a much easier route is to recognize that since (X, Y) has a uniform distribution, the problem reduces to finding areas. And then it is clear from the picture that the desired probability is 1/2. ∎

Problems involving two or more continuous random variables will require multiple integrals. Setting these up correctly can be challenging. As an example, consider the probability $P(X < Y)$ from the last Example 6.20, but where (X, Y) has joint density

$$f(x, y) = \frac{3x^2 y}{64}, \quad \text{for } 0 < x < 2, \ 0 < y < 4.$$

The distribution here is not uniform so the problem cannot be reduced to finding areas. The graph of the density function, as shown at the top of Figure 6.12, is a surface over the rectangle $[0, 2] \times [0, 4]$.

The event $\{X < Y\}$ determines the shaded region at the bottom of Figure 6.12. For solving problems such as this, the most important step is to draw the picture. The desired probability is the integral of the density $f(x, y)$ over this shaded region.

In setting up the double integral be aware of *all* constraints on each variable. If, say, x is chosen as the variable for the outer integral, then the limits of integration for the outer integral will not depend on y. Otherwise, y will appear in the final answer.

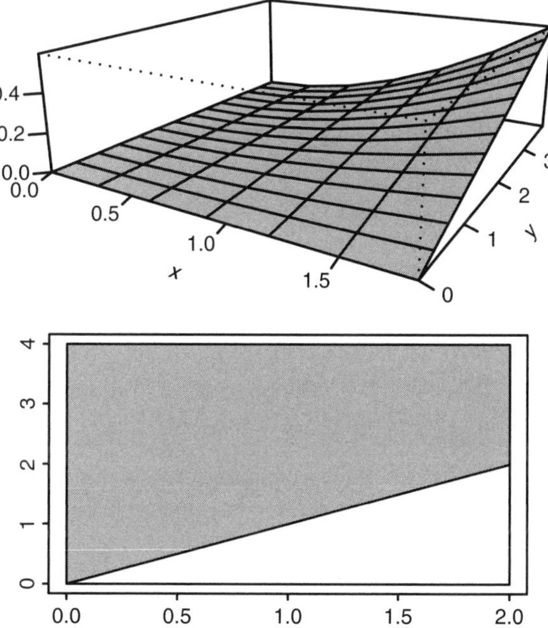

FIGURE 6.12: Top: joint density $f(x, y) = 3x^2y/64$. Bottom: domain of joint density. Shaded region shows event $\{X < Y\}$.

To set up the double integral for $P(X < Y)$, the limits for the outer integral are $0 < x < 2$. For the inner integral, the constraints on y are $0 < y < 4$ and $y > x$ which together give $x < y < 4$. Thus,

$$P(X < Y) = \int_{x=0}^{2} \int_{y=x}^{4} \frac{3x^2y}{64}\, dy\, dx.$$

Notice that we have explicitly written x and y in the limits of integration to help keep track of which variable corresponds to which integral.

The multiple integral can also be set up with y on the outside integral and x on the inside. If y is the first variable, then the limits for the outer integral are $0 < y < 4$. For the inner integral, the constraints on x are $x < y$ and $0 < x < 2$. This is equivalent to $0 < x < \min(y, 2)$. The limits of integration for x depend on y. We need to break up the outer integral into two parts. For $0 < y < 2$, x ranges from 0 to y. For $2 < y < 4$, x ranges from 0 to 2. This gives

$$P(X < Y) = \int_{y=0}^{2} \int_{x=0}^{y} \frac{3x^2y}{64}\, dx\, dy + \int_{y=2}^{4} \int_{x=0}^{2} \frac{3x^2y}{64}\, dx\, dy.$$

In both cases, the final answer is $P(X < Y) = 17/20$.

Example 6.21 The joint density of X and Y is

$$f(x,y) = \frac{1}{x^2y^2}, \quad x > 1, y > 1.$$

Find $P(X \geq 2Y)$.

See Figure 6.13. The constraints on the variables are $x > 1$, $y > 1$, and $x \geq 2y$. Setting up the multiple integral with y as the outside variable gives

$$\begin{aligned} P(X \geq 2Y) &= \int_{y=1}^{\infty} \int_{x=2y}^{\infty} \frac{1}{x^2y^2} \, dy \, dx = \int_{y=1}^{\infty} \frac{1}{y^2} \left(\frac{-1}{x} \right) \Big|_{2y}^{\infty} \, dy \\ &= \int_{y=1}^{\infty} \frac{1}{y^2} \left(\frac{1}{2y} \right) dy = \left(\frac{-1}{4y^2} \right) \Big|_{1}^{\infty} = \frac{1}{4}. \end{aligned}$$

To set up the integral with x on the outside consider the constraints $x > 1, x \geq 2y$, and $y > 1$. Since the minimum value of y is 1 and $x \geq 2y$, we must have $x \geq 2$ for the outer integral. For the inside variable, we have $y > 1$ and $y \leq x/2$. That is, $1 < y \leq x/2$. This gives

$$\begin{aligned} P(X \geq 2Y) &= \int_{x=2}^{\infty} \int_{y=1}^{x/2} \frac{1}{x^2y^2} \, dy \, dx = \int_{x=2}^{\infty} \frac{1}{x^2} \left(\frac{-1}{y} \right) \Big|_{1}^{x/2} \, dx \\ &= \int_{x=2}^{\infty} \frac{1}{x^2} \left(1 - \frac{2}{x} \right) dx = \left(\frac{-1}{x} + \frac{1}{x^2} \right) \Big|_{2}^{\infty} = \frac{1}{2} - \frac{1}{4} = \frac{1}{4}. \end{aligned}$$ ■

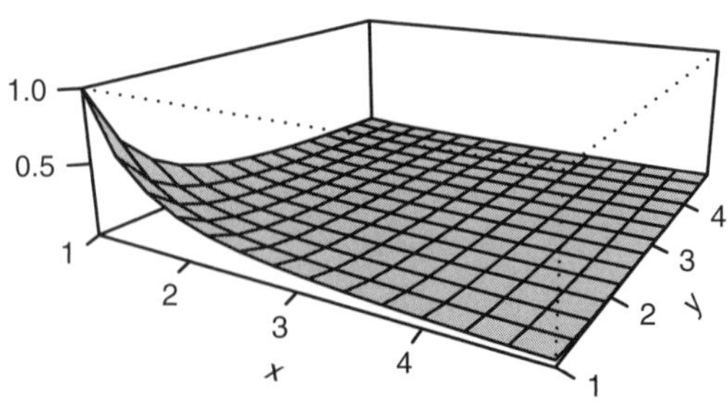

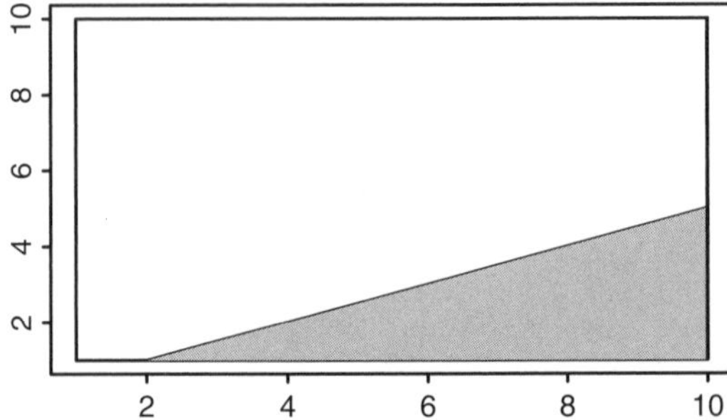

FIGURE 6.13: Top: joint density. Bottom: shaded region shows part of event $\{X \geq 2Y\}$.

Example 6.22 There are many applications in reliability theory of two-unit systems in which the lifetimes of the two units are described by a joint probability distribution. See Harris (1968) for examples. Suppose such a system depends on components A and B whose respective lifetimes X and Y are jointly distributed with density function

$$f(x, y) = e^{-y}, \quad \text{for } 0 < x < y < \infty.$$

Find the probability (i) that component B lasts at least three time units longer than component A; (ii) that both components last for at least two time units.

(i) The desired probability is

$$\begin{aligned} P(Y > X + 3) &= \int_{x=0}^{\infty} \int_{y=x+3}^{\infty} e^{-y} \, dy \, dx \\ &= \int_{x=0}^{\infty} e^{-(x+3)} \, dx = e^{-3} = 0.050. \end{aligned}$$

(ii) The desired probability is

$$\begin{aligned} P(X > 2, Y > 2) &= \int_{y=2}^{\infty} \int_{x=2}^{y} e^{-y} \, dx \, dy \\ &= \int_{y=2}^{\infty} (y-2) e^{-y} \, dy = e^{-2} = 0.135. \end{aligned}$$

■

The joint density of X and Y captures all the "probabilistic information" about X and Y. In principle, it can be used to find any probability which involves these variables. From the joint density the marginal densities are obtained by integrating out the extra variable. (In the discrete case, we sum over the other variable.)

MARGINAL DISTRIBUTIONS FROM JOINT DENSITIES

$$f_X(x) = \int_{-\infty}^{\infty} f(x, y) \, dy \quad \text{and} \quad f_Y(y) = \int_{-\infty}^{\infty} f(x, y) \, dx.$$

To see why these equations hold, consider

$$\{X \le x\} = \{X \le x, -\infty < Y < \infty\}.$$

Hence,

$$P(X \le x) = P(X \le x, -\infty < Y < \infty) = \int_{-\infty}^{x} \left[\int_{-\infty}^{\infty} f(s, y) \, dy \right] \, ds.$$

Differentiating with respect to x, applying the fundamental theorem of calculus, gives

$$f_X(x) = \int_{-\infty}^{\infty} f(x, y) \, dy,$$

and similarly for $f_Y(y)$.

Example 6.23 Consider the joint density function from the last Example 6.22

$$f(x, y) = e^{-y}, \quad \text{for } 0 < x < y < \infty.$$

Find the marginal densities of X and Y.

The marginal density of X is

$$f_X(x) = \int_{-\infty}^{\infty} f(x, y)\, dy = \int_{x}^{\infty} e^{-y}\, dy = e^{-x}, \quad \text{for } x > 0.$$

Note that once we integrate out the y variable, the domain of the x variable is all positive real numbers. From the form of the density function we see that X has an exponential distribution with parameter $\lambda = 1$.

The marginal density of Y is

$$f_Y(y) = \int_{0}^{y} e^{-y}\, dx = ye^{-y}, \quad \text{for } y > 0.$$

■

Computing expectations of functions of two or more continuous random variables should offer no surprises as we use the continuous form of the "law of the unconscious statistician."

EXPECTATION OF FUNCTION OF JOINTLY DISTRIBUTED RANDOM VARIABLES

If X and Y have joint density f, and $g(x, y)$ is a function of two variables, then

$$E[g(X, Y)] = \int_{-\infty}^{\infty} \int_{-\infty}^{\infty} g(x, y) f(x, y)\, dx\, dy.$$

The expected product of random variables X and Y is thus

$$E[XY] = \int_{-\infty}^{\infty} \int_{-\infty}^{\infty} xyf(x, y)\, dx\, dy.$$

Example 6.24 Suppose (X, Y) is uniformly distributed on the circle of radius 1 centered at the origin. Find the expected distance $D = \sqrt{X^2 + Y^2}$ to the origin.

Let C denote the circle. Since the area of C is π, the joint density function of X and Y is

$$f(x, y) = \frac{1}{\pi}, \quad \text{for } (x, y) \in C,$$

and 0, otherwise. This gives

$$\begin{aligned}E[D] &= E[\sqrt{X^2+Y^2}]\\ &= \int_{-\infty}^{\infty}\int_{-\infty}^{\infty}\sqrt{x^2+y^2}f(x,y)\,dx\,dy = \iint_C \sqrt{x^2+y^2}\frac{1}{\pi}\,dx\,dy.\end{aligned}$$

Changing to (r,θ) polar coordinates gives

$$E[D] = \int_0^{2\pi}\int_0^1 \sqrt{r^2}\left(\frac{1}{\pi}\right) r\,dr\,d\theta = \int_0^{2\pi}\frac{1}{3\pi}\,d\theta = \frac{2}{3}.$$ ■

6.8 INDEPENDENCE

As in the discrete case, if random variables X and Y are independent, then for all A and B,

$$P(X \in A, Y \in B) = P(X \in A)P(Y \in B).$$

In particular, for all x and y,

$$F(x,y) = P(X \le x, Y \le y) = P(X \le x)P(Y \le y) = F_X(x)F_Y(y).$$

Take derivatives with respect to x and y on both sides of this equation. This gives the following characterization of independence for continuous random variables in terms of probability density functions.

INDEPENDENCE AND DENSITY FUNCTIONS

Continuous random variables X and Y are independent if and only if their joint density function is the product of their marginal densities. That is,

$$f(x,y) = f_X(x)f_Y(y), \quad \text{for all } x, y.$$

More generally, if $X_1,\ldots,X_n$ are jointly distributed with joint density function f, then the random variables are mutually independent if and only if

$$f(x_1,\ldots,x_n) = f_{X_1}(x_1)\cdots f_{X_n}(x_n), \quad \text{for all } x_1,\ldots,x_n.$$

Example 6.25 Alex and Michael are at the airport terminal buying tickets. Alex is in the regular ticket line where the waiting time A has an exponential distribution with mean 10 minutes. Michael is in the express line, where the waiting time M has an exponential distribution with mean 5 minutes. Waiting times for the two lines are independent. What is the probability that Alex gets to the ticket counter before Michael?

The desired probability is $P(A < M)$. By independence, the joint density of A and M is

$$f(a,m) = f_A(a)f_M(m) = \frac{1}{10}e^{-a/10}\frac{1}{5}e^{-m/5}, \text{ for } a > 0, m > 0.$$

Then,

$$\begin{aligned}
P(A < M) &= \iint_{\{(a,m):a<m\}} f(a,m)\,da\,dm \\
&= \int_0^\infty \int_0^m \frac{1}{50}e^{-a/10}e^{-m/5}\,da\,dm \\
&= \frac{1}{50}\int_0^\infty e^{-m/5}\left(-10e^{-a/10}\right)\Big|_0^\infty\,dm \\
&= \frac{1}{5}\int_0^\infty e^{-m/5}\left(1 - e^{-m/10}\right)\,dm \\
&= \frac{1}{5}\left(\frac{5}{3}\right) = \frac{1}{3}.
\end{aligned}$$

■

Example 6.26 Let X, Y, and Z be i.i.d. random variables with common marginal density $f(t) = 2t$, for $0 < t < 1$, and 0, otherwise. Find $P(X < Y < Z)$.

Here we have three jointly distributed random variables. Results for joint distributions of three or more random variables are natural extensions of the two variable case. By independence, the joint density of (X, Y, Z) is,

$$f(x,y,z) = f_X(x)f_Y(y)f_Z(z) = (2x)(2y)(2z) = 8xyz,$$

for $0 < x, y, z < 1$.

To find $P(X < Y < Z)$ integrate the joint density function over the three-dimensional region $S = \{(x,y,z) : x < y < z\}$.

$$\begin{aligned}
P(X < Y < Z) &= \iiint_S f(x,y,z)\,dx\,dy\,dz \\
&= \int_0^1 \int_0^z \int_0^y 8xyz\,dx\,dy\,dz \\
&= \int_0^1 \int_0^z 8yz\left(\frac{y^2}{2}\right)\,dy\,dz \\
&= \int_0^1 4z\left(\frac{z^4}{4}\right)\,dz = \frac{1}{6}.
\end{aligned}$$

An alternate solution appeals to symmetry. There are $3! = 6$ ways to order X, Y, and Z. Since the marginal densities are all the same, the probabilities for each ordered relationship are the same. And thus

$$P(X < Y < Z) = P(X < Z < Y) = \cdots = P(Z < Y < X) = \frac{1}{6}.$$

Note that in this last derivation we have not used the specific form of the density function f. The result holds for *any* three continuous i.i.d. random variables.

By extension, if $X_1, \ldots, X_n$ are continuous i.i.d. random variables then for any permutation $(i_1, \ldots, i_n)$ of $\{1, \ldots, n\}$,

$$P(X_{i_1} < \cdots < X_{i_n}) = \frac{1}{n!}.$$

■

If X and Y are independent then their joint density function factors into a product of two functions—one that depends on x and one that depends on y. Conversely, suppose the joint density of X and Y has the form $f(x, y) = g(x)h(y)$ for some functions g and h. Then it follows that X and Y are independent with $f_X(x) \propto g(x)$ and $f_Y(y) \propto h(y)$. We leave the proof to the reader.

Example 6.27 (i) The joint density of X and Y is

$$f(x, y) = 6e^{-(2x+3y)}, \quad \text{for } x, y > 0.$$

The joint density can be written as $f(x, y) = g(x)h(y)$, where $g(x) = \text{constant} \times e^{-2x}$, for $x > 0$; and $h(y) = \text{constant} \times e^{-3y}$, for $y > 0$. It follows that X and Y are independent. Furthermore, $X \sim \text{Exp}(2)$ and $Y \sim \text{Exp}(3)$. This becomes clear if we write $g(x) = 2e^{-2x}$ for $x > 0$, and $h(y) = 3e^{-3y}$ for $y > 0$.

(ii) Consider the joint density function

$$f(x, y) = 15e^{-(2x+3y)}, \quad \text{for } 0 < x < y.$$

Superficially the joint density appears to have a similar form as the density in (i). However, the constraints on the domain $x < y$ show that X and Y are not independent. The domain constraints are part of the function definition. We cannot factor the joint density into a product of two functions which only depend on x and y, respectively.

■

6.8.1 Accept–Reject Method

Rectangles are "nice" sets because their areas are easy to compute. If (X, Y) is uniformly distributed on the rectangle $[a, b] \times [c, d]$, then the joint pdf of (X, Y) is

$$f(x, y) = \frac{1}{(b-a)(d-c)}, \quad \text{for } a < x < b \quad \text{and} \quad c < y < d.$$

The density factors as

$$f(x, y) = f_X(x) f_Y(y),$$

where f_X is the uniform density on (a, b), and f_Y is the uniform density on (c, d).

This observation suggests how to simulate a uniformly random point in the rectangle: Generate a uniform number X in (a, b). Independently generate a uniform number Y in (c, d). Then (X, Y) gives the desired uniform point on the rectangle.

In R, type

```
> c(runif(1,a,b),runif(1,c,d))
```

to generate $(X, Y) \sim \text{Unif}([a, b] \times [c, d])$.

How can we generate a point uniformly distributed on some "complicated" set S? The idea is intuitive. Assume S is bounded, and enclose the set in a rectangle.

1. Generate a uniformly random point in the rectangle.
2. If the point is contained in S *accept* it as the desired point.
3. If the point is not contained in S, *reject* it and generate another point and keep doing so until a point is contained in S. When it does, *accept* that point.

We show that the accepted point is uniformly distributed in S. This method is known as the *accept–reject method.*

Let R be a rectangle that encloses S. Let $\boldsymbol{R}$ be a point uniformly distributed in R. And let $\boldsymbol{S}$ be the point obtained by the accept–reject method.

Proposition 6.2. *The random point* **S** *generated by the accept–reject method is uniformly distributed on* S.

Proof. To show that $\boldsymbol{S} \sim \text{Unif}(S)$, we need to show that for all $A \subseteq S$,

$$P(\boldsymbol{S} \in A) = \frac{\text{Area}\,(A)}{\text{Area}\,(S)}.$$

Let $A \subseteq S$. If $\boldsymbol{S} \in A$ then necessarily a point $\boldsymbol{R}$ was accepted. Furthermore, $\boldsymbol{R} \in A$. That is,

$$\begin{aligned} P(\boldsymbol{S} \in A) &= P(\boldsymbol{R} \in A | \boldsymbol{R} \text{ is accepted}) \\ &= \frac{P(\boldsymbol{R} \in A, \boldsymbol{R} \text{ is accepted})}{P(\boldsymbol{R} \text{ is accepted})} \\ &= \frac{P(\boldsymbol{R} \in A)}{P(\boldsymbol{R} \text{ is accepted})} \\ &= \frac{\text{Area}\,(A)/\text{Area}\,(R)}{\text{Area}\,(S)/\text{Area}\,(R)} = \frac{\text{Area}\,(A)}{\text{Area}\,(S)}. \end{aligned}$$

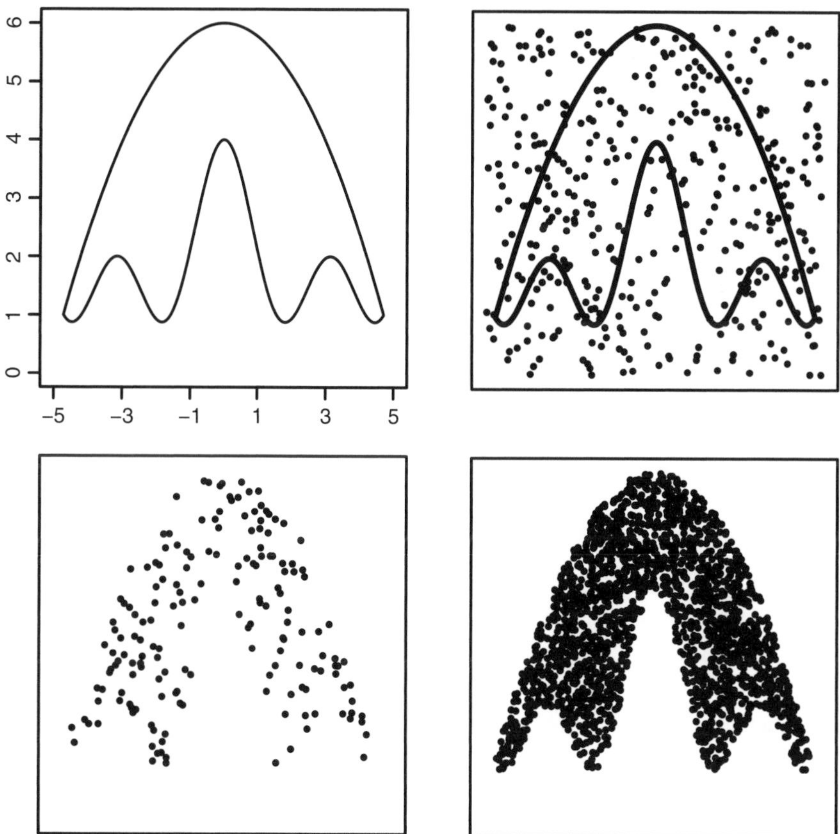

FIGURE 6.14: Accept–reject method for simulating points uniformly distributed in the top-left set S. In the top-right, 400 points are generated uniformly on the rectangle. Points inside S are accepted in the bottom left. In bottom right, the accept–reject method is used with an initial 5000 points of which 1970 are accepted.

Example 6.28 See Figure 6.14. The shape in the top left is the region S between the functions

$$f_1(x) = \frac{-20x^2}{9\pi^2} + 6 \text{ and } f_2(x) = \cos x + \cos 2x + 2.$$

The region is contained in the rectangle $[-5, 5] \times [0, 6]$. Points are generated uniformly on the rectangle with the commands

```
> x <- runif(1,-5,5)
> y <- runif(1,0,6)
```

and accepted if they fall inside S. The accepted points are uniformly distributed on S. See the script **AcceptReject1.R**.

The accept–reject method for generating uniform points in planar regions, extends in a natural way to three (and higher) dimensions. ■

Example 6.29 Let (X, Y, Z) be a point uniformly distributed on the sphere of radius 1 centered at the origin. Estimate the mean and standard deviation for the distance from the point to the origin.

Let $D = \sqrt{X^2 + Y^2 + Z^2}$ be the distance from (X, Y, Z) to the origin. We will simulate D by evaluating the distance function for a uniformly random point in the sphere.

Enclose the sphere in a cube of side length 2 centered at the origin. The command

```
> pt <- runif(3,-1,1)
```

generates a point uniformly distributed in the cube. A point (x, y, z) is contained in the unit sphere if $x^2 + y^2 + z^2 < 1$. The command

```
> if ((pt[1]^2 + pt[2]^2 + pt[3]^2) < 1) 1 else 0
```

checks whether `pt` lies in the sphere or not.

Here is the simulation. The points generated in the sphere are shown in Figure 6.15. To estimate the mean distance to the origin simply compute the distance to the origin for each of the simulated points and take their average. Similarly for the standard deviation.

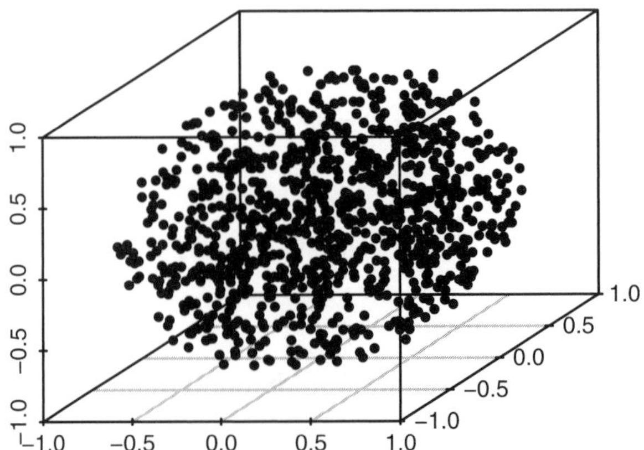

FIGURE 6.15: One thousand points generated in the unit sphere using the accept–reject method.

```
# Distance in unit sphere
> n <- 10000
> mat <- matrix(rep(0,3*n),nrow=n) # initialize n-x-3 matrix
> i <- 1
> while (i <= n) {
     pt <- runif(3,-1,1)
     if ((pt[1]^2 + pt[2]^2 + pt[3]^2) < 1 ) {
         mat[i,] <- pt
         i <- i+1          } }
> d <-sqrt(mat[,1]^2 + mat[,2]^2 + mat[,3]^2)
> mean(d)
[1] 0.7516139
> sd(d)
[1] 0.1934755
```

The exact theoretical values for the mean and standard deviation of distance are $E[D] = 3/4$ and $\text{SD}[D] = \sqrt{3/80} = 0.194\ldots$ ■

6.9 COVARIANCE, CORRELATION

For jointly distributed continuous random variables, the covariance and correlation are defined as for discrete random variables.

COVARIANCE

Let X and Y be jointly distributed continuous random variables with joint density function f. Let $\mu_X = E[X]$ and $\mu_Y = E[Y]$. The *covariance of X and Y* is

$$\begin{aligned}\text{Cov}(X,Y) &= E[(X-\mu_X)(Y-\mu_Y)] \\ &= \int_{-\infty}^{\infty}\int_{-\infty}^{\infty}(x-\mu_X)(y-\mu_Y)f(x,y)\,dx\,dy.\end{aligned}$$

For all jointly distributed random variables—continuous and discrete—we have $\text{Cov}(X,Y) = E[XY] - E[X]E[Y]$. We remind the reader

$$\text{Corr}(X,Y) = \frac{\text{Cov}(X,Y)}{\text{SD}[X]\text{SD}[Y]}.$$

Example 6.30 A point (X, Y) is uniformly distributed on the triangle with vertices (0,0), (1,0), and (1,1). Find the covariance and correlation between X and Y.

The area of the triangle is 1/2 and is described by the constraints $0 < y < x < 1$. The joint density function is thus

$$f(x, y) = 2, \quad \text{for } 0 < y < x < 1.$$

We have

$$E[XY] = \int_0^1 \int_0^x 2xy \, dy \, dx = \int_0^1 2x \left(\frac{x^2}{2}\right) dx = \int_0^1 x^3 \, dx = \frac{1}{4}.$$

For the marginal density of X, integrate out the y variable, giving

$$f_X(x) = \int_y f(x, y) \, dy = \int_0^x 2 \, dy = 2x, \quad \text{for } 0 < x < 1.$$

Thus,

$$E[X] = \int_0^1 x(2x) \, dx = \frac{2}{3}.$$

For the Y variable,

$$f_Y(y) = \int_y^1 2 \, dx = 2(1 - y), \quad \text{for } 0 < y < 1,$$

and

$$E[Y] = \int_0^1 y2(1 - y) \, dy = \frac{1}{3}.$$

This gives

$$\text{Cov}(X, Y) = E[XY] - E[X]E[Y] = \frac{1}{4} - \left(\frac{2}{3}\right)\left(\frac{1}{3}\right) = \frac{1}{36}.$$

For the correlation, we need the marginal standard deviations.

$$E[X^2] = \int_0^1 x^2(2x) \, dx = \frac{1}{2},$$

giving

$$V[X] = E[X^2] - E[X]^2 = \frac{1}{2} - \left(\frac{2}{3}\right)^2 = \frac{1}{18} = 0.0278.$$

Check that $V[Y] = 1/18$. This gives

$$\text{Corr}(X, Y) = \frac{\text{Cov}(X, Y)}{\text{SD}[X]\text{SD}[Y]} = \frac{1/36}{1/18} = \frac{1}{2}.$$

R: SIMULATION OF COVARIANCE, CORRELATION

The accept–reject method is used to simulate uniform points in the triangle. The sample covariance and correlation of the simulated points, computed with the R commands `cov(x,y)` and `cor(x,y)`, give Monte Carlo estimates of the theoretical covariance and correlation.

```
> xsim <-c()
> ysim <- c()
> for (i in 1:10000) {
   x <- runif(1)
   y <- runif(1)
   if (y < x) {
   xsim <- c(xsim,x)
   ysim <- c(ysim,y) }}
> cov(xsim,ysim)
[1] 0.02775561
> cor(xsim,ysim)
[1] 0.501157
```

■

6.10 FUNCTIONS OF RANDOM VARIABLES II

6.10.1 Maximums and Minimums

A scientist is monitoring four similar experiments and waiting for a chemical reaction to occur in each. Suppose the time until the reaction has some probability distribution and is modeled with random variables T_1, T_2, T_3, T_4, which represent the respective reaction time for each experiment. Of interest may be the time until the *first* reaction occurs. This is the *minimum* of the T_i's. The time until the *last* reaction is the *maximum* of the T_i's.

Maximums and minimums of collections of random variables arise frequently in applications. The key to working with them are the following algebraic relations.

INEQUALITIES FOR MAXIMUMS AND MINIMUMS

Let $x_1, \ldots, x_n, s, t$ be arbitrary numbers.

1. All of the x_k's are greater than or equal to s if and only if the *minimum* of the x_k's is greater than or equal to s. That is,

$$x_1 \geq s, \ldots, x_n \geq s \Leftrightarrow \min(x_1, \ldots, x_n) \geq s. \tag{6.6}$$

2. All of the x_k's are less than or equal to t if and only if the *maximum* of the x_k's is less than or equal to t. That is,

$$x_1 \leq t, \ldots, x_n \leq t \Leftrightarrow \max(x_1, \ldots, x_n) \leq t. \tag{6.7}$$

The same results hold if partial inequalities are replaced with strict inequalities.

Example 6.31 The reliability of three lab computers is being monitored. Let X_1, X_2, X_3, be the respective times until their systems crash and need to reboot. The "crash times" are independent of each other and have exponential distributions with respective parameters $\lambda_1 = 2$, $\lambda_2 = 3$, and $\lambda_3 = 5$. Let M be the time that the first computer crashes. Find the distribution of M.

The time of the first crash is the minimum of X_1, X_2, X_3. Using the above result for minimums, consider $P(M > m)$. For $m > 0$,

$$\begin{aligned} P(M > m) &= P(\min(X_1, X_2, X_3) > m) \\ &= P(X_1 > m, X_2 > m, X_3 > m) \\ &= P(X_1 > m)P(X_2 > m)P(X_3 > m) \\ &= e^{-2m}e^{-3m}e^{-5m} = e^{-10m}, \end{aligned}$$

where the third equality is from independence of the X_i's. This gives $F_M(m) = P(M \leq m) = 1 - e^{-10m}$, which is the cdf of an exponential random variable with parameter $\lambda = 10$. The minimum has an exponential distribution. See Figure 6.16 for a comparison of simulated distributions of X_1, X_2, X_3, and M. ■

This example illustrates a general result—the distribution of the minimum of independent exponential random variables is exponential.

MINIMUM OF INDEPENDENT EXPONENTIAL DISTRIBUTIONS

Let $X_1, \ldots, X_n$ be independent random variables with $X_k \sim \text{Exp}(\lambda_k)$ for $k = 1, \ldots, n$. Then,

$$\min(X_1, \ldots, X_n) \sim \text{Exp}(\lambda_1 + \cdots + \lambda_n).$$

Example 6.32 After a severe storm, an insurance company expects many claims for damages. An actuary has modeled the company's payout per claim, in thousands of dollars, with the probability density function

$$f(x) = \frac{x^2}{9}, \quad 0 < x < 3.$$

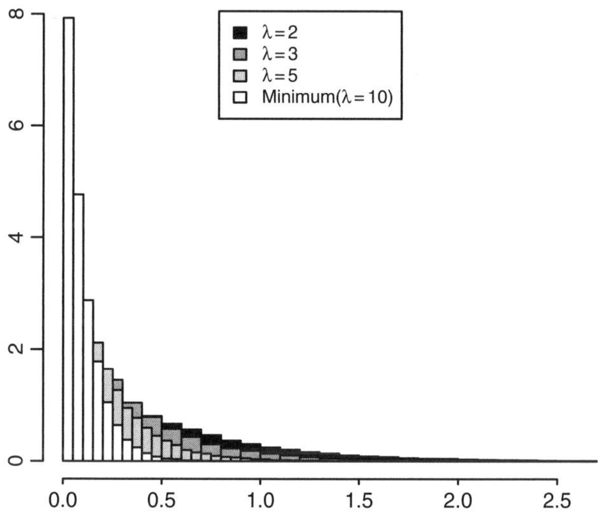

FIGURE 6.16: Simulated distribution of three independent exponential random variables and their minimum.

Suppose $X_1, \ldots, X_n$ represent payouts from n independent claims. Find the expected value of the maximum payout.

Let $M = \max(X_1, \ldots, X_n)$. First find the cdf of M. For $0 < m < 3$,

$$\begin{aligned} F_M(m) &= P(M \leq m) = P(\max(X_1, \ldots, X_n) \leq m) \\ &= P(X_1 \leq m, \ldots, X_n \leq m) \\ &= P(X_1 \leq m) \cdots P(X_n \leq m) \\ &= [P(X_1 \leq m)]^n = [F(m)]^n . \end{aligned}$$

Differentiating with respect to m gives

$$\begin{aligned} f_M(m) &= n\,[F(m)]^{n-1}\,f(m) \\ &= n\left[\int_0^m \frac{t^2}{9}\,dt\right]^{n-1}\frac{m^2}{9} = n\left[\frac{m^3}{27}\right]^{n-1}\frac{m^2}{9} \\ &= \frac{n}{9(27^{n-1})}m^{3n-1} = \frac{3nm^{3n-1}}{27^n}, \quad \text{for } 0 < m < 3. \end{aligned}$$

The expectation with respect to this density is

$$\begin{aligned} E[M] &= \int_0^3 m\frac{3nm^{3n-1}}{27^n}\,dm = \frac{3n}{27^n}\int_0^3 m^{3n}\,dm \\ &= \frac{3n}{27^n}\frac{3^{3n+1}}{3n+1} = \frac{9n}{3n+1}. \end{aligned}$$

Observe that expected maximum payout approaches \$3000 as the number of claims n gets large. ■

6.10.2 Sums of Random Variables

We have already seen several applications of sums of random variables. Let X and Y be independent continuous variables with respective densities f and g. We give a general expression for the density function of $X+Y$. Since $f(x,y) = f(x)g(y)$, the cdf of $X+Y$ is

$$\begin{aligned} P(X+Y \le t) &= \int_{-\infty}^{\infty} \int_{-\infty}^{t-y} f(x,y)\,dx\,dy \\ &= \int_{-\infty}^{\infty} \left(\int_{-\infty}^{t-y} f(x)\,dx \right) g(y)\,dy. \end{aligned}$$

Differentiating with respect to t and applying the fundamental theorem of calculus gives

$$f_{X+Y}(t) = \int_{-\infty}^{\infty} f(t-y)g(y)\,dy. \tag{6.8}$$

This is called the *convolution* of the densities of X and Y. Compare to the discrete formula for the probability mass function of a sum given in Equation 4.7.

Example 6.33 Sum of independent exponentials. Once a web page goes on-line, suppose the times between successive hits received by the page are independent and exponentially distributed with parameter λ. Find the density function of the time that the second hit is received.

Let X be the time that the first hit is received. Let Y be the additional time until the second hit. The time that the second hit is received is $X+Y$. We find the density of $X+Y$ two ways: (i) by using the convolution formula and (ii) from first principles and setting up the multiple integral.

(i) Suppose the density of X is f. The random variables X and Y have the same distribution and thus the convolution formula gives

$$f_{X+Y}(t) = \int_{-\infty}^{\infty} f(t-y)f(y)\,dy.$$

Consider the domain constraints on the densities and the limits of integration. Since the exponential density is equal to 0 for negative values, the integrand expression $f(t-y)f(y)$ will be positive when $t-y>0$ and $y>0$. That is, $0<y<t$. Hence for $t>0$,

$$\begin{aligned} f_{X+Y}(t) &= \int_{-\infty}^{\infty} f(t-y)f(y)\,dy \\ &= \int_0^t (\lambda e^{-\lambda(t-y)})(\lambda e^{-\lambda y})\,dy \\ &= \lambda^2 e^{-\lambda t} \int_0^t dy = \lambda^2 t e^{-\lambda t}. \end{aligned}$$

(ii) First find the cdf of $X + Y$. For $t > 0$,

$$\begin{aligned}
P(X + Y \le t) &= \iint_{\{(x,y):x+y\le t\}} f(x, y)\, dx\, dy \\
&= \int_0^t \int_0^{t-y} \lambda e^{-\lambda x} \lambda e^{-\lambda y}\, dx\, dy \\
&= \int_0^t \lambda e^{-\lambda y}(1 - e^{-\lambda(t-y)})\, dy \\
&= \int_0^t \lambda \left(e^{-\lambda y} - e^{-\lambda t}\right)\, dy \\
&= 1 - e^{-\lambda t} - \lambda t e^{-\lambda t}.
\end{aligned}$$

Differentiating with respect to t gives

$$f_{X+Y}(t) = \lambda e^{-\lambda t} - \lambda \left(e^{-\lambda t} - \lambda t e^{-\lambda t}\right) = \lambda^2 t e^{-\lambda t}, \quad \text{for } t > 0.$$

The density function of $X+Y$ is the density of a *gamma distribution*, which arises for sums of independent exponential random variables. The distribution is discussed in Chapter 7. ■

Example 6.34 Sum of independent uniforms. Let X and Y be i.i.d. random variables uniformly distributed on $(0,1)$. Find the density of $X + Y$.

Let f be the density function for the uniform distribution. We use the convolution formula. Since f is equal to 0 outside the interval (0,1), the integrand expression $f(t - y)f(y)$ will be positive when both factors are positive giving $0 < t - y < 1$ and $0 < y < 1$, or $t - 1 < y < t$ and $0 < y < 1$. We can write these two conditions as $\max(0, t - 1) < y < \min(1, t)$.

Since $X + Y$ takes values between 0 and 2, we consider two cases:
(i) For $0 < t < 1$,

$$\max(0, t - 1) = 0 < y < \min(1, t) = t$$

and thus $0 < y < t$. In that case,

$$f_{X+Y}(t) = \int_0^t f(t - y)f(y)\, dy = \int_0^t dy = t.$$

(ii) For $1 \le t \le 2$,

$$\max(0, t - 1) = t - 1 < y < \min(1, t) = 1$$

and thus $t - 1 < y < 1$. In that case,

$$f_{X+Y}(t) = \int_{t-1}^1 f(t - y)f(y)\, dy = \int_{t-1}^1 dy = 2 - t.$$

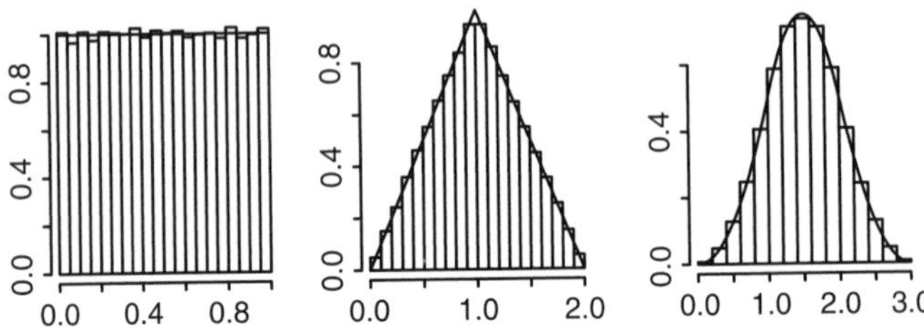

FIGURE 6.17: Distributions from left to right: uniform, sum of two independent uniforms, sum of three independent uniforms. Histogram is from 100,000 trials. Curve is the theoretical density.

Together, we get the *triangular density*

$$f_{X+Y}(t) = \begin{cases} t, & \text{if } 0 < t \leq 1 \\ 2 - t, & \text{if } 1 < t \leq 2, \end{cases}$$

and 0, otherwise. See Figure 6.17 for the graph of the triangular density and the density of the sum of three independent uniform random variables. ■

There are several equivalent ways to tackle this problem. A different approach using geometry is introduced next.

6.11 GEOMETRIC PROBABILITY

In this section we use geometry to solve probability problems. Geometric methods are a powerful tool for working with independent uniform random variables.

If X and Y are independent and each uniformly distributed on a bounded interval, then (X, Y) is uniformly distributed on a rectangle. Often problems involving two independent uniform random variables can be recast in two-dimensions where they can be approached geometrically.

Example 6.35 When Angel meets Lisa. Angel and Lisa are planning to meet at a coffee shop for lunch. Let M be the event that they meet. Each will arrive at the coffee shop at some time uniformly distributed between 1:00 and 1:30 p.m. independently of each other. When each arrives they will wait for the other person for 5 minutes, but then leave if the other person does not show up. What is the probability that they will meet?

Let A and L denote their respective arrival times in minutes from one o'clock. Then both A and L are uniformly distributed on $(0, 30)$. They will meet if they both arrive within 5 minutes of each other. Let M be the event that they meet. Then $M = \{|A - L| < 5\}$ and

$$P(M) = P(|A - L| < 5) = P(-5 < A - L < 5).$$

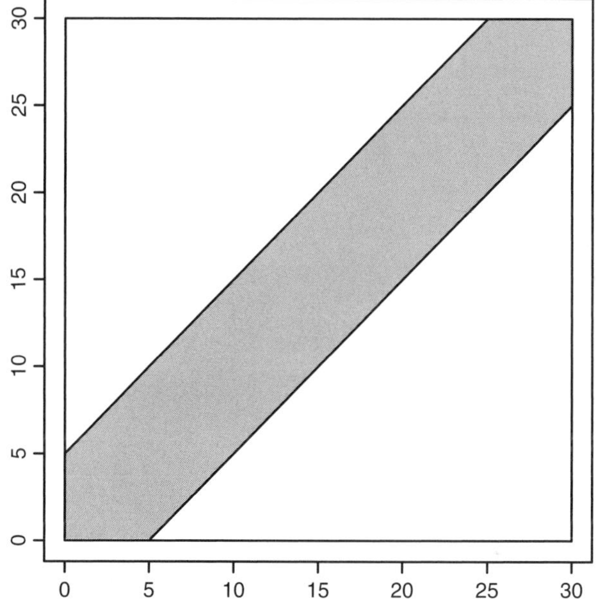

FIGURE 6.18: The shaded region is the event that Angel and Lisa meet.

The analytic approach to solving this probability involves a tricky double integral and working with the joint density function. We take a geometric approach.

Represent the arrival times A and L as a point (A, L) in a $[0{,}30] \times [0{,}30]$ square in the a-l plane. Since A and L are independent, the point (A, L) is uniformly distributed on the square. The event that Angel and Lisa meet corresponds to the region five units above and below the line $L = A$. See Figure 6.18.

The probability $P(M)$ is the area of the shaded region as a proportion of the area of the square. The area of the shaded region is best found by computing the area of the non-shaded region, which consists of two 25×25 triangles. Since the area of the square is $30 \times 30 = 900$, this gives

$$P(M) = \frac{900 - 625}{900} = \frac{11}{36} = 0.306. \qquad \blacksquare$$

We can use geometric methods not only to find probabilities but to derive cdfs and density functions as well.

Example 6.36 Let X and Y be independent and uniformly distributed on $(0, a)$. Let $Z = |X - Y|$ be their absolute difference. Find the density and expectation of Z.

The cdf of Z is $F(z) = P(Z \leq z) = P(|X - Y| \leq z)$, for $0 < z < a$. Observe that in the last example we found this probability for Angel and Lisa with $a = 30$ and $z = 5$. Generalizing the results there with a $[0, a] \times [0, a]$ square gives

$$P(|X - Y| \leq z) = \frac{a^2 - (a - z)^2}{a^2}.$$

Differentiating gives the density of Z

$$f_Z(z) = \frac{2(a - t)}{a^2}, \quad \text{for } 0 < t < a.$$

The expectation of the absolute difference of X and Y is

$$E[Z] = \int_0^a t\frac{2(a - t)}{a^2}\, dt = \frac{2}{a^2}\int_0^a (at - t^2)\, dt = \frac{2}{a^2}\left(\frac{at^2}{2} - \frac{t^3}{3}\right)\Bigg|_0^a = \frac{a}{3}. \quad ■$$

Example 6.37 A random point (X, Y) is uniformly distributed on the triangle with vertices at (0,0), (1,0), and (1,1). Find the distribution of the X-coordinate.

In Example 6.30 we gave the analytic solution using multiple integrals. Here we use geometry. First draw the picture. (See Figure 6.19.) For $0 < x < 1$, the event $\{X \leq x\}$ consists of the shaded region, with area $x^2/2$. The area of the large triangle is 1/2. Thus

$$P(X \leq x) = \frac{x^2/2}{1/2} = x^2.$$

Differentiating gives $f(x) = 2x$, for $0 < x < 1$.

Note that the marginal distribution of X is *not* uniform as the x-coordinate of the triangle is more likely to be large than small. ■

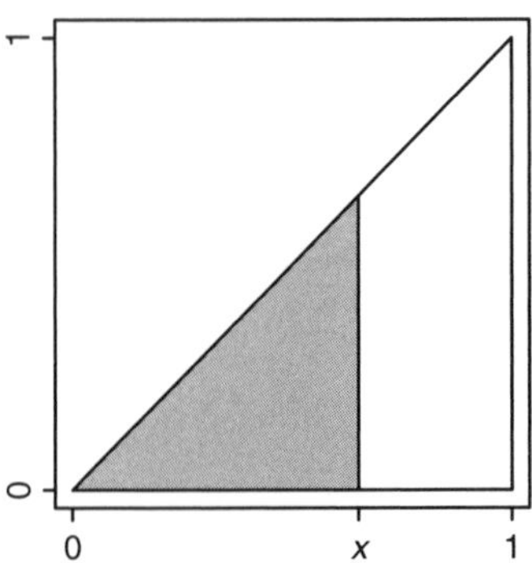

FIGURE 6.19: Let (X, Y) be uniformly distributed on the triangle. The shaded region is the event $\{X \leq x\}$.

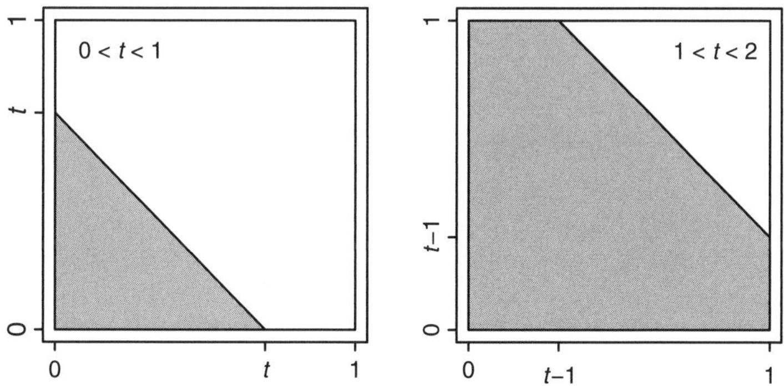

FIGURE 6.20: Region $\{X + Y \leq t\}$ for $0 < t < 1$ and $1 < t < 2$.

Example 6.38 Sum of uniforms revisited. The distribution of $X + Y$ for independent random variables uniformly distributed on $(0,1)$ was derived with the convolution formula in Exercise 6.34. Here we do it geometrically.

For $0 < t < 2$, consider $P(X + Y \leq t)$. The region $x + y \leq t$ in the $[0, 1] \times [0, 1]$ square is the region below the line $y = t - x$. The y-intercept of the line is $y = t$, which is inside the square, if $t < 1$, and outside the square, if $t > 1$. (See Figure 6.20.) The shaded region is the event $\{X + Y \leq t\}$ for different choices of t.

For $0 < t < 1$, the area of the shaded region is $t^2/2$. For $1 < t < 2$, the area of the shaded region is $1 - (2 - t)^2/2$. Thus,

$$P(X + Y \leq t) = \begin{cases} 0, & \text{if } t \leq 0 \\ t^2/2, & \text{if } 0 < t \leq 1 \\ 1 - (2-t)^2/2, & \text{if } 1 < t \leq 2 \\ 1, & \text{if } t \geq 2. \end{cases}$$

Differentiating gives

$$f_{X+Y}(t) = \begin{cases} t, & \text{if } 0 < t \leq 1 \\ 2 - t, & \text{if } 1 < t \leq 2, \end{cases}$$

and 0, otherwise. ■

Example 6.39 Buffon's needle. The most famous problem in geometric probability is Buffon's needle problem, introduced in 1733 by Georges-Louis Leclerc, Comte de Buffon, who posed the question:

> Suppose we have a floor made of parallel strips of wood, each the same width, and we drop a needle onto the floor. What is the probability that the needle will lie across a line between strips?

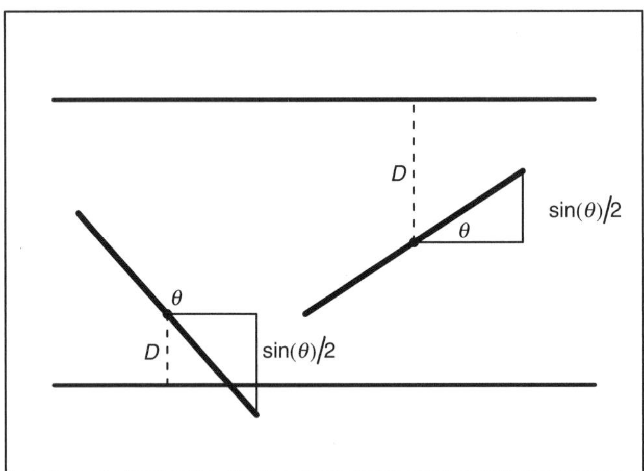

FIGURE 6.21: Geometry of Buffon's needle problem. The needle intersects a line if $\sin(\theta)/2 > D$.

Assume that the lines between strips of wood are one unit apart. and the length of the needle also has length one. It should be clear that we only need to consider what happens on one strip of wood.

Parametrize the needle's position with two numbers. Let D be the distance between the center of the needle and the closest line. Let Θ be the angle that the needle makes with that line. Then all positions of the needle can be described by (D, Θ), with $0 < D < 1/2$ and $0 < \Theta < \pi$.

Suppose $\Theta = \theta$. Consider a right triangle whose hypotenuse is the half of the needle closest to the nearest line and with two vertices at the center and endpoint of the needle. The length of the hypotenuse is 1/2. The triangle's angle at the needle's center will either be θ, if $0 < \theta < \pi/2$, or $\pi - \theta$, if $\pi/2 < \theta < \pi$. In either case, the side of the right triangle opposite that angle has length $(\sin\theta)/2$. See Figure 6.21.

Considering different cases shows that the needle intersects a line if and only if the distance of that side is greater than D. Thus the event that the needle crosses a line is equal to the event that $\sin(\Theta)/2 > D$ and

$$P\,(\text{Needle crosses line}) = P\left(\frac{\sin\Theta}{2} > D\right).$$

The needle's position (D, θ) can be expressed as a point in the $[0, \pi] \times [0, 1/2]$ rectangle in the θ-d plane. The event $\{\sin\Theta/2 > D\}$ is the region under the curve $d = (\sin\theta)/2$. (See Figure 6.22.)

The area of this region is

$$\int_0^{\pi} \frac{\sin\theta}{2}\,d\theta = \left.\frac{-\cos\theta}{2}\right|_0^{\pi} = 1.$$

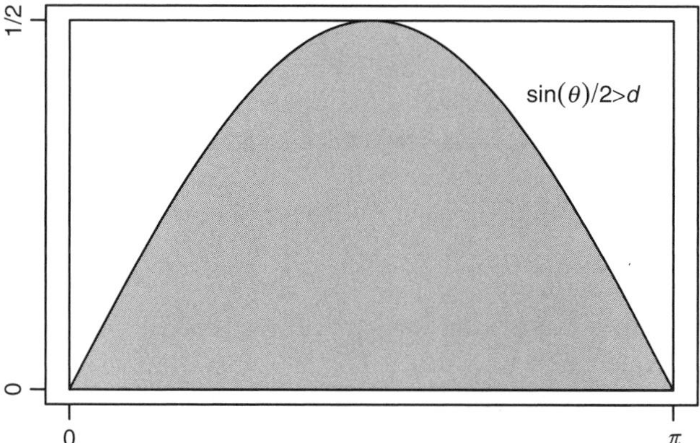

FIGURE 6.22: Buffon's needle problem is solved by finding the area under the curve $d = \sin(\theta)/2$ as a proportion of the area of the $[0, \pi] \times [0, 1/2]$ rectangle.

As a fraction of the area of the rectangle $\pi/2$ this gives

$$P\,(\text{Needle crosses line}) = \frac{1}{\pi/2} = \frac{2}{\pi} = 0.6366\ldots$$

Although the number π, pervasive throughout mathematics, is not a physical constant, here we have a physical method to simulate π: drop needles on a wooden floor and count the number of times a line is crossed. For large n,

$$P\,(\text{Needle crosses line}) \approx \frac{\text{Number of needle crosses}}{n},$$

which gives

$$\pi = \frac{2}{P(\text{Needles crosses line})} \approx \frac{2n}{\text{Number of needle crosses}}.$$ ■

Ants, fish, and noodles

- In "Ants estimate area using Buffon's needle," Mallon and Franks (2000) show a remarkable connection between the Buffon needle problem and how ants measure the size of their potential nest sites. Size assessment is made by individual ant scouts who are able to detect intersections of their tracks along potential nest sites. The authors argue that the ants use a "Buffon needle algorithm" to assess nest areas.

- Power plants use large volumes of water for cooling. Water is usually drawn into the plant by means of intake pumps which have large impeller blades. Sometimes the water contains small larval fish. In "Larval fish, power plants and Buffon's needle," Eby and Beauchamp (1977) extend the classic Buffon needle problem to estimate the probability of a larval fish being killed by an impeller blade of a pump in a power plant.
- Among the many generalizations of Buffon's needle, our favorite is "Buffon's noodle." In this version, the needle is allowed to be a curve. Although the probability that the noodle crosses the line depends on the shape of the curve, remarkably the expected number of crossings does not. That is, the expected number of lines which the noodle crosses is equal to the expected number of lines that a straight needle will cross. (See Ramaley 1969.)

Steinhaus, with his predilection for metaphors, used to quote a Polish proverb, 'Forturny kolem sie tocza' (Luck runs in circles), to explain why π, so intimately connected with circles, keeps cropping up in probability theory and statistics, the two disciplines that deal with randomness and luck.

—Mark Kac

6.12 SUMMARY

Continuous probability is introduced in this chapter. In the continuous setting, the sample space is uncountable. Integrals replace sums and density functions replace probability mass functions. Many concepts first introduced for discrete probability, such as expectation, variance, joint and conditional distributions, extend naturally to the continuous framework. The density function plays a similar role as that of the discrete probability mass function: we integrate the density to compute probabilities. However, unlike the pmf, the density is not a probability.

For all random variables X, discrete and continuous, there is a cumulative distribution function $F(x) = P(X \leq x)$. For continuous random variables, the density function is the derivative of the cdf. This important property is used extensively in finding distributions of continuous variables.

The continuous uniform and exponential distributions are introduced in this chapter. The uniform distribution on an interval (a, b) has density function constant on that interval. The exponential distribution is the only continuous distribution that is memoryless.

For two or more jointly distributed random variables, there is a joint density function. Probabilities which involve multiple random variables will require multiple integrals. For independent random variables the joint density $f(x, y)$ is a product of the marginal densities $f_X(x)f_Y(y)$. Many examples are given in this chapter for finding the density of a function of random variables, for instance, maximums and minimums, and sums of independent variables.

When X and Y are independent and uniformly distributed, the random pair (X, Y) is uniformly distributed on a rectangle in the plane. Often, such problems can be treated geometrically. The chapter concludes with several examples of geometric probability.

- **Continuous random variable:** A random variable which takes values in a continuous set.
- **Probability density function:** A function f is the density function of a continuous random variable if
  1. $f(x) \geq 0$ for all x.
  2. $\int_{-\infty}^{\infty} f(x)\,dx = 1$.
  3. For all $S \subseteq \mathbb{R}$, $P(X \in S) = \int_S f(x)\,dx$.
- **Cumulative distribution function:** The cdf of X is $F(x) = P(X \leq x)$, defined for all real x.
- **Pdf and cdf:** $F'(x) = f(x)$.
- **Properties of cdf:**
  1. $\lim_{x\to\infty} F(x) = 1$.
  2. $\lim_{x\to-\infty} F(x) = 0$.
  3. $F(x)$ is right-continuous at all x.
  4. $F(x)$ is an increasing function of x.
- **Expectation:** $E[X] = \int_{-\infty}^{\infty} xf(x)\,dx$.
- **Variance:** $V[X] = \int_{-\infty}^{\infty} (x - E[X])^2 f(x)\,dx$.
- **Law of unconscious statistician:** If g is a function, then

$$E[g(X)] = \int_{-\infty}^{\infty} g(x)f(x)\,dx.$$

- **Uniform distribution:** A continuous random variable X is uniformly distributed on (a, b) if the density function of X is

$$f(x) = \frac{1}{b-a}, \text{ for } a < x < b,$$

  and 0, otherwise.
- **Uniform setting:** The uniform distribution arises as a model for equally likely outcomes. Properties of the continuous uniform distribution include:
  1. $E[X] = (b + a)/2$.
  2. $V[X] = (b - a)^2/12$.
  3. $F(x) = P(X \leq x) = (x - a)/(b - a)$, if $a < x < b$, 0, if $x \leq a$, and 1, if $x \geq b$.

- **Exponential distribution:** The distribution of X is exponential with parameter $\lambda > 0$ if the density of X is $f(x) = \lambda e^{-\lambda x}$, for $x > 0$.
- **Exponential setting:** The exponential distribution is often used to model arrival times—the time until some event occurs, such as phone calls, traffic accidents, component failures, etc. Properties of the exponential distribution include:
  1. $E[X] = 1/\lambda$.
  2. $V[X] = 1/\lambda^2$.
  3. $F(x) = P(X \leq x) = 1 - e^{-\lambda x}$.
  4. The exponential distribution is the only continuous distribution which is memoryless.
- **Inverse transform method:** If the cdf F of a random variable X is invertible, and $U \sim$ Unif(0,1), then $F^{-1}(U)$ has the same distribution as X. This gives a method for simulating X.
- **Joint probability density function:** For jointly continuous random variables the joint density $f(x,y)$ has similar properties as the univariate density function:
  1. $f(x,y) \geq 0$ for all x and y
  2. $\int_{-\infty}^{\infty}\int_{-\infty}^{\infty} f(x,y) = 1$
  3. For all $S \subseteq \mathbb{R}^2$, $P((X,Y) \in S) = \iint\limits_S f(x,y)\,dx\,dy$.
- **Joint cumulative distribution function:** $F(x,y) = P(X \leq x, Y \leq y)$, defined for all real x and y.
- **Joint cdf and joint pdf:** $\frac{\partial^2}{\partial x \partial y} F(x,y) = f(x,y)$.
- **Expectation of function of two random variables:** If $g(x,y)$ is a function of two variables, then $E[g(X,Y)] = \int_{-\infty}^{\infty}\int_{-\infty}^{\infty} g(x,y)f(x,y)\,dx\,dy$.
- **Independence:** If X and Y are jointly continuous and independent, with marginal densities f_X and f_Y, respectively, then the joint density of X and Y is $f(x,y) = f_X(x)f_Y(y)$.
- **Accept–reject method:** Suppose S is a bounded set in the plane. The method gives a way to simulate from the uniform distribution on S. Enclose S in a rectangle R. Generate a point uniformly distributed in R. If the point is in S, "accept"; if the point is not in R, "reject" and try again. The first accepted point will be uniformly distributed on S.
- **Convolution:** If X and Y are continuous and independent, with respective marginal densities f_X and f_Y, then the density of $X+Y$ is given by the convolution formula

$$f_{X+Y}(t) = \int_{-\infty}^{\infty} f_X(t-y)f_Y(y)\,dy.$$

- **Covariance:**

$$\begin{aligned}\text{Cov}(X,Y)&E[(X-E[X])(Y-E[Y])]\\ &= \int_{-\infty}^{\infty}\int_{\infty}^{\infty}(x-E[X])(y-E[Y])\,dx\,dy.\end{aligned}$$

- **Problem solving strategies:**
  1. **Densities of functions of random variables:** Suppose $Y = g(X)$. To find the density of Y, start with the cdf $P(Y \leq y) = P(g(X) \leq y)$. Obtain an expression of the form $P(X \leq h(y))$ for some function h. Differentiate to obtain the desired density. Most likely you will need to apply the chain rule.
  2. **Setting up multiple integrals:** Many continuous problems involving two random variables or functions of random variables will involve multiple integrals. Make sure to define the limits of integration carefully. Several examples in the book, such as $P(X < Y)$ for different distributions of X and Y, are good problems to practice on.
  3. **Geometric probability:** Geometric probability is a powerful method for solving problems involving independent uniform random variables. If $X \sim \text{Unif}(a, b)$ and $Y \sim \text{Unif}(c, d)$ are independent, then (X, Y) is uniformly distributed on the rectangle $[a, b] \times [c, d]$. Finding uniform probabilities in two-dimensions reduces to finding areas. Try to find a geometrically-based solution for these types of problems.

EXERCISES

Density, cdf, expectation, variance

6.1 A random variable X has density function

$$f(x) = ce^x, \quad \text{for } -2 < x < 2.$$

(a) Find c.

(b) Find $P(X < -1)$.

(c) Find $E[X]$.

6.2 A random variable X has density function proportional to x^{-5} for $x > 1$.

(a) Find the constant of proportionality.

(b) Find and graph the cdf of X.

(c) Use the cdf to find $P(2 < X < 3)$.

(d) Find the mean and variance of X.

6.3 The cumulative distribution function for a random variable X is

$$F(x) = \begin{cases} 0, & \text{if } x \leq 0 \\ \sin x, & \text{if } 0 < x \leq \pi/2 \\ 1, & \text{if } x > \pi/2. \end{cases}$$

(a) Find $P(0.1 < X < 0.2)$.

(b) Find $E[X]$.

6.4 The Laplace distribution, also known as the double exponential distribution, has density function proportional to $e^{-|x|}$ for all real x. Find the mean and variance of the distribution.

6.5 The random variable X has density f that satisfies

$$f(x) \propto \frac{1}{1+x^2}$$

on the real numbers.

(a) Find $P(X > 1)$.

(b) Show that the expectation of X does not exist.

6.6 Show that

$$f(x) = e^x e^{-e^x}, \quad \text{for all } x,$$

is a probability density function. If X has such a density function find the cdf of X.

6.7 Let $X \sim \text{Unif}(a, b)$. Find a general expression for the *kth moment* $E[X^k]$.

6.8 An isosceles right triangle has side length uniformly distributed on (0,1). Find the expectation and variance of the length of the hypotenuse.

6.9 Suppose $f(x)$ and $g(x)$ are probability density functions. Under what conditions on the constants α and β will the function $\alpha f(x) + \beta g(x)$ be a probability density function?

6.10 Some authors take the following as the definition of continuous random variables: A random variable is continuous if the cumulative distribution function $F(x)$ is continuous for all real x. Show that if X is a discrete random variable, then the cdf of X is not continuous.

6.11 For continuous random variable X and constants a and b, prove that $E[aX + b] = aE[X] + b$.

Exponential distribution

6.12 It is 9:00 p.m. The time until Joe receives his next text message has an exponential distribution with mean 5 minutes.

(a) Find the probability that he will not receive a text in the next 10 minutes.

(b) Find the probability that the next text arrives between 9:07 and 9:10 p.m.

(c) Find the probability that a text arrives before 9:03 p.m.

(d) A text has not arrived for 5 minutes. Find the probability that none will arrive for 7 minutes.

6.13 Let $X \sim \text{Exp}(\lambda)$. Suppose $0 < s < t$. Since X is memoryless, is it true that $\{X > s + t\}$ are $\{X > t\}$ are independent events?

6.14 Derive the mean of the exponential distribution with parameter λ.

6.15 Derive the variance of the exponential distribution with parameter λ.

6.16 Let X have an exponential distribution conditioned to be greater than 1. That is, for $t > 1$, $P(X \leq t) = P(Y \leq t|Y > 1)$, where $Y \sim \text{Exp}(\lambda)$.

(a) Find the density of X.

(b) Find $E[X]$.

6.17 The time each student takes to finish an exam has an exponential distribution with mean 45 minutes. In a class of 10 students, what is the probability that at least one student will finish in less than 20 minutes? Assume students' times are independent.

6.18 For a continuous random variable X, the number m such that

$$P(X \leq m) = \frac{1}{2}$$

is called the *median* of X.

(a) Find the median of an $\text{Exp}(\lambda)$ distribution.

(b) Give a simplified expression for the difference between the mean and the median. Is the difference positive or negative?

6.19 Find the probability that an exponential random variable is within two standard deviations of the mean. That is, compute

$$P(|X - \mu| \leq 2\sigma),$$

where $\mu = E[X]$ and $\sigma = \text{SD}[X]$.

6.20 Solve these integrals without calculus. (Hint: Think exponential distribution.)

(a) $\int_0^\infty e^{-3x/10}\, dx$

(b) $\int_0^\infty te^{-4t}\, dt$

(c) $\int_0^\infty z^2 e^{-2z}\, dz$

6.21 Suppose $X \sim \text{Exp}(\lambda)$. Find the density of $Y = cX$ for $c > 0$. Describe the distribution of Y.

Functions of random variables

6.22 Suppose $U \sim \text{Unif(0,1)}$. Find the density of

$$Y = \tan\left(\pi U - \frac{\pi}{2}\right).$$

6.23 Let $X \sim \text{Unif}(0,1)$. Find $E\left[e^X\right]$ two ways:

(a) By finding the density of e^X and then computing the expectation with respect to that distribution.

(b) By using the law of the unconscious statistician.

6.24 The density of a random variable X is given by

$$f(x) = \begin{cases} 3x^2, & \text{for } 0 < x < 1 \\ 0, & \text{otherwise.} \end{cases}$$

Let $Y = e^X$.

(a) Find the density function of Y

(b) Find $E[Y]$ two ways: (i) Using the density of Y and (ii) Using the density of X.

6.25 Let X have a Cauchy distribution. That is, the density of X is

$$f(x) = \frac{1}{\pi(1+x^2)}, \quad -\infty < x < \infty.$$

Show that $1/X$ has a Cauchy distribution.

6.26 **Extreme value distribution.** Suppose $X_1, \ldots, X_n$ is an independent sequence of exponential random variables with parameter $\lambda = 1$. Let

$$Z = \max(X_1, \ldots, X_n) - \log n.$$

(a) Show that the cdf of Z is

$$F_Z(z) = \left(1 - \frac{e^{-z}}{n}\right)^n, \quad z > -\log n.$$

(b) Show that for all z,

$$F_Z(z) \to e^{-e^{-z}} \text{ as } n \to \infty.$$

The limit is a probability distribution called an *extreme value distribution*. It is used in many fields which model extreme values, such as hydrology (intense rainfall), actuarial science, and reliability theory.

(c) Suppose the times between heavy rainfalls are independent and have an exponential distribution with mean 1 month. Find the probability that in the next 10 years, the maximum time between heavy rainfalls is greater than 3 months in duration.

6.27 Let $X \sim \text{Unif}(a, b)$. Suppose Y is a linear function of X. That is $Y = mX + n$. Where m and n are constants. Assume also that $m > 0$. Show that Y is uniformly distributed on the interval $(ma + n, mb + n)$.

6.28 Suppose X has density function

$$f(x) = \frac{e^4}{e^4 - 1}|x|e^{-x^2/2}, \quad \text{for } -2 < x < 2.$$

Find the density of $Y = X^2$.

6.29 If r is a real number, the *ceiling of* r, denoted $\lceil r \rceil$, is the smallest integer not less than r. For instance, $\lceil 0.25 \rceil = 1$ and $\lceil 4 \rceil = 4$. Suppose $X \sim \text{Exp}(\lambda)$. Let $Y = \lceil X \rceil$. Show that Y has a geometric distribution.

Joint distributions

6.30 The joint density of X and Y is

$$f(x, y) = \frac{2x}{9y}, \quad \text{for } 0 < x < 3, 1 < y < e.$$

(a) Are X and Y independent?

(b) Find the joint cumulative distribution function.

(c) Find $P(1 < X < 2, Y > 2)$.

6.31 The joint density of X and Y is

$$f(x, y) = 2e^{-(x+2y)}, \quad \text{for } x > 0, y > 0.$$

(a) Find the joint cumulative distribution function.

(b) Find the cumulative distribution function of X.

(c) Find $P(X < Y)$.

6.32 The joint density of X and Y is

$$f(x, y) = ce^{-2y}, \quad \text{for } 0 < x < y < \infty.$$

(a) Find c.

(b) Find the marginal densities of X and Y. Do you recognize either of these distributions?

(c) Find $P(Y < 2X)$.

6.33 Suppose the joint density function of X and Y is

$$f(x, y) = 2e^{-6x}, \quad \text{for } x > 0, 1 < y < 4.$$

By noting the form of the joint density and without doing any calculations show that X and Y are independent. Describe their marginal distributions.

6.34 The time until the light in Bob's office fails is exponentially distributed with mean 2 hours. The time until the computer crashes in Bob's office is exponentially distributed with mean 3 hours. Failure and crash times are independent.

(a) Find the probability that neither the light nor computer fail in the next 2 hours.

(b) Find the probability that the computer crashes at least 1 hour after the light fails.

6.35 Let (X, Y, Z) be uniformly distributed on a three-dimensional box with side lengths 3, 4, and 5. Find $P(X < Y < Z)$.

6.36 A stick of unit length is broken into two pieces. The break occurs at a location uniformly at random on the stick. What is the expected length of the longer piece?

6.37 In Example 6.33 we found the distribution of the sum of two i.i.d. exponential variables with parameter λ. Call the sum X. Let Y be a third independent exponential variable with parameter λ. Use the convolution formula 6.8 to find the sum of three independent exponential random variables by finding the distribution of $X + Y$.

6.38 Suppose B and C are independent random variables each uniformly distributed on $(0,1)$. Find the probability that the roots of the equation

$$x^2 + Bx + C = 0$$

are real.

6.39 Let X be a continuous random variable with cdf F. Since $F(x)$ is a function of x, $F(X)$ is a random variable which is a function of X. Suppose F is invertible. Find the distribution of $F(X)$.

6.40 Suppose X has density function

$$f(x) = \frac{1}{(1+x)^2}, \quad \text{for } x > 0.$$

Show how to use the inverse transform method to simulate X.

6.41 Tom and Danny each pick uniformly random real numbers between 0 and 1. Find the expected value of the smaller number.

6.42 Let $X_1, \ldots, X_n$ be an i.i.d. sequence of Uniform (0,1) random variables. Let $M = \max(X_1, \ldots, X_n)$.

(a) Find the density function of M.

(b) Find $E[M]$ and $V[M]$.

6.43 Let X_1 and X_2 be independent exponential random variables with parameter λ. Show that $Y = |X_1 - X_2|$ is also exponentially distributed with parameter λ.

6.44 Suppose X and Y are i.i.d. exponential random variables with $\lambda = 1$. Find the density of X/Y and use it to compute $P(X/Y < 1)$.

6.45 Suppose (X, Y) are distributed uniformly on the circle of radius 1 centered at the origin. Find the marginal densities of X and Y. Are X and Y independent?

6.46 Suppose X and Y have joint probability density $f(x, y) = g(x)h(y)$ for some functions g and h which depend only on x and y, respectively. Show that X and Y are independent with $f_X(x) \propto g(x)$ and $f_Y(y) \propto h(y)$.

6.47 Consider the following attempt at generating a point uniformly distributed in the circle of radius 1 centered at the origin. In polar coordinates, pick R uniformly at random on (0,1). Pick Θ uniformly at random on $(0, 2\pi)$, independently of R. Show that this method does *not* work. That is show (R, Θ) is not uniformly distributed on the circle.

6.48 Let $A = UV$, where U and V are independent and uniformly distributed on (0,1).

(a) Find the density of A.

(b) Find $E[A]$ two ways: (i) using the density of A and (ii) not using the density of A.

6.49 See the joint density in Exercise 6.32. Find the covariance of X and Y.

6.50 Let X, Y, and Z be i.i.d. random variables uniformly distributed on (0,1). Find the density of $X + Y + Z$.

Geometric probability

6.51 Suppose X and Y are independent random variables, each uniformly distributed on (0,2).

(a) Find $P(X^2 < Y)$.

(b) Find $P(X^2 < Y | X + Y < 2)$.

6.52 Suppose (X, Y) is uniformly distributed on the region in the plane between the curves $y = \sin x$ and $y = \cos x$, for $0 < x < \pi/2$. Find $P(Y > 1/2)$.

6.53 Suppose (X, Y, Z) is uniformly distributed on the sphere of radius 1 centered at the origin. Find the probability that (X, Y, Z) is contained in the inscribed cube.

6.54 Solve Buffon's needle problem for a "short" needle. That is, suppose the length of the needle is $x < 1$.

6.55 Suppose you use Buffon's needle problem to simulate π. Let n be the number of needles you drop on the floor. Let X be the number of needles that cross a line. Find the distribution, expectation and variance of X.

6.56 Suppose X and Y are independent random variables uniformly distributed on (0,1). Use geometric arguments to find the density of $Z = X/Y$.

Simulation and R

6.57 Simulate the expected length of the hypotenuse of the isosceles right triangle in Exercise 6.8.

6.58 Let X and Y be independent exponential random variables with parameter $\lambda = 1$. Simulate $P(X/Y < 1)$. (See Exercise 6.44.)

6.59 Use the accept–reject method to simulate points uniformly distributed on the circle of radius 1 centered at the origin. Use your simulation to approximate the expected distance of a point inside the circle to the origin. (See Example 6.24.)

6.60 Let $R \sim \text{Unif}(1, 4)$. Let A be the area of the circle of radius R. Use R to simulate R. Simulate the mean and pdf of A and compare to the exact results. Create one graph with both the theoretical density and the simulated distribution.

6.61 See the R script **AcceptReject1.R**. Make up your own interesting shape S and use the accept–reject method to generate uniformly distributed points in S.

6.62 Let X be a random variable with density function $f(x) = 1/x^2$, for $1 < x < 4$, and 0, otherwise. Simulate $E[X]$ using the inverse transform method. Compare to the exact value.

6.63 Write an R script to estimate π using Buffon's needle problem. How many simulation iterations do you need to perform to be reasonably confident that your estimation is good within two significant digits? That is, $\pi \approx 3.14$.

6.64 Let $X_1, \ldots, X_n$ be independent random variables each uniformly distributed on $[-1, 1]$. Let $p_n = P(X_1^2 + \cdots + X_n^2 < 1)$. Conduct a simulation study to approximate p_n for increasing values of n.

For $n = 2$, p_2 is the probability that a point uniformly distributed on the square $[-1, 1] \times [-1, 1]$ falls in the inscribed circle of radius 1 centered at the origin. For $n = 3$, p_3 is the probability that a point uniformly distributed on the cube $[-1, 1] \times [-1, 1] \times [-1, 1]$ falls in the inscribed sphere of radius 1 centered at the origin. For $n > 3$, you are in higher dimensions estimating the probability that a point in a "hypercube" falls within the inscribed "hypersphere." What happens when n gets large?

7

CONTINUOUS DISTRIBUTIONS

> At this point an enigma presents itself which in all ages has agitated inquiring minds. How can it be that mathematics, being after all a product of human thought which is independent of experience, is so admirably appropriate to the objects of reality?
>
> —Albert Einstein

7.1 NORMAL DISTRIBUTION

> I know of scarcely anything so apt to impress the imagination as the wonderful form of cosmic order expressed by "the law of error." The law would have been personified by the Greeks and deified, if they had known of it. It reigns with severity in complete self-effacement amidst the wildest confusion. The huger the mob and the greater the anarchy the more perfect is its sway. Let a large sample of chaotic elements be taken and marshalled in order of their magnitudes, and then, however wildly irregular they appeared, an unexpected and most beautiful form of regularity proves to have been present all along.
>
> —Sir Francis Galton

The "law of error" in the quotation is now known as the normal distribution. It is perhaps the most important distribution in statistics, is ubiquitous as a model for natural phenomenon, and arises as the limit for many random processes and distributions throughout probability and statistics. It is sometimes called the Gaussian distribution, after Carl Friedrich Gauss, one of the greatest mathematicians in history, who discovered its utility as a model for astronomical measurement errors.

Probability: With Applications and R, First Edition. Robert P. Dobrow.

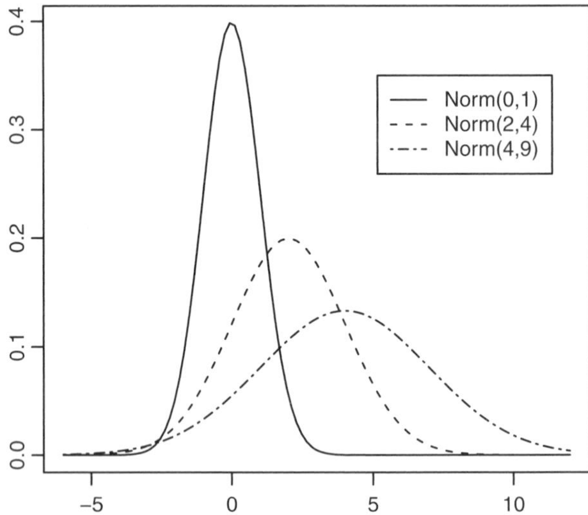

FIGURE 7.1: Three normal distributions.

Adolphe Quetelet, the father of quantitative social science, was the first to apply it to human measurements, including his detailed study of the chest circumferences of 5738 Scottish soldiers. Statistician Karl Pearson penned the name "normal distribution" in 1920, although he did admit that it "had the disadvantage of leading people to believe that all other distributions of frequency are in one sense or another *abnormal.*"

NORMAL DISTRIBUTION

A random variable X has the *normal distribution with parameters* μ *and* σ^2, if the density function of X is

$$f(x) = \frac{1}{\sigma\sqrt{2\pi}} e^{-\frac{(x-\mu)^2}{2\sigma^2}}, \quad -\infty < x < \infty.$$

We write $X \sim \text{Norm}(\mu, \sigma^2)$.

The shape of the density curve is the famous "bell curve" (see Fig. 7.1). The parameters μ and σ^2 are, respectively, the mean and variance of the distribution. The density curve is symmetric about the line $x = \mu$ and takes values in the set of all real numbers. The inflection points, where the curvature of the density function changes, occur one standard deviation unit from the mean, that is, at the points $x = \mu \pm \sigma$.

Surprisingly there is no closed form expression for the normal cumulative distribution function

$$F(x) = \int_{-\infty}^{x} \frac{1}{\sigma\sqrt{2\pi}} e^{-\frac{(t-\mu)^2}{2\sigma^2}}\, dt.$$

The integral has no antiderivative that is expressible with elementary functions. So numerical methods must be used to find normal probabilities.

The R commands for working with the normal distribution are

R: NORMAL DISTRIBUTION

```
> dnorm(x,μ,σ)
> pnorm(x,μ,σ)
> rnorm(x,μ,σ)
```

Default values for parameters are $\mu = 0$ and $\sigma = 1$.

Note that R uses the standard deviation σ in specifying the normal distribution, not the variance σ^2 as we do.

Although in general it is not possible to get exact closed form expressions for normal probabilities of the form

$$P(a < X < b) = \int_{a}^{b} f(x)\, dx,$$

it is possible with basic calculus to show that the density integrates to 1 on $(-\infty, \infty)$. It requires working in two dimensions with polar coordinates. Write

$$I = \int_{-\infty}^{\infty} \frac{1}{\sigma\sqrt{2\pi}} e^{-\frac{(t-\mu)^2}{2\sigma^2}}\, dt.$$

Change variables by setting $x = (t - \mu)/\sigma$ to get

$$I = \frac{1}{\sqrt{2\pi}} \int_{-\infty}^{\infty} e^{-\frac{x^2}{2}}\, dx.$$

Consider

$$I^2 = \frac{1}{2\pi} \int_{-\infty}^{\infty} \int_{-\infty}^{\infty} e^{-\frac{x^2+y^2}{2}}\, dx\, dy.$$

Work in polar coordinates, setting $x^2 + y^2 = r^2$ and $dx\, dy = r\, dr\, d\theta$. Then

$$I^2 = \frac{1}{2\pi} \int_{0}^{\infty} \int_{0}^{2\pi} re^{-\frac{r^2}{2}}\, dr\, d\theta = \int_{0}^{\infty} re^{-\frac{r^2}{2}}\, dr.$$

Solve this integral with the substitution $z = r^2/2$ and $dz = r\,dr$, giving

$$I^2 = \int_0^\infty e^{-z}\,dz = 1,$$

and thus $I = 1$.

7.1.1 Standard Normal Distribution

For real μ and $\sigma > 0$, suppose $X \sim \text{Norm}(\mu, \sigma^2)$. Define the *standardized* random variable

$$Z = \frac{X - \mu}{\sigma} = \left(\frac{1}{\sigma}\right) X - \frac{\mu}{\sigma}.$$

We show that Z is normally distributed with mean 0 and variance 1.

For real z,

$$P(Z \leq z) = P\left(\frac{X - \mu}{\sigma} \leq z\right) = P(X \leq \mu + \sigma z).$$

Differentiating with respect to z, using the chain rule, gives

$$f_Z(z) = \sigma f_X(\mu + \sigma z) = \sigma \left(\frac{1}{\sigma\sqrt{2\pi}} e^{-\frac{((\mu+\sigma z)-\mu)^2}{2\sigma^2}}\right) = \frac{1}{\sqrt{2\pi}} e^{-\frac{z^2}{2}},$$

which is the density of a normal distribution with mean 0 and variance 1. We call this the *standard normal distribution* and reserve the letter Z for a standard normal random variable.

Thus, any normal random variable $X \sim \text{Norm}(\mu, \sigma^2)$ can be transformed to a standard normal variable Z by the linear function $Z = (1/\sigma)X - \mu/\sigma$. In fact, *any* linear function of a normal random variable is normally distributed. We leave the proof of that result to the exercises.

LINEAR FUNCTION OF NORMAL RANDOM VARIABLE

Let X be normally distributed with mean μ and variance σ^2. For constants $a \neq 0$ and b, the random variable $Y = aX + b$ is normally distributed with mean

$$E[Y] = E[aX + b] = aE[X] + b = a\mu + b$$

and variance

$$V[Y] = V[aX + b] = a^2V[X] = a^2\sigma^2.$$

That is, $aX + b \sim \text{Norm}(a\mu + b, a^2\sigma^2)$.

The ability to transform any normal distribution to a standard normal distribution by a change of variables makes it possible to simplify many computations. Most statistics textbooks include tables of standard normal probabilities, and problems involving normal probabilities are usually solved by first standardizing the variables to work with the standard normal distribution. Today this is somewhat outdated due to technology.

Since a standard normal distribution has mean 0 and standard deviation 1,

$$P(|Z| \leq z) = P(-z \leq Z \leq z)$$

is the probability that Z is within z standard deviation units from the mean. It is also the probability that *any* normal random variable $X \sim \text{Norm}(\mu, \sigma^2)$ is within z standard deviation units from the mean since

$$P(|X - \mu| \leq \sigma z) = P\left(\left|\frac{X - \mu}{\sigma}\right| \leq z\right) = P(|Z| \leq z).$$

The probability that a normal random variable is within one, two, and three standard deviations from the mean, respectively, is found in R.

```
> pnorm(1)-pnorm(-1)
[1] 0.6826895
> pnorm(2)-pnorm(-2)
[1] 0.9544997
> pnorm(3)-pnorm(-3)
[1] 0.9973002
```

This gives the so-called "68-95-99.7 rule": For the normal distribution, the probability of being within one, two, and three standard deviations from the mean is, respectively, about 68, 95, and 99.7%. This rule can be a valuable tool for doing "back of the envelope" calculations when technology is not available.

Example 7.1 Babies' birth weights are normally distributed with mean $\mu = 120$ and standard deviation $\sigma = 20$ ounces. What is the probability that a random baby's birth weight will be greater than 140 ounces?

Let X represent a baby's birth weight. Then $X \sim \text{Norm}(120, 20^2)$. The desired probability is

$$P(X > 140) = P\left(\frac{X-\mu}{\sigma} > \frac{140-120}{20}\right) = P(Z > 1).$$

The weight $140 = 120 + 20$ is one standard deviation above the mean. As 68% of the probability mass is within one standard deviation of the mean, the remaining 32% is evenly divided between the outer two halves of the distribution. The desired probability is about 0.16. ■

7.1.2 Normal Approximation of Binomial Distribution

One of the first uses of the normal distribution was to approximate binomial probabilities for large n. Before the use of computers or technology, calculations involving binomial coefficients with large factorials were extremely hard to compute. The approximation made it possible to numerically solve many otherwise intractable problems and also gave theoretical insight into the importance of the normal distribution.

If $X \sim \text{Binom}(n, p)$ and n is large, then the distribution of X is approximately normal with mean np and variance $np(1-p)$. Equivalently, the standardized random variable

$$\frac{X - np}{\sqrt{np(1-p)}}$$

has an approximate standard normal distribution.

The result, discovered by Abraham de Moivre in early 1700s and generalized by Laplace 100 years later, says that if X is a binomial random variable with parameters n and p, then for $a < b$,

$$\lim_{n\to\infty} P\left(a \leq \frac{X-np}{\sqrt{np(1-p)}} \leq b\right) = \frac{1}{\sqrt{2\pi}} \int_a^b e^{-x^2/2}\, dx.$$

We see the normal approximation in Figure 7.2. The six binomial distributions all have parameter $p = 0.10$ with $n = 4, 10, 25, 50, 100, 500$. Each super-imposed curve is a normal density with mean np and variance $np(1-p)$.

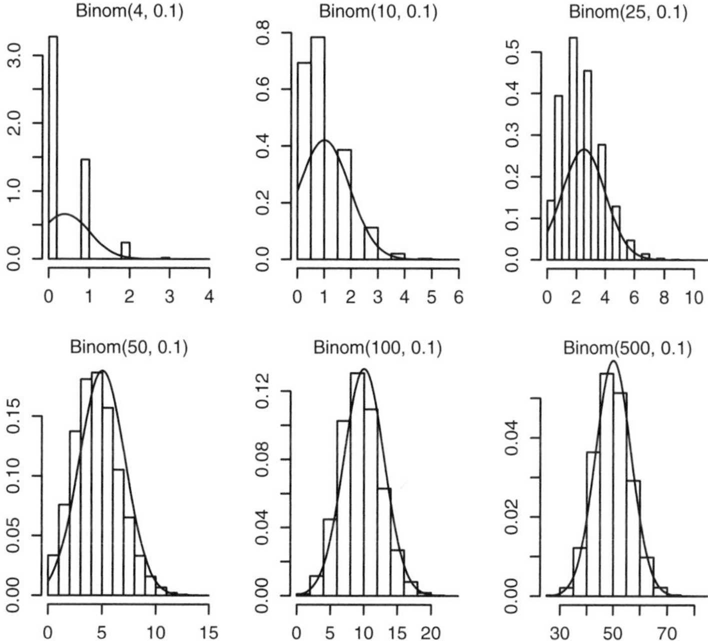

FIGURE 7.2: Normal approximation of the binomial distribution. For fixed p, as n gets large the binomial distribution tends to a normal distribution.

Example 7.2 In 600 rolls of a die, what is the probability of rolling between 90 and 110 fours?

Let X be the number of fours in 600 rolls. Then $X \sim \text{Binom}(600, 1/6)$. Let Y be a normal random variable with mean $np = 600(1/6) = 100$ and variance $np(1-p) = 500/6$. The normal approximation of the binomial gives

$$P(90 \leq X \leq 110) \approx P(90 \leq Y \leq 110) = 0.7267.$$

```
> pnorm(110,100,sqrt(500/6))-pnorm(90,100,sqrt(500/6))
[1] 0.7266783
```

An alternate derivation works with the standardized random variable to obtain

$$\begin{aligned} P(90 \leq X \leq 110) &= P\left(\frac{90-100}{\sqrt{500/6}} \leq \frac{X-np}{\sqrt{np(1-p)}} \leq \frac{110-100}{\sqrt{500/6}}\right) \\ &\approx P(-1.095 \leq Z \leq 1.095) \\ &= F_Z(1.095) - F_Z(-1.095) = 0.7267. \end{aligned}$$

Compare the normal approximation with the exact binomial probability

$$P(90 \leq X \leq 100) = \sum_{k=90}^{110} \binom{600}{k} \left(\frac{1}{6}\right)^{90} \left(\frac{5}{6}\right)^{110} = 0.7501.$$

```
> pbinom(110,600,1/6)-pbinom(89,600,1/6)
[1] 0.7501249
```

■

Continuity correction. Accuracy can often be improved in the normal approximation by accounting for the fact that we are using the area under a continuous density curve to approximate a discrete sum.

If X is a discrete random variable taking integer values, then for integers $a < b$ the probabilities

$$P(a \leq X \leq b),\ P\left(a - 1/2 \leq X \leq b + 1/2\right), \quad \text{and} \quad P(a - 1 < X < b + 1)$$

are all equal to each other. However, if Y is a continuous random variable with density f, then the corresponding probabilities

$$P(a \leq Y \leq b),\ P\left(a - 1/2 \leq Y \leq b + 1/2\right), \quad \text{and} \quad P(a - 1 < Y < b + 1)$$

are all different, as the integrals

$$\int_a^b f(y)\,dy, \quad \int_{a-1/2}^{b+1/2} f(y)\,dy, \quad \text{and} \quad \int_{a-1}^{b-1} f(y)\,dy$$

have different intervals of integration.

The *continuity correction* uses the middle integral for the approximation. That is,

$$P(a \leq X \leq b) = P(a - 1/2 \leq X \leq b + 1/2) \approx \int_{a-1/2}^{b+1/2} f(y)\,dy,$$

taking the limits of integration out one-half unit to the left of a and to the right of b.

For example, suppose the normal distribution is used to approximate $P(11 \leq X \leq 13)$, where X is a binomial random variable with parameters $n = 20$ and $p = 0.5$. Let Y be a normal random variable with the same mean and variance as X (see Fig. 7.3). The curve in the graphs is part of the normal density. The rectangles are part of the probability mass function of X.

The light gray area represents the binomial probability $P(11 \leq X \leq 13)$. The dark area in the first panel represents the normal probability $P(11 \leq Y \leq 13)$. However, since the interval of integration is $[11, 13]$, the area under the normal density curve does not capture the area of the binomial probability to the left of 11 and to the right of 13. In the second panel, the dark area represents $P(10.5 \leq Y \leq 13.5)$, and

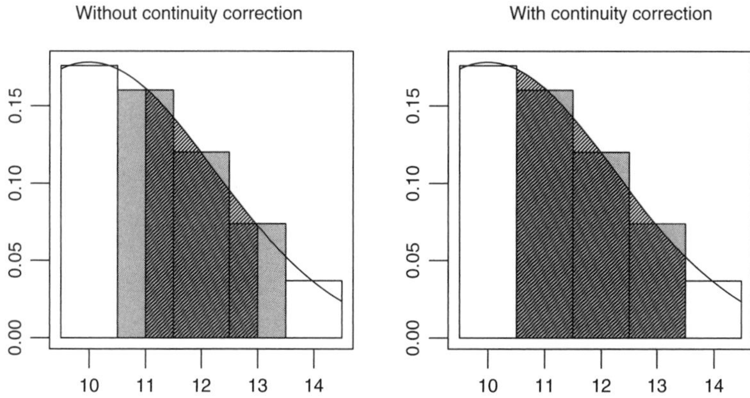

FIGURE 7.3: Approximating the binomial probability $P(11 \leq X \leq 13)$ using a normal distribution with and without the continuity correction.

visually we can see that this will give a better approximation of the desired binomial probability.

In fact, the exact binomial probability is $P(11 \leq X \leq 13) = 0.3542$. Compare to the normal approximations $P(11 \leq Y \leq 13) = 0.2375$ and $P(10.5 \leq Y \leq 13.5) = 0.3528$ to see the significant improvement using the continuity correction.

```
> dbinom(11,20,0.5)+dbinom(12,20,0.5)+dbinom(13,20,0.5)
[1] 0.3542423
> pnorm(13,10,sqrt(5))-pnorm(11,10,sqrt(5))
[1] 0.2375042
> pnorm(13.5,10,sqrt(5))-pnorm(10.5,10,sqrt(5))
[1] 0.3527692
```

Example 7.3 Dice, continued. In Example 7.2, applying the continuity correction gives

$$\begin{aligned} P(90 \leq X \leq 110) &= P(89.5 \leq X \leq 110.5) \\ &= P\left(\frac{89.5 - 100}{9.13} \leq \frac{X - np}{\sqrt{np(1-p)}} \leq \frac{110.5 - 100}{9.13}\right) \\ &\approx P(-1.15 \leq Z \leq 1.15) \\ &= F_Z(1.095) - F_Z(-1.095) = 0.740, \end{aligned}$$

which improves upon the previous estimate 0.727. The exact answer is 0.7501. ■

Quantiles. It is hard to escape taking standardized tests in the United States. Typically after taking such a test when your scores are reported, they usually include your percentile ranking. If you scored at the 90th percentile, also called the 90th quantile, this means that you scored as well, or better, than 90% of all test takers.

QUANTILE

Definition 7.1. *Let* $0 < p < 1$. *If* X *is a continuous random variable, then the* p*th quantile is the number* q *that satisfies*

$$P(X \leq q) = p/100.$$

That is, the p*th quantile separates the bottom* p *percent of the probability mass from the top* $(1 - p)$ *percent.*

The R command for finding quantiles of a probability distribution is obtained by prefacing the distribution name with the letter `q`. Normal quantiles are found with the command `qnorm`. To find the 25th quantile of the standard normal distribution, type

```
> qnorm(0.25)
[1] -0.6744898
```

As seen in the definition, the quantile function is inverse to the cumulative distribution function. The cdf evaluated at a quantile returns the original $p/100$.

```
> pnorm(-.6744899)
[1] 0.25
```

Example 7.4 The 68-95-99.7 rule says that the area under the standard normal curve between $z = -2$ and $z = 2$ is about 0.95. The area to the left of $z = -2$ is about 0.025. Thus, the total area to the left of $z = 2$ is $0.95 + 0.025 = 0.975$. Hence, $q = 2$ is the approximate 97.5th quantile of the standard normal distribution.

In R, we find the exact 97.5th quantile of a standard normal distribution.

```
> qnorm(0.975)
[1] 1.959964
```

■

Example 7.5 **IQR.** The *interquartile range (IQR)* of a distribution is the difference between the 75th and 25th quantiles, also called the third and first quartiles. If $X \sim \text{Norm}(\mu, \sigma)$, find the IQR of X.

Let q_{75} be the 75th quantile of the given normal distribution. Since

$$0.75 = P(X \leq q_{75}) = P\left(Z \leq \frac{q_{75} - \mu}{\sigma}\right),$$

$(q_{75} - \mu)/\sigma$ is the 75th quantile of the standard normal distribution. In R, we find

```
> qnorm(0.75)
[1] 0.6744898.
```

Thus, $(q_{75} - \mu)/\sigma = 0.6745$, which gives $q_{75} = \mu + 0.6745\sigma$. The normal distribution is symmetric about μ. It follows that the 25th quantile is $q_{25} = \mu - 0.6745\sigma$. The interquartile range is

$$\text{IQR} = q_{75} - q_{25} = (\mu + 0.6745\sigma) - (\mu - 0.6745\sigma) = 1.35\sigma.$$

Many statistical software programs will flag an observation as an *outlier* by the so-called $1.5 \times$ IQR rule: an observation is labeled an outlier if it is more than $1.5 \times$ IQR units above the 75th quantile, or $1.5 \times$ IQR units below the 25th quantile. Since

$$q_{75} + (1.5)\text{IQR} = (\mu + 0.6745\sigma) + (1.5)(1.35\sigma) = \mu + 2.7\sigma$$

and

$$q_{25} - (1.5)\text{IQR} = (\mu - 0.6745\sigma) - (1.5)(1.35\sigma) = \mu - 2.7\sigma,$$

an observation X from a normal distribution would be labeled an outlier if it is more than 2.7 standard deviations from the mean. The probability of this occurring is

$$P(|Z| > 2.7) = 2P(Z > 2.7) = 0.007.$$

While this probability is relatively small, it also means that in a dataset of 1000 observations taken from a normal distribution, the software would label about seven points as outliers. ■

Example 7.6 Let X be the number of heads in 10,000 coin tosses. We expect $X \approx 5000$. What range of values of X are typically observed with high probability, say 0.99? More precisely, what number t satisfies

$$P(5000 - t \leq X \leq 5000 + t) = 0.99?$$

Since $X \sim \text{Binom}(10,000, 0.5)$, the distribution of X is approximated by a normal distribution with mean 5000 and standard deviation $\sqrt{10{,}000(0.5)(0.5)} = 50$. Since n is fairly large, we will ease notation and omit the continuity correction, as it would have a negligible effect. We have

$$
\begin{aligned}
P(5000 - t \leq X \leq 5000 + t) &= P\left(\frac{-t}{50} \leq \frac{X - 5000}{50} \leq \frac{t}{50}\right) \\
&\approx P\left(\frac{-t}{50} \leq Z \leq \frac{t}{50}\right) \\
&= F_Z\left(\frac{t}{50}\right) - F_Z\left(\frac{-t}{50}\right) \\
&= F_Z\left(\frac{t}{50}\right) - \left[1 - F_Z\left(\frac{t}{50}\right)\right] \\
&= 2F_Z\left(\frac{t}{50}\right) - 1,
\end{aligned}
$$

where the next-to-last equality is from the symmetry of the standard normal density about 0. Setting the last expression equal to 0.99 gives

$$
F_Z\left(\frac{t}{50}\right) = P\left(Z \leq \frac{t}{50}\right) = \frac{1 + 0.99}{2} = 0.995. \tag{7.1}
$$

Thus, $t/50$ is the 99.5th quantile of the standard normal distribution. We find

```
> qnorm(0.995)
[1] 2.575829
```

This gives $t = 50(2.576) = 128.8 \approx 130$. In 10,000 coin tosses, we expect to observe 5000 ± 130 heads with high probability. A claim of tossing a fair coin 10,000 times and obtaining less than 4870 or more than 5130 heads would be highly suspect. ■

10,000 coin flips

Mathematician John Kerrich actually flipped 10,000 coins when he was interned in a Nazi prisoner of war camp during World War II. Kerrich was visiting Copenhagen at the start of the war, just when the German army was occupying Denmark. The data from Kerrich's coin tossing experiment are given in Freedman et al. (2007). Kerrich's results are in line with expected numbers. He got 5067 heads. In Chapter 9, we use Kerrich's data to illustrate the law of large numbers.

Kerrich spent much of his time in the war camp conducting experiments to demonstrate laws of probability. He is reported to have used ping-pong balls to illustrate Bayes formula.

7.1.3 Sums of Independent Normals

The normal distribution has the property that the sum of independent normal random variables is normally distributed. We show this is true for the sum of two standard normal variables.

Let X and Y be independent standard normal variables. To derive the density function of $X + Y$, use the convolution formula of Equation 6.8. For real t,

$$\begin{aligned} f_{X+Y}(t) &= \int_{-\infty}^{\infty} f(t-y)f(y)\,dy \\ &= \int_{-\infty}^{\infty} \frac{1}{\sqrt{2\pi}} e^{-\frac{(t-y)^2}{2}} \frac{1}{\sqrt{2\pi}} e^{-\frac{y^2}{2}}\,dy \\ &= \frac{1}{2\pi}\int_{-\infty}^{\infty} e^{-\frac{(2y^2-2yt+t^2)}{2}}\,dy. \end{aligned} \tag{7.2}$$

Consider the exponent of e in the last integral. Write

$$2y^2 - 2yt + t^2 = 2\left(y^2 - yt + \frac{t^2}{4} + \frac{t^2}{4}\right) = 2\left(y - \frac{t}{2}\right)^2 + \frac{t^2}{2}.$$

This gives

$$\begin{aligned} f_{X+Y}(t) &= \frac{1}{2\pi}\int_{-\infty}^{\infty} e^{-\left(y-\frac{t}{2}\right)^2} e^{-\frac{t^2}{4}}\,dy \\ &= \frac{1}{\sqrt{2\pi}\sqrt{2}} e^{-\frac{t^2}{4}} \int_{-\infty}^{\infty} \frac{\sqrt{2}}{\sqrt{2\pi}} e^{-\left(y-\frac{t}{2}\right)^2}\,dy \\ &= \frac{1}{\sqrt{2\pi}\sqrt{2}} e^{-t^2/4}, \end{aligned}$$

where the last equality follows because the integrand is the density of a normal distribution with mean $t/2$ and variance $1/2$, and thus integrates to one. The final expression is the density of a normal distribution with mean 0 and variance 2. That is, $X + Y \sim$ Norm(0,2).

The general result stated next can be derived with a messier integral and mathematical induction. It is also found by using moment generating functions as described in Section 9.5.

SUM OF INDEPENDENT NORMAL RANDOM VARIABLES IS NORMAL

Let $X_1, \ldots, X_n$ be a sequence of independent normal random variables with

$$X_k \sim \text{Norm}(\mu_k, \sigma_k^2), \text{ for } k = 1, \ldots, n.$$

Then

$$X_1 + \cdots + X_n \sim \text{Norm}(\mu_1 + \cdots + \mu_n, \sigma_1^2 + \cdots + \sigma_n^2).$$

Example 7.7 The mass of a cereal box is normally distributed with mean 385 g and standard deviation 5 g. What is the probability that 10 boxes will contain less than 3800 g of cereal?

Let $X_1, \ldots, X_{10}$ denote the respective cereal weights in each of the 10 boxes. Then $T = X_1 + \cdots + X_{10}$ is the total weight. We have that T is normally distributed with mean $E[T] = 10(385) = 3850$ and variance $V[T] = 10(5^2) = 250$.

The desired probability $P(T < 3800)$ can be found two ways. The most direct route uses technology.

```
> pnorm(3800,3850,sqrt(250))
[1] 0.0007827011
```

An alternate derivation, which gives more insight, is to standardize T, giving

$$P(T < 3800) = P\left(\frac{T-\mu}{\sigma} < \frac{3800-3850}{\sqrt{250}}\right) = P(Z < -3.16).$$

We see that 3800 g is a little more than three standard deviations below the mean. We can use software to find the exact probability. By the 68-95-99.7 rule, the probability is less than $0.003/2 = 0.0015$. ∎

Example 7.8 According to the National Center for Health Statistics, the mean male adult height in the United States is $\mu_M = 69.2$ inches with standard deviation $\sigma_M = 2.8$ inches. The mean female adult height is $\mu_F = 63.6$ inches with standard deviation $\sigma_F = 2.5$ inches. Assume male and female heights are normally distributed and independent of each other. If a man and woman are chosen at random, what is the probability that the woman will be taller than the man?

Let M and F denote male and female height, respectively. The desired probability is $P(F > M) = P(F - M > 0)$. One approach to find this probability is to obtain the joint density of F and M and then integrate over the region $\{(f,m) : f > m\}$.

But a much easier approach is to recognize that since F and M are independent normal variables, $F - M$ is normally distributed with mean

$$E[F-M] = E[F] - E[M] = 63.6 - 69.2 = -5.6$$

and variance

$$V[F-M] = V[F] + V[M] = (2.5)^2 + (2.8)^2 = 14.09.$$

This gives

$$P(F > M) = P(F - M > 0) = P\left(Z > \frac{0-(-5.6)}{\sqrt{14.09}}\right)$$
$$= P(Z > 1.492) = 0.068.$$

∎

Averages. Averages of i.i.d. random variables figure prominently in probability and statistics. If $X_1, \ldots, X_n$ is an i.i.d. sequence (not necessarily normal), let $S_n = X_1 + \cdots + X_n$. The average of the X_i's is S_n/n. Results for averages are worthwhile to highlight and remember.

AVERAGES OF i.i.d. RANDOM VARIABLES

Let $X_1, \ldots, X_n$ be an i.i.d. sequence of random variables with common mean μ and variance σ^2. Let $S_n = X_1 + \cdots + X_n$. Then

$$E\left[\frac{S_n}{n}\right] = \mu \text{ and } V\left[\frac{S_n}{n}\right] = \frac{\sigma^2}{n}.$$

If the X_i's are normally distributed, then $S_n/n \sim \text{Norm}(\mu, \sigma^2/n)$.

The mean and variance of the average are

$$E\left[\frac{S_n}{n}\right] = E\left[\frac{X_1 + \cdots + X_n}{n}\right] = \frac{1}{n}(\mu + \cdots + \mu) = \frac{n\mu}{n} = \mu$$

and

$$V\left[\frac{S_n}{n}\right] = V\left[\frac{X_1 + \cdots + X_n}{n}\right] = \frac{1}{n^2}\left(\sigma^2 + \cdots + \sigma^2\right) = \frac{n\sigma^2}{n^2} = \frac{\sigma^2}{n}.$$

Example 7.9 Averages are better than single measurements. The big metal spring scale at the fruit stand is not very precise and has a significant measurement error. When measuring fruit, the measurement error is the difference between the true weight and what the scale says. Measurement error is often modeled with a mean 0 normal distribution.

Suppose the scale's measurement error M is normally distributed with $\mu = 0$ and $\sigma = 2$ ounces. If a piece of fruit's true weight is w, then the *observed weight* of the fruit is what the customer sees on the scale—the sum of the true weight and the measurement error. Let X be the observed weight. Then $X = w + M$. Since w is a constant, X is normally distributed with mean $E[X] = E[w+M] = w+E[M] = w$, and variance $V[X] = V[w + M] = V[M] = 4$.

When a shopper weighs their fruit, the probability that the observed measurement is within 1 ounce of the true weight is

$$
\begin{aligned}
P(|X-w| \leq 1) &= P(-1 \leq X-w \leq 1) \\
&= P\left(\frac{-1}{2} \leq \frac{X-w}{\sigma} \leq \frac{1}{2}\right) \\
&= P\left(\frac{-1}{2} \leq Z \leq \frac{1}{2}\right) = 0.383.
\end{aligned}
$$

Equivalently, there is a greater than 60% chance that the scale will show a weight that is off by more than 1 ounce.

However, a savvy shopper decides to take n independent measurements $X_1, \ldots, X_n$ and rely on the average $S_n/n = (X_1 + \cdots + X_n)/n$. Since $S_n/n \sim \text{Norm}\,(w, 4/n)$, the probability that the average measurement is within 1 ounce of the true weight is

$$
\begin{aligned}
P\left(|S_n/n - w| \leq 1\right) &= P(-1 \leq S_n/n - w \leq 1) \\
&= P\left(\frac{-1}{\sqrt{4/n}} \leq \frac{S_n/n - w}{\sigma/\sqrt{n}} \leq \frac{1}{\sqrt{4/n}}\right) \\
&= P\left(\frac{-\sqrt{n}}{2} \leq Z \leq \frac{\sqrt{n}}{2}\right) \\
&= 2F\left(\frac{\sqrt{n}}{2}\right) - 1.
\end{aligned}
$$

If the shopper wants to be "95% confident" that the average is within 1 ounce of the true weight, that is, $P(|S_n/n - w| \leq 1) = 0.95$, how many measurements should they take? Solve $0.95 = 2F\left(\sqrt{n}/2\right) - 1$ for n. We have

$$
F\left(\frac{\sqrt{n}}{2}\right) = \frac{1+0.95}{2} = 0.975.
$$

The 97.5th quantile of the standard normal distribution is 1.96. Thus, $\sqrt{n}/2 = 1.96$ and $n = (2 \times 1.96)^2 = 7.68$. The shopper should take eight measurements. ■

Example 7.10 Darts. Stern (1997) collected data from 590 throws at a dart board, measuring the distance between the dart and the bullseye. To create a model of dart throws, it is assumed that the horizontal and vertical errors H and V are independent random variables normally distributed with mean 0 and variance σ^2. Let T be the distance from the dart to the bullseye. Then $T = \sqrt{H^2 + V^2}$ is the radial distance. We find the distribution of T.

The radial distance is a function of two independent normals. The joint density of H and V is the product of their marginal densities. That is,

$$f(h,v) = \left(\frac{1}{\sqrt{2\pi}\sigma}e^{-h^2/2\sigma^2}\right)\left(\frac{1}{\sqrt{2\pi}\sigma}e^{-v^2/2\sigma^2}\right) = \frac{1}{2\pi\sigma^2}e^{-(h^2+v^2)/2\sigma^2}.$$

For $t > 0$,

$$\begin{aligned} P(T \leq t) &= P\left(\sqrt{H^2+V^2} \leq t\right) = P(H^2+V^2 \leq t^2) \\ &= \int_{-t}^{t}\int_{-\sqrt{t^2-h^2}}^{\sqrt{t^2-h^2}} \frac{1}{2\pi\sigma^2}e^{-(h^2+v^2)/2\sigma^2}\,dh\,dv. \end{aligned}$$

Changing to polar coordinates gives

$$\begin{aligned} P(T \leq t) &= \frac{1}{2\pi\sigma^2}\int_0^t\int_0^{2\pi} e^{-r^2/2\sigma^2} r\,d\theta\,dr \\ &= \frac{1}{\sigma^2}\int_0^t re^{-r^2/2\sigma^2}\,dr = 1 - e^{-t^2/2\sigma^2}. \end{aligned}$$

Differentiating with respect to t gives

$$f_T(t) = \frac{t}{\sigma^2}e^{-\frac{t^2}{2\sigma^2}}, \quad \text{for } t > 0.$$

This is the density of a *Weibull distribution* often used in reliability theory and industrial engineering.

In the darts model, the parameter σ is a measure of players' accuracy. Several statistical techniques can be used to estimate σ from the data. Estimates for two professional darts players were made at $\sigma \approx 13.3$. For this level of accuracy, the probability of missing the bullseye by more than 40 mm (about 1.5 inches) is

$$P(T > 40) = \int_{40}^{\infty} \frac{t}{(13.3)^2}e^{-\frac{t^2}{2(13.3)^2}} = 0.011. \quad \blacksquare$$

FIGURE 7.4: A German 10-mark note honoring the contributions of Carl Gauss. There one finds a bell curve representing the Gaussian normal distribution.

7.2 GAMMA DISTRIBUTION

The gamma distribution is a family of positive, continuous distributions with two parameters. The density curve can take a wide variety of shapes, which allows the distribution to be used to model variables that exhibit skewed and nonsymmetric behavior. See Fig. 7.5.

> **GAMMA DISTRIBUTION**
>
> A random variable X has a *gamma distribution with parameters* a > 0 *and* $\lambda > 0$ if the density function of X is
>
> $$f(x) = \frac{\lambda^a x^{a-1} e^{-\lambda x}}{\Gamma(a)}, \quad \text{for} x > 0,$$
>
> where
>
> $$\Gamma(a) = \int_0^\infty t^{a-1} e^{-t} dt.$$
>
> We write $X \sim \text{Gamma}(a, \lambda)$.

The function Γ is the *gamma function.* The function is continuous, defined by an integral, and arises in many applied settings. Observe that $\Gamma(1) = 1$. Integration by parts (see Exercise 7.18) gives

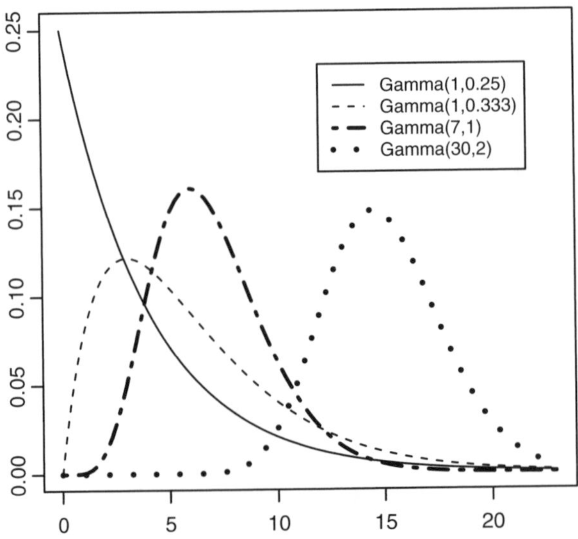

FIGURE 7.5: Four gamma distributions.

$$\Gamma(x) = (x-1)\Gamma(x-1), \quad \text{for all } x. \tag{7.3}$$

If x is a positive integer, unwinding Equation 7.3 shows that

$$\begin{aligned}\Gamma(x) &= (x-1)\Gamma(x-1)\\ &= (x-1)(x-2)\Gamma(x-2)\\ &= (x-1)(x-2)\cdots(1)\\ &= (x-1)!.\end{aligned}$$

We have always found it remarkable that the "most discrete of all functions"—the factorial function—is contained in the continuous gamma function.

If the first parameter a of the gamma distribution is equal to one, the gamma density function reduces to an exponential density. The gamma distribution is a two-parameter generalization of the exponential distribution.

In many applications, exponential random variables are used to model *interarrival times* between events, such as the times between successive highway accidents, component failures, telephone calls, or bus arrivals. The nth occurrence, or time of the nth arrival, is the sum of n interarrival times. The gamma distribution arises, as we will see, since the sum of n i.i.d. exponential random variables has a gamma distribution.

SUM OF INDEPENDENT EXPONENTIALS HAS GAMMA DISTRIBUTION

Let $E_1, \ldots, E_n$ be an i.i.d. sequence of exponential random variables with parameter λ. Let

$$S = E_1 + \cdots + E_n.$$

Then S has a gamma distribution with parameters n and λ.

Zwahlen et al. (2000) explores the process by which at-risk individuals take and retake HIV tests. The time between successive retaking of the HIV test is modeled with an exponential distribution. The time that an individual retakes the test for the nth time is fitted to a gamma distribution with parameters n and λ. Men and women retake the test at different rates, and statistical methods are used to estimate λ for men and women and for different populations. Understanding the process of test-taking can help public health officials treat at-risk populations more effectively.

It was shown in the last chapter (Example 6.33) that the sum of 2 i.i.d. exponential random variables X and Y with parameter λ has density $f_{X+Y}(t) = \lambda^2 te^{-\lambda t}$, for $t > 0$. This is a gamma distribution with parameters $n = 2$ and λ.

The general result for the sum of n independent exponentials follows from the convolution formula and mathematical induction. We do not assume knowledge of mathematical induction in this book, but interested readers can see the derivation at the end of this section. The result will also be shown using moment generating functions in Section 9.5.

Example 7.11 The times between insurance claims following a natural disaster are modeled as i.i.d. exponential random variables. Claims arrive during the first few days at a rate of about four claims per hour. (i) What is the probability that the 100th insurance claim will not arrive during the first 24 hours? (ii) The probability is at least 95% that the 100th claim will arrive before what time?

Let S_{100} be the time of the 100th claim (in hours). Then S_{100} has a gamma distribution with parameters $n = 100$ and $\lambda = 4$. (i) The desired probability is $P(S_{100} > 24)$. In R, type

```
> 1-pgamma(24,100,4)
[1] 0.6450564
```

(ii) Solve $P(S_{100} \leq t) \geq 0.95$. Find the 95th quantile of S_{100}. Type

```
> qgamma(0.95,100,4)
[1] 29.24928
```

The probability is at least 95% that the 100th claim will arrive before time $t = 29.25$, which is 5 hours and 15 minutes into the second day. ■

R : SIMULATING THE GAMMA DISTRIBUTION FROM A SUM OF EXPONENTIALS

We simulate the sum of 20 independent exponential random variables with $\lambda = 2$ and compare to the gamma density with $a = 20$ and $\lambda = 2$ (see Fig. 7.6).

```
> simlist <- replicate(10000,sum(rexp(20,2)))
> hist(simlist,prob=T)
> curve(dgamma(x,20,2),0,20,add=T)
```

Expectation and variance. For the gamma distribution with parameters a and λ, the expectation μ and variance σ^2 are

$$\mu = \frac{a}{\lambda} \quad \text{and} \quad \sigma^2 = \frac{a}{\lambda^2}.$$

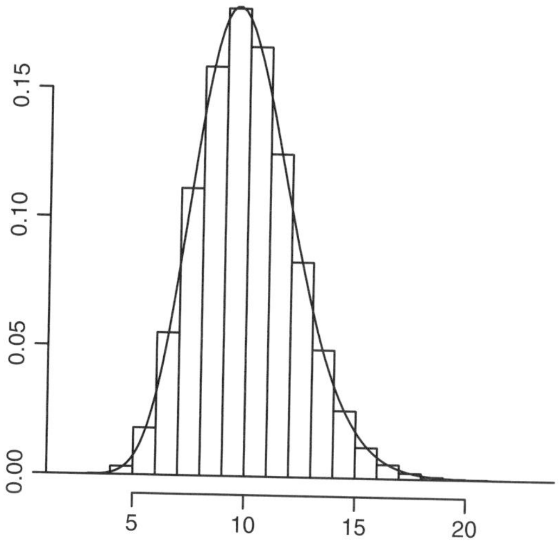

FIGURE 7.6: The histogram is from simulating the sum of 20 exponential variables with $\lambda = 2$. The curve is the density of a gamma distribution with parameters $a = 20$ and $\lambda = 2$.

These should not be surprising. If a is a positive integer, consider a sum of a i.i.d. exponential random variables with parameter λ. Each exponential variable has expectation $1/\lambda$, and the sum has a gamma distribution. Similarly, the variance of each exponential variable is $1/\lambda^2$, and the variance of the sum is a/λ^2. The expectation and variance for general a are derived in the next section.

7.2.1 Probability as a Technique of Integration

Many integrals can be solved by recognizing that the integrand is proportional to a known probability density.

Example 7.12 Solve

$$I = \int_0^\infty e^{-3x/5}\, dx.$$

This is an "easy" integral with a substitution. But before solving it with calculus, recognize that the integrand is "almost" an exponential density. It is proportional to an exponential density with $\lambda = 3/5$. Putting in the necessary constant to make it so gives

$$I = \frac{5}{3}\int_0^\infty \frac{3}{5} e^{-3x/5}\, dx = \frac{5}{3},$$

since the integrand is now a probability density, which integrates to one. ■

Example 7.13 Solve

$$I = \int_{-\infty}^{\infty} e^{-x^2}\, dx. \tag{7.4}$$

The integral looks like a normal density function. Make it so. Since the exponential factor of the normal density has the form $e^{-(x-\mu)^2/2\sigma^2}$, in order to match up with the integrand take $\mu = 0$ and $2\sigma^2 = 1$, that is, $\sigma^2 = 1/2$. Reexpressing the integral as a normal density gives

$$I = \int_{-\infty}^{\infty} e^{-x^2}\, dx = \sqrt{\pi} \int_{-\infty}^{\infty} \frac{1}{\sqrt{2\pi}\sqrt{1/2}} e^{-\frac{x^2}{2(1/2)}}\, dx = \sqrt{\pi},$$

where the last equality follows since the final integrand is a normal density with $\mu = 0$ and $\sigma^2 = 1/2$ and thus integrates to one. ■

Example 7.14 Meteorologists and hydrologists use probability to model numerous quantities related to rainfall. *Rainfall intensity*, as discussed in Watkins (), (http://www.sjsu.edu/faculty/watkins/raindrop.htm.) is equal to the raindrop volume times the intensity of droplet downfall. Let R denote rainfall intensity. (The units are usually millimeters per hour.) A standard model is to assume that the intensity of droplet downfall k (number of drops per unit area per hour) is constant and that the radius of a raindrop D has an exponential distribution with parameter λ. This gives $R = 4k\pi D^3/3$. The expected rainfall intensity is

$$E[R] = E\left[\frac{4k\pi D^3}{3}\right] = \frac{4k\pi}{3} E[D^3] = \frac{4k\pi}{3} \int_0^{\infty} x^3 \lambda e^{-x\lambda}\, dx.$$

The integral can be solved by three applications of integration by parts. But avoid the integration by recognizing that the integrand is proportional to a gamma density with parameters $a = 4$ and λ. The integral is equal to

$$\int_0^{\infty} \lambda x^3 e^{-\lambda x}\, dx = \frac{\Gamma(4)}{\lambda^3} \int_0^{\infty} \frac{\lambda^4 x^3 e^{-\lambda x}}{\Gamma(4)}\, dx = \frac{\Gamma(4)}{\lambda^3} = \frac{6}{\lambda^3}.$$

The expected rainfall intensity is

$$E[R] = \frac{4k\pi}{3}\left(\frac{6}{\lambda^3}\right) = \frac{8k\pi}{\lambda^3}.$$

■

Expectation of gamma distribution. The expectation of the gamma distribution is obtained similarly. Let $X \sim \text{Gamma}(a, \lambda)$. Then

$$E[X] = \int_0^{\infty} x \frac{\lambda^a x^{a-1} e^{-\lambda x}}{\Gamma(a)}\, dx = \int_0^{\infty} \frac{\lambda^a x^a e^{-\lambda x}}{\Gamma(a)}\, dx.$$

Because of the $x^a = x^{(a+1)-1}$ term in the integrand, make the integrand look like the density of a gamma distribution with parameters $a+1$ and λ. This gives

$$\int_0^\infty \frac{\lambda^a x^a e^{-\lambda x}}{\Gamma(a)}\,dx = \frac{\Gamma(a+1)}{\lambda\Gamma(a)}\int_0^\infty \frac{\lambda^{a+1} x^a e^{-\lambda x}}{\Gamma(a+1)}\,dx = \frac{\Gamma(a+1)}{\lambda\Gamma(a)},$$

since the last integral integrates to one. From the relationship in Equation 7.3 for the gamma function, we have

$$E[X] = \frac{\Gamma(a+1)}{\lambda\Gamma(a)} = \frac{a\Gamma(a)}{\lambda\Gamma(a)} = \frac{a}{\lambda}.$$

We invite the reader to use this technique to derive the variance of the gamma distribution $V[X] = a/\lambda^2$.

7.2.2 Sum of Independent Exponentials*

Although the result in this section assumes knowledge of mathematical induction, the interested reader who has not had induction may still gain from studying the proof. Let $X_1, X_2, \ldots$ be an i.i.d. sequence of independent exponential random variables with common parameter λ. Let

$$S_n = X_1 + \cdots + X_n.$$

We show that the distribution of S_n is a gamma distribution with parameters $a = n$ and λ. That is,

$$f_{S_n}(x) = \frac{\lambda^n x^{n-1} e^{-\lambda x}}{\Gamma(n)} = \lambda e^{-\lambda x}\frac{(\lambda x)^{n-1}}{(n-1)!}, \quad \text{for } x > 0.$$

Observe that the result holds for $n = 1$ since $S_1 = X_1$. We also showed in Example 6.33 that the result holds for $n = 2$.

The induction step of the proof assumes that the result is true for all values less than n. Now consider

$$S_n = X_1 + \cdots + X_{n-1} + X_n = S_{n-1} + X_n.$$

Since S_{n-1} only depends on $X_1, \ldots, X_{n-1}$, the random variables S_{n-1} and X_n are independent. For $x > 0$, the convolution formula for independent random variables gives

$$
\begin{aligned}
f_{S_n}(x) &= \int_{-\infty}^{\infty} f_{X_n}(x-y) f_{S_{n-1}}(y)\, dy \\
&= \int_0^x \left(\lambda e^{-\lambda(x-y)}\right)\left(\lambda e^{-\lambda y}\frac{(\lambda y)^{n-2}}{(n-2)!}\right) dy \\
&= e^{-\lambda x}\frac{\lambda^n}{(n-2)!}\int_0^x y^{n-2}\, dy \\
&= e^{-\lambda x}\frac{\lambda^n}{(n-2)!}\frac{x^{n-1}}{n-1} \\
&= \lambda e^{-\lambda x}\frac{(\lambda x)^{n-1}}{(n-1)!},
\end{aligned}
$$

where the second equality uses the induction hypothesis.

7.3 POISSON PROCESS

Consider a process whereby "events"—also called "points" or "arrivals"—occur randomly in time or space. Examples include phone calls throughout the day, car accidents along a stretch of highway, component failures, service times, and radioactive particle emissions.

For many applications, it is reasonable to model the times between successive events as memoryless (e.g., a phone call doesn't "remember" when the last phone call took place). Model these interarrival times as an independent sequence $E_1, E_2, \ldots$ of exponential random variables with parameter λ, where E_k is the time between the $(k-1)$st and kth arrival. Set $S_0 = 0$ and let

$$S_n = E_1 + \cdots + E_n,$$

for $n = 1, 2, \ldots$. Then S_n is the time of the nth arrival and $S_n \sim \text{Gamma}(n, \lambda)$. The sequence $S_0, S_1, S_2, \ldots$ is the arrival sequence.

For each time $t \geq 0$, let N_t be the number of arrivals that occur up through time t. Then for each t, N_t is a discrete random variable. We now show that N_t has a Poisson distribution. The collection of N_t's forms a random process called a *Poisson process with parameter* λ. It is an example of what is called a *stochastic process*. Formally, a stochastic process is a collection of random variables defined on a common sample space.

Throughout this book, we have considered sequences of random variables as models for random processes. However, the collection of N_t's in a Poisson process is not a sequence, because N_t is defined for all real t. The N_t's form an uncountable collection of random variables. We write $(N_t)_{t\geq 0}$ to denote a Poisson process.

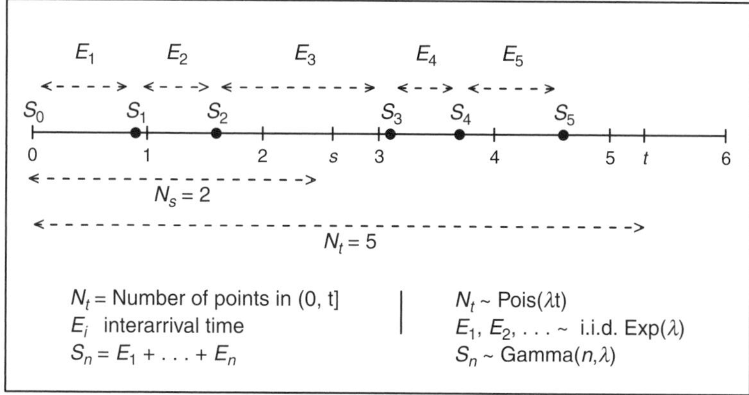

FIGURE 7.7: Poisson process: relationships of underlying random variables.

To understand the Poisson process, it is important to understand the relationship between the interarrival times (E_k's), the arrival times (S_n's), and the number of arrivals (N_t) (see Fig. 7.7).

N_t has a Poisson distribution. Although the *time* of arrivals is continuous, the *number* of arrivals is discrete. We show that N_t, the number of arrivals up to time t, has a Poisson distribution with parameter λt.

DISTRIBUTION OF N_t FOR POISSON PROCESS WITH PARAMETER λ

Let $(N_t)_{t\geq 0}$ be a Poisson process with parameter λ. Then

$$P(N_t = k) = \frac{e^{-\lambda t}(\lambda t)^k}{k!}, \text{ for } k = 0, 1, \ldots.$$

Consider the event $\{N_t = k\}$ that there are k arrivals in $(0, t]$. This occurs if and only if (i) the kth arrival occurrs by time t and (ii) the $(k + 1)$st arrival occurs after time t. That is, $\{N_t = k\} = \{S_k \leq t, S_{k+1} > t\}$. Since

$$S_{k+1} = E_1 + \cdots + E_k + E_{k+1} = S_k + E_{k+1},$$

we have

$$P(N_t = k) = P(S_k \leq t, S_k + E_{k+1} > t).$$

The sum S_k is a function of $E_1, \ldots, E_k$, which are independent of E_{k+1}. Thus S_k and E_{k+1} are independent random variables and the joint density of (S_k, E_{k+1}) is the product of their marginal densities. That is,

$$f_{S_k,E_{k+1}}(x,y) = f_{S_k}(x)f_{E_{k+1}}(y) = \frac{\lambda^k x^{k-1} e^{-\lambda x}}{(k-1)!} \lambda e^{-\lambda y}, \quad \text{for } x, y > 0.$$

The region defined by

$$\{S_k \le t, S_k + E_{k+1} > t\} = \{S_k \le t, E_{k+1} > t - S_k\}$$

is the set of all (x, y) such that $0 < x \le t$ and $y > t - x$. This gives

$$\begin{aligned}
P(N_t = k) &= P(S_k \le t, S_k + E_{k+1} > t) \\
&= \int_0^t \int_{t-x}^{\infty} \frac{\lambda^k x^{k-1} e^{-\lambda x}}{(k-1)!} \lambda e^{-\lambda y} \, dy \, dx \\
&= \int_0^t \frac{\lambda^k x^{k-1} e^{-\lambda x}}{(k-1)!} \left(\int_{t-x}^{\infty} \lambda e^{-\lambda y} \, dy \right) dx \\
&= \int_0^t \frac{\lambda^k x^{k-1} e^{-\lambda x}}{(k-1)!} e^{-(t-x)\lambda} \, dx \\
&= \frac{e^{-t\lambda} \lambda^k}{(k-1)!} \int_0^t x^{k-1} \, dx \\
&= \frac{e^{-t\lambda} (\lambda t)^k}{k!}, \quad \text{for } k = 0, 1, \ldots,
\end{aligned}$$

which establishes the result.

Example 7.15 Marketing. A commonly used model in marketing is the so-called NBD model, introduced in late 1950s by Andrew Ehrenberg (1959) and still popular today. (NBD stands for negative binomial distribution, but we will not discuss the role of that distribution in our example.) Individual customers' purchasing occasions are modeled as a Poisson process. Different customers purchase at different rates λ and often the goal is to estimate such λ. Many empirical studies show a close fit to a Poisson process.

Suppose the time scale for such a study is in days. We assume that Ben's purchases form a Poisson process with parameter $\lambda = 0.5$. Consider the following questions of interest.

1. What is the average rate of purchases?
2. What is the probability that Ben will make at least three purchases within the next 7 days?
3. What is the probability that his 10th purchase will take place within the next 20 days?
4. What is the expected number of purchases Ben will make next month?
5. What is the probability that Ben will not buy anything for the next 5 days given that he has not bought anything for 2 days?

Solutions:

1. Ben purchases at the rate of $\lambda = 0.5$ items per day.
2. The desired probability is $P(N_7 \geq 3)$, where N_7 has a Poisson distribution with parameter $\lambda t = (0.5)7 = 3.5$. This gives

$$\begin{aligned}P(N_7 \geq 3) &= 1 - P(N_7 \leq 2)\\ &= 1 - e^{-7/2} - \frac{7e^{-7/2}}{2} - \frac{49e^{-7/2}}{8} = 0.679.\end{aligned}$$

3. The time of Ben's 10th purchase is S_{10}. The desired probability is $P(S_{10} \leq 20)$. Since $S_{10} \sim \text{Gamma}(10, 0.5)$, we have

$$P(S_{10} \leq 20) = \int_0^{20} \frac{(1/2)^{10} x^9 e^{-x/2}}{\Gamma(10)}\, dx = 0.542.$$

4. The expectation is $E[N_{30}]$. The variable N_{30} has a Poisson distribution with parameter $(0.5)30 = 15$. Thus, the expected number of purchases next month is $E[N_{30}] = 15$.
5. If we consider the present time as $t = 0$, then the time of Ben's next purchase is E_1. The desired probability is $P(E_1 > 5 | E_1 > 2)$. By the memoryless property of the exponential distribution, this is equal to $P(E_1 > 3) = e^{-3\lambda} = e^{-1.5} = 0.223$. ∎

Properties of the Poisson process. Suppose calls come in to a call center starting at 8 a.m. according to a Poisson process with parameter λ. Jack and Jill work at the center. Jill gets in to work at 8 a.m. Jack does not start until 10 a.m.

If we consider the process of phone call arrivals that Jack sees starting at 10 a.m., that arrival process is also a Poisson process with parameter λ. Because of the memoryless property of the exponential distribution, the process started at 10 a.m. is a translated version of the original process, shifted over 2 hours (see Fig. 7.8).

The number of calls that Jack sees in the hour between 11 a.m. and noon has the same distribution as the number of calls that Jill sees between 9 and 10 a.m. (We did not say that the number of calls between 10 and 11 a.m. is *equal* to the number of calls between 8 and 9 a.m. Rather that their *distributions* are the same.) This is the *stationary increments property* of a Poisson process and is a consequence of memorylessness. The distribution of the number of arrivals in an interval only depends on the *length* of the interval, not on the location of the interval.

In addition, whether or not calls come in between 8 and 9 a.m. has no influence on the distribution of calls between 10 and 11 a.m. The number of calls in each disjoint interval is independent of each other. This is called the *independent increments property* of a Poisson process.

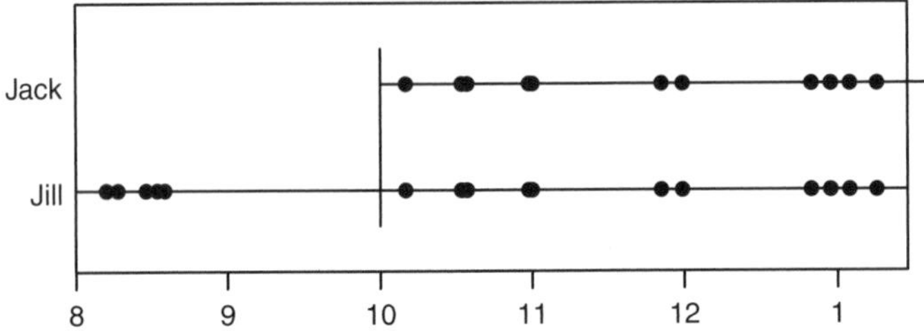

FIGURE 7.8: The calls that Jack sees starting at 10 a.m. have the same probabilistic behavior as the calls that Jill sees starting at 8 a.m. Both are Poisson processes with the same parameter.

PROPERTIES OF POISSON PROCESS

Let $(N_t)_{t\geq 0}$ be a Poisson process with parameter λ.

Independent increments

If $0 < q < r < s < t$, then $N_r - N_q$ and $N_t - N_s$ are independent random variables. That is,

$$P(N_r - N_q = m, N_t - N_s = n) = P(N_r - N_q = m)P(N_t - N_s = n),$$

for all $m, n = 0, 1, 2, \ldots$.

Stationary increments

For all $0 < s < t$, $N_{t+s} - N_t$ and N_s have the same distribution. That is,

$$P(N_{t+s} - N_t = k) = P(N_s = k) = \frac{e^{-\lambda s}(\lambda s)^k}{k!}, \quad \text{for } k = 0, 1, 2, \ldots.$$

Example 7.16 Starting at 6:00 a.m., birds perch on a power line according to a Poisson process with parameter $\lambda = 8$ birds per hour.

1. Find the probability that at most two birds arrive between 7 and 7:15 a.m.

   We start time at time $t = 0$ and keep track of hours after 6:00 a.m. The desired probability is $P(N_{1.25} - N_1 \leq 2)$. By the stationary increments property, this is equal to the probability that at most two birds perch during the first 15 minutes (one-fourth of an hour). The desired probability is

$$\begin{aligned}P(N_{1.25} - N_1 \leq 2) &= P(N_{0.25} \leq 2)\\ &= e^{-8/4}\left[1 + 2 + 2^2/2\right]\\ &= 5e^{-2} = 0.677.\end{aligned}$$

2. Between 7 and 7:30 a.m, five birds arrive on the power line. What is the expectation and standard deviation for the number of birds that arrive from 3:30 to 4:00 p.m?

   By the independent increments property, the number of birds that perch in the afternoon is independent of the number that perch in the morning. By the stationary increments property, the number of birds within each half-hour period has a Poisson distribution with parameter $\lambda/2 = 4$. Hence, the desired expectation is four birds and the standard deviation is two birds. ■

Example 7.17 The number of goals scored during a soccer match is modeled as a Poisson process with parameter λ. Suppose five goals are scored in the first t minutes. Find the probability that two goals are scored in the first s minutes.

The event that two goals are scored in the time interval $[0, s]$ and five goals are scored in $[0, t]$ is equal to the event that two goals are scored in $[0, s]$ and three goals are scored in $(s, t]$. Thus,

$$\begin{aligned}P(N_s = 2|N_t = 5) &= \frac{P(N_s = 2, N_t = 5)}{P(N_t = 5)}\\ &= \frac{P(N_s = 2, N_t - N_s = 3)}{P(N_t = 5)}\\ &= \frac{P(N_s = 2)P(N_t - N_s = 3)}{P(N_t = 5)}\\ &= \frac{P(N_s = 2)P(N_{t-s} = 3)}{P(N_t = 5)}\\ &= \frac{(e^{-\lambda s}(\lambda s)^2/2!)(e^{-\lambda(t-s)}(t-s)^3/3!)}{e^{-\lambda t}(\lambda t)^5/5!}\\ &= \binom{5}{2}\frac{s^2(t-s)^3}{t^5} = \binom{5}{2}\left(\frac{s}{t}\right)^2\left(1-\frac{s}{t}\right)^3,\end{aligned}$$

where the third equality is from independent increments and the fourth equality is from stationary increments.

The final expression is a binomial probability. Extrapolating from this example we obtain a general result: Let $(N_t)_{t\geq 0}$ be a Poisson process with parameter λ. For $0 < s < t$, the conditional distribution of N_s given $N_t = n$ is binomial with parameters n and $p = s/t$. ■

R: SIMULATING A POISSON PROCESS

One way to simulate a Poisson process is to first simulate exponential interarrival times and then construct the arrival sequence. To generate n arrivals of a Poisson process with parameter λ, type

```
> inter <- rexp(n,λ).
> arrive <- cumsum(inter)
```

The `cumsum` command returns a cumulative sum from the interarrival times. The vector `arrive` contains the times of successive arrivals. The last arrival time is

```
> last <- tail(arrive,1)
```

For $t \leq$ `last`, the number of arrivals up to time t N_t is simulated by

```
> sum(arrive <= t)
```

For example, consider a Poisson process with parameter $\lambda = 1/2$. We simulate the probability $P(N_5 - N_2 = 1)$:

```
> n <- 10000
> simlist <- numeric(n)
> for (i in 1:n) {
  inter <- rexp(30,1/2)
  arrive <- cumsum(inter)
  nt <- sum( 2 <= arrive & arrive <= 5)
  simlist[i] <- if (nt ==1) 1 else 0 }
> mean(simlist)
[1] 0.3335
```

As $P(N_5 - N_2 = 1) = P(N_3 = 1)$, here is the exact answer:

```
> dpois(1,3/2)
[1] 0.3346952
```

Example 7.18 Waiting time paradox. The following classic is from Feller (1968). Buses arrive at a bus stop according to a Poisson process with parameter λ. The expected time between buses is $1/\lambda$. Suppose you get to the bus stop at noon. How long can you expect to wait for a bus? Here are two possible answers.

1. The memoryless property of the Poisson process means that the distribution of your waiting time should not depend on when you arrive at the bus stop. Thus, your expected waiting time is $1/\lambda$.

2. You arrive at some time between two consecutive buses. By symmetry your expected waiting time should be half the expected time between two consecutive buses, that is, $1/(2\lambda)$.

We explore the issue with a simulation, assuming that buses arrive on average every 20 minutes and $\lambda = 1/20$. Assume you arrive at the bus stop at time $t = 200$. The variable `wait` is your waiting time. The simulation repeats the experiment 10,000 times.

R: WAITING TIME PARADOX

```
> mytime <- 200  # arbitrary time you get to bus stop
> lambda <- 1/20
> simlist <- numeric(10000)
> for (i in 1:10000) {
   arrivals <- cumsum(rexp(30, lambda))
   wait <- arrivals[arrivals > mytime][1] - mytime
   simlist[i] <- wait  }
> mean(simlist)
[1] 19.83012
  ...


> mean(simlist)
[1] 21.6133
  ...


> mean(simlist)
[1] 20.03134
```

The simulation suggests that the expected waiting time is $20 = 1/\lambda$ minutes. Here is the fallacy of the second approach. Buses arrive on average, say, every 20 minutes. But the time between buses is random, buses may arrive one right after the other, or there may be a long time between consecutive buses. When you get to the bus stop, it is more likely that you get there during a longer interval between buses rather than a shorter interval.

This phenomenon is known as *length-biased* or *size-biased* sampling. If you reach into a bag containing pieces of string of different lengths and pick one "at random," you tend to pick a longer rather than shorter piece. Bus interarrival time intervals are analogous to pieces of string.

Here is another example of size-biased sampling. Suppose you want to estimate how much time people spend working out at the gym. If you go to the gym to ask people at random how long they work out, you are likely to get a biased estimate that is too big. You are more likely to sample someone who works out a lot. The people who rarely go to the gym are likely not to be there when you go there.

In the bus waiting problem, it turns out that the expected length of an interarrival time interval *which contains a fixed time* t is actually about $2/\lambda$, which is twice as large as the expected interarrival time. To verify, we modify the simulation code, keeping track of bus arrival times before and after the time we get to the bus stop.

R: WAITING TIME SIMULATION CONTINUED

```
> mytime <- 200
> for (i in 1:1000) {
   arrivals <- cumsum(rexp(30, lambda))
   indx <- which(arrivals>mytime)[1]
   lengthintrvl <- arrivals[indx] - arrivals[indx-1]
   simlist[i] <- lengthintrv    }
> mean(simlist)
[1] 40.71362
 ...

 > mean(simlist)
 [1] 40.89851
 ...

 > mean(simlist)
 [1] 39.85409
```

So the second answer was not entirely wrong. Symmetry says that your expected waiting time should be half the expected time between the two buses that arrive before and after the time you get there. This gives the expected waiting time for a bus as $(1/2)(2/\lambda) = 1/\lambda$.

Finally, there is nothing special about the time $t = 200$. The result holds for *any* fixed time. ∎

7.4 BETA DISTRIBUTION

The beta distribution generalizes the uniform distribution. A beta random variable takes values between zero and one. The distribution is parametrized by two positive numbers a and b (see Fig. 7.9 for examples).

A random variable X has a *beta distribution with parameters* $a > 0$ *and* $b > 0$ if the density function of X is

$$f(x) = \frac{\Gamma(a+b)}{\Gamma(a)\Gamma(b)} x^{a-1}(1-x)^{b-1}, \quad 0 < x < 1.$$

We write $X \sim \text{Beta}(a, b)$.

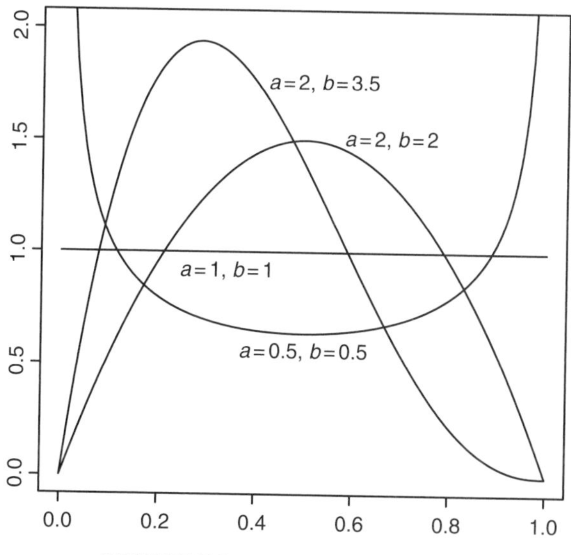

FIGURE 7.9: Four beta distributions.

When $a = b = 1$, the beta distribution reduces to the uniform distribution on $(0, 1)$.

Using integration by parts, one can show that

$$\int_0^1 x^{a-1}(1-x)^{b-1}\,dx = \frac{\Gamma(a)\Gamma(b)}{\Gamma(a+b)}, \tag{7.5}$$

for $a, b > 0$, and thus the beta density integrates to one.

To find the expectation of a beta distribution, use the methods of Section 7.2.1. Let $X \sim \text{Beta}(a, b)$. Then

$$E[X] = \int_0^1 x\frac{\Gamma(a+b)}{\Gamma(a)\Gamma(b)}x^{a-1}(1-x)^{b-1}\,dx = \frac{\Gamma(a+b)}{\Gamma(a)\Gamma(b)}\int_0^1 x^a(1-x)^{b-1}\,dx.$$

The integrand is proportional to a beta density with parameters $a+1$ and b. This gives

$$\begin{aligned}
E[X] &= \frac{\Gamma(a+b)}{\Gamma(a)\Gamma(b)}\int_0^1 x^a(1-x)^{b-1}\,dx \\
&= \frac{\Gamma(a+b)}{\Gamma(a)\Gamma(b)}\left(\frac{\Gamma(a+1)\Gamma(b)}{\Gamma(a+1+b)}\right)\int_0^1 \frac{\Gamma(a+1+b)}{\Gamma(a+1)\Gamma(b)}x^a(1-x)^{b-1}\,dx \\
&= \frac{\Gamma(a+b)}{\Gamma(a)\Gamma(b)}\left(\frac{\Gamma(a+1)\Gamma(b)}{\Gamma(a+1+b)}\right) = \frac{a}{a+b},
\end{aligned}$$

using the property $\Gamma(x+1) = x\Gamma(x)$ of the gamma function.

We leave it to the reader to derive the variance of the beta distribution

$$V[X] = \frac{ab}{(a+b)^2(a+b+1)}.$$

Since probabilities and proportions are numbers between zero and one, the beta distribution is widely used to model an unknown probability or proportion p. For instance, Chia and Hutchinson (1991) model the fraction of daylight hours not receiving bright sunshine with a beta distribution. Gupta and Nadarajan (2004) is a book-length treatment of the many and diverse applications of the beta distribution.

Example 7.19 The proportion of his daily study time that Jack devotes to probability is modeled with a density function proportional to $x^3(1-x)$, for $0 < x < 1$. What is the probability that Jack will spend more than half his study time tomorrow on probability?

Let X be the proportion of Jack's study time spent on probability. The given density function is that of a beta distribution with parameters $a = 4$ and $b = 2$. The desired probability is

$$P(X > 1/2) = \int_{1/2}^{1} \frac{\Gamma(6)}{\Gamma(4)\Gamma(2)} x^3(1-x)\,dx = \int_{1/2}^{1} 20x^3(1-x)\,dx = \frac{3}{16}.$$

In R, type

```
> 1-pbeta(1/2,4,2)
[1] 0.8125
```

■

Example 7.20 Order statistics. Given a sequence of random variables $X_1, \ldots, X_n$, order the values from smallest to largest. Let $X_{(k)}$ denote the kth largest value. That is,

$$X_{(1)} \leq \cdots \leq X_{(n)}.$$

The minimum of the n variables is $X_{(1)}$ and the maximum is $X_{(n)}$. The random variables $(X_{(1)}, \ldots, X_{(n)})$ are called the *order statistics*.

For instance, take a sample $X_1, \ldots, X_5$, where the X_i's are uniformly distributed on (0,1).

```
> runif(5)
[1] 0.1204 0.6064 0.5070 0.6805 0.1185
```

Then

$$(X_{(1)}, X_{(2)}, X_{(3)}, X_{(4)}, X_{(5)}) = (0.1185, 0.1204, 0.5070, 0.6064, 0.6805).$$

Consider the order statistics for n i.i.d. random variables $X_1, \ldots, X_n$ uniformly distributed on (0,1). We show that for each $k = 1, \ldots, n$, the kth order statistic $X_{(k)}$ has a beta distribution.

For $0 < x < 1$, $X_{(k)}$ is less than or equal to x if and only if at least k of the X_i's are less than or equal to x. Since the X_i's are independent, and $P(X_i \leq x) = x$, the number of X_i's less than or equal to x has a binomial distribution with parameters n and $p = x$. Letting Z be a random variable with such a distribution gives

$$P(X_{(k)} \leq x) = P(Z \geq k) = \sum_{i=k}^{n} \binom{n}{i} x^i (1-x)^{n-i},$$

Differentiate to find the density of $X_{(k)}$. Consider the case $k < n$.

$$\begin{aligned}
f_{X_{(k)}}(x) &= \sum_{i=k}^{n} \binom{n}{i} \left[i x^{i-1}(1-x)^{n-i} - (n-i)(1-x)^{n-i-1} x^i \right] \\
&= \sum_{i=k}^{n} \frac{n}{i!(n-i)!} i x^{i-1}(1-x)^{n-i} \\
&\quad - \sum_{i=k}^{n} \frac{n!}{i!(n-i)!}(n-i) x^i (1-x)^{n-i-1} \\
&= \sum_{i=k}^{n} \frac{n!}{(i-1)!(n-i)!} x^{i-1}(1-x)^{n-i} \\
&\quad - \sum_{i=k}^{n-1} \frac{n!}{i!(n-i-1)!} x^i (1-x)^{n-i-1} \\
&= \sum_{i=k}^{n} \frac{n!}{(i-1)!(n-i)!} x^{i-1}(1-x)^{n-i} \\
&\quad - \sum_{i=k+1}^{n} \frac{n!}{(i-1)!(n-i)!} x^{i-1}(1-x)^{n-i} \\
&= \frac{n!}{(k-1)!(n-k)!} x^{k-1}(1-x)^{n-k}.
\end{aligned}$$

This gives the density function of a beta distribution with parameters $a = k$ and $b = n - k + 1$.

For the case $k = n$, the order statistic is the maximum. The density function for the maximum of n independent uniforms is $f_{X_{(n)}}(x) = nx^{n-1}$, for $0 < x < 1$. This is a beta distribution with parameters $a = n$ and $b = 1$. In both cases, we have the following result.

DISTRIBUTION OF ORDER STATISTICS

Let $X_1, \ldots, X_n$ be an i.i.d. sequence uniformly distributed on (0,1). Let $X_{(1)}, \ldots, X_{(n)}$ be the corresponding order statistics. For each $k = 1, \ldots, n$,

$$X_{(k)} \sim \text{Beta}(k, n - k + 1).$$

■

Example 7.21 Simulating beta random variables. We show how to simulate beta random variables when the parameters a and b are integers. Suppose $X_1, \ldots, X_{a+b-1}$ are i.i.d. uniform (0,1) random variables. Then

$$X_{(a)} \sim \text{Beta}(a, (b + a - 1) - a + 1) = \text{Beta}(a, b).$$

To simulate a Beta(a, b) random variable, generate $a + b - 1$ uniform (0,1) variables and choose the ath largest.

This is not the most efficient method for simulating beta variables, but it works fine for small and even moderate values of a and b. And it is easily coded. In R, type

```
> sort(runif(a+b-1))[a].
```

■

Beta distributions are often used to model continuous random variables that take values in some bounded interval. The beta distribution can be extended to the interval (s, t) by the scale change

$$Y = (t - s)X + s.$$

If X has a beta distribution on $(0, 1)$, then Y has an extended beta distribution on (s, t). We leave the derivation to the exercises.

7.5 PARETO DISTRIBUTION, POWER LAWS, AND THE 80-20 RULE

The normal, exponential, gamma, Poisson, and geometric distributions all have probability functions that contain an exponential factor of the form a^{-x}, where a is a positive constant. What this means is that the probability of a "large" outcome far from the mean in the "tail" of the distribution is essentially negligible. For instance, the probability that a normal distribution takes a value greater than six standard deviations from the mean is about two in a billion.

In many real-world settings, however, variables take values over a wide range covering several orders of magnitude. For instance, although the average size of U.S. cities is about 6000, there are American cities whose populations are a thousand times that much.

Tens of thousands of earthquakes occur every day throughout the world, measuring less than 3.0 on the Richter magnitude scale. Yet each year there are about a hundred earthquakes of magnitude between 6.0 and 7.0, and one or two greater than 8.0.

The Pareto distributions is an example of a "power–law" distribution whose density function is proportional to the power function x^{-a}. Such distributions have "heavy tails," which give nonnegligible probability to extreme values. The Pareto distribution has been used to model population, magnitude of earthquakes, size of meteorites, maximum one-day rainfalls, price returns of stocks, Internet traffic, and wealth in America.

The distribution was discovered by the Italian economist Vilfredo Pareto who used it to model income distribution in Italy at the beginning of the twentieth century.

PARETO DISTRIBUTION

A random variable X has a *Pareto distribution with parameters* $m > 0$ *and* $a > 0$ if the density function of X is

$$f(x) = a\frac{m^a}{x^{a+1}}, \quad \text{for } x > m.$$

The distribution takes values above a minimum positive number m. We write $X \sim \text{Pareto}(m, a)$.

Example 7.22 Consider a Pareto distribution with parameters $m = 1$ and $a = 3$. The expectation and variance of this distribution are $\mu = 3/2$ and $\sigma^2 = 3/4$. In Table 7.1, we compare the tail probability $P(|X - \mu| > k\sigma)$ of falling more than k standard deviations from the mean for this distribution and for a normal distribution with the same mean and variance.

The data highlight the heavy tail property of the Pareto distribution. In a billion observations of a normally distributed random variable, you would expect about $1.973 \times 10^{-9} \times 10^9 = 1.973 \approx 2$ observations greater than six standard deviations from the mean. But for a Pareto distribution with the same mean and

TABLE 7.1: Comparison of tail probabilities for normal and Pareto distributions.

	$P(\lvert X - \mu\rvert > k\sigma)$	
k	Normal(3/2, 3/4)	Pareto(1, 3)
1	3.173×10^{-1}	7.549×10^{-2}
2	4.550×10^{-2}	2.962×10^{-2}
3	2.700×10^{-3}	1.453×10^{-2}
4	6.334×10^{-5}	8.175×10^{-3}
5	5.733×10^{-7}	5.046×10^{-3}
6	1.973×10^{-9}	3.331×10^{-3}
7	2.560×10^{-12}	2.312×10^{-3}

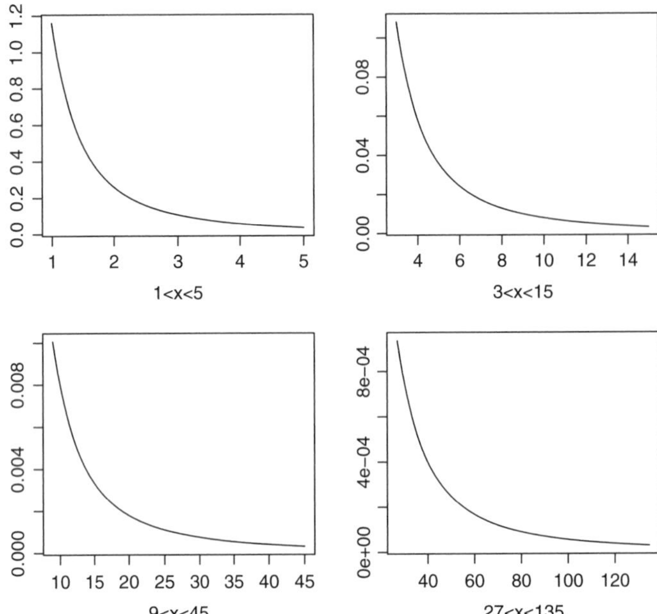

FIGURE 7.10: Scale-invariance of the Pareto distribution. The density curve is for a Pareto distribution with $m = 1$ and $a = 1.16$. The curve is shown on intervals of the form $c < x < 5c$, for $c = 1, 3, 9, 27$.

variance, you would expect to see about $3.31 \times 10^{-3} \times 10^9 = 3,310,000$ such outcomes. ■

The Pareto distribution is characterized by a special *scale-invariance* property. Intuitively this means that the shape of the distribution does not change by changing the scale of measurements. For instance, it is reasonable to model income with a scale-invariant distribution since the distribution of income does not change if units are measured in dollars, or converted to euros or to yen. You can see this phenomenon in Figure 7.10. The plots show the density curve for a Pareto distribution with $m = 1$ and $a = 1.16$ over four different intervals of the form $c < x < 5c$. The shape of the curve does not change for any value of c.

SCALE-INVARIANCE

Definition 7.2. *A probability distribution with density* f *is scale-invariant if for all positive constants* c*,*

$$f(cx) = g(c)f(x),$$

where g *is some function that does not depend on* x.

For the Pareto distribution,

$$f(cx) = a\frac{m^a}{(cx)^{a+1}} = \left(\frac{1}{c^{a+1}}\right) a\frac{m^a}{x^{a+1}} = \frac{1}{c^{a+1}} f(x).$$

Hence, the distribution is scale-invariant.

Newman (2005) is an excellent source for background on the Pareto distribution. He shows that a power–law distribution is the only probability distribution that is scale-invariant.

The 80-20 rule. Ever notice that it seems to take 20% of your time to do 80% of your homework? Or you wear 20% of your clothes 80% of the time? The numbers 80 and 20 seem to be empirically verified by many real-life phenomenon. Pareto observed that 80% of the land in Italy was owned by 20% of the people. The so-called 80-20 rule is characteristic of the Pareto distribution.

Suppose income is modeled with a Pareto distribution. Further suppose that income follows an 80-20 rule with 80% of the wealth owned by 20% of the people. How should the parameter a be determined?

Let X be a random variable with the given Pareto distribution. Then the proportion of the population with income larger than x is

$$P(X > x) = \int_x^\infty a\frac{m^a}{t^{a+1}}\,dt = \left(\frac{m}{x}\right)^a, \quad \text{for } x > m.$$

Solving $0.20 = (m/x)^a$ gives $x = m5^{1/a}$.

The proportion of total income in the hands of those people is estimated in economics by the *Lorenz function*

$$L(x) = \frac{\int_x^\infty tf(t)\,dt}{\int_m^\infty tf(t)\,dt} = \left(\frac{m}{x}\right)^{a-1}.$$

This gives

$$L\left(m5^{1/a}\right) = 5^{(1-a)/a}.$$

If 80% of the income is owned by 20% of the people, then $0.80 = 5^{(1-a)/a}$, which gives $\log 4/5 = (1-a)/a \log 5$, and thus $a = \log 5/\log 4 = 1.16$.

7.5.1 Simulating the Pareto Distribution

R does not have a built-in function to simulate from the Pareto distribution. To simulate a Pareto random variable, we use the inverse transform method, introduced in Section 6.6.1. Let $X \sim \text{Pareto}(m, a)$. Then

$$F_X(x) = 1 - \left(\frac{m}{x}\right)^a$$

with inverse function

$$F^{-1}(u) = \frac{m}{(1-u)^{1/a}}.$$

If $U \sim$ Unif(0,1), then $m/(1-U)^{1/a}$ has the desired Pareto distribution. And we can even do a little better. Since $1 - U$ is also uniformly distributed on (0,1), simplify the simulation formula and use $m/U^{1/a}$ as an observation from a Pareto(m, a) distribution.

To illustrate the 80-20 rule for income, we conduct a simulation study, simulating from a Pareto distribution with parameters $m = 1$ and $a = \log 5/\log 4$. See the script file **Pareto8020.R**.

R: SIMULATING THE 80-20 RULE

What percent of the population owns 80% of the income?

```
# Pareto8020.R
> m <- 1
> a <- log(5)/log(4)
> simlist <- m/runif(100000)^(1/a)
> totalwealth <- sum(simlist)
> totalwealth80 <- 0.80*totalwealth   # 80% of wealth
> indx <- which(cumsum(simlist) > totalwealth80)[1]
> 1-indx/100000   # % of population who own 80% of
income
[1] 0.20846
```

7.6 SUMMARY

Several families of continuous distributions and random processes are introduced in this chapter. The normal distribution is perhaps the most important distribution in statistics. Many key properties of the distribution are presented, such as (i) linear functions of normal random variables are normal; (ii) sums of independent normal variables are normally distributed; and (iii) the so-called "68-95-99.7 rule," which quantifies the probability that a normal variable is within one, two, and three standard deviations from the mean. The normal approximation of the binomial distribution is discussed, as well as the continuity correction used when a continuous density curve is used to approximate a discrete probability. Quantiles are introduced, as well as general results for sums and averages of i.i.d. random variables.

The gamma distribution arises as the sum of i.i.d. exponential random variables. This flexible two-parameter family of distributions generalizes the exponential distribution. The Poisson process, a continuous time stochastic (random) process, is presented. The process sees the interrelationship between gamma, exponential, and Poisson random variables. The process arises as a model for the "arrival" of "events" (e.g., phone calls at a call center, accidents on the highway, and goals at a soccer game) in continuous time. The key properties of a Poisson process—stationary and independent increments—are explained.

Also introduced in this chapter is the beta distribution, a two-parameter family of distributions on $(0,1)$ that generalizes the uniform distribution. Finally, the Pareto distribution, part of a larger class of scale-invariant, power–law distributions, is presented. Connections are drawn with the so-called 80-20 rule.

- **Normal distribution:** A random variable X is normally distributed with parameters μ and σ^2 if the density of X is

$$f(x) = \frac{1}{\sigma\sqrt{2\pi}} e^{-(x-\mu)^2/2\sigma^2}, \text{ for } -\infty < x < \infty.$$

  Parameters μ and σ^2 are, respectively, the mean and variance of the distribution.
- **Normal setting:** The bell-shaped, symmetric distribution arises as a simple model for many complex phenomena. It has been used to model measurement error, standardized test scores, and human characteristics like height and weight.
- **68-95-99.7 rule:** For any normal distribution, the probability that an outcome is within one, two, and three standard deviations from the mean is, respectively, about 68, 95, and 99.7%.
- **Standard normal:** A normal random variable with mean $\mu = 0$ and variance $\sigma^2 = 1$ is said to have a standard normal distribution.
- **Linear function:** If $X \sim \text{Norm}(\mu, \sigma^2)$, then for constants $a \neq 0$ and b, the random variable $aX + b$ is normally distributed with mean $a\mu + b$ and variance $a^2\sigma^2$. It follows that $(X - \mu)/\sigma \sim \text{Norm}(0,1)$.
- **Normal approximation of the binomial distribution:** If $X \sim \text{Binom}(n, p)$, and n is large, then X has an approximate normal distribution with mean $E[X] = np$ and variance $V[X] = np(1-p)$. In particular, for $a < b$,

$$\lim_{n\to\infty} P\left(a \leq \frac{X - np}{\sqrt{np(1-p)}} \leq b\right) = \frac{1}{\sqrt{2\pi}} \int_a^b e^{-x^2/2}\, dx.$$

- **Continuity correction:** Suppose X is a discrete random variable and we wish to approximate $P(a \leq X \leq b)$ using a continuous distribution whose density function is f. Then take

$$P(a \le X \le b) \approx \int_{a-1/2}^{b+1/2} f(x)\,dx.$$

- **Quantiles:** If X is a continuous random variable and $0 < p < 1$, then the pth quantile is the number q that satisfies $P(X \le q) = p/100$.
- **Averages of i.i.d. random variables.** Suppose $X_1, \ldots, X_n$ are i.i.d. with mean μ and variance σ^2. Let $S_n = X_1 + \cdots + X_n$. Then S_n/n is the average and
  1. $E[S_n/n] = \mu$.
  2. $V[S_n/n] = \sigma^2/n$.
  3. If the X_i's are normally distributed, then so is S_n/n. That is, $S_n/n \sim \text{Norm}(\mu, \sigma^2/n)$.
- **Gamma distribution:** A random variable X has a gamma distribution with parameters a and λ if the density of X is

$$f(x) = \frac{\lambda^a x^{a-1} e^{-\lambda x}}{\Gamma(a)}, \quad \text{for } x > 0,$$

  where $\Gamma(a) = \int_0^\infty t^{a-1} e^{-t}\,dt$ is the gamma function.
- **Gamma setting:** The distribution is a flexible, two-parameter distribution defined on the positive reals. A sum of n i.i.d. exponential random variables with parameter λ has a gamma distribution with parameters n and λ. The gamma distribution generalizes the exponential distribution (with $a = 1$).
- **Gamma distribution properties:** If $X \sim \text{Gamma}(a, \lambda)$, then
  1. $E[X] = a/\lambda$.
  2. $V[X] = a/\lambda^2$.
- **Poisson process:** This is a model for the distribution of "events" or "arrivals" in space and time. Times between arrivals are modeled as i.i.d. exponential random variables with parameter λ. Let N_t be the number of arrivals up to time t, defined for all nonnegative t. Then $N_t \sim \text{Pois}(\lambda t)$. The collection of random variables $(N_t)_{t \ge 0}$ is a Poisson process with a parameter λ.
- **Properties of Poisson process:**
  1. **Stationary increments:** For $0 < s < t$, the distribution of $N_{t+s} - N_t$, the number of arrivals between times t and $t + s$, only depends on the length of the interval $(t + s) - t = s$. That is, the distribution of $N_{t+s} - N_t$ is the same as the distribution of N_s. Hence,

$$P(N_{t+s} - N_t = k) = \frac{e^{-\lambda s}(\lambda s)^k}{k!}, \quad \text{for } k = 0, 1, \ldots.$$

  2. **Independent increments:** For $0 < a < b < c < d$, $N_b - N_a$ and $N_d - N_c$ are independent random variables. The number of arrivals in disjoint intervals is an independent random variable.

- **Beta distribution:** A random variable X has a beta distribution with parameters $a > 0$ and $b > 0$ if the density of X is

$$f(x) = \frac{\Gamma(a+b)}{\Gamma(a)\Gamma(b)} x^{a-1}(1-x)^{b-1}, \text{ for } 0 < x < 1.$$

- **Beta setting:** The distribution is a flexible two-parameter family of distributions on $(0,1)$. It is often used to model an unknown proportion or probability. The distribution generalizes the uniform distribution (with $a = b = 1$).
- **Beta distribution properties:** If $X \sim \text{Beta}(a, b)$, then
  1. $E[X] = a/(a + b)$.
  2. $V[X] = ab/((a + b)^2(a + b + 1))$.
- **Pareto distribution:** A random variable X has a Pareto distribution with parameters $m > 0$ and $a > 0$ if the density of X is

$$f(x) = a\frac{m^a}{x^{a+1}}, \text{ for } x > m.$$

- **Pareto setting:** The distribution is used to model heavily skewed data such as population and income. It is an example of a "power–law" distribution with the so-called heavy tails.
- **Scale-invariance:** A distribution with density function f is scale-invariant if for all constants c, $f(cx) = g(c)f(x)$, where g is some function that does not depend on x. The Pareto distribution is scale-invariant.

EXERCISES

Normal distribution

7.1 Let $X \sim \text{Norm}(-4, 25)$. Find the following probabilities without using software:

(a) $P(X > 6)$.

(b) $P(-9 < X < 1)$.

(c) $P(\sqrt{X} > 1)$.

7.2 Let $X \sim \text{Norm}(4, 4)$. Find the following probabilities using R:

(a) $P(|X| < 2)$.

(b) $P\left(e^X < 1\right)$.

(c) $P(X^2 > 3)$.

7.3 Babies' birth weights are normally distributed with mean 120 ounces and standard deviation 20 ounces. *Low birth weight* is an important indicator of a newborn baby's chances of survival. One definition of low birth weight is that it is the fifth percentile of the weight distribution.

(a) Babies who weigh less than what amount would be considered low birth weight?

(b) *Very low birth weight* is used to described babies who are born weighing less than 52 ounces. Find the probability that a baby is born with very low birth weight.

(c) Given that a baby is born with low birth weight, what is the probability that they have very low birth weight?

7.4 An elevator's weight capacity is 1000 pounds. Three men and three women are riding the elevator. Adult male weight is normally distributed with mean 172 pounds and standard deviation 29 pounds. Adult female weight is normally distributed with mean 143 pounds and standard deviation 29 pounds. Find the probability that the passengers' total weight exceeds the elevator's capacity.

7.5 The rates of return of ten stocks are normally distributed with mean $\mu = 2$ and standard deviation $\sigma = 4$ (units are percentages). Rates of return are independent from stock to stock. Each stock sells for \$10 a share. Amy, Bill, and Carrie have \$100 each to spend on their stock portfolio. Amy buys one share of each stock. Bill buys two shares from five different companies. Carrie buys ten shares from one company. For each person, find the probability that their average rate of return is positive.

7.6 The two main standardized tests in the United States for high school students are the ACT and SAT. ACT scores are normally distributed with mean 18 and standard deviation 6. SAT scores are normally distributed with mean 500 and standard deviation 100. Suppose Jill takes an SAT exam and scores 680. Jack plans to take the ACT. What score does Jack need to get so that his standardized score is the same as Jill's standardized score? What percentile do they each score at?

7.7 The SAT exam is a composite of three exams—in reading, math, and writing. For 2011 college-bound high school seniors, Table 7.2 gives the mean and standard deviation of the three exams. The data are from the College Board (2011).

Let R, M, and W denote the reading, math, and writing scores. The College Board also estimates the following correlations between the three exams:

$$\text{Corr}(R, M) = \text{Corr}(W, M) = 0.72, \quad \text{Corr}(R, W) = 0.84.$$

(a) Find the mean and standard deviation of the total composite SAT score $T = R + M + W$.

(b) Assume total composite score is normally distributed. Find the 80th and 90th percentiles.

(c) Laura took the SAT. Find the probability that the average of her three exam scores is greater than 600.

TABLE 7.2: SAT statistics for 2011 college-bound seniors.

	Reading	Math	Writing
Mean	497	514	489
SD	114	117	113

7.8 Let $X \sim \text{Norm}(\mu, \sigma^2)$. Show that $E[X] = \mu$.

7.9 Find all the inflection points of a normal density curve. Show how this information can be used to draw a normal curve given values of μ and σ.

7.10 Suppose $X_i \sim \text{Norm}(i, i)$ for $i = 1, 2, 3, 4$. Further assume all the X_i's are independent. Find $P(X_1 + X_3 < X_2 + X_4)$.

7.11 Let $X_1, \ldots, X_n$ be an i.i.d. sequence of normal random variables with mean μ and variance σ^2. Let $S_n = X_1 + \cdots + X_n$.

(a) Suppose $\mu = 5$ and $\sigma^2 = 1$. Find $P(|X_1 - \mu| \leq \sigma)$.

(b) Suppose $\mu = 5$, $\sigma^2 = 1$, and $n = 9$. Find $P(|S_n/n - \mu| \leq \sigma)$.

7.12 Let $X \sim \text{Norm}(\mu, \sigma^2)$. Suppose $a \neq 0$ and b are constants. Show that $Y = aX + b$ is normally distributed.

7.13 Let $X_1, \ldots, X_n$ be i.i.d. normal random variables with mean μ and standard deviation σ. Recall that $\bar{X} = (X_1 + \cdots + X_n)/n$ is the sample average. Let

$$S = \frac{1}{n-1} \sum_{i=1}^{n} (X_i - \bar{X})^2.$$

Show that $E[S] = \sigma^2$. In statistics, S is called the *sample variance*.

7.14 Let $Z \sim \text{Norm}(0,1)$. Find $E[|Z|]$.

7.15 If Z has a standard normal distribution, find the density of Z^2.

7.16 If X and Y are independent standard normal random variables, show that $X^2 + Y^2$ has an exponential distribution.

7.17 If X, Y, Z are independent standard normal random variables, find the distribution of $X^2 + Y^2 + Z^2$. (Hint: Think spherical coordinates.)

Gamma distribution, Poisson process

7.18 Using integration by parts, show that the gamma function satisfies

$$\Gamma(a) = (a-1)\Gamma(a-1).$$

7.19 Let $X \sim \text{Gamma}(a, \lambda)$. Find $V[X]$ using the methods of Section 7.2.1.

7.20 May is tornado season in Oklahoma. According to the National Weather Service, the rate of tornados in Oklahoma in May is 21.7 per month, based on data from 1950 to the present. Assume tornados follow a Poisson process.

(a) What is the probability that next May there will be more than 25 tornados?

(b) What is the probability that the 10th tornado in May occurs before May 15?

(c) What is the expectation and standard deviation for the number of tornados during 7 days in May?

(d) What is the expected number of days between tornados in May? What is the standard deviation?

7.21 Starting at 9 a.m., students arrive to class according to a Poisson process with parameter $\lambda = 2$ (units are minutes). Class begins at 9:15 a.m. There are 30 students.

(a) What is the expectation and variance of the number of students in class by 9:15 a.m.?

(b) Find the probability there will be at least 10 students in class by 9:05 a.m.

(c) Find the probability that the last student who arrives is late.

(d) Suppose exactly six students are late. Find the probability that exactly 15 students arrived by 9:10 a.m.

(e) What is the expected time of arrival of the seventh student who gets to class?

7.22 Solve the following integrals without calculus by recognizing the integrand as related to a known probability distribution and making the necessary substitution(s).

(a)

$$\int_{-\infty}^{\infty} e^{-(x+1)^2/18} \, dx$$

(b)

$$\int_{-\infty}^{\infty} x e^{-(x-1)^2/2} \, dx$$

(c)

$$\int_{0}^{\infty} x^4 e^{-2x} \, dx$$

(d)

$$\int_{0}^{\infty} x^s e^{-tx} \, dx,$$

for positive integer s and positive real t.

7.23 A *spatial Poisson process* is a model for the distribution of points in two-dimensional space. For a set $A \subseteq \mathbb{R}^2$, let N_A denote the number of points in A. The two defining properties of a spatial Poisson process with parameter λ are

1. If A and B are disjoint sets, then N_A and N_B are independent random variables.
2. For all $A \subseteq \mathbb{R}^2$, N_A has a Poisson distribution with parameter $\lambda|A|$ for some $\lambda > 0$, where $|A|$ denotes the area of A. That is,

$$P(N_A = k) = \frac{e^{-\lambda|A|}(\lambda|A|)^k}{k!}, \ k = 0, 1, \ldots.$$

Consider a spatial Poisson process with parameter λ. Let $\boldsymbol{x}$ be a fixed point in the plane.

(a) Find the probability that there are no points of the spatial process that are within two units distance from $\boldsymbol{x}$. (Draw the picture.)

(b) Let X be the distance between $\boldsymbol{x}$ and the nearest point of the spatial process. Find the density of X. (Hint: Find $P(X > x)$.)

7.24 Suppose $(N_t)_{t\geq 0}$ is a Poisson process with parameter $\lambda = 1$.

(a) Find $P(N_3 = 4)$.

(b) Find $P(N_3 = 4, N_5 = 8)$.

(c) Find $P(N_3 = 4|N_5 = 8)$.

(d) Find $P(N_3 = 4, N_5 = 8, N_6 = 10)$.

7.25 Suppose $(N_t)_{t\geq 0}$ is a Poisson process with parameter λ. Find $P(N_s = k|N_t = n)$ when $s > t$.

7.26 The number of accidents on a highway is modeled as a Poisson process with parameter λ. Suppose exactly one accident has occurred by time t. If $0 < s < t$, find the probability that accident occurred by time s.

7.27 Suppose $(N_t)_{t\geq 0}$ is a Poisson process with parameter λ. For $s < t$, find $\text{Cov}(N_s, N_t)$.

7.28 If X has a gamma distribution and c is a positive constant, show that cX has a gamma distribution. Find the parameters.

Beta distribution

7.29 A density function is proportional to $f(x) = x^3(1-x)^7$, for $0 < x < 1$.

(a) Find the constant of proportionality.

(b) Find the mean and variance of the distribution.

(c) Use R to find $P(X > 0.5)$.

7.30 A random variable X has density function proportional to

$$f(x) = \frac{1}{\sqrt{x(1-x)}}, \quad \text{for } 0 < x < 1.$$

Use R to find $P(1/8 < X < 1/4)$.

7.31 **Order statistics.** Suppose $X_1, \ldots, X_{100}$ are independent and uniformly distributed on (0,1).

(a) Find the probability the 25th smallest variable is less than 0.20.

(b) Find $E[X_{(95)}]$ and $V[X_{(95)}]$.

(c) The *range* of a set of numbers is the difference between the maximum and the minimum. Find the expected range.

7.32 If n is odd, the *median* of a list of n numbers is the middle value. Suppose a sample of size $n = 13$ is taken from a uniform distribution on (0,1). Let M be the median. Find $P(M > 0.55)$.

7.33 Let $X \sim$ Beta(a, b). For $s < t$, let $Y = (t-s)X+s$. Then Y has an extended beta distribution on (s, t). Find the density function of Y.

7.34 Find the variance of a beta distribution with parameters a and b.

7.35 Let $X \sim$ Beta(a, b). Find the distribution of $Y = 1 - X$.

7.36 Let $X \sim$ Beta(a, b), for $a > 1$. Find $E[1/X]$.

Pareto, scale-invariant distribution

7.37 Let $X \sim$ Pareto(m, a).

(a) For what values of a does the mean exist? For such a, find $E[X]$.

(b) For what values of a does the variance exist? For such a, find $V[X]$.

7.38 In a population, suppose personal income above $15,000 has a Pareto distribution with $a = 1.8$ (units are $10,000). Find the probability that a randomly chosen individual has income greater than $60,000.

7.39 In Newman (2005), the population of U.S. cities larger than 40,000 is modeled with a Pareto distribution with $a = 2.30$. Find the probability that a random city's population is greater than $k = 3, 4, 5$, and 6 standard deviations above the mean.

7.40 Let $X \sim$ Beta$(a, 1)$. Show that $Y = 1/X$ has a Pareto distribution.

7.41 Let $X \sim$ Exp(a). Let $Y = me^X$. Show that $Y \sim$ Pareto(m, a).

7.42 Zipf's law is a discrete distribution related to the Pareto distribution. If X has a Zipf's law distribution with parameters $s > 0$ and $n \in \{1, 2, \ldots\}$, then

$$P(X = k) = \frac{1/k^s}{\sum_{i=1}^{n}(1/i^s)}, \quad \text{for } k = 1, \ldots, n.$$

The distribution is used to model frequencies of words in languages.

(a) Show that Zipf's law is scale-invariant.

(b) Assume there are a million words in the English language and word frequencies follow Zipf's law with $s = 1$. The three most frequently occurring words are, in order, "the," "of," and "and." What does Zipf's law predict for their relative frequencies?

7.43 The "99-10" rule on the Internet says that 99% of the content generated in Internet chat rooms is created by 10% of the users. If the amount of chat room content has a Pareto distribution, find the value of the parameter a.

Simulation and R

7.44 Conduct a simulation study to illustrate that sums of independent normal random variables are normal. In particular, let $X_1, \ldots, X_{30}$ be normally distributed with $\mu = 1$ and $\sigma^2 = 4$. Simulate $X_1 + \cdots + X_{30}$. Plot a histogram estimate of the distribution of the sum together with the exact density function of a normally distributed random variable with mean 30 and variance 120.

7.45 Suppose phone calls arrive at a Help Desk according to a Poisson process with parameter $\lambda = 10$. Show how to simulate the arrival of phone calls.

8

CONDITIONAL DISTRIBUTION, EXPECTATION, AND VARIANCE

Conditioning is a must for martial artists.

—Bruce Lee

At this crossroad, we bring together several important ideas in both discrete and continuous probability. We focus on conditional distributions and conditional expectation. We will introduce conditional density functions, extend the law of total probability, present problems with both discrete and continuous components, and expand our available tools for computing probabilities.

8.1 CONDITIONAL DISTRIBUTIONS

Conditional distributions for discrete variables were introduced in Chapter 4. For continuous variables, the conditional density function plays the analogous role to the conditional probability mass function.

CONDITIONAL DENSITY FUNCTION

If X and Y are jointly continuous random variables, the conditional density of Y given $X = x$ is

$$f_{Y|X}(y|x) = \frac{f(x,y)}{f_X(x)}, \tag{8.1}$$

for $f_X(x) > 0$.

Probability: With Applications and R, First Edition. Robert P. Dobrow.

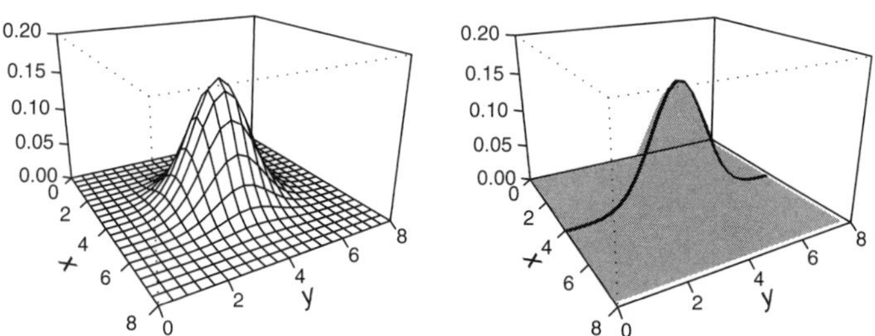

FIGURE 8.1: Left: graph of a joint density function. Right: conditional density of Y given $X = 4$.

The conditional density function is in fact a valid probability density function, as it is nonnegative and integrates to 1. For a given x,

$$\int_{-\infty}^{\infty} f_{Y|X}(y|x)\,dy = \int_{-\infty}^{\infty} \frac{f(x,y)}{f_X(x)}\,dy = \frac{1}{f_X(x)} \int_{-\infty}^{\infty} f(x,y)\,dy = \frac{f_X(x)}{f_X(x)} = 1.$$

Geometrically, the conditional density of Y given $X = x$ is a one-dimensional "slice" of the two-dimensional joint density along the line $X = x$, but "renormalized" so that the resulting curve integrates to 1 (see Fig. 8.1).

Conditional densities are used to compute conditional probabilities. For $A \subseteq \mathbb{R}$,

$$P(Y \in A|X = x) = \int_A f_{Y|X}(y|x)\,dy.$$

When working with conditional density functions $f_{Y|X}(y|x)$, it is most important to remember that the conditioning variable x is treated as *fixed*. The conditional density function is a function of its first argument y.

Example 8.1 Random variables X and Y have joint density

$$f(x,y) = \frac{y}{4x}, \quad \text{for } 0 < y < x < 4.$$

Find $P(Y < 1|X = x)$ and $P(Y < 1|X = 2)$.

The desired probability is

$$P(Y < 1|X = x) = \int_0^1 f_{Y|X}(y|x)\,dy.$$

To find the conditional density of Y given $X = x$, we first need the marginal density of X

$$f_X(x) = \int_0^x \frac{y}{4x}\,dy = \frac{1}{4x}\left(\frac{y^2}{2}\right)\Big|_0^x = \frac{x}{8}, \quad \text{for } 0 < x < 4.$$

The conditional density is

$$f_{Y|X}(y|x) = \frac{f(x,y)}{f_X(x)} = \frac{y/(4x)}{x/8} = \frac{2y}{x^2}, \quad \text{for } 0 < y < x.$$

This gives,

$$P(Y < 1|X = x) = \int_0^1 \frac{2y}{x^2}\, dy = \frac{1}{x^2}$$

and $P(Y < 1|X = 2) = 1/4$. ∎

Example 8.2 Random variables X and Y have joint density function

$$f(x,y) = e^{-x^2 y}, \text{ for } x > 1, y > 0.$$

Find and describe the conditional distribution of Y given $X = x$.

The marginal density of X is

$$f_X(x) = \int_0^\infty f(x,y)\, dy = \int_0^\infty e^{-x^2 y}\, dy = \frac{1}{x^2}, \quad \text{for } x > 1.$$

The conditional density function is

$$f_{Y|X}(y|x) = \frac{f(x,y)}{f_X(x)} = \frac{e^{-x^2 y}}{1/x^2} = x^2 e^{-x^2 y}, \quad \text{for } y > 0.$$

In the conditional density function, x^2 is treated as a constant. Thus, this is the density of an exponential distribution with parameter x^2. That is, the conditional distribution of Y given $X = x$ is exponential with parameter x^2. ∎

Example 8.3 Let X and Y be uniformly distributed in the circle of radius one centered at the origin. Find (i) the marginal distribution of X and (ii) the conditional distribution of Y given $X = x$.

(i) The area of the circle is π. The equation of the circle is $x^2 + y^2 = 1$. The joint density of X and Y is

$$f(x,y) = \frac{1}{\pi}, \quad \text{for } -1 \le x \le 1, -\sqrt{1-x^2} \le y \le \sqrt{1-x^2},$$

and 0, otherwise. Integrating out the y term gives the marginal density of X

$$f_X(x) = \int_{-\sqrt{1-x^2}}^{\sqrt{1-x^2}} \frac{1}{\pi}\, dy = \frac{2\sqrt{1-x^2}}{\pi}, \quad \text{for } -1 \le x \le 1. \tag{8.2}$$

(ii) The conditional density of Y given $X = x$ is

$$f_{Y|X}(y|x) = \frac{f(x,y)}{f_X(x)} = \frac{1/\pi}{2\sqrt{1-x^2}/\pi}$$
$$= \frac{1}{2\sqrt{1-x^2}}, \quad \text{for } -\sqrt{1-x^2} < y < \sqrt{1-x^2}.$$

The function $f_{Y|X}(y|x)$ does not depend on y. As x is treated as a constant, the conditional distribution is uniform on the interval $(-\sqrt{1-x^2}, \sqrt{1-x^2})$.

Observe that while the conditional distribution of Y given $X = x$ is uniform, the marginal distribution of X is not. The marginal distribution of X is the distribution of the X-coordinate of a point (X, Y) in the circle. Points tend to be closer to the origin than to the outside of the circle as described by the half-circular marginal density. ■

Example 8.4 Rachel is working on a project that has many tasks to complete, including tasks A and B. Let X be the proportion of the project time she spends on task A. Let Y be the proportion of time she spends on B. The joint density of X and Y is

$$f(x,y) = 24xy, \quad \text{for } 0 < x < 1,\ 0 < y < 1 - x.$$

If the fraction of the time Rachel spends on task A is x, find the probability that she spends at least half the time on task B.

The total fraction of the time that Rachel spends on both tasks A and B must be less than 1. The desired probability, for $x < 1/2$, is $P(Y > 1/2|X = x)$. (If $x \geq 1/2$, then $P(Y > 1/2|X = x) = 0$.)

The marginal density of X is

$$f_X(x) = \int_0^{1-x} 24xy\, dy = 12x(1-x)^2, \quad \text{for } 0 < x < 1.$$

The conditional density of Y given $X = x$ is

$$f_{Y|X}(y|x) = \frac{f(x,y)}{f_X(x)} = \frac{24xy}{12x(1-x)^2} = \frac{2y}{(1-x)^2}, \quad \text{for } 0 < y < 1 - x.$$

For $0 < x < 1/2$,

$$P(Y > 1/2|X = x) = \int_{1/2}^{1-x} \frac{2y}{(1-x)^2}\, dy = 1 - \frac{1}{4(1-x)^2}.$$ ■

Many random experiments are performed in stages. Consider the following two-stage, *hierarchical* model. Ben picks a number X uniformly distributed in (0,1). Ben shows his number $X = x$ to Lily, who picks a number Y uniformly distributed in $(0, x)$. The conditional distribution of Y given $X = x$ is uniform on $(0, x)$ and thus the conditional density is

$$f_{Y|X}(y|x) = \frac{1}{x}, \quad \text{for } 0 < y < x.$$

If we know that Ben picked 2/3, then the probability that Lily's number is greater than $1/3$ is

$$P(Y > 1/3|X = 2/3) = \int_{1/3}^{2/3} \frac{3}{2}\, dy = \frac{1}{2}.$$

On the other hand, suppose we only see the second stage of the experiment. If Lily's number is 1/3, what is the probability that Ben's original number is greater than 2/3? The desired probability is $P(X > 2/3|Y = 1/3)$. This requires the conditional density of X given $Y = y$.

We are given the conditional density $f_{Y|X}(y|x)$ and we want to find the "inverse" conditional density $f_{X|Y}(x|y)$. This has the flavor of Bayes formula used to invert conditional probabilities. We appeal to the continuous version of Bayes formula. By rearranging the conditional density formula,

$$f_{X|Y}(x|y) = \frac{f(x,y)}{f_Y(y)} = \frac{f_{Y|X}(y|x)f_X(x)}{f_Y(y)}.$$

with denominator equal to

$$f_Y(y) = \int_{-\infty}^{\infty} f(t,y)\, dt = \int_{-\infty}^{\infty} f_{Y|X}(y|t)f_X(t)\, dt.$$

This gives the continuous version of Bayes formula.

BAYES FORMULA

Let X and Y be jointly distributed continuous random variables. Then

$$f_{X|Y}(x|y) = \frac{f_{Y|X}(y|x)f_X(x)}{\int_{-\infty}^{\infty} f_{Y|X}(y|t)f_X(t)\, dt}. \tag{8.3}$$

Using Bayes formula on Ben and Lily's problem, we find the conditional density of X given $Y = y$

$$f_{X|Y}(x|y) = \frac{(1/x)(1)}{\int_y^1 (1/t)(1)\, dt} = \frac{-1}{x \ln y}, \quad \text{for } y < x < 1.$$

The desired probability is

$$P(X > 2/3|Y = 1/3) = \int_{2/3}^{1} \frac{-1}{x \ln 1/3}\,dx = (\ln 3)(\ln 3/2) = 0.445.$$

Example 8.5 The time between successive tsunamis in the Caribbean is modeled with an exponential distribution. See Parsons and Geist (2008) for the use of probability in predicting tsunamis. The parameter value of the exponential distribution is unknown and is itself modeled as a random variable Λ uniformly distributed on (0, 1). Suppose the time X between the last two consecutive tsunamis was 2 years. Find the conditional density of Λ given $X = 2$. (We use the notation Λ since it is the Greek capital letter lambda for λ.)

The conditional distribution of X given $\Lambda = \lambda$ is exponential with parameter λ. That is, $f_{X|\Lambda}(x|\lambda) = \lambda e^{-x\lambda}$, for $x > 0$. The (unconditional) density of Λ is $f_\Lambda(\lambda) = 1$, for $0 < \lambda < 1$. By Bayes formula,

$$\begin{aligned} f_{\Lambda|X}(\lambda|x) &= \frac{f_{X|\Lambda}(x|\lambda)f_\Lambda(\lambda)}{\int_{-\infty}^{\infty} f_{X|\Lambda}(x|t)f_\Lambda(t)\,dt} \\ &= \frac{\lambda e^{-x\lambda}}{\int_0^1 te^{-tx}\,dt} \\ & \frac{\lambda e^{-x\lambda}}{(1 - e^{-x} - xe^{-x})/x^2}, \quad \text{for } 0 < \lambda < 1. \end{aligned}$$

The conditional density of Λ given $X = 2$ is

$$\begin{aligned} f_{\Lambda|X}(\lambda|2) &= \frac{\lambda e^{-2\lambda}}{(1 - e^{-2} - 2e^{-2})/4} \\ &= \frac{4\lambda e^{-2\lambda}}{1 - 3e^{-2}}, \quad \text{for } 0 < \lambda < 1. \end{aligned}$$

Of interest to researchers is estimating λ given the observed data $X = 2$. One approach is to find the value of λ that maximizes the conditional density function (see Fig. 8.2). The density is maximized at $\lambda = 1/2$.

In hindsight, this is not surprising. A natural estimate for λ is 2 given that the mean of an exponential distribution with parameter $\lambda = 1/2$ is 2, the value of X. ■

Motivating the definition. For jointly continuous random variables X and Y, and $A \subseteq R$, the probability $P(Y \in A|X = x)$ is found by integrating the conditional density function. Although the probability is conditional, we *cannot* use the conditional probability formula

$$P(Y \in A|X = x) = \frac{P(Y \in A, X = x)}{P(X = x)},$$

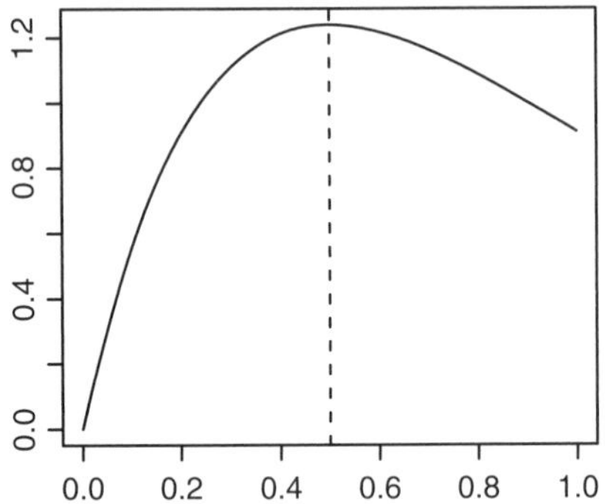

FIGURE 8.2: Graph of conditional density of Λ given $X = 2$. The density function takes its largest value at $\lambda = 1/2$.

since if X is continuous both numerator and denominator are equal to 0, and the conditional probability is undefined.

However, when we try to divide 0 by 0, there often is a limit lurking in the background. To motivate the definition of the conditional density function in Equation 8.1, we have

$$\begin{aligned}
P(Y \in A|X = x) &= \lim_{\epsilon \to 0} P(Y \in A|x \le X \le x + \epsilon) \\
&= \lim_{\epsilon \to 0} \frac{P(Y \in A, x \le X \le x + \epsilon)}{P(x \le X \le x + \epsilon)} \\
&= \lim_{\epsilon \to 0} \frac{\int_x^{x+\epsilon} \int_A f(t, y)\, dy\, dt}{\int_x^{x+\epsilon} f_X(t)\, dt} \\
&= \frac{\lim_{\epsilon \to 0} \frac{1}{\epsilon} \int_x^{x+\epsilon} \int_A f(t, y)\, dy\, dt}{\lim_{\epsilon \to 0} \frac{1}{\epsilon} \int_x^{x+\epsilon} f_X(t)\, dt} \\
&= \frac{\int_A f(x, y)\, dy}{f_X(x)} = \int_A \left[\frac{f(x, y)}{f_X(x)} \right] dy \\
&= \int_A f_{Y|X}(y|x)\, dy, \quad \text{for all } A \subseteq \mathbb{R}.
\end{aligned}$$

8.2 DISCRETE AND CONTINUOUS: MIXING IT UP

Hopefully the reader has appreciated the close similarity between results for discrete and continuous random variables. In an advanced probability course, using the tools of real analysis and measure theory, the two worlds are unified. A new type of integral—called a Lebesgue integral—is used for *both* continuous and discrete random variables, and for random variables that exhibit properties of both. There is no need to treat discrete and continuous problems as two separate categories.

Such a discussion is beyond the scope of this book. However, we can, and will, define joint and conditional distributions of random variables where one variable is discrete and the other is continuous.

To illustrate, here is a "continuous-discrete" two-stage random experiment. Pick a number U uniformly at random between 0 and 1. Given $U = u$, consider a "biased" coin whose probability of heads is equal to u. Flip such a coin n times and let H be the number of heads.

The conditional distribution of H given $U = u$ is a binomial distribution, with parameters n and u. The conditional probability mass function of H given $U = u$ is

$$P(H = h|U = u) = \binom{n}{h} u^h (1-u)^{n-h}, \quad h = 0, \ldots, n.$$

What is the *unconditional* distribution of H?

The joint density of H and U is

$$\begin{aligned} f(h, u) &= P(H = h|U = u) f_U(u) \\ &= \binom{n}{h} u^h (1-u)^{n-h}, \quad h = 0, \ldots, n, \ 0 < u < 1. \end{aligned} \tag{8.4}$$

The function contains a discrete component and a continuous component. To compute probabilities, we sum *and* integrate:

$$P(H \in A, U \in B) = \sum_{h \in A} \int_B f(h, u)\, du = \sum_{h \in A} \int_B \binom{n}{h} u^h (1-u)^{n-h}\, du.$$

The marginal distribution of H is obtained by integrating out u in the joint density, giving

$$P(H = h) = \int_0^1 \binom{n}{h} u^h (1-u)^{n-h}\, du = \binom{n}{h} \int_0^1 u^h (1-u)^{n-h}\, du.$$

The integral can be solved using integration by parts. If you read Section 7.4 on the beta distribution, recognize that the integrand is proportional to a beta density with parameters $a = h + 1$ and $b = n - h + 1$. Hence, the integral is equal to

$$\frac{\Gamma(h+1)\Gamma(n-h+1)}{\Gamma(n+2)} = \frac{h!(n-h)!}{(n+1)!}.$$

This gives

$$P(H = h) = \binom{n}{h} \frac{h!(n-h)!}{(n+1)!} = \frac{1}{n+1}, \quad \text{for } h = 0, \ldots, n.$$

We see that H is uniformly distributed on the integers $\{0, \ldots, n\}$.

Another quantity of interest in this example is the conditional density of U given $H = h$. For $h = 0, \ldots, n$,

$$f_{U|H}(u|h) = \frac{f(h,u)}{f_H(h)} = \frac{\binom{n}{h} u^h (1-u)^{n-h}}{1/(n+1)}$$
$$= \frac{(n+1)!}{h!(n-h)!} u^h (1-u)^{n-h}, \quad 0 < u < 1.$$

The density almost looks like a binomial probability expression. But h is fixed and the density is a continuous function of u. The conditional distribution of U given $H = h$ is a beta distribution with parameters $a = h + 1$ and $b = n - h + 1$.

The example we have used to motivate this discussion is a case of the *beta-binomial model*, an important model in Bayesian statistics and numerous applied fields.

■ **Example 8.6 Will the sun rise tomorrow?** Suppose an event, such as the daily rising of the sun, has occurred n times without fail. What is the probability that it will occur again?

In 1774, Laplace formulated his *law of succession* to answer this question. In modern notation, let S be the unknown probability that the sun rises on any day. And let X be the number of days the sun has risen among the past n days. If $S = s$, and we assume successive days are independent, then the conditional distribution of X is binomial with parameters n and s, and $P(X = n|S = s) = s^n$. Laplace assumed in the absence of any other information that the "unknown" sun rising probability S was uniformly distributed on $(0, 1)$.

Using Bayes formula, the conditional density of the sun rising probability, given that the sun has risen for the past n days, is

$$f_{S|X}(s|n) = \frac{P(X = n|S = s) f_S(s)}{\int_0^1 P(X = n|S = t) f_S(t)\, dt}$$
$$= \frac{s^n}{\int_0^1 t^n\, dt} = (n+1)s^n.$$

Laplace computed the mean sun rising probability with respect to this conditional density

$$\int_0^1 s(n+1)s^n \, ds = \frac{n+1}{n+2}.$$

He argued that the sun has risen for the past 5000 years, or 1,826,213 days. And thus the probability that the sun will rise tomorrow is

$$\frac{1,826,214}{1,826,215} = 0.9999994524\ldots.$$

With such "certainty," hopefully you will sleep better tonight! ■

Example 8.7 The following two-stage "exponential-Poisson" setting has been used to model traffic flow on networks and highways. Consider e-mail traffic during a 1-hour interval. Suppose the unknown rate of e-mail traffic Λ has an exponential distribution with parameter one. Further suppose that if $\Lambda = \lambda$, the number of e-mail messages M that arrive during the hour has a Poisson distribution with parameter 100λ. Find the probability mass function of M.

The conditional distribution of M given $\Lambda = \lambda$ is Poisson with parameter 100λ. The joint density of M and Λ is

$$\begin{aligned} f(m,\lambda) = f_{M|\Lambda}(m|\lambda) f_\Lambda(\lambda) &= \frac{e^{-100\lambda}(100\lambda)^m}{m!} e^{-\lambda} \\ &= \frac{e^{-101\lambda}(100\lambda)^m}{m!}, \quad \text{for } \lambda > 0 \text{ and } m = 0,1,2,\ldots. \end{aligned}$$

Integrating the "mixed" joint density with respect to λ gives the discrete probability mass function

$$\begin{aligned} P(M=m) = \frac{1}{m!}\int_0^\infty e^{-101\lambda}(100\lambda)^m \, d\lambda &= \frac{1}{m!}\left(\frac{m!}{101}\left(\frac{100}{101}\right)^m\right) \\ &= \left(1 - \frac{1}{101}\right)^{m-1} \frac{1}{101}, \quad \text{for } m = 1,2,\ldots \end{aligned}$$

The number of e-mail messages that arrive during the hour has a geometric distribution with parameter $p = 1/101$. ■

R: SIMULATING EXPONENTIAL-POISSON TRAFFIC FLOW MODEL

The following code simulates the joint distribution of Λ and M in the two-stage traffic flow model. Letting n be the number of trials in the simulation, the data are stored in an $n \times 2$ matrix `simarray`. Each row consists of an outcome of (Λ, M).

```
> n <- 100
> simarray <- matrix(0,n,2)
> for (i in 1:n) {
  simarray[i,1] <- rexp(1,1)
  simarray[i,2] <- rpois(1,100*simarray[i,1]) }
```

Marginal distributions of Λ and M are simulated by simply taking the respective first and second columns of the `simarray` matrix.

See Figure 8.3 for graphs of the simulated joint distribution of (Λ, M) and the marginal distributions. Compare the outcomes of M with the geometric distribution with parameter $p = 1/101$.

Here are the R commands for generating the Figure 8.3 graphs.

```
> par(mfrow=c(2,2))
> plot(simarray,xlab=expression(Lambda),ylab="M",
    pch=20, main=expression(paste("Joint distribution
    of  ", Lambda," and M")))
> hist(simarray[,1],xlab="",ylab="",
  main=expression(paste("Marginal distribution of ",
  Lambda)),prob=T)
> curve(dexp(x,1),0,5,add=T)
> hist(simarray[,2],xlab="",ylab="",
  main="Marginal distribution of M")
> hist(rgeom(n,1/101),xlab="",ylab="",
  main="Geometric(1/101) distribution",breaks=15)
```

8.3 CONDITIONAL EXPECTATION

Uncertainty and expectation are the joys of life.

—William Congreve

A *conditional expectation* is an expectation computed with respect to a conditional distribution. We write $E[Y|X = x]$ is for the *conditional expectation of Y given $X = x$*. Our treatment of conditional expectation combines both discrete and continuous settings.

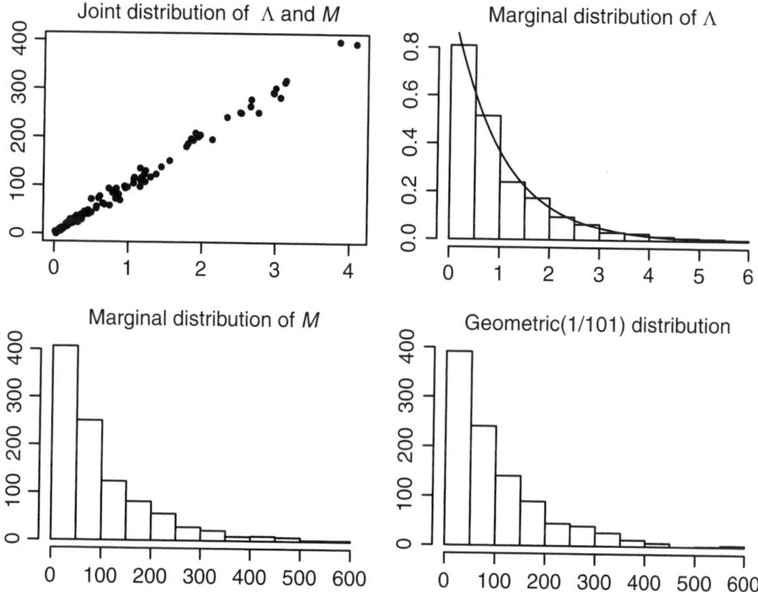

FIGURE 8.3: Simulations of (i) joint distribution of (Λ, M), (ii) marginal distribution of Λ together with exponential density curve ($\lambda = 1$), (iii) marginal distribution of M, and (iv) geometric distribution ($p = 1/101$).

CONDITIONAL EXPECTATION OF Y GIVEN $X = x$

$$E[Y|X = x] = \begin{cases} \sum_y yP(Y = y|X = x), & \text{discrete} \\ \int_y y f_{Y|X}(y|x)\,dy, & \text{continuous.} \end{cases}$$

Most important is that $E[Y|X = x]$ is a *function* of x.

Example 8.8 Random variables X and Y have joint probability mass function described by the following table. Find $E[Y|X = x]$.

		Y = **1**	**2**
X	x_1	0.1	0.2
	x_2	0.3	0.4

We treat $X = x_1$ and $X = x_2$ separately. The conditional probability mass function of of Y given $X = x_1$, is

$$P(Y = k|X = x_1) = \begin{cases} 1/3, & \text{if } k = 1 \\ 2/3, & \text{if } k = 2. \end{cases}$$

Thus,

$$E[Y|X = x_1] = 1\left(\frac{1}{3}\right) + 2\left(\frac{2}{3}\right) = \frac{5}{3}.$$

Similarly, the conditional pmf of Y given $X = x_2$,

$$P(Y = k|X = x_2) = \begin{cases} 3/7, & \text{if } k = 1 \\ 4/7, & \text{if } k = 2, \end{cases}$$

which gives

$$E[Y|X = x_2] = 1\left(\frac{3}{7}\right) + 2\left(\frac{4}{7}\right) = \frac{11}{7}.$$

Writing the conditional expectation $E[Y|X = x]$ as a function of x gives

$$E[Y|X = x] = \begin{cases} 5/3, & \text{if } x = x_1 \\ 11/7, & \text{if } x = x_2. \end{cases}$$ ■

Example 8.9 Suppose X and Y have joint density

$$f(x, y) = \frac{2}{xy}, \quad \text{for } 1 < y < x < e.$$

Find $E[Y|X = x]$.

The marginal density of X is

$$f_X(x) = \int_1^x \frac{2}{xy}\, dy = \frac{2 \ln x}{x}, \quad \text{for } 1 < x < e.$$

The conditional density of Y given $X = x$ is

$$f_{Y|X}(y|x) = \frac{f(x, y)}{f_X(x)} = \frac{2/xy}{2 \ln x/x} = \frac{1}{y \ln x}, \quad \text{for } 1 < y < x.$$

The conditional expectation is

$$E[Y|X = x] = \int_1^x y f_{Y|X}(y|x)\, dy = \int_1^x \frac{y}{y \ln x}\, dy = \frac{x - 1}{\ln x}.$$ ■

Example 8.10 Suppose X and Y have joint density function

$$f(x,y) = e^{-y},\ 0 < x < y < \infty.$$

Find the conditional expectations (i) $E[X|Y = y]$ and (ii) $E[Y|X = x]$.

(i) The marginal density of Y is

$$f_Y(y) = \int_{-\infty}^{\infty} f(x,y)\,dx = \int_0^y e^{-y}\,dx = ye^{-y},\ y > 0.$$

This gives

$$f_{X|Y}(x|y) = \frac{f(x,y)}{f_Y(y)} = \frac{e^{-y}}{ye^{-y}} = \frac{1}{y},\ 0 < x < y.$$

Remember that y is fixed. The conditional distribution of X given $Y = y$ is uniform on the interval $(0, y)$. It immediately follows that the conditional expectation of X given $Y = y$ is the midpoint of the interval $(0, y)$. That is, $E[X|Y = y] = y/2$.

(ii) The marginal density of X is

$$f_X(x) = \int_x^{\infty} e^{-y}\,dy = e^{-x},\ x > 0.$$

This gives the conditional density function

$$f_{Y|X}(y|x) = \frac{f(x,y)}{f_X(x)} = \frac{e^{-y}}{e^{-x}} = e^{-(y-x)},\ y > x. \tag{8.5}$$

The function looks like an exponential density, but it is "shifted over" x units. Let Z be an exponential random variable with parameter $\lambda = 1$. Then the conditional distribution of Y given $X = x$ is the same distribution as that of $Z + x$. In particular, for $y > x$,

$$P(Z + x \le y) = P(Z \le y - x) = 1 - e^{-(y-x)}.$$

Differentiating with respect to y shows that

$$f_{Z+x}(y) = e^{-(y-x)},\ \text{ for } y > x.$$

Since the two distributions are the same, their expectations are the same. This gives

$$E[Y|X = x] = E[Z + x] = E[Z] + x = 1 + x.$$

Or, if you prefer the integral,

$$E[Y|X = x] = \int_{y=x}^{\infty} ye^{-(y-x)}\, dy = 1 + x. \qquad \blacksquare$$

8.3.1 From Function to Random Variable

We are about to take a big leap in our treatment of conditional expectation. Please make sure that your seat belt is securely fastened.

In the previous section, we emphasized that $E[Y|X = x]$ is a function of x. Temporarily write this function with functional notation as $f(x) = E[Y|X = x]$. Since f is a function, we can define a random variable $f(X)$. What exactly is $f(X)$? When X takes the value x, the random variable $f(X)$ takes the value $f(x) = E[Y|X = x]$.

We give a new name to $f(X)$ and call it $E[Y|X]$, the *conditional expectation of Y given X*.

> **CONDITIONAL EXPECTATION *E*[*Y*|*X*]**
>
> For jointly distributed random variables X and Y, the *conditional expectation of Y given X*, denoted $E[Y|X]$, is a random variable, which
>
> 1. Is a function of X
> 2. Is equal to $E[Y|X = x]$ when $X = x$

Example 8.11 Evan picks a random number X uniformly distributed on (0,1). If Evan picks x, he shows it to Kaitlyn who picks a number Y uniformly distributed on $(0, x)$. Find the conditional expectation of Y given X.

The conditional distribution of Y given $X = x$ is provided explicitly. Since the conditional distribution is uniform on $(0, x)$, it follows immediately that $E[Y|X = x] = x/2$, the midpoint of the interval. This holds for all $0 < x < 1$, and thus $E[Y|X] = X/2$. ■

There is much to be learned from this simple, two-stage experiment. We will return to it again.

As much as they look the same, there is a fundamental difference between the conditional expectations $E[Y|X = x]$ and $E[Y|X]$. The former is a function of x. Its domain is a set of real numbers. The function can be evaluated and graphed. For instance, in the last example, $E[Y|X = x] = x/2$ is a linear function of x with slope 1/2. On the other hand, $E[Y|X]$ is a random variable. As such, it has a probability distribution. And thus it makes sense to take *its* expectation with respect to that distribution.

The expectation of a conditional expectation might be a lot to chew on. But it leads to one of the most important results in probability.

LAW OF TOTAL EXPECTATION

For a random variable Y and any random variable X defined jointly with Y, the expectation of Y is equal to the expectation of the conditional expectation of Y given X. That is,

$$E[Y] = E[E[Y|X]]. \tag{8.6}$$

We prove this for the discrete case and leave the continuous case as an exercise. Since $E[Y|X]$ is a random variable that is a function of X, it will be helpful to write it explicitly as $E[Y|X] = f(X)$, where $f(x) = E[Y|X = x]$. When we take the expectation $E[E[Y|X]] = E[f(X)]$ of this function of a random variable, we use the law of the unconscious statistician:

$$\begin{aligned} E[E[Y|X]] = E[f(X)] &= \sum_x f(x)P(X = x) \\ &= \sum_x E[Y|X = x]P(X = x) \\ &= \sum_x \left(\sum_y yP(Y = y|X = x) \right) P(X = x) \\ &= \sum_y y \sum_x P(Y = y|X = x)P(X = x) \\ &= \sum_y y \sum_x P(X = x, Y = y) \\ &= \sum_y yP(Y = y) = E[Y]. \end{aligned}$$

The fifth equality is achieved by changing the order of the double summation.

Example 8.12 Angel will harvest T tomatoes in her vegetable garden, where T has a Poisson distribution with parameter λ. Each tomato is checked for defects. The chance that a tomato has defects is p. Assume that having defects or not is independent from tomato to tomato. Find the expected number of defective tomatoes.

Let X be the number of defective tomatoes. Intuitively, the expected number of defective tomatoes is $E[X] = pE[T] = p\lambda$.

Rigorously, observe that the conditional distribution of X given $T = n$ is binomial with parameters n and p. Thus, $E[X|T = n] = pn$. Since this is true for all n, $E[X|T] = pT$. By the law of total expectation,

$$E[X] = E[E[X|T]] = E[pT] = pE[T] = p\lambda. \qquad \blacksquare$$

Example 8.13 When Katie goes to the gym, she will either run, bicycle, or row. She will choose one of the aerobic activities with respective probabilities 0.5, 0.3, and 0.2. And having chosen an activity the amount of time (in minutes) she spends exercising is exponentially distributed with respective parameters 0.05, 0.025, and 0.01. Find the expectation of Katie's exercise time.

Let T be her exercise time, and let A be a random variable that takes values 1, 2, and 3 corresponding to her choice of running, bicycling, and rowing. The conditional distribution of exercise time given her choice is exponentially distributed. Thus, the conditional expectation of exercise time given her choice is the reciprocal of the corresponding parameter value. By the law of total expectation,

$$\begin{aligned} E[T] &= E[E[T|A]] = \sum_{a=1}^{3} E[T|A=a]P(A=a) \\ &= \frac{1}{0.05}(0.5) + \frac{1}{0.025}(0.3) + \frac{1}{0.01}(0.2) = 42 \text{ minutes.} \end{aligned}$$

■

Example 8.14 **At the gym, continued.** When Tyler goes to the gym, he has similar habits as Katie, except, whenever he chooses rowing he only rows for 10 minutes, stops to take a drink of water, and starts all over, choosing one of the three activities at random as if he had just walked in the door. Find the expectation of Tyler's exercise time.

The conditional expectation of Tyler's exercise time given that he chooses running or bicycling is the same as Katie's. The conditional expectation given that he picks rowing is $E[T|A = 3] = 10 + E[T]$ since after 10 minutes of rowing, Tyler's subsequent exercise time has the same distribution as if he had just walked into the gym and started anew. His expected exercise time is

$$\begin{aligned} E[T] &= E[E[T|A]] \\ &= E[T|A=1]P(A=1) + E[T|A=2]P(A=2) \\ &\quad + E[T|A=3]P(A=3) \\ &= \frac{0.5}{0.05} + \frac{0.3}{0.025} + (E[T]+10)(0.2) \\ &= 24 + E[T](0.2). \end{aligned}$$

Solving for $E[T]$ gives $E[T] = 24/(0.8) = 30$ minutes. See **Exercise.R** for a simulation.

■

In the last example, observe carefully the difference between $E[T]$, $E[T|A = a]$, and $E[T|A]$. The first is a number, the second is a function of a, and the third is a random variable.

Since conditional expectations *are* expectations, they have the same properties as regular expectations, such as linearity. The law of the unconscious statistician also applies. In addition, we highlight two new properties specific to conditional expectation.

PROPERTIES OF CONDITIONAL EXPECTATION

1. (Linearity) For constants a, b, and random variables Y and Z,

$$E[aY + bZ|X] = aE[Y|X] + bE[Z|X].$$

2. (Law of the unconscious statistician) If g is a function, then

$$E[g(Y)|X = x] = \begin{cases} \sum_y g(y)P(Y = y|X = x) & \text{discrete} \\ \int_y g(y) f_{Y|X}(y|x)\,dy & \text{continuous} \end{cases}$$

3. (Independence) If X and Y are independent, then

$$E[Y|X] = E[Y].$$

4. (Y function of X) If $Y = g(X)$ is a function of X, then

$$E[Y|X] = E[g(X)|X] = g(X) = Y. \tag{8.7}$$

We prove the last two properties. For Property 3, we show the discrete case. If X and Y are independent,

$$E[Y|X = x] = \sum_y yP(Y = y|X = x) = \sum_y yP(Y = y) = E[Y],$$

for all x. Thus, $E[Y|X] = E[Y]$.

For Property 4, let $Y = g(X)$ for some function g, then

$$E[Y|X = x] = E[g(X)|X = x] = E[g(x)|X = x] = g(x),$$

where the last equality follows since the expectation of a constant is that constant. Since $E[Y|X = x] = g(x)$ for all x, we have that $E[Y|X] = g(X) = Y$.

Example 8.15 Let X and Y be independent Poisson random variables with respective parameters $\lambda_X > \lambda_Y$. Let U be uniformly distributed on (0,1) and independent of X and Y. The conditional expectation $E[UX+(1-U)Y|U]$ is a random variable. Find its distribution, mean, and variance.

We have

$$\begin{aligned}
E[UX + (1-U)Y|U] &= E[UX|U] + E[(1-U)Y|U] \\
&= UE[X|U] + (1-U)E[Y|U] \\
&= UE[X] + (1-U)E[Y] \\
&= U\lambda_X + (1-U)\lambda_Y \\
&= \lambda_Y + (\lambda_X - \lambda_Y)U,
\end{aligned}$$

where the first equality uses linearity, the second equality uses Property 4, and the third equality uses independence. Since U is uniformly distributed on (0,1), it follows that the conditional expectation is uniformly distributed on (λ_Y, λ_X).

We could use the law of total expectation for the mean, but this is not necessary. Appealing to results for the uniform distribution, the mean of the conditional expectation is the midpoint $(\lambda_X + \lambda_Y)/2$; the variance is $(\lambda_Y - \lambda_X)^2/12$. ■

Example 8.16 Suppose X and Y have joint density

$$f(x,y) = xe^{-y}, \ \ 0 < x < y < \infty.$$

Find $E\left[e^{-Y}|X\right]$.

We will need the conditional density of Y given $X = x$. The marginal density of X is

$$f_X(x) = \int_x^\infty xe^{-y}\, dy = xe^{-x}, \quad \text{for } x > 0.$$

The conditional density of Y given $X = x$ is

$$f_{Y|X}(y|x) = \frac{f(x,y)}{f_X(x)} = \frac{xe^{-y}}{xe^{-x}} = e^{-(y-x)}, \quad \text{for } y > x.$$

This gives

$$\begin{aligned}
E\left[e^{-Y}|X = x\right] &= \int_{-\infty}^\infty e^{-y} f_{Y|X}(y|x)\, dy = \int_x^\infty e^{-y}e^{-(y-x)}\, dy \\
&= e^x \int_x^\infty e^{-2y}\, dy = e^x\left(\frac{e^{-2x}}{2}\right) = \frac{e^{-x}}{2},
\end{aligned}$$

for all $x > 0$. Thus, $E\left[e^{-Y}|X\right] = e^{-X}/2$. ■

Example 8.17 Let $X_1, X_2, \ldots$ be an i.i.d. sequence of random variables with common mean μ. Let $S_n = X_1 + \cdots + X_n$, for each $n = 1, 2, \ldots$ Find $E[S_m|S_n]$ for (i) $m \leq n$, and (ii) $m > n$.

(i) For $m \leq n$,

$$E[S_m|S_n] = E[X_1 + \cdots + X_m|S_n] = \sum_{i=1}^{m} E[X_i|S_n] = mE[X_1|S_n], \tag{8.8}$$

where the second equality is from linearity of conditional expectation, and the last equality is because of symmetry, since all the X_i's are identically distributed. Since S_n is, of course, a function of S_n, we have that

$$S_n = E[S_n|S_n] = E[X_1 + \cdots X_n|S_n] = \sum_{i=1}^{n} E[X_i|S_n] = nE[X_1|S_n].$$

Thus, $E[X_1|S_n] = S_n/n$. With Equation 8.8 this gives

$$E[S_m|S_n] = mE[X_1|S_n] = \left(\frac{m}{n}\right) S_n.$$

(ii) For $m > n$,

$$E[S_m|S_n] = E[S_n + X_{n+1} + \cdots + X_m|S_n] = E[S_n|S_n] + \sum_{i=n+1}^{m} E[X_i|S_n] = S_n + \sum_{i=n+1}^{m} E[X_i] = S_n + (m-n)\mu,$$

where μ is the common mean of the X_i's. The last equality is because for $i > n$, X_i is independent of $(X_1, \ldots, X_n)$ and thus X_i is independent of S_n.

The amounts that a waiter earns every day in tips form an i.i.d. sequence with mean \$50. Given that he earns \$1400 in tips during 1 month (30 days), what is his expected gain in tips for the first week? By (i),

$$E[S_7|S_{30} = 1400] = (7/30)(1400) = \$326.67.$$

On the other hand, if the waiter makes \$400 in tips in the next week, how much can he expect to earn in tips in the next month? By (ii),

$$E[S_{30}|S_7 = 40] = 40 + (30-7)\mu = 40 + 23(50) = \$1190.$$ ■

8.3.2 Random Sum of Random Variables

Sums of random variables where the *number* of summands is also a random variable arise in numerous applications. Van Der Lann and Louter (1986) use such a model to study the total cost of damage from traffic accidents.

Let X_k be the amount of damage from an individual's kth traffic accident. It is assumed that $X_1, X_2, \ldots$ is an i.i.d. sequence of random variables with common mean μ. Furthermore, the number of accidents N for an individual driver is a random variable with mean λ. It is often assumed that N has a Poisson distribution.

The total cost of damages is

$$X_1 + \cdots + X_N = \sum_{k=1}^{N} X_k.$$

The number of summands is random. To find the expected total cost, it is *not* correct to write $E\left[\sum_{k=1}^{N} X_k\right] = \sum_{k=1}^{N} E[X_k]$, assuming that linearity of expectation applies. Linearity of expectation does not apply here because the number of summands is random, not fixed. (Observe that the equation does not even make sense as the left-hand side is a number and the right-hand side is a random variable.)

To find the expectation of a random sum, condition on the number of summands N. Let $T = X_1 + \cdots + X_N$. By the law of total expectation, $E[T] = E[E[T|N]]$. To find $E[T|N]$, consider

$$\begin{aligned} E[T|N = n] &= E\left[\sum_{k=1}^{N} X_k | N = n\right] = E\left[\sum_{k=1}^{n} X_k | N = n\right] \\ &= E\left[\sum_{k=1}^{n} X_k\right] = \sum_{k=1}^{n} E[X_k] = n\mu. \end{aligned}$$

The third equality follows because N is independent of the X_i's. The equality holds for all n, thus $E[T|N] = N\mu$. By the law of total expectation,

$$E[T] = E[E[T|N]] = E[N\mu] = \mu E[N] = \mu\lambda.$$

The result is intuitive. The expected total cost is the product of the expected number of accidents times the expected cost per accident.

8.4 COMPUTING PROBABILITIES BY CONDITIONING

When we first introduced indicator variables, we showed that probabilities can actually be treated as expectations, since for any event A, $P(A) = E[I_A]$, where

I_A is an indicator random variable. Applying the law of total expectation gives

$$P(A) = E[I_A] = E[E[I_A|X]].$$

If X is continuous with density f_X,

$$P(A) = E[E[I_A|X]] = \int_{-\infty}^{\infty} E[I_A|X=x]\, f_X(x)\, dx$$
$$= \int_{-\infty}^{\infty} P(A|X=x) f_X(x)\, dx.$$

This is the continuous version of the law of total probability, a powerful tool for finding probabilities by conditioning.

For instance, consider $P(X < Y)$, where X and Y are continuous. Treating $\{X < Y\}$ as the event A, and conditioning on Y, gives

$$P(X < Y) = \int_{-\infty}^{\infty} P(X < Y|Y = y) f_Y(y)\, dy = \int_{-\infty}^{\infty} P(X < y|Y = y) f_Y(y)\, dy.$$

If X and Y are independent,

$$P(X < Y) = \int_{-\infty}^{\infty} P(X < y|Y = y) f_Y(y)\, dy$$
$$= \int_{-\infty}^{\infty} P(X < y) f_Y(y)\, dy$$
$$= \int_{-\infty}^{\infty} F_X(y) f_Y(y)\, dy.$$

Example 8.18 The times F and Z that Lief and Liz arrive to class have exponential distributions with respective parameters λ_F and λ_Z. If their arrival times are independent, the probability that Liz arrives before Lief is

$$P(Z < F) = \int_0^{\infty} \left(1 - e^{-\lambda_Z y}\right) \lambda_F e^{-\lambda_F y}\, dy$$
$$= 1 - \lambda_Z \int_0^{\infty} e^{-y(\lambda_Z + \lambda_F)}\, dy$$
$$= 1 - \frac{\lambda_F}{\lambda_Z + \lambda_F} = \frac{\lambda_Z}{\lambda_Z + \lambda_F}.$$

■

Example 8.19 The density function of X is $f(x) = xe^{-x}$, for $x > 0$. Given $X = x$, Y is uniformly distributed on $(0, x)$. Find $P(Y < 2)$.

Observe that

$$P(Y < 2|X = x) = \begin{cases} 1, & \text{if } 0 < x \leq 2 \\ 2/x, & \text{if } x > 2. \end{cases}$$

By conditioning on X,

$$
\begin{aligned}
P(Y < 2) &= \int_0^\infty P(Y < 2|X = x) f_X(x)\,dx \\
&= \int_0^2 P(Y < 2|X = x) x e^{-x}\,dx + \int_2^\infty P(Y < 2|X = x) x e^{-x}\,dx \\
&= \int_0^2 x e^{-x}\,dx + \int_2^\infty \left(\frac{2}{x}\right) x e^{-x}\,dx \\
&= (1 - 3e^{-2}) + 2e^{-2} = 1 - e^{-2} = 0.8647.
\end{aligned}
$$

■

Example 8.20 We capture the convolution formula Equation 6.8 for the sum of independent random variables $X + Y$ by conditioning on one of the variables.

$$
\begin{aligned}
P(X + Y \le t) &= \int_{-\infty}^\infty P(X + Y \le t|Y = y) f_Y(y)\,dy \\
&= \int_{-\infty}^\infty P(X + y \le t|Y = y) f_Y(y)\,dy \\
&= \int_{-\infty}^\infty P(X \le t - y) f_Y(y)\,dy,
\end{aligned}
$$

where the last equality is because of independence. Differentiating with respect to t gives the density of $X + Y$

$$
f_{X+Y}(t) = \int_{-\infty}^\infty f(t - y) f(y)\,dy.
$$

■

The following example treats a random variable that has both discrete and continuous components. It arises naturally in many applications, including modeling insurance claims.

Example 8.21 Bob's insurance will pay for a medical expense subject to a \$100 deductible. Suppose the amount of the expense is exponentially distributed with parameter λ. Find (i) the distribution of the amount of the insurance company's payment and (ii) the expected payout.

(i) Let M be the amount of the medical expense and let X be the company's payout. Then

$$
X = \begin{cases} M - 100, & \text{if } M > 100 \\ 0, & \text{if } M \le 100, \end{cases}
$$

where $M \sim \text{Exp}(\lambda)$.

The random variable X has both discrete and continuous components. Observe that $X = 0$ if and only if $M \leq 100$. Thus,

$$P(X = 0) = P(M \leq 100) = \int_0^{100} \lambda e^{-\lambda m}\, dm = 1 - e^{-\lambda 100}.$$

For $x > 0$, we find the cdf of X by conditioning on M.

$$\begin{aligned}
P(X \leq x) &= \int_0^{\infty} P(X \leq x | M = m)\lambda e^{-\lambda m}\, dm \\
&= \int_0^{100} P(0 \leq x | M = m)\lambda e^{-\lambda m}\, dm \\
&\quad + \int_{100}^{\infty} P(M - 100 \leq x | M = m)\lambda e^{-\lambda m}\, dm \\
&= \int_0^{100} \lambda e^{-\lambda m}\, dm + \int_{100}^{\infty} P(M \leq x + 100)\lambda e^{-\lambda m}\, dm \\
&= 1 - e^{-\lambda 100} + \int_{100}^{x+100} \lambda e^{-\lambda m}\, dm \\
&= 1 - e^{-\lambda 100} + \left(e^{-100\lambda} - e^{-\lambda(100+x)}\right) \\
&= 1 - e^{-\lambda(100+x)}.
\end{aligned}$$

Thus, the cdf of X is

$$P(X \leq x) = \begin{cases} 0, & \text{if } x < 0 \\ 1 - e^{-\lambda(100+x)}, & \text{if } x \geq 0. \end{cases}$$

The cdf is not continuous and has a jump discontinuity at $x = 0$ (see Fig. 8.4 for the graph of the cdf).

(ii) For the expected payout $E[X]$, we apply the law of total expectation, giving

$$\begin{aligned}
E[X] = E[E[X|M]] &= \int_0^{\infty} E[X|M = m]\lambda e^{-\lambda m}\, dm \\
&= \int_{100}^{\infty} E[M - 100 | M = m]\lambda e^{-\lambda m}\, dm \\
&= \int_{100}^{\infty} (m - 100)\lambda e^{-\lambda m}\, dm \\
&= \frac{e^{-\lambda 100}}{\lambda}.
\end{aligned}$$

■

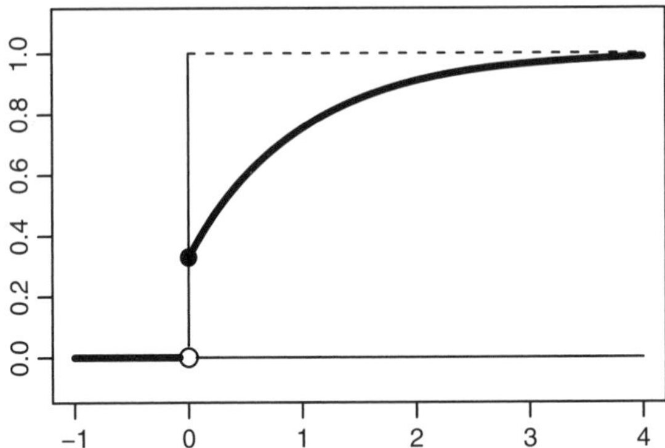

FIGURE 8.4: Cumulative distribution function of insurance payout has jump discontinuity at $x = 0$. The distribution has both discrete and continuous components.

8.5 CONDITIONAL VARIANCE

Conceptually, the conditional variance is derived in a way similar to that of the conditional expectation. It is a variance taken with respect to a conditional distribution. The conditional variance $V[Y|X = x]$ is defined as

CONDITIONAL VARIANCE OF Y GIVEN $X = x$

$$V[Y|X=x] = \begin{cases} \sum_y (y - E[Y|X=x])^2 P(Y=y|X=x), & \text{discrete} \\ \int_y (y - E[Y|X=x])^2 f_{Y|X}(y|x)\, dy, & \text{continuous.} \end{cases}$$

Compare with the unconditional variance formula. Note the conditional expectation $E[Y|X = x]$ takes the place of the unconditional expectation $E[Y]$ in the usual variance formula.

Example 8.22 Recall the two-stage uniform model in Example 8.11. Let $X \sim \text{Unif}(0, 1)$. Conditional on $X = x$, let $Y \sim \text{Unif}(0, x)$. Find the conditional variance $V[Y|X = x]$.

From the defining formula,

$$V[Y|X=x] = \int_0^x (y - E[Y|X=x])^2 \frac{1}{x}\, dy = \int_0^x \frac{(y - x/2)^2}{x}\, dy = \frac{x^2}{12}. \quad \blacksquare$$

Properties for the regular variance transfer to the conditional variance.

PROPERTIES OF CONDITIONAL VARIANCE

1. $$V[Y|X=x] = E[Y^2|X=x] - (E[Y|X=x])^2$$
2. For constants a and b,
$$V[aY+b|X=x] = a^2 V[Y|X=x].$$
3. If Y and Z are independent, then
$$V[Y+Z|X=x] = V[Y|X=x] + V[Z|X=x].$$

As with the development of conditional expectation, we define the *conditional variance* $V[Y|X]$ as the random variable that is a function of X which takes the value $V[Y|X=x]$ when $X=x$. For instance, in Example 8.22, $V[Y|X=x] = x^2/12$, for $0 < x < 1$. And thus $V[Y|X] = X^2/12$.

While the conditional expectation formula $E[Y] = E[E[Y|X]]$ may get credit for one of the most important formulas in this book, the following variance formula is perhaps the most aesthetically pleasing.

LAW OF TOTAL VARIANCE

$$V[Y] = E[V[Y|X]] + V[E[Y|X]].$$

The proof is easier than you might think. We have that

$$\begin{aligned} E[V[Y|X]] &= E\left[E[Y^2|X] - (E[Y|X])^2\right] \\ &= E\left[E[Y^2|X]\right] - E\left[(E[Y|X])^2\right] \\ &= E[Y^2] - E\left[(E[Y|X])^2\right]. \end{aligned}$$

And

$$\begin{aligned} V[E[Y|X]] &= E\left[(E[Y|X])^2\right] - (E[E[Y|X]])^2 \\ &= E\left[(E[Y|X])^2\right] - (E[Y])^2 \end{aligned}$$

Thus,

$$
\begin{aligned}
&E[V[Y|X]] + V[E[Y|X]] \\
&\quad = \left(E[Y^2] - E\left[(E[Y|X])^2\right]\right) + \left(E\left[(E[Y|X])^2\right] - (E[Y])^2\right) \\
&\quad = E[Y^2] - (E[Y])^2 = V[Y].
\end{aligned}
$$

Example 8.23 We continue with the two-stage random experiment of first picking X uniformly in (0,1) and, if $X = x$, then picking Y uniformly in $(0, x)$. Find the expectation and variance of Y.

In Examples 8.11 and 8.22, we found that $E[Y|X] = X/2$ and $V[Y|X] = X^2/12$. By the law of total expectation,

$$E[Y] = E[E[Y|X]] = E\left[\frac{X}{2}\right] = \frac{1}{2}E[X] = \frac{1}{4}.$$

By the law of total variance,

$$
\begin{aligned}
V[Y] &= E[V[Y|X]] + V[E[Y|X]] \\
&= E\left[\frac{X^2}{12}\right] + V\left[\frac{X}{2}\right] \\
&= \frac{1}{12}E[X^2] + \frac{1}{4}V[X] \\
&= \frac{1}{12}\left(\frac{1}{3}\right) + \frac{1}{4}\left(\frac{1}{12}\right) = \frac{7}{144} = 0.04861.
\end{aligned}
$$

The nature of the hierarchical experiment makes it easy to simulate.

R: SIMULATION OF TWO-STAGE UNIFORM EXPERIMENT

```
> simlist <- replicate(100000,runif(1,0,runif(1,0,1)))
> mean(simlist)
[1] 0.2495215
> var(simlist)
[1] 0.04867183
```

■

Example 8.24 During his 4 years at college, Danny takes N exams, where N has a Poisson distribution with parameter λ. On each exam, he scores an A with probability p, independently of any other test. Let Z be the number of A's he receives. Find the correlation between N and Z.

We first find the covariance $\text{Cov}(N, Z) = E[NZ] - E[N]E[Z]$. Conditional on $N = n$, the number of Danny's A's has a binomial distribution with parameters n and p. Thus, $E[Z|N = n] = np$ and $E[Z|N] = Np$. This gives

$$E[Z] = E[E[Z|T]] = E[Np] = pE[N] = p\lambda.$$

By conditioning on N,

$$\begin{aligned} E[NZ] &= E[E[NZ|N]] \\ &= E[NE[Z|N] = E[N(pN)] \\ &= pE[N^2] = p(\lambda + \lambda^2). \end{aligned}$$

The second equality is because of Property 4 in Equation 8.7. We have

$$\text{Cov}(N, Z) = E[NZ] - E[N]E[Z] = p(\lambda + \lambda^2) - (p\lambda)\lambda = p\lambda.$$

To find the correlation, we need the standard deviations. The conditional variance is $V[Z|N = n] = np(1 - p)$ and thus $V[Z|N] = Np(1 - p)$. By the law of total variance,

$$\begin{aligned} V[Z] &= E[V[Z|N]] + V[E[Z|N]] \\ &= E[p(1 - p)N] + V[pN] = p(1 - p)E[N] + p^2V[N] \\ &= p(1 - p)\lambda + p^2\lambda = p\lambda. \end{aligned}$$

Thus, $SD[N]SD[Z] = \sqrt{\lambda}\sqrt{p\lambda} = \lambda\sqrt{p}$. This gives

$$\text{Corr}(N, Z) = \frac{\text{Cov}(N, Z)}{\text{SD}[N]\text{SD}[Z]} = \frac{\lambda p}{\lambda\sqrt{p}} = \sqrt{p}.$$

See the script file **CorrTest.R** for a simulation. The graph in Figure 8.5 was created with parameters $\lambda = 20$ and $p = 0.60$. The correlation between N and Z is $\sqrt{0.60} = 0.775$. ■

Example 8.25 Random sums continued. The number of customers N who come in every day to Alice's Restaurant has mean and variance μ_N and σ_N^2, respectively. Customers each spend on average μ_C dollars with variance σ_C^2. Customers' spending is independent of each other and of N. Find the mean and standard deviation of customers' total spending.

Let $C_1, C_2, \ldots$ be the amounts each customer spends at the restaurant. Then the total spending is $T = C_1 + \cdots + C_N$, a random sum of random variables.

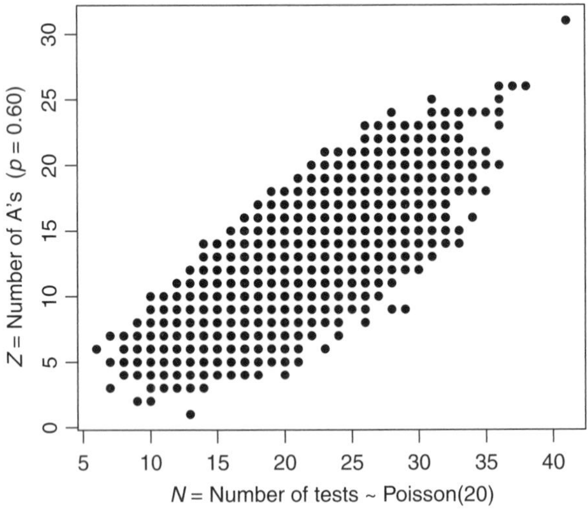

FIGURE 8.5: Number of exams versus number of A's. Correlation is 0.775.

In Section 8.3.2, we showed that

$$E[T] = E[C_1]E[N] = \mu_C \mu_N.$$

For the variance, condition on N. By the law of total variance,

$$\begin{aligned} V[T] &= V\left[\sum_{k=1}^{N} C_k\right] \\ &= E\left[V\left[\sum_{k=1}^{N} C_k \middle| N\right]\right] + V\left[E\left[\sum_{k=1}^{N} C_k \middle| N\right]\right]. \end{aligned}$$

We have that

$$\begin{aligned} V\left[\sum_{k=1}^{N} C_k \middle| N = n\right] &= V\left[\sum_{k=1}^{n} C_k \middle| N = n\right] \\ &= V\left[\sum_{k=1}^{n} C_k\right] = \sum_{k=1}^{n} V[C_k] \\ &= n\sigma_C^2. \end{aligned}$$

The second equality is because N and the C_k's are independent. The third equality is because all of the C_k's are independent. This gives

$$V\left[\sum_{k=1}^{N} C_k \middle| N\right] = \sigma_C^2 N.$$

From results for conditional expectation, we have that

$$E\left[\sum_{k=1}^{N} C_k \middle| N\right] = E[C_1]N = \mu_C N.$$

This gives

$$\begin{aligned} V[T] &= E\left[V\left[\sum_{k=1}^{N} C_k \middle| N\right]\right] + V\left[E\left[\sum_{k=1}^{N} C_k \middle| N\right]\right] \\ &= E[\sigma_C^2 N] + V[\mu_C N] \\ &= \sigma_C^2 E[N] + (\mu_C)^2 V[N] \\ &= \sigma_C^2 \mu_N + (\mu_C)^2 \sigma_N^2. \end{aligned}$$

Suppose, on average, $\lambda = 100$ customers arrive each day. Customers each spend on average \$14 with standard deviation \$2. Then total spending at Alice's Restaurant has mean $E[T] = (100)(14) = \$1400$ and standard deviation

$$\text{SD}[T] = \sqrt{(2^2)(100) + (14^2)(100)} = \sqrt{20,000} = \$141.42.$$

R: TOTAL SPENDING AT ALICE'S RESTAURANT

```
# Alice.R
> n = 50000
> simlist <- numeric(n)
> for (i in 1:n) {
   N <- rpois(1,100) # Number of customers
   cust <- rnorm(N,14,2)
   total <-sum(cust)
   simlist[i] <- total  }
> mean(simlist)
[1] 1400.176
> sd(simlist)
[1] 141.3404
```

■

8.6 SUMMARY

Conditional distribution, expectation, and variance are the focus of this chapter. Many results are given for both discrete and continuous settings. For jointly distributed continuous random variables, the conditional density function is a density function defined with respect to a conditional distribution. The continuous version of Bayes formula is presented. Conditional distributions arise naturally in two-stage, hierarchical models, with several related examples given throughout the chapter.

Conditional expectation is given an extensive treatment. A conditional expectation is an expectation with respect to a conditional distribution. Similarly for conditional variance, we first focus on the conditional expectation of Y given $X = x$ $E[Y|X = x]$, which is a function of x. We then introduce $E[Y|X]$, the conditional expectation of Y given a random variable X. The law of total expectation, a central result in probability, is presented. From the law of total expectation, we show how to compute continuous probabilities by conditioning on random variables.

The last section presents the law of total variance.

- **Conditional density function:** The conditional density of Y given $X = x$ is

$$f_{Y|X}(y|x) = \frac{f(x, y)}{f_X(x)},$$

for $f_X(x) > 0$.

- **Continuous Bayes formula:**

$$f_{X|Y}(x|y) = \frac{f_{Y|X}(y|x) f_X(x)}{\int_{-\infty}^{\infty} f_{Y|X}(y|t) f_X(t)\, dt}.$$

- **Conditional expectation of Y given $X = x$:**

$$E[Y|X = x] = \begin{cases} \int_y y f_{Y|X}(y|x)\, dy, & \text{continuous} \\ \sum_y y P(Y = y|X = x), & \text{discrete.} \end{cases}$$

- **Conditional expectation of Y given X:** $E[Y|X]$ is a random variable and a function of X. When $X = x$, the random variable $E[Y|X]$ takes the value $E[Y|X = x]$.
- **Law of total expectation:** $E[Y] = E[E[Y|X]]$.
- **Properties of conditional expectation:**

1. For constants a and b,

$$E[aY + bZ|X] = aE[Y|X] + bE[Z|X].$$

2. If g is a function,

$$E[g(Y)|X = x] = \begin{cases} \sum_y g(y)P(Y = y|X = x), & \text{discrete} \\ \int_y g(y)f_{Y|X}(y|x)\,dy, & \text{continuous} \end{cases}$$

3. If X and Y are independent, then $E[Y|X] = E[Y]$.
4. If $Y = g(X)$ is a function of X, then $E[Y|X] = Y$.

- **Law of total probability (continuous version):**

$$P(A) = \int_{-\infty}^{\infty} P(A|X = x)f_X(x)\,dx.$$

- **Conditional variance:**

$$V[Y|X = x] = \begin{cases} \sum_y (y - E[Y|X = x])^2 P(Y = y|X = x), & \text{discrete} \\ \int_y (y - E[Y|X = x])^2 f_{Y|X}(y|x)\,dy, & \text{continuous.} \end{cases}$$

- **Properties of conditional variance:**
1. $V[Y|X = x] = E[Y^2|X = x] - (E[Y|X = x])^2$.
2. For constants a and b, $V[aY + b|X = x] = a^2 V[Y|X = x]$.
3. If Y and Z are independent,

$$V[Y + Z|X = x] = V[Y|X = x] + V[Z|X = x].$$

- **Law of total variance:** $V[Y] = E[V[Y|X]] + V[E[Y|X]]$.

EXERCISES

Conditional distributions

8.1 Let X and Y have joint density

$$f(x, y) = 12x(1 - x), \quad \text{for } 0 < x < y < 1.$$

(a) Find the marginal densities of X and Y.

(b) Find the conditional density of Y given $X = x$. Describe the conditional distribution.

(c) Find $P(X < 1/4|Y = 0.5)$

8.2 Let $X \sim \text{Unif}(0, 2)$. If $X = x$, let Y be uniformly distributed on $(0, x)$.

(a) Find the joint density of X and Y.

(b) Find the marginal densities of X and Y.

(c) Find the conditional densities of X and Y.

8.3 Let X and Y have joint density

$$f(x, y) = 4e^{-2x}, \quad \text{for } 0 < y < x < \infty.$$

Find $P(1 < Y < 2|X = 3)$.

8.4 Let X and Y be uniformly distributed on the triangle with vertices (0,0), (1,0), and (1,1).

(a) Find the joint and marginal densities of X and Y.

(b) Find the conditional density of Y given $X = x$. Describe the conditional distribution.

8.5 Let X and Y be i.i.d. exponential random variables with parameter λ. Find the conditional density function of $X + Y$ given $X = x$. Describe the conditional distribution.

8.6 Let X and Y have joint density

$$f(x, y) = \sqrt{\frac{2}{\pi}} y e^{-xy} e^{-y^2/2}, \quad \text{for } x > 0, y > 0.$$

(a) By examining the joint density function can you guess the conditional density of X given $Y = y$? (Treat y as a constant.) Confirm your guess.

(b) Similarly, first conjecture about the conditional distribution of Y given $X = x$. Find the conditional density.

(c) Find $P(X < 1|Y = 1)$.

8.7 Let $Z \sim \text{Gamma}(a, \lambda)$, where a is an integer. Conditional on $Z = z$, let $X \sim \text{Pois}(z)$. Show that X has a negative binomial distribution with the following interpretation. For $k = 0, 1, \ldots$, $P(X = k)$ is equal to the number of failures before a successes in a sequence of i.i.d. Bernoulli trials with success parameter $\lambda/(1 + \lambda)$.

Conditional expectation

8.8 Explain carefully the difference between $E[Y|X]$ and $E[Y|X = x]$

8.9 Show that $E[E[Y|X]] = E[Y]$ in the continuous case.

8.10 On one block of "Eat Street" in downtown Minneapolis there are 10 restaurants to choose from. The waiting time for each restaurant is exponentially distributed with respective parameters $\lambda_1, \ldots, \lambda_{10}$. Bob will decide to eat at restaurant i with probability p_i for $i = 1, \ldots, 10$ ($p_1 + \cdots + p_{10} = 1$). What is Bob's expected waiting time?

8.11 Let X and Y have joint density function

$$f(x, y) = e^{-x(y+1)},\ x > 0, 0 < y < e - 1.$$

(a) Find and describe the conditional distribution of X given $Y = y$.

(b) Find $E[X|Y = y]$ and $E[X|Y]$.

(c) Find $E[X]$ in two ways: (i) using the law of total expectation; (ii) using the distribution of X.

8.12 Let X and Y have joint density

$$f(x, y) = x + y,\ 0 < x < 1, 0 < y < 1.$$

Find $E[X|Y = y]$.

8.13 Let $P(X = 0, Y = 0) = 0.1$, $P(X = 0, Y = 1) = 0.2$, $P(X = 1, Y = 0) = 0.3$, and $P(X = 1, Y = 1) = 0.4$. Show that

$$E[X|Y] = \frac{9 - Y}{12}.$$

8.14 Let A and B be events such that $P(A) = 0.3$, $P(B) = 0.5$, and $P(AB) = 0.2$.

(a) Find $E[I_A|I_B = 0]$ and $E[I_A|I_B = 1]$.

(b) Show that $E[E[I_A|I_B]] = E[I_A]$.

8.15 Let X and Y be independent and uniformly distributed on (0,1). Let $M = \min(X, Y)$ and $N = \max(X, Y)$.

(a) Find the joint density of M and N. (Hint: For $0 < m < n < 1$, show that $M \leq m$ and $N > n$ if and only if either $\{X \leq m \text{ and } Y > n\}$ or $\{Y \leq m \text{ and } X > n\}$.)

(b) Find the conditional density of N given $M = m$ and describe the conditional distribution.

8.16 Given an event A, define the conditional expectation of Y given A as

$$E[Y|A] = \frac{E[YI_A]}{P(A)}, \tag{8.9}$$

where I_A is the indicator random variable.

(a) Let $Y \sim \text{Exp}(\lambda)$. Find $E[Y|Y > 1]$.

(b) An insurance company has a \$250 deductible on a claim. Suppose C is the amount of damages claimed by a customer. Let X be the amount that the insurance company will pay on the claim. Suppose C has an exponential distribution with mean 300. That is,

$$X = \begin{cases} 0 & \text{if } C \leq 250 \\ C - 250 & \text{if } C > 250. \end{cases}$$

Find $E[X]$, the expected payout by the insurance company.

8.17 Let X_1, X_2 be the rolls of two four-sided tetrahedron dice. Let $S = X_1 + X_2$ be the sum of the dice. Let $M = \max(X_1, X_2)$ be the largest of the two numbers rolled. Find the following:

(a) $E[X_1|X_2]$

(b) $E[X_1|S]$

(c) $E[M|X_1 = x]$

(d) $E[X_1X_2|X_1]$

8.18 A *random walker* starts at one vertex of a triangle, moving left or right with probability 1/2 at each step. The triangle is *covered* when the walker visits all three vertices. Find the expected number of steps for the walker to cover the triangle.

8.19 Repeat the last exercise. Only this time at each step the walker either moves left, moves right, or stays put, each with probability 1/3. Staying put counts as one step.

Computing probabilities with conditioning

8.20 Let $X \sim \text{Unif}(0, 1)$. If $X = x$, then $Y \sim \text{Exp}(x)$. Find $P(Y > 1)$ by conditioning on X.

8.21 Let $X \sim \text{Unif}(0, 1)$. If $X = x$, then $Y \sim \text{Unif}(x)$.

(a) Find $P(Y < 1/4)$ by conditioning on X.

(b) Find $P(Y < 1/4)$ by using the marginal density of Y.

(c) Approximate $P(Y < 1/4)$ by simulation.

8.22 Suppose the density of X is proportional to $x^2(1 - x)$ for $0 < x < 1$. If $X = x$, then $Y \sim \text{Binom}(10, x)$. Find $P(Y = 6)$ by conditioning on X.

8.23 Suppose X and Y are independent and positive random variables with density functions f_X and f_Y, respectively. Use conditioning to find a general expression for the density function of

(a) XY

(b) X/Y

(c) $X - Y$

8.24 Let X and Y be independent uniform random variables on (0,1). Find the density function of $Z = X/Y$. Show that the mean of Z does not exist.

Conditional variance

8.25 A biased coin has heads probability p. Let $N \sim \text{Pois}(\lambda)$. If $N = n$, we will flip the coin n times. Let X be the number of heads.

(a) Use the law of total expectation to find $E[X]$.

(b) Use the law of total variance to find $V[X]$.

8.26 Suppose Λ is an exponential random variable with mean 1. Conditional on $\Lambda = \lambda$, N is a Poisson random variable with parameter λ.

(a) Find $E[N|\Lambda]$ and $V[N|\Lambda]$

(b) Use the law of total expectation to find $E[N]$.

(c) Use the law of total variance to find $V[N]$.

(d) Find the probability mass function of N.

(e) Find $E[N]$ and $V[N]$ again using the pmf of N.

8.27 Tom tosses 100 coins. Let H be the number of heads he gets. For each head that he tosses, Tom will get a reward. The amount of each reward is normally distributed with $\mu = 5$ and $\sigma^2 = 1$. (The units are in dollars, and Tom might get a negative "reward.") Individual rewards are independent of each other and independent of H. Find the expectation and standard deviation of Tom's total reward.

8.28 The joint density of X and Y is

$$f(x, y) = xe^{-3xy}, \ \ 1 < x < 4, \ y > 0.$$

(a) Describe the marginal distribution of X.

(b) Describe the conditional distribution of Y given $X = x$.

(c) Find $E[Y|X]$.

(d) Use the law of total expectation to show $E[Y] = \ln 4/9$.

(e) Find $E[XY]$ by conditioning on X.

(f) Find $\text{Cov}(X, Y)$.

8.29 The number of deaths by horsekick per army corps has a Poisson distribution with mean λ. However, λ varies from corps unit to corps unit and can be thought of as a random variable. Determine the mean and variance of the number of deaths by horsekick when $\Lambda \sim \text{Unif}(0, 3)$.

8.30 Revisit Exercise 8.13. Find $V[X|Y = 0]$ and $V[X|Y = 1]$. Find a general expression for $V[X|Y]$ as a function of Y.

8.31 Let $X_1, \ldots, X_n$ be i.i.d. random variables with mean μ and variance σ^2. Let $\bar{X} = (X_1 + \cdots + X_n)/n$ be the average.

(a) Find $E[\bar{X}|X_1]$.

(b) Find $V[\bar{X}|X_1]$.

8.32 If X and Y are independent, does $V[Y|X] = V[Y]$?

8.33 If $Y = g(X)$ is a function of X, what is $V[Y|X]$?

Simulation and R

8.34 Suppose X has a Poisson distribution whose parameter value is the outcome of an independent exponential random variable with parameter μ. (i) Write an R function `rexppois(k,`μ`)` for simulating k copies of such a random variable. (ii) Use your function to estimate $E[X]$ for the case $\mu = 1$ and match up with the result of Example 8.7.

8.35 Write an R script to simulate Katie and Tyler's gym activities as described in Examples 8.13 and 8.14. Estimate their expected exercise time.

8.36 Let N be a Poisson random variable with parameter λ. Write a one-line R command for simulating $Z = \sum_{i=1}^{N} X_i$, where $X_1, X_2, \ldots$ are i.id. normal random variables with parameters μ and σ^2. Now show how to write a one-line command for estimating $E[Z]$.

8.37 See the previous exercise. Suppose $\lambda = 10$ and $\mu = \sigma = 1$. It might be tempting to believe that the distribution of Z is normal with mean 10 and variance 20. (Why might someone make this mistake?) Give visual evidence that this is not the case by simulating the distribution of Z and superimposing a normal density curve with mean 10 and variance 20.

9

LIMITS

We must learn our limits.

—Blaise Pascal

We study limits in probability to understand the long-term behavior of random processes and sequences of random variables. Limits can lead to simplified formulas for otherwise intractable probability models, and they may give insight into complex problems. We have already seen some limit results—the normal approximation, and the Poisson approximation, of the binomial distribution.

The use of simulation to approximate the probability of an event A is justified by one of the most important limit results in probability—the law of large numbers. A consequence of the law of the large numbers is that in repeated trials of a random experiment the proportion of trials in which A occurs converges to $P(A)$, as the number of trials goes to infinity. This is often described as the relative frequency interpretation of probability.

The earliest version of the law of large numbers was discovered at the beginning of the eighteenth century by James Bernoulli, who called it his "golden theorem."

Let $X_1, X_2, \ldots$ be an independent and identically distributed sequence of random variables with finite expectation μ. For $n = 1, 2, \ldots$, let

$$S_n = X_1 + \cdots + X_n.$$

Probability: With Applications and R, First Edition. Robert P. Dobrow.

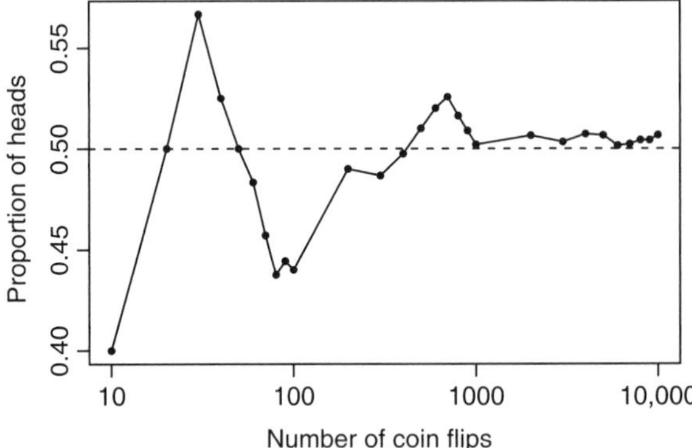

FIGURE 9.1: Mathematician John Kerrich tossed 10,000 coins when he was interned in a prisoner of war camp during World War II. His results illustrate the law of large numbers.

The law of large numbers says that the average S_n/n converges to μ, as $n \to \infty$.

In statistics, where $X_1, \ldots, X_n$ might represent data from a random sample taken from a population, this says that the sample mean converges to the population mean.

Recall mathematician John Kerrich's experiment tossing 10,000 coins when interned in a prisoner of war camp during World War II, described in Section 7.1. His results are shown in Figure 9.1, which graphs the proportion of heads he got for 10–10,000 tosses. The data, given in Freedman et al. (2007), illustrate the law of large numbers. Wide fluctuations in possible outcomes at the beginning of the trials eventually settle down and reach an equilibrium at the mean of the distribution. In this example, $X_1, X_2, \ldots, X_{10{,}000}$ represents independent coin tosses, and S_n/n is the proportion of heads in n tosses, with $\mu = 1/2$.

To see how the law of large numbers leads to the relative frequency interpretation of probability, consider a random experiment and some event A. Repeat the experiment many times, obtaining a sequence of outcomes, and let

$$X_k = \begin{cases} 1, & \text{if } A \text{ occurs on the } k\text{th experiment} \\ 0, & \text{otherwise,} \end{cases}$$

for $k = 1, 2, \ldots$. The X_k's form an i.i.d. sequence with mean $\mu = E[X_k] = P(A)$. The average $S_n/n = (X_1 + \cdots + X_n)/n$ is the proportion of n trials in which A occurs. The law of large numbers says that $S_n/n \to \mu = P(A)$, as $n \to \infty$. That is, the proportion of n trials in which A occurs converges to $P(A)$.

THE "LAW OF AVERAGES" AND A RUN OF BLACK AT THE CASINO

The law of large numbers is sometimes confused with the so-called "law of averages," also called the "gamblers fallacy." This says that outcomes of a random experiment will "even out" within a small sample. Sportscasters love to cite the "law of averages" to explain why players break out of a "slump" (although interestingly they do not ever cite it to explain why players break out of a "streak").

Darrel Huff, in his excellent and amusing book *How to take a chance* (1964), tells the story of a run on black at roulette in a Monte Carlo casino in 1913:

> Black came up a record 26 times in succession. Except for the question of the house limit, if a player had made a one-louis ($4) bet when the run started and pyramided for precisely the length of the run on black, he could have taken away 268 million dollars. What actually happened was a near-panicky rush to bet on red, beginning about the time black had come up a phenomenal 15 times... Players doubled and tripled their stakes [believing] that there was not a chance in a million of another repeat. In the end the unusual run enriched the Casino by some millions of franc.

James Bernoulli's discovery marked a historic shift in the development of probability theory. Previously most probability problems that were studied involved games of chance where outcomes were equally likely. The law of large numbers gave rigorous justification to the fact that *any* probability could be computed, or at least estimated, with a large enough sample size of repeated trials.

Today there are two versions of the law of large numbers, each based on different ways to define what it means for a sequence of random variables to *converge*. What we now know as the *weak law of large numbers* is the "golden theorem" discovered by Bernoulli some 300 years ago.

9.1 WEAK LAW OF LARGE NUMBERS

> This is therefore the problem that I now wish to publish here, having considered it closely for a period of twenty years, and it is a problem of which the novelty as well as the high utility together with its grave difficulty exceed in value all the remaining chapters of my doctrine.
>
> —James Bernoulli, Ars Conjectandi (The Art of Conjecturing) IV

The weak law of large numbers says that for any $\epsilon > 0$ the sequence of probabilities

$$P\left(\left|\frac{S_n}{n} - \mu\right| < \epsilon\right) \to 1, \quad \text{as } n \to \infty.$$

That is, the probability that S_n/n is arbitrarily close to μ converges to 1. Equivalently,

$$P\left(\left|\frac{S_n}{n} - \mu\right| \geq \epsilon\right) \to 0, \quad \text{as } n \to \infty.$$

To understand the result, it is helpful to work out a specific example in detail. Consider i.i.d. coin flips, that is, Bernoulli trials with $p = \mu = 1/2$. Let $\epsilon > 0$. Then

$$\begin{aligned} P\left(\left|\frac{S_n}{n} - \frac{1}{2}\right| < \epsilon\right) &= P\left(-\epsilon < \frac{S_n}{n} - \frac{1}{2} < \epsilon\right) \\ &= P\left(\frac{n}{2} - n\epsilon < S_n < \frac{n}{2} + n\epsilon\right). \end{aligned}$$

The sum $S_n = X_1 + \cdots + X_n$ has a binomial distribution with parameters n and $1/2$. We find the probability $P(|S_n/n - \mu| < \epsilon)$ in R. An illustration of the limiting behavior, with $\epsilon = 0.01$, is shown in Figure 9.2.

R: WEAK LAW OF LARGE NUMBERS

The function `wlln(`n`,`ϵ`)` computes the probability $P(|S_n/n - \mu| < \epsilon)$ for $\mu = p = 1/2$.

```
> p <- 1/2
> wlln <- function(n,eps) {
    pbinom(n*p+n*eps,n,p)-pbinom(n*p-n*eps,n,p) }
> wlln(100,0.1)
[1] 0.9539559
> wlln(100,0.01)
[1] 0.1576179
> wlln(1000,0.01)
[1] 0.4726836
> wlln(10000,0.01)
[1] 0.9544943
```

9.1.1 Markov and Chebyshev Inequalities

Bernoulli's original proof of the weak law of large numbers is fairly complicated and technical. A much simpler proof was discovered in the mid-1800s based on what is now called Chebyshev's inequality. The use of inequalities to bound probabilities is a topic of independent interest.

Theorem 9.1. Markov's inequality. *Let* X *be a nonnegative random variable with finite expectation. Then for all* $\epsilon > 0$,

$$P(X \geq \epsilon) \leq \frac{E[X]}{\epsilon}.$$

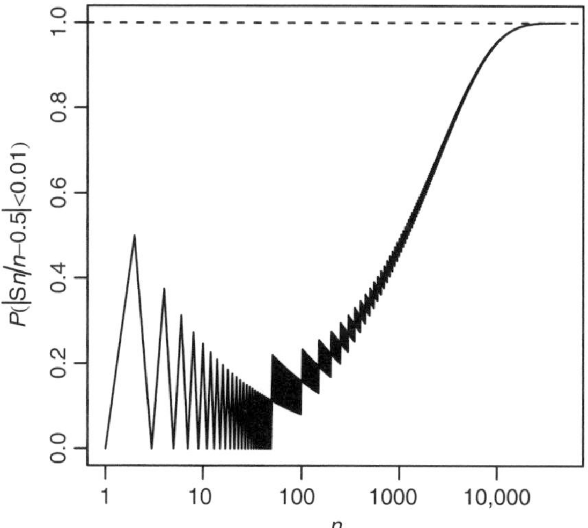

FIGURE 9.2: An illustration of the weak law of large numbers for Bernoulli coin flips with $\epsilon = 0.01$.

Proof. We give the proof for the case when X is continuous with density function f, and invite the reader to show it for the discrete case. We have

$$E[X] = \int_0^\infty x f(x)\,dx \geq \int_\epsilon^\infty x f(x)\,dx \geq \int_\epsilon^\infty \epsilon f(x)\,dx = \epsilon P(X \geq \epsilon).$$

The first inequality is a consequence of the fact that the integrand is nonnegative and $\epsilon > 0$. Rearranging gives the result.

Example 9.1 Let $\epsilon = kE[X] = k\mu$ in Markov's inequality for positive integer k. Then

$$P(X \geq k\mu) \leq \frac{\mu}{k\mu} = \frac{1}{k}.$$

For instance, the probability that a nonnegative random variable is at least twice its mean is at most 1/2. ■

Corollary 9.2. *If f is an increasing positive function, then*

$$P(X \geq \epsilon) = P(f(X) \geq f(\epsilon)) \leq \frac{E[f(X)]}{f(\epsilon)}.$$

By careful choice of the function f, one can often improve the Markov inequality upper bound.

Theorem 9.3. Chebyshev's inequality. *Let X be a random variable (not necessarily positive) with finite mean* μ *and variance* σ^2. *Then for all* $\epsilon > 0$,

$$P(|X-\mu| \geq \epsilon) \leq \frac{\sigma^2}{\epsilon^2}.$$

Proof. Let $f(x) = x^2$ on $(0,\infty)$. By Corollary 9.2, applied to the nonnegative random variable $|X-\mu|$,

$$P(|X-\mu| \geq \epsilon) = P(|X-\mu|^2 \geq \epsilon^2) \leq \frac{E[(X-\mu)^2]}{\epsilon^2} = \frac{\sigma^2}{\epsilon^2}.$$

Example 9.2 Let X be an exponential random variable with mean and variance equal to 1. Consider $P(X \geq 4)$. By Markov's inequality,

$$P(X \geq 4) \leq \frac{1}{4} = 0.25.$$

To bound $P(X \geq 4)$ using Chebyshev's inequality, we have

$$P(X \geq 4) = P(X-1 \geq 3) = P(|X-1| \geq 3) \leq \frac{1}{9} = 0.111. \tag{9.1}$$

We see the improvement of Chebyshev's bound over Markov's bound.

In fact, $P(X \geq 4) = e^{-4} = 0.0183$. So both bounds are fairly crude. However, the power of Markov's and Chebyshev's inequalities is that they apply to *all* random variables without regard to distribution.

Observe that the second equality in Equation 9.1 holds since

$$\begin{aligned}\{|X-1| \geq 3\} &= \{X-1 \geq 3 \text{ or } X-1 \leq -3\}\\ &= \{X-1 \geq 3 \text{ or } X \leq -2\} = \{X-1 \geq 3\},\end{aligned}$$

since X is a positive random variable. However, in general, for a random variable Y and constant c,

$$P(Y \geq c) \leq P(|Y| \geq c),$$

since $\{Y \geq c\}$ implies $\{|Y| \geq c\}$. ■

Example 9.3 Let $\epsilon = k\sigma$ in Chebyshev's inequality for positive integer k. The probability that X is within k standard deviations of the mean is

$$P(|X-\mu| < k\sigma) = 1 - P(|X-\mu| \geq k\sigma) \geq 1 - \frac{\sigma^2}{k^2\sigma^2} = 1 - \frac{1}{k^2}.$$

With $k = 2$, the probability that *any* random variable is within two standard deviations from the mean is at least 75%. The probability that any random variable is within $k = 3$ standard deviations from the mean is at least 88.89%.

For a normally distributed random variable, these probabilities are, respectively, 95 and 99.7%. But again, the utility of the inequalities is that they apply to *all* random variables. ■

Example 9.4 Can Chebyshev's inequality, as a general bound for all random variables, be improved upon? For $k \geq 1$, define a random variable X that takes values $-1, 0$, and 1, with

$$P(X = -1) = P(X = 1) = \frac{1}{2k^2} \quad \text{and} \quad P(X = 0) = 1 - \frac{1}{k^2}.$$

Observe that $\mu = E[X] = 0$ and $\sigma^2 = V[X] = 1/k^2$. Chebyshev's inequality gives

$$P(|X - \mu| \geq k\sigma) \leq \frac{1/k^2}{k^2\sigma^2} = \frac{1}{k^2}.$$

An exact calculation finds

$$P(|X - \mu| \geq k\sigma) = P(|X| \geq 1) = P(X = 1) + P(X = -1) = \frac{1}{k^2}.$$

Thus, this random variable X achieves the Chebyshev bound. This shows that for a general bound that applies to *all* distributions, Chebyshev's bound is best possible; it cannot be improved upon. ■

The proof of the weak law of large numbers is remarkably easy with Chebyshev's inequality.

Theorem 9.4. Weak law of large numbers. *Let $X_1, X_2, \ldots$ be an i.i.d. sequence of random variables with finite mean μ and variance σ^2. For $n = 1, 2, \ldots$, let $S_n = X_1 + \cdots + X_n$. Then*

$$P\left(\left|\frac{S_n}{n} - \mu\right| \geq \epsilon\right) \to 0,$$

as $n \to \infty$.

Proof. We have

$$E\left[\frac{S_n}{n}\right] = \mu \quad \text{and} \quad V\left[\frac{S_n}{n}\right] = \frac{\sigma^2}{n}.$$

Let $\epsilon > 0$. By Chebyshev's inequality,

$$P\left(\left|\frac{S_n}{n} - \mu\right| \geq \epsilon\right) \leq \frac{\sigma^2}{n\epsilon^2} \to 0,$$

as $n \to \infty$.

Remarks:

1. The requirement that the X_i's have finite variance $\sigma^2 < \infty$ is not necessary for the weak law of large numbers to hold. We include it only to simplify the proof.
2. The type of convergence stated in the weak law is known as *convergence in probability*. More generally, say that a sequence of random variables $X_1, X_2, \ldots$ converges in probability to a random variable X, if, for all $\epsilon > 0$,

$$P(|X_n - X| \geq \epsilon) \to 0,$$

as $n \to \infty$. We write $X_n \xrightarrow{p} X$. Thus, the weak law of large numbers says that S_n/n converges in probability to μ.

3. In statistics, a sequence of estimators $\widehat{p_1}, \widehat{p_2}, \ldots$ for an unknown parameter p is called *consistent* if $\widehat{p_n} \xrightarrow{p} p$, as $n \to \infty$. For instance, the sample mean $(X_1 + \cdots + X_n)/n$ is a consistent estimator for the population mean μ.

The weak law of large numbers says that for large n, the average S_n/n is with high probability close to μ. For instance, in a million coin flips we find, using the R function `wlln(n,ε)` given earlier, that the probability that the proportion of heads is within one-one thousandth of the mean is about 0.954.

```
> wlln(1000000,0.001)
[1] 0.9544997
```

However, the weak law does *not* say that as you continue flipping more coins your sequence of coin flips will *stay* close to 1/2. You might get a run of "bad luck" with a long sequence of tails, which will temporarily push the average S_n/n below 1/2, further away than 0.001.

Recall that if a sequence of numbers $x_1, x_2, \ldots$ converges to a limit x, then eventually, for n sufficiently large, the terms $x_n, x_{n+1}, x_{n+2} \ldots$ will *all* be arbitrarily close to x. That is, for any $\epsilon > 0$, there is some index N such that $|x_n - x| \leq \epsilon$ for all $n \geq N$.

The weak law of large numbers, however, does not say that the sequence of averages $S_1/1, S_2/2, S_3/3, \ldots$ behaves like this. It does not say that having come close to μ with high probability the sequence of averages will always *stay* close to μ.

It might seem that since the terms of the sequence $S_1/1, S_2/2, S_3/3, \ldots$ are random variables it would be unlikely to guarantee such strong limiting behavior. And yet the remarkable strong law of large numbers says exactly that.

9.2 STRONG LAW OF LARGE NUMBERS

The Strong Hotel has infinitely many rooms. In each room, a guest is flipping coins—forever. Each guest generates an infinite sequence of zeros and ones. We are interested in the limiting behavior of the sequences in each room. In six of the rooms, we find the following outcomes:

Room 1: 0, 1, 1, 0, 0, 0, 0, 0, 0, 0, 0, 1, 0, 0, 0, 0, 0, 1, 1, 0, 1, 0, 1, 0, 1, 1, 0, ...

Room 2: 1, 0, 0, 0, 0, 1, 0, 1, 1, 1, 1, 0, 1, 1, 1, 0, 1, 1, 1, 0, 0, 1, 0, 0, 0, 1, 0, ...

Room 3: 0, 1, 1, 0, 1, 0, 1, 1, 0, 1, 1, 1, 1, 0, 1, 1, 1, 1, 0, 0, 0, 1, 0, 0, 0, 1, 1, ...

Room 4: 1, 1, 1, 0, 0, 0, 1, 0, 1, 1, 1, 0, 1, 1, 0, 0, 1, 0, 1, 1, 1, 1, 1, 0, 1, 1, 1, ...

Room 5: 1, 0, 0, 1, 0, 0, 1, 0, 0, 1, 0, 0, 1, 0, 0, 1, 0, 0, 1, 0, 0, 1, 0, 0, 1, 0, 0, ...

Room 6: 1, 1, 1, 1, 1, 1, 1, 1, 1, 1, 1, 1, 1, 1, 1, 1, 1, 1, 1, 1, 1, 1, 1, 1, 1, 1, 1, ...

While the sequences in the first four rooms do not reveal any obvious pattern, in room 5, heads appears once every three flips. And in room 6, the guest seems to be continually flipping heads. The sequence of partial averages appears to be converging to 1/3 in room 5, and to 1 in room 6. One can imagine many rooms in which the guest will create sequences whose partial averages do not converge to 1/2.

However, the strong law of large numbers says that in *virtually every room of the hotel* the sequence of averages will converge to 1/2. And not only will these averages get arbitrarily close to 1/2 after a very long time, but each will stay close to 1/2 for all the remaining terms of the sequence. Those sequences whose averages converge to 1/2 constitute a set of "probability 1." And those sequences whose averages do not converge to 1/2 constitute a set of "probability 0."

Theorem 9.5. Strong law of large numbers. *Let* $X_1, X_2, \ldots$ *be an i.i.d. sequence of random variables with finite mean* μ. *For* $n = 1, 2, \ldots$, *let* $S_n = X_1 + \cdots + X_n$. *Then*

$$P\left(\lim_{n\to\infty} \frac{S_n}{n} = \mu\right) = 1. \tag{9.2}$$

We say that S_n/n *converges to* μ *with probability* 1.

The proof of the strong law is beyond the scope of this book. However, we will spend some time explaining what it says and means, as well as giving examples of its use.

Implicit in the statement of the theorem is the fact that the set

$$\left\{\lim_{n\to\infty}\frac{S_n}{n}=\mu\right\}$$

is an event on some sample space. Otherwise, it would not make sense to take its probability. In order to understand the strong law, we need to understand the sample space and the probability function defined on that sample space.

In the case where $X_1, X_2, \ldots$ is an i.i.d. sequence of Bernoulli trials representing coin flips, the sample space Ω is the set of all infinite sequences of zeros and ones. A simple outcome $\omega \in \Omega$ is an infinite sequence.

It can be shown (usually in a more advanced analysis course) that Ω is uncountable. The set can be identified (put in one-to-one correspondence) with the set of real numbers in the interval (0,1). Every real number r between 0 and 1 has a binary base two expansion of the form

$$r=\frac{x_1}{2}+\frac{x_2}{4}+\cdots+\frac{x_k}{2^k}+\cdots, \tag{9.3}$$

where the x_k's are 0 or 1. For every $r \in (0{,}1)$, the correspondence yields a 0–1 sequence $(x_1, x_2, \ldots) \in \Omega$. Conversely, every coin-flipping sequence of zeros and ones $(x_1, x_2, \ldots)$ yields a real number $r \in (0{,}1)$.

The construction of the "right" probability function on Ω is also beyond the scope of this book, and one of the important topics in an advanced probability course based on measure theory. But it turns out that the desired probability function is equivalent to the uniform distribution on (0,1).

As an illustration, consider the probability that in an infinite sequence of coin flips the first two outcomes are both tails. With our notation, we have that $X_1 = X_2 = 0$. Sequences of zeros and ones in which the first two terms are zero yield binary expansions of the form

$$\frac{0}{2}+\frac{0}{4}+\frac{x_3}{8}+\frac{x_4}{16}+\frac{x_5}{32}+\cdots,$$

where the x_k's are 0 or 1 for $k \geq 3$. The set of of all such resulting real numbers gives the interval $(0, 1/4)$. And if $U \sim$ Unif(0,1), then $P(0 < U < 1/4) = 1/4$, the probability that the first two coins come up tails.

For a 0–1 sequence $\omega \in \Omega$, write $\omega = (\omega_1, \omega_2, \ldots)$. The set $\{S_n/n \to \mu\}$ is the set of all sequences ω with the property that $(\omega_1 + \cdots + \omega_n)/n \to \mu$, as $n \to \infty$. This set is in fact an event, that is, it is contained in Ω, and thus we can take its probability. And the strong law of large numbers says that this probability is equal to 1.

Remarks:

1. The type of convergence described in the strong law of large numbers is known as *almost sure convergence.* More generally, we say that a sequence

of random variables $X_1, X_2, \ldots$ converges almost surely to a random variable X if $P(X_n \to X) = 1$. We write $X_n \overset{a.s.}{\to} X$. The strong law says that S_n/n converges almost surely to μ.

2. Almost sure convergence is a stronger form of convergence than convergence in probability. Sequences of random variables that converge almost surely also converge in probability. However, the converse is not necessarily true.
3. The set of 0–1 sequences whose partial averages do not converge to 1/2 is very large. In fact, it is uncountable. Nevertheless, the strong law of large numbers asserts that such sequences constitute a set of probability 0. In that sense, the set of such sequences is very small!

The strong law of large numbers is illustrated in Figure 9.3, which shows the characteristic convergence of S_n/n for nine sequences of 1000 coin flips. Wide fluctuations at the beginning of the sequence settle down to an equilibrium very close to μ.

It is interesting to compare such convergence with what happens for a sequence of random variables in which the expectation does not exist, and for which the strong law of large numbers does not apply. The Cauchy distribution, with density $f(x) = 1/(\pi(1 + x^2))$, for all real x, provides such an example. The distribution is symmetric about zero, but does not have finite expectation. Observe the much different behavior of the sequence of averages in Figure 9.4. Several sequences display erratic behavior with no settling down to an equilibrium as in the finite expectation case.

See Exercise 9.35 for the R code for creating the graphs in Figure 9.3 and Figure 9.4. The code is easily modified to show the strong law for i.i.d. sequences with other distributions.

■ **Example 9.5** The **Weierstrass approximation theorem** says that a continuous function f on [0,1] can be approximated arbitrarily closely by polynomials. In particular, for $0 \leq p \leq 1$,

$$\sum_{k=0}^{n} f\left(\frac{k}{n}\right)\binom{n}{k}p^k(1-p)^{n-k} \to f(p),$$

as $n \to \infty$. The nth degree polynomial function of p on the left-hand side is known as the *Bernstein polynomial.*

The result seems to have no connection to probability. However, there is a probabilistic proof based on the law of large numbers. We give the broad strokes.

Let $X_1, X_2, \ldots$ be an independent sequence of Bernoulli random variables with parameter p. Then by the strong law of large numbers,

$$\frac{X_1 + \cdots + X_n}{n} \to p, \quad \text{as } n \to \infty,$$

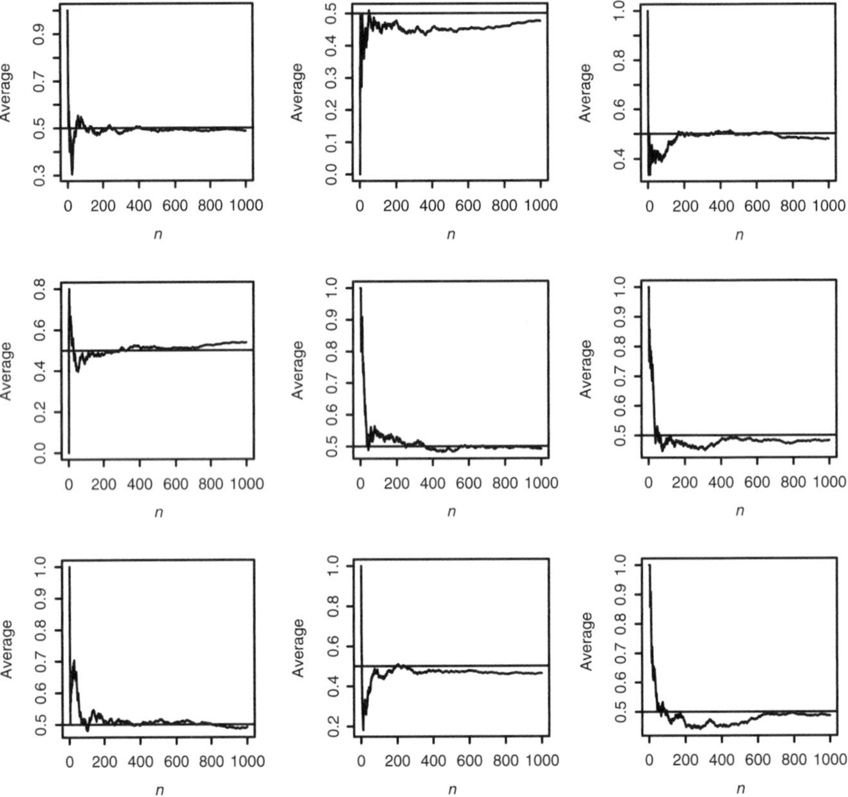

FIGURE 9.3: Nine realizations of 1000 coin flips.

with probability 1. If f is a continuous function, then with probability 1,

$$f\left(\frac{X_1 + \cdots + X_n}{n}\right) \to f(p).$$

It also follows that

$$E\left[f\left(\frac{X_1 + \cdots + X_n}{n}\right)\right] \to E[f(p)] = f(p), \quad \text{as } n \to \infty.$$

Since $X_1 + \cdots + X_n \sim \text{Binom}(n, p)$, the left-hand side is

$$E\left[f\left(\frac{X_1 + \cdots + X_n}{n}\right)\right] = \sum_{k=0}^{n} f\left(\frac{k}{n}\right)\binom{n}{k} p^k (1-p)^{n-k},$$

which gives the result. ∎

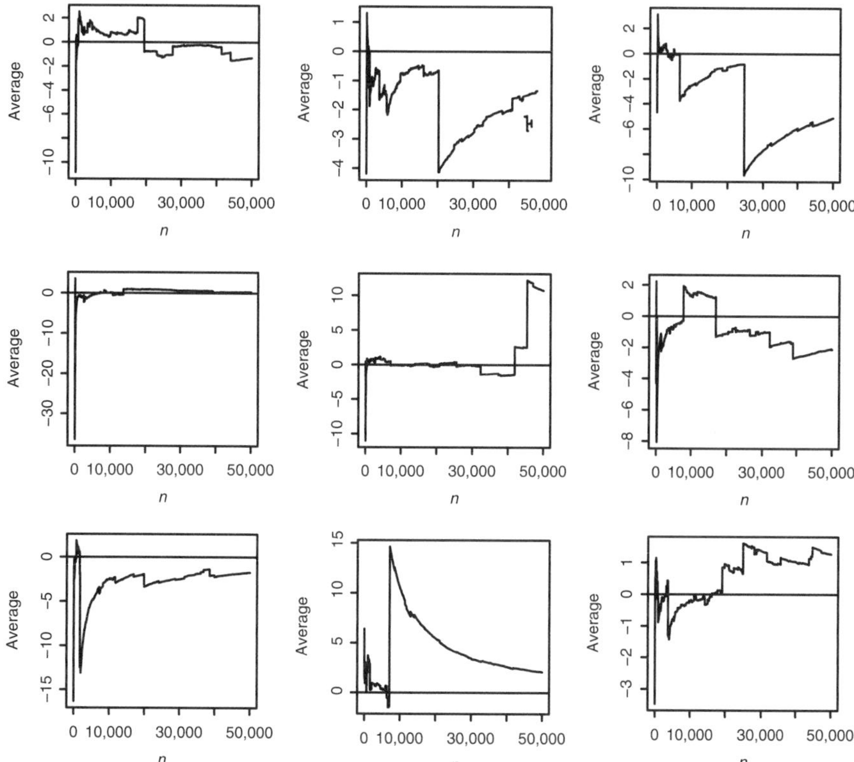

FIGURE 9.4: Sequences of averages ($n = 50,000$) for the Cauchy distribution whose expectation does not exist. Observe the erratic behavior for some sequences.

Consider the problem of estimating the area of a "complicated" set. Here is a Monte Carlo method based on the accept–reject method introduced in Section 6.8.1.

Let S be a bounded set in the plane. Suppose R is a rectangle that encloses S. Let $X_1, \ldots, X_n$ be i.i.d. points uniformly distributed in R. Define

$$I_k = \begin{cases} 1, & \text{if } X_k \in S \\ 0, & \text{otherwise,} \end{cases}$$

for $k = 1, \ldots, n$. The I_k's form an i.i.d. sequence. Their common expectation is

$$E[I_k] = P(X_k \in S) = \frac{\text{Area }(S)}{\text{Area }(R)}.$$

The proportion of the n points that lie in S is

$$\frac{I_1 + \cdots + I_n}{n} \to \frac{\text{Area }(S)}{\text{Area }(R)}, \quad \text{as } n \to \infty,$$

with probability 1. Thus for large n,

$$\text{Area }(S) \approx \left(\frac{I_1 + \cdots + I_n}{n}\right) \text{Area }(R).$$

The area of S is approximately equal to the area of the rectangle R times the proportion of the n points that fall in S.

Example 9.6 Estimating the area of the United States. What is the area of the continental United States? The distance from a line parallel to the northernmost point in the continental United States (Lake of the Woods, Minnesota) to a line parallel to the southernmost point (Ballast Key, Florida) is 1720 miles. From the westernmost point (Cape Alava, Washington) to the easternmost point (West Quoddy Head, Maine) is 3045 miles. The continental United States fits inside a rectangle that is $1720 \times 3045 = 5{,}237{,}400$ square miles.

We enclosed a map of the continental United States inside a comparable rectangle. One thousand points uniformly distributed in the rectangle were generated and we counted 723 in the map (see Fig. 9.5). The area of the continental United States is estimated as $(0.723)(5{,}237{,}400) = 3{,}786{,}640$ square miles.

The area of the continental United States is in fact 3,718,710 square miles. ■

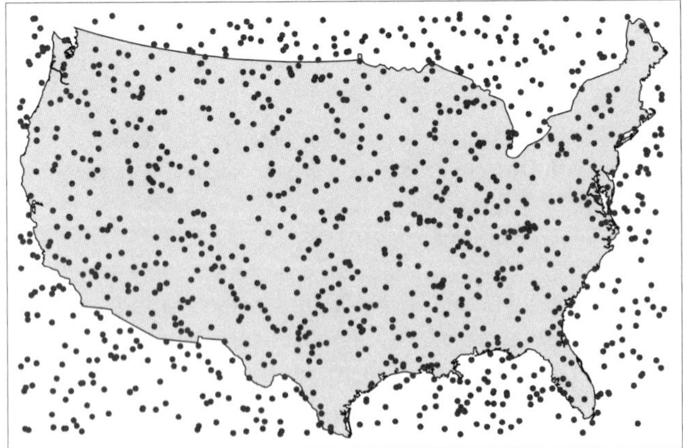

FIGURE 9.5: Using probability and random numbers to estimate the area of the United States.

9.3 MONTE CARLO INTEGRATION

Let f be a continuous function on $(0,1)$. Consider solving the integral

$$I = \int_0^1 f(x)\, dx,$$

assuming the integral converges. If f does not have a known antiderivative, solving the integral may be hard and require numerical approximation. Although the problem does not seem to have any connection with probability, observe that we can write the integral as the expectation of a random variable. In particular,

$$\int_0^1 f(x)\,dx = E[f(X)],$$

where X is uniformly distributed on $(0,1)$. Writing the integral as an expectation allows for using the law of large numbers to approximate the integral.

Let $X_1, X_2, \ldots$ be an i.i.d. sequence of uniform (0,1) random variables. Then $f(X_1),\ f(X_2), \ldots$ is also an i.i.d. sequence with expectation $E[f(X)]$. By the strong law of large numbers, with probability 1,

$$\frac{f(X_1) + \cdots + f(X_n)}{n} \to E[f(X)] = I, \quad \text{as } n \to \infty.$$

This gives a recipe for a Monte Carlo approximation of the integral.

MONTE CARLO INTEGRATION

Let f be a continuous function on $(0,1)$. The following Monte Carlo algorithm gives an approximation for the integral

$$I = \int_0^1 f(x)\,dx.$$

1. Generate n uniform (0,1) random variables $X_1, \ldots, X_n$.
2. Evaluate f at each X_i, giving $f(X_1), \ldots,\ f(X_n)$.
3. Take the average as a Monte Carlo approximation of the integral. That is,

$$I \approx \frac{f(X_1) + \cdots + f(X_n)}{n}.$$

Example 9.7 Solve

$$\int_0^1 (\sin x)^{\cos x}\,dx.$$

Here $f(x) = (\sin x)^{\cos x}$. The integral has no simple antiderivative. We use Monte Carlo approximation with $n = 10{,}000$.

R: MONTE CARLO INTEGRATION

```
> f <- function(x) sin(x)^cos(x)
> n <- 10000
> simlist <- f(runif(n))
> mean(simlist)
[1] 0.4998372
```

The solution can be "checked" using numerical integration. The R command `integrate` numerically solves the integral within a margin of error.

```
> integrate(f,0,1)
0.5013249 with absolute error < 1.1e-09
```

■

The method we have given can be used, in principle, to approximate *any* convergent integral on $(0,1)$ using uniform random variables. However, the Monte Carlo method can be used in a wider context, with different ranges of integration, types of distributions, and with either integrals or sums.

Example 9.8 Solve

$$\int_0^\infty x^{-x}\,dx.$$

Since the range of integration is $(0,\infty)$, we cannot express the integral as a function of a uniform random variable on $(0,1)$. However, write

$$\int_0^\infty x^{-x}\,dx = \int_0^\infty x^{-x}e^x e^{-x}\,dx = \int_0^\infty \left(\frac{e}{x}\right)^x e^{-x}\,dx.$$

Letting $f(x) = (e/x)^x$, the integral is equal to $E[f(X)]$, where X has an exponential distribution with parameter $\lambda = 1$. For Monte Carlo integration, generate exponential random variables, evaluate using f, and take the resulting average.

R: MONTE CARLO INTEGRATION

```
> f <- function(x) (exp(1)/x)^x
> n <- 10000
> simlist <- f(rexp(n,1))
> mean(simlist)
[1] 1.996624
```

Here is a check by numerical integration:

```
> g <- function(x) x^(-x)
> integrate(g,0,Inf)
1.995456 with absolute error < 0.00016
```

■

Monte Carlo approximation is not restricted to integrals.

Example 9.9 Solve

$$\sum_{k=1}^{\infty} \frac{\log k}{3^k}.$$

Write

$$\sum_{k=1}^{\infty} \frac{\log k}{3^k} = \sum_{k=1}^{\infty} \frac{\log k}{2} \left(\frac{1}{3}\right)^{k-1} \left(\frac{2}{3}\right) = \sum_{k=1}^{\infty} \frac{\log k}{2} P(X = k),$$

where X has a geometric distribution with parameter $p = 2/3$. The last expression is equal to $E[f(X)]$, where $f(x) = \log x/2$.

R: MONTE CARLO SUMMATION

```
> f <- function(x) log(x)/2
> n <- 10000
> simlist <- f(rgeom(n,2/3) + 1)
> mean(simlist)
[1] 0.1474316
```

The exact sum of the series, to seven decimal places, is 0.1452795. ■

Example 9.10 Solve

$$I = \int_{-\infty}^{\infty} \int_{-\infty}^{\infty} \int_{-\infty}^{\infty} \sin(x^2 + 2y - z) e^{-(x^2+y^2+z^2)/2} \, dx \, dy \, dz.$$

The multiple integral can be expressed as an expectation with respect to a joint distribution. Let (X, Y, Z) be independent standard normal random variables. The joint density of (X, Y, Z) is

$$f(x, y, z) = \frac{1}{2\pi\sqrt{2\pi}} e^{-(x^2+y^2+z^2)/2}.$$

Write the multiple integral as $I = 2\pi\sqrt{2\pi} E[g(X, Y, Z)]$, where $g(x, y, z) = \sin(x^2 + 2y - z)$. Let $(X_i, Y_i, Z_i), i = 1, \ldots, n$ be an i.i.d. sample from the joint normal distribution. Then

$$I = 2\pi\sqrt{2\pi}E[g(X,Y,Z)] \approx \frac{2\pi\sqrt{2\pi}}{n}\sum_{i=1}^{n} g(X_i, Y_i, Z_i).$$

R: MONTE CARLO INTEGRATION

```
> n <- 1000000
> simlist <- numeric(n)
> for (i in 1:n) {
  vect <- rnorm(3)
  x <- vect[1]
  y <- vect[2]
  z <- vect[3]
  simlist[i] <- sin(x^2+2*y-z) }
> 2*pi*sqrt(2*pi)*mean(simlist)
[1] 0.4489869
```

A symbolic mathematical software system took 3 minutes to solve the integral giving the exact answer

$$I = 2\pi^{3/2}e^{-5/2}\sqrt{\frac{1}{\sqrt{5}} - \frac{1}{5}} = 0.454522\ldots. \quad ■$$

Is Monte Carlo a practical method for solving integrals? For a single integral, probably not, as there are many deterministic numerical integration methods that give excellent approximations to desired levels of accuracy.

However in high dimensions, with large multiple integrals, numerical techniques break down and are not able to give accurate results. It is not uncommon in statistics, biology, physics, and many fields to encounter intractable integrals involving hundreds, even thousands, of variables. For such problems, randomized methods are the only way to go. Monte Carlo integration, often using sophisticated methods involving random processes called Markov chains, is the only practical way in which such integrals can be solved. According to the prominent mathematician and probabilist Persi Diaconis (2008), the use of simulation for such problems has "revolutionized applied mathematics." We introduce this revolutionary methodology, called Markov chain Monte Carlo, in Section 10.6.

9.4 CENTRAL LIMIT THEOREM

> Everyone believes in the law of errors, the experimenters because they think it is a mathematical theorem, the mathematicians because they think it is an experimental fact.
>
> —Gabriel Lipman in conversation with Henri Poincaré

The next theorem rivals the law of large numbers in importance. It gives insight into the behavior of sums of random variables, it is fundamental to much of statistical inference, and it quantifies the size of the error in using Monte Carlo methods to approximate integrals, expectations, and probabilities.

CENTRAL LIMIT THEOREM

Let $X_1, X_2, \ldots$ be an i.i.d. sequence of random variables with finite mean μ and variance σ^2. For $n = 1, 2, \ldots$, let $S_n = X_1 + \cdots + X_n$. Then the distribution of the standardized random variable $(S_n/n - \mu)/(\sigma/\sqrt{n})$ converges to a standard normal distribution in the following sense. For all t,

$$P\left(\frac{S_n/n - \mu}{\sigma/\sqrt{n}} \leq t\right) \to P(Z \leq t), \text{ as } n \to \infty,$$

where $Z \sim \text{Norm}(0,1)$.

We first illustrate the central limit theorem with a simulation experiment. Let $X_1, X_2, \ldots,$ denote i.i.d. Bernoulli trials with $p = 1/2$ (e.g., fair coin flips). The mean and variance of the X_i's are, respectively, $\mu = 1/2$ and $\sigma^2 = 1/4$. Consider the sequence

$$\frac{S_n/n - \mu}{\sigma/\sqrt{n}} = \frac{S_n/n - 1/2}{1/(2\sqrt{n})} = 2\sqrt{n}\left(\frac{S_n}{n} - \frac{1}{2}\right), \quad n = 1, 2, \ldots \tag{9.4}$$

Observe that by the law of large numbers, $(S_n/n - 1/2) \to 0$, as $n \to \infty$, since $S_n/n \to 1/2$. Thus on the right-hand side of Equation 9.4, we take a sequence that converges to 0 and multiply it by $2\sqrt{n}$, a sequence that diverges to infinity. This gives the indeterminate form $\infty \cdot 0$, and *a priori* we do not know whether the resulting sequence will converge to 0, tend to infinity, or do something "in between" like converge to a constant. Furthermore, the terms of the sequence are random variables.

In the simulation code given next, the function `cltsequence(n)` simulates terms of this sequence. Choose a large value of n, enter values repeatedly at your keyboard, and it becomes apparent that the sequence does not converge to 0, nor diverge to infinity, but does "something else." When we repeat many times, graph the distribution of outcomes with a histogram, and superimpose a normal density curve on top of the graph, we obtain an excellent fit to a standard normal distribution, as seen in Figure 9.6.

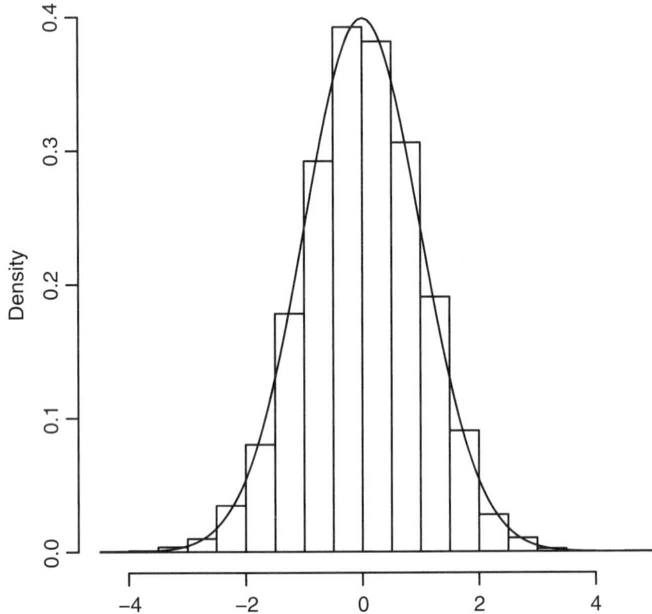

FIGURE 9.6: Histogram of $(S_n/n-\mu)/(\sigma/\sqrt{n})$ from an underlying sequence of $n = 1000$ Bernoulli trials. Density curve is standard normal.

R: SIMULATION EXPERIMENT

```
> cltsequence <- function(n)
    (mean(rbinom(n,1,1/2))-1/2)*(2*sqrt(n))
> cltsqnce(1000)
[1] 0.5059644
> cltsqnce(1000)
[1] 0.3794733
> sqnce(1000)
[1] 1.328157
> sqnce(1000)
[1] -0.7589466
> simlist <- replicate(10000,cltsqnce(1000))
> hist(simlist,prob=T)
> curve(dnorm(x),-4,4,add=T)
```

The astute reader may notice that since a sum of Bernoulli trials has a binomial distribution, what we have illustrated is just the normal approximation of the binomial distribution, discussed in Section 7.1.2.

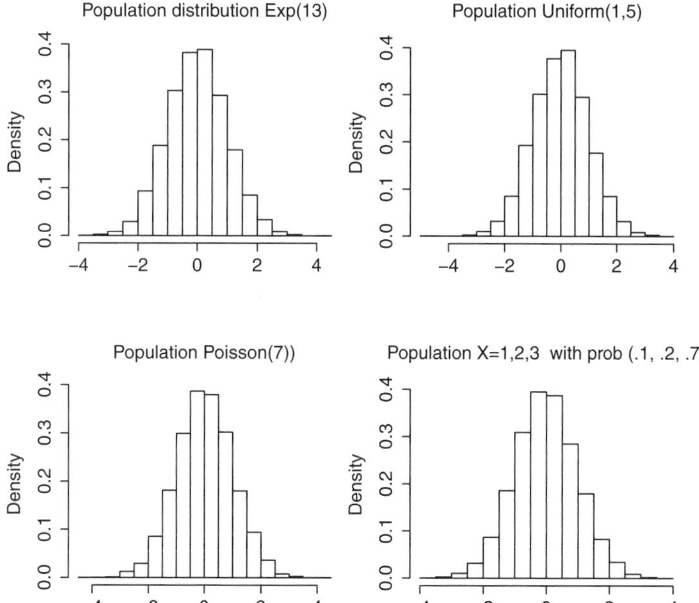

FIGURE 9.7: The simulated distribution of $(S_n/n - \mu)/(\sigma/\sqrt{n})$ for four population distributions ($n = 1000$).

True enough. However, the power of the central limit theorem is that it applies to *any* population distribution with finite mean and variance. In Figure 9.7, observe the results from simulation experiments with four different population distributions: (i) exponential with $\lambda = 13$; (ii) uniform on (1,5); (iii) Poisson with $\lambda = 7$; and (iv) a discrete distribution that takes values 1, 2, and 3, with respective probabilities 0.1, 0.2, and 0.7. In every case, the distribution of the standardized random variable $(S_n/n - \mu)/(\sigma/\sqrt{n})$ tends to a standard normal distribution as n tends to infinity.

Remarks:

1. If the distribution of the X_i's is normal, then by results for sums of independent normal random variables, the distributions of S_n and S_n/n are *exactly* normal. The specialness of the central limit theorem is that it applies to *any* distribution of the X_i's (with finite mean and variance).
2. There are several equivalent ways to formulate the central limit theorem. For large n, the sum $S_n = X_1 + \cdots + X_n$ is approximately normal with mean $n\mu$ and variance $n\sigma^2$. Also, the average S_n/n is approximately normal with mean μ and variance σ^2/n.

EQUIVALENT EXPRESSIONS FOR THE CENTRAL LIMIT THEOREM

If $X_1, X_2, \ldots$ satisfies the assumptions of the central limit theorem, then for large n,

$$X_1 + \cdots + X_n \approx \text{Norm}(n\mu, n\sigma^2)$$

and

$$\frac{X_1 + \cdots + X_n}{n} \approx \text{Norm}(\mu, \sigma^2/n).$$

3. The type of convergence described in the central limit theorem is called *convergence in distribution.* We say that random variables $X_1, X_2, \ldots$ *converge in distribution to* X, if for all t, $P(X_n \leq t) \to P(X \leq t)$, as $n \to \infty$.

Example 9.11 Customers at a popular restaurant are waiting to be served. Waiting times are independent and exponentially distributed with mean $1/\lambda = 30$ minutes. If 16 customers are waiting what is the probability that their average wait is less than 25 minutes?

The average waiting time is $S_{16}/16$. Since waiting time is exponentially distributed, the mean and standard deviation of individual waiting time is $\mu = \sigma = 30$. By the central limit theorem,

$$P(S_{16}/16 < 25) = P\left(\frac{S_{16}/16 - \mu}{\sigma/\sqrt{n}} < \frac{25 - 30}{7.5}\right)$$
$$\approx P(Z < -0.667) = 0.252.$$

For this example, we can compare the central limit approximation with the exact probability. The distribution of S_{16}, the sum of independent exponentials, is a gamma distribution with parameters 16 and 1/30. This gives $P(S_{16}/16 < 25) = P(S_{16} < 400)$. In R,

```
> pgamma(400,16,1/30)
[1] 0.2666045
> pnorm(-0.6667)
[1] 0.2524819
```

■

Example 9.12 More than three million high school students took an AP exam in 2011. The grade distribution for the exams is given in Table 9.1.

In one high school, a sample of 30 AP exam scores was taken. What is the probability that the average score is above 3?

TABLE 9.1: Grade distribution for AP exams.

1	2	3	4	5
0.214	0.211	0.236	0.195	0.144

Assume scores are independent. The desired probability is $P(S_{30}/30 > 3)$. The expectation of exam score is

$$1(0.214) + 2(0.211) + 3(0.236) + 4(0.195) + 5(0.144) = 2.844.$$

The mean of the squared scores is

$$1(0.214)^2 + 2(0.211)^2 + 3(0.236)^2 + 4(0.195)^2 + 5(0.144)^2 = 9.902,$$

giving the standard deviation of exam score to be $\sqrt{9.902 - (2.844)^2} = 1.347$. By the central limit theorem,

$$\begin{aligned} P(S_{30}/30 > 3) &= P\left(\frac{S_{30}/30 - \mu}{\sigma/\sqrt{n}} > \frac{3 - 2.844}{1.347/\sqrt{30}}\right) \\ &\approx P(Z > 0.634) = 0.263. \end{aligned}$$

■

Example 9.13 Random walk. Random walks are fundamental models in physics, ecology, and numerous fields. They have been popularized in finance with applications to the stock market. A particle starts at the origin on the integer number line. At each step, the particle moves left or right with probability 1/2. Find the expectation and standard deviation of the distance of the walk from the origin after n steps.

A random walk process is constructed as follows. Let $X_1, X_2, \ldots$ be an independent sequence of random variables taking values ± 1 with probability 1/2 each. The X_i's represent the individual steps of the random walk. For $n = 1, 2, \ldots$, let $S_n = X_1 + \cdots + X_n$ be the position of the walk after n steps. The random walk process is the sequence $(S_1, S_2, S_3, \ldots)$. Four "paths" for such a process, each taken for $10,000$ steps, are shown in Figure 9.8.

The X_i's have mean 0 and variance 1. Thus, $E[S_n] = 0$ and $V[S_n] = n$. By the central limit theorem, for large n the distribution of S_n is approximately normal with mean 0 and variance n.

After n steps, the random walk's distance from the origin is $|S_n|$. Using the normal approximation, the expected distance from the origin is

$$\begin{aligned} E[|S_n|] &\approx \int_{-\infty}^{\infty} |t| \frac{1}{\sqrt{2\pi n}} e^{-t^2/2n}\, dt = \frac{2}{\sqrt{2\pi n}} \int_0^{\infty} t e^{-t^2/2n}\, dt \\ &= \frac{2n}{\sqrt{2\pi n}} = \sqrt{\frac{2}{\pi}} \sqrt{n} \approx (0.80)\sqrt{n}. \end{aligned}$$

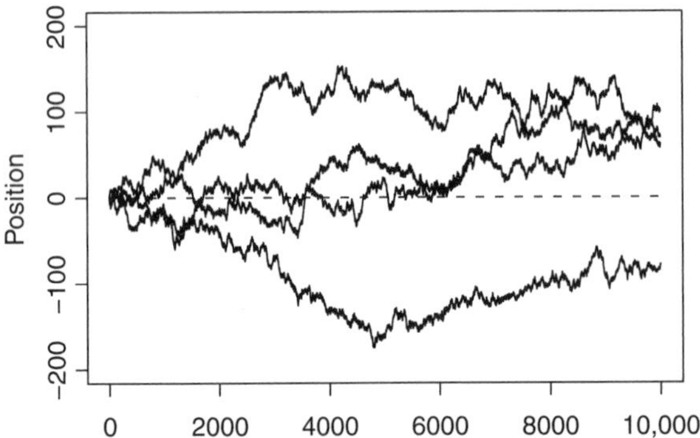

FIGURE 9.8: Four random walk paths of 10,000 steps. The horizontal axis represents the number of steps n. The vertical axis is position.

For the standard deviation of distance, $E[|S_n|^2] = E[S_n^2] = n$. Thus,

$$V[|S_n|] = E[|S_n^2|] - E[|S_n|]^2 \approx n - \frac{2n}{\pi} = n\left(\frac{\pi - 2}{\pi}\right),$$

giving

$$\text{SD}[|S_n|] \approx \sqrt{\frac{\pi - 2}{\pi}}\sqrt{n} \approx (0.60)\sqrt{n}.$$

For instance, after 10,000 steps, a random walk is about $(0.80)\sqrt{10,000} = 80$ steps from the origin give or take about $(0.60)\sqrt{10,000} = 60$ steps.

R: RANDOM WALK DISTANCE FROM ORIGIN

Here we simulate the expectation and standard deviation of a random walk's distance from the origin. Also included is the code for generating the random walk graphs in Figure 9.8.

```
> simlist <- numeric(5000)
> for (i in 1:5000) {
      rw <- sample(c(-1,1),10000, replace=T)
      simlist[i] <- abs(tail(cumsum(rw),1))   }
> mean(simlist)
[1] 80.152
> sd(simlist)
[1] 59.82924

# random walk path
```

```
> steps <- 10000
> rw <- sample(c(-1,1),steps,replace=T)
> plot(c(1,1),type="n",xlim=c(0,steps),
  ylim=c(-200,200),xlab="",ylab="Position")
> lines(cumsum(rw),type="l")
```

■

The central limit theorem requires that n be "large" for the normal approximation to hold. But how large? In statistics, empirical evidence based on the behavior of many actual datasets suggests that $n \approx 30$–40 is usually "good enough" for the central limit theorem to take effect. Heavily skewed distributions may require larger n. However, if the population distribution is fairly symmetric with no extreme values, even relatively small values of n may work reasonably well.

Example 9.14 With just six dice and the central limit theorem one can obtain a surprisingly effective method for simulating normal random variables.

Let $S_6 = X_1 + \cdots + X_6$ be the sum of six dice throws. Then $E[S_6] = 6(3.5) = 21$. And $V[S_6] = 6V[X_1] = 6(35/12) = 35/2$, with $\text{SD}[S_6] = \sqrt{35/2} \approx 4.1833$.

Roll six dice and take $(S_6 - 21)/4.1833$ as a simulation of one standard normal random variable. Here are some results.

R: SUM OF SIX DICE ARE CLOSE TO NORMAL

```
> (sum(sample(1:6,6,rep=T))-21)/4.1833
[1] -0.2390457
> (sum(sample(1:6,6,rep=T))-21)/4.1833
[1] -1.434274
> (sum(sample(1:6,6,rep=T))-21)/4.1833
[1] 0.2390457
> (sum(sample(1:6,6,rep=T))-21)/4.1833
[1] 1.195229
> (sum(sample(1:6,6,rep=T))-21)/4.1833
[1] 0.7171372
```

To explore the distribution of the standardized sum, we simulate 10,000 replications. The normal distribution satisfies the "68-95-99.7 rule." Counting the number of observations within 1, 2, and 3 standard deviations, respectively, gives fairly close agreement with the normal distribution to within about 0.02.

```
> simlist <- replicate(10000,
   (sum(sample(1:6,6,rep=T))-21)/4.1833)
> sum(-1 <= simlist & simlist <= 1)/10000
[1] 0.7087
> sum(-2 <= simlist & simlist <= 2)/10000
```

```
[1] 0.9596
> sum(-3 <= simlist & simlist <= 3)/10000
[1] 0.9985
```

See the comparison in Figure 9.9 between the distribution of the standardized sum of six dice throws and the standard normal density curve.

■

9.4.1 Central Limit Theorem and Monte Carlo

In broad strokes, the law of large numbers asserts that $S_n/n \approx \mu$, when n is large. The central limit theorem states that for large n, $(S_n/n - \mu)/(\sigma/\sqrt{n}) \approx Z$, where Z is a standard normal random variable. Equivalently,

$$\frac{S_n}{n} \approx \mu + \frac{\sigma}{\sqrt{n}} Z.$$

Thus, the central limit theorem can be seen as giving the second-order term in the approximation of μ by S_n/n. This suggests that the error $|S_n/n - \mu|$ in the approximation decreases on the order of $1/\sqrt{n}$, as $n \to \infty$.

What this means is that increasing the number of trials in a simulation by a factor of $n = 100$ will roughly decrease the error by a factor of $1/\sqrt{n} = 1/10$. In other words, every 100-fold increase in the number of Monte Carlo trials improves accuracy in the simulation by about one significant digit.

See Table 9.2 to observe the accuracy of Monte Carlo approximation for $n = 10^1$, 10^3, 10^5, and 10^7. Each simulation increases the number of trials by 100. For each value of n, the simulation is repeated 12 times. With just 10 trials, the results are not uniformly accurate for even one digit. With 1000 trials, the results are uniformly

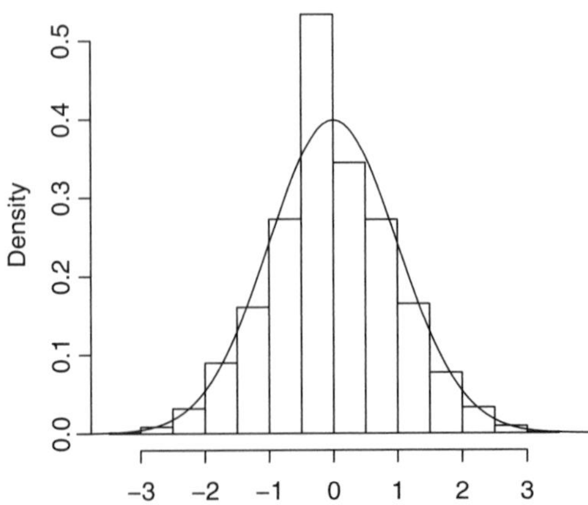

FIGURE 9.9: The normalized sum of just six dice throws comes close the normal distribution.

TABLE 9.2: Monte Carlo approximation of the mean of a uniform (0,1) distribution. Compare the precision for $n = 10^1, 10^3, 10^5, 10^7$. Each simulation is repeated 12 times.

	Number of trials			
Outcome	10	1000	100,000	10,000,000
1	0.7999731	0.4887204	0.5000470	0.4999549
2	0.6176654	0.4922006	0.5004918	0.5000157
3	0.5858701	0.5034127	0.4999026	0.4999761
4	0.5304252	0.4914003	0.5003041	0.5000184
5	0.5602126	0.5070693	0.4997541	0.4999820
6	0.4448233	0.5097890	0.4992946	0.4998743
7	0.3839213	0.4936749	0.5024522	0.5001255
8	0.5434252	0.5010844	0.4997956	0.5000897
9	0.3328748	0.5147627	0.4988693	0.4999183
10	0.5452109	0.4917534	0.5016036	0.5001748
11	0.4943713	0.5178478	0.4999336	0.5001578
12	0.5573252	0.5001346	0.5025030	0.4998666

accurate in the first digit. That is, each approximation is equal to 0.5 when rounded to one digit. With 100,000 trials, the results are precise to two digits. And with 10 million trials, all of the approximations round to 0.500 to three digits.

9.5 MOMENT-GENERATING FUNCTIONS

For $k = 1, 2, \ldots$, the *kth moment* of a random variable X is $E[X^k]$. For instance, the first moment of X is the expectation $E[X]$. The concept is related to the notion of moment in physics. The moment-generating function (mgf), as the name suggests, can be used to generate the moments of a random variable.

MOMENT-GENERATING FUNCTION

Definition 9.1. *Let X be a random variable. The mgf of X is the real-valued function*

$$m(t) = E[e^{tX}]$$

defined for all real t when this expectation exists.

Moments of X are obtained from the mgf by successively differentiating $m(t)$ and evaluating at $t = 0$. We have

$$m'(t) = \frac{d}{dt}E[e^{tX}] = E\left[\frac{d}{dt}e^{tX}\right] = E[Xe^{tX}]$$

and $m'(0) = E[X]$.

Taking second derivatives gives

$$m''(t) = \frac{d}{dt}m'(t) = \frac{d}{dt}E[Xe^{tX}] = E\left[\frac{d}{dt}Xe^{tX}\right] = E\left[X^2e^{tX}\right]$$

and $m''(0) = E[X^2]$.

In general, the kth derivative of the mgf gives

$$m^{(k)}(0) = E[X^k], \quad \text{for } k = 1, 2, \ldots$$

Remarks:

1. To define the mgf it suffices that the expectation $E[e^{tX}]$ exists for values of t in an *interval* that, contains zero.
2. For some distributions, the expectation $E[e^{tX}]$ is not finite. This is true for the Cauchy distribution and the t-distribution, used in statistics. However, another type of generating function called the *characteristic function* can be used. The characteristic function $E[e^{itX}]$ is defined for *all* distributions and requires the use of complex (imaginary) numbers.
3. In physics, the concept of moments is used to describe physical properties, with the first moment roughly equal to the center of mass. For a probability distribution, the first moment is the "center" of probability mass. Similar analogies can be made with higher moments.

■ **Example 9.15 Poisson distribution.** Let $X \sim \text{Pois}(\lambda)$. The mgf of X is

$$m(t) = E\left[e^{tX}\right] = \sum_{k=0}^{\infty} e^{tk}\frac{e^{-\lambda}\lambda^k}{k!} = e^{-\lambda}\sum_{k=0}^{\infty}\frac{(\lambda e^t)^k}{k!} = e^{-\lambda}e^{\lambda e^t} = e^{\lambda(e^t-1)}.$$

Differentiating gives

$$m'(t) = \lambda e^t e^{\lambda(e^t-1)}$$

with

$$m'(0) = \lambda = E[X];$$

and

$$m''(t) = (\lambda e^t)\left(\lambda e^t + 1\right)\left(e^{\lambda(e^t-1)}\right)$$

with

$$m''(0) = \lambda^2 + \lambda = E[X^2].$$

■

Example 9.16 Let X be a random variable taking values -1, 0, and 1, with

$$P(X=-1)=p, \ \ P(X=0)=q, \ \text{ and } \ P(X=1)=r,$$

with $p+q+r=1$. Find the mgf and use it to find the mean and variance of X.

The mgf of X is

$$m(t)=E[e^{tX}]=pe^{-t}+q+re^{t},$$

with

$$m'(t)=-pe^{-t}+re^{t} \ \text{ and } \ m''(t)=pe^{-t}+re^{t}.$$

Thus,

$$E[X]=m'(0)=r-p,$$

and

$$V[X]=E[X^2]-E[X]^2=m''(0)-[m'(0)]^2=(r+p)-(r-p)^2. \qquad \blacksquare$$

Example 9.17 **Standard normal distribution.** Let $X \sim$ Norm(0,1). The mgf of X is

$$\begin{aligned} m(t)=E\left[e^{tX}\right] &= \int_{-\infty}^{\infty} e^{tx}\frac{1}{\sqrt{2\pi}}e^{-x^2/2}\,dx \\ &= \frac{1}{\sqrt{2\pi}}\int_{-\infty}^{\infty} e^{-(x^2-2tx)/2}\,dx \\ &= e^{t^2/2}\frac{1}{\sqrt{2\pi}}\int_{-\infty}^{\infty} e^{-(x-t)^2/2}\,dx \\ &= e^{t^2/2}, \end{aligned}$$

as the last integral gives the density of a normal distribution with mean t and variance one. Check that $m'(0)=0=E[X]$ and $m''(0)=1=E[X^2]$. $\blacksquare$

Here are three important properties of the mgf.

PROPERTIES OF MOMENT GENERATING FUNCTIONS

1. If X and Y are independent random variables, then the mgf of their sum is the product of their mgfs.
   That is,

$$\begin{aligned} m_{X+Y}(t) &= E\left[e^{t(X+Y)}\right] \\ &= E\left[e^{tX}e^{tY}\right] = E\left[e^{tX}\right]E\left[e^{tY}\right] \\ &= m_X(t)m_Y(t). \end{aligned}$$

2. Let X be a random variable and c a constant. Then

$$m_{cX}(t) = E\left[e^{t(cX)}\right] = E\left[e^{(tc)X}\right] = m_X(ct).$$

3. Moment-generating functions uniquely determine the underlying probability distribution. That is, if two random variables have the same mgf, then they have the same probability distribution.

Example 9.18 Sum of independent Poissons is Poisson. Let X and Y be independent Poisson random variables with respective parameters λ_1 and λ_2. Then

$$m_{X+Y}(t) = m_X(t)m_Y(t) = e^{\lambda_1(e^t-1)}e^{\lambda_2(e^t-1)} = e^{(\lambda_1+\lambda_2)(e^t-1)},$$

which is the mgf of a Poisson random variable with parameter $\lambda_1 + \lambda_2$.

This constitutes a proof that the sum of independent Poisson random variables is Poisson, a result derived by other means in Section 4.4.1. ■

This last example gives a taste of the power of mgfs. Many results involving sums and limits of random variables can be proven using them.

Example 9.19 Sum of i.i.d. exponentials is gamma. We have shown that the sum of i.i.d. exponential random variables has a gamma distribution. Here we prove the result using the mgfs. First find the mgf of the exponential distribution.

Let $X \sim \text{Exp}(\lambda)$. The mgf of X is

$$\begin{aligned} m_X(t) = E\left[e^{tX}\right] &= \int_{-\infty}^{\infty} e^{tx}\lambda e^{-\lambda x}\,dx \\ &= \frac{\lambda}{\lambda - t}\int_{-\infty}^{\infty}(\lambda - t)e^{-(\lambda-t)x}\,dx \\ &= \frac{\lambda}{\lambda - t}, \end{aligned}$$

defined for all $0 < t < \lambda$.

Now find the mgf of the gamma distribution. Let $Z \sim \text{Gamma}(a, \lambda)$. The mgf of Z is

$$\begin{aligned} m_Z(t) = E\left[e^{tZ}\right] &= \int_0^\infty e^{tz} \frac{\lambda^a z^{a-1} e^{-\lambda z}}{\Gamma(a)} \, dz \\ &= \int_0^\infty \frac{\lambda^a z^{a-1} e^{-(\lambda - t)z}}{\Gamma(a)} \, dz \\ &= \left(\frac{\lambda}{\lambda - t}\right)^a \int_0^\infty \frac{(\lambda - t)^a z^{a-1} e^{-(\lambda - t)z}}{\Gamma(a)} \, dz \\ &= \left(\frac{\lambda}{\lambda - t}\right)^a. \end{aligned}$$

That is, $m_Z(t) = [m_X(t)]^a$. For positive integer a, this is the mgf of the sum of a independent exponential random variables with parameter λ. ∎

The moment-generating functions play a significant role in establishing limit theorems for random variables because of the following result, which we state without proof.

Theorem 9.7. Continuity theorem. *Let* $X_1, X_2, \ldots$ *be a sequence of random variables with corresponding mgfs* $m_{X_1}(t), m_{X_2}(t), \ldots$*. Further suppose that for all* t,

$$m_{X_n}(t) \to m_X(t), \quad \text{as } n \to \infty,$$

where $m_X(t)$ *is the mgf of a random variable X. Then*

$$P(X_n \le x) \to P(X \le x), \quad \text{as } n \to \infty,$$

at each x *where* $P(X \le x)$ *is continuous.*

Thus, convergence of the mgfs corresponds to convergence in distribution for random variables.

9.5.1 Proof of Central Limit Theorem

Here we prove the central limit theorem using moment generating functions. We show that the cumulative distribution function of $(S_n - n\mu)/(\sigma\sqrt{n})$ converges to the cdf of the standard normal distribution.

Proof. Let $X_1, X_2, \ldots$ be an i.i.d. sequence of random variables with finite mean μ and variance σ^2. First suppose that $\mu = 0$ and $\sigma^2 = 1$. Then $S_n/n = (X_1 + \cdots + X_n)/n$ has mean 0 and variance $1/n$. We need to show that the mgf of $(S_n/n)/(1/\sqrt{n}) = S_n/\sqrt{n}$ converges to the mgf of Z, where $Z \sim \text{Norm}(0,1)$.

Let m be the common mgf of the X_i's. By Properties 1 and 2 for the mgfs,

$$m_{S_n/\sqrt{n}}(t) = \left[m\left(\frac{t}{\sqrt{n}}\right)\right]^n.$$

The mgf of the standard normal distribution is $m_Z(t) = e^{t^2/2}$. We need to show that for all t,

$$m\left(\frac{t}{\sqrt{n}}\right)^n \to e^{t^2/2}, \quad \text{as } n \to \infty.$$

Equivalently, we take logarithms and show that

$$\ln\left(m\left(\frac{t}{\sqrt{n}}\right)^n\right) \to \frac{t^2}{2}, \quad \text{as } n \to \infty.$$

Consider the limit, as $n \to \infty$, of $\ln m(t/\sqrt{n})^n = n \ln m(t/\sqrt{n})$. Assuming continuity of the mgf,

$$\lim_{n\to\infty} \ln\left(m\left(\frac{t}{\sqrt{n}}\right)\right) = \ln\left(m\left(\lim_{n\to\infty}\frac{t}{\sqrt{n}}\right)\right) = \ln m(0) = \ln 1 = 0.$$

Thus, $\lim_{n\to\infty} n \ln m(t/\sqrt{n})$ is an indeterminate form of type $\infty \cdot 0$. Apply l'Hôpital's rule two times, but first make a change of variables letting $\epsilon = t/\sqrt{n}$, so $n = t^2/\epsilon^2$, giving

$$\begin{aligned}
\lim_{n\to\infty} n \ln m(t/\sqrt{n}) &= t^2 \lim_{\epsilon\to 0} \frac{\ln m(\epsilon)}{\epsilon^2} \\
&= t^2 \lim_{\epsilon\to 0} \frac{m'(\epsilon)}{m(\epsilon)2\epsilon} \\
&= \lim_{\epsilon\to 0} t^2 \frac{m''(\epsilon)}{m'(\epsilon)2\epsilon + 2m(\epsilon)} \\
&= t^2 \frac{m''(0)}{2m(0)} = \frac{t^2}{2}.
\end{aligned}$$

This proves the theorem for the case when $\mu = 0$ and $\sigma^2 = 1$.

For the general case, let $X_i^* = (X_i - \mu)/\sigma$. Then $X_i = \sigma X_i^* + \mu$ and

$$S_n = \sum_{i=1}^{n} X_i = \sum_{i=1}^{n} (\sigma X_i^* + \mu) = \sigma S_n^* + n\mu,$$

where $S_n^* = X_1^* + \cdots + X_n^*$. This gives

$$\frac{S_n/n - \mu}{\sigma/\sqrt{n}} = \frac{S_n^*}{\sqrt{n}}.$$

Observe that S_n^* is the sum of i.i.d. random variables with mean 0 and variance 1. Thus, the mgf of $S_n^*/\sqrt{n}$ converges to the standard normal mgf, and the result is shown.

9.6 SUMMARY

This chapter introduces the important limit theorems of probably, two inequalities for bounding general probabilities, and mgfs. For i.i.d. random variables $X_1, X_2, \ldots$, with finite mean μ, the law of large numbers says that the sequence of averages $S_n/n = (X_1 + \cdots + X_n)/n$ converges to μ, as n tends to infinity. There are two versions of the limit result, the weak law and the strong law, based on different ways in which a sequence of random variables can converge to a limit. The proof of the weak law of large numbers uses Markov's and Chebyshev's inequalities, which bound probabilities of the form $P(X \geq x)$. The strong law of large numbers asserts that S_n/n converges to μ "almost surely," or with probability 1. The use of Monte Carlo simulation to approximate integrals and sums is shown.

The central limit theorem is discussed and proved in this chapter. The theorem says that for large n, the standardized averages $(S_n/n-\mu)/(\sigma/\sqrt{n})$ have an approximate standard normal distribution. The result is remarkable in that it applies to all population distributions with finite mean and variance.

The last section of this chapter introduces the mgfs that are useful devices for finding the moments of a probability distribution and for proving results about sums of independent random variables as well as limit theorems. The central limit theorem is proven using the mgfs.

For the results given next, let $X_1, X_2, \ldots$ be an i.i.d. sequence of random variables with finite mean μ and variance σ^2. Let $S_n = X_1 + \cdots + X_n$.

- **Weak law of large numbers:** For all $\epsilon > 0$,

$$P\left(|S_n/n - \mu| < \epsilon\right) \to 1, \quad \text{as } n \to \infty.$$

- **Markov's inequality:** For nonnegative random variable X with finite expectation, $P(X \geq \epsilon) \leq E[X]/\epsilon$.

- **Chebyshev's inequality:** If the mean μ and variance σ^2 of X are finite, then $P(|X - \mu| \geq \epsilon) \leq \sigma^2/\epsilon^2$.
- **Strong law of large numbers:** $P(\lim_{n\to\infty} S_n/n = \mu) = 1$.
- **Monte Carlo integration:** An integral such as $I = \int_0^1 f(x)\,dx$ can be written as the expectation of a random variable $E[f(X)]$, where $X \sim \text{Unif}(0,1)$. If $X_1, X_2, \ldots$ are i.i.d. random variables uniformly distributed on $(0,1)$, then

$$\frac{f(X_1) + \cdots + f(X_n)}{n} \approx E[f(X)] = I.$$

- **Central limit theorem:** For all t,

$$P\left(\frac{S_n/n - \mu}{\sigma/\sqrt{n}} \leq t\right) \to P(Z \leq t), \quad \text{as } n \to \infty,$$

where $Z \sim \text{Norm}(0, 1)$.
- **Moments:** The kth moment of X is $E[X^k]$.
- **Moment-generating function:** $m(t) = E[e^{tX}]$ defined for real t when the expectation exists.
- **Properties of the moment-generating functions:**
  1. $E[X^k] = m^{(k)}(0)$.
  2. If X and Y are independent, then $m_{X+Y}(t) = m_X(t)m_Y(t)$.
  3. For constant c, $m_{cX}(t) = m_X(ct)$.
  4. If two random variables have the same mgf, they have the same distribution.
  5. Let $X_1, X_2, \ldots$ be a sequence of random variables. If $m_{X_n}(t) \to m_X(t)$, as $n \to \infty$, where $m_X(t)$ is the mgf of X, then $P(X_n \leq x) \to P(X \leq x)$, as $n \to \infty$.

EXERCISES

Law of Large Numbers

9.1 Describe in your own words the law of large numbers.

9.2 Your roommate missed probability class again. Explain to him/her the difference between the weak and strong laws of large numbers.

9.3 Let S be the sum of 100 fair dice rolls. Use (i) Markov's inequality and (ii) Chebyshev's inequality to bound $P(S \geq 380)$.

9.4 Find the best value of c so that $P(X \geq 5) \leq c$ using Markov's and Chebyshev's inequalities, filling in the subsequent table. Compare with the exact probabilities.

Distribution	Markov	Chebyschev	Exact Probability
Pois(2)			
Exp(1/2)			
Norm(2, 4)			
Geom(1/2)			

9.5 Let X be a positive random variable with $\mu = 50$ and $\sigma^2 = 25$.
(a) What can you say about $P(X \geq 60)$ using Markov's inequality?
(b) What can you say about $P(X \geq 60)$ using Chebyshev's inequality?

9.6 Prove Markov's inequality for the discrete case.

9.7 Let X be a positive random variable. Show that for all c,

$$P(\log X \geq c) \leq \mu e^{-c}.$$

9.8 The expected sum of two fair dice is 7; the variance is 35/6. Let X be the sum after rolling n pairs of dice. Use Chebyshev's inequality to find z such that

$$P(|X - 7n| < z) \geq 0.95.$$

In 10,000 rolls of two dice there is at least a 95% chance that the sum will be between what two numbers?

9.9 Show how to use Monte Carlo techniques to approximate the following integrals and sums.

(a)
$$I = \int_0^1 \sin(x) e^{-x^2}\, dx.$$

(b)
$$I = \int_0^\infty \sin(x) e^{-x^2}\, dx.$$

(c)
$$I = \int_{-\infty}^\infty \log(x^2) e^{-x^2}\, dx.$$

(d)
$$S = \sum_{k=1}^\infty \frac{\sin k}{2^k}.$$

(e)

$$S = \sum_{k=0}^{\infty} \frac{\cos \cos k}{k!}.$$

9.10 Make up your own "hard" integral to solve using Monte Carlo approximation. Do the same with a "hard" sum.

Central Limit Theorem

9.11 The local farm packs its tomatoes in crates. Individual tomatoes have mean weight of 10 ounces and standard deviation 3 ounces. Find the probability that a crate of 50 tomatoes weighs between 480 and 510 ounces.

9.12 The waiting time on the cashier's line at the school cafeteria is exponentially distributed with mean 2 minutes. Use the central limit theorem to find the approximate probability that the average waiting time is more than 2.5 minutes for a group of 20 people. Use R and compare with the exact probability.

9.13 Recall the game of roulette and the casino's fortunes when a player places a "red" bet. (See Example 4.28). For one \$1 red bet, let G be the casino's gain. Then $P(G = 1) = 20/38$ and $P(G = -1) = 18/38$. Suppose in 1 month, one million red bets are placed. Let T be the casino's total gain. Find $P(50,000 < T < 55,000)$.

9.14 A baseball player has a batting average of 0.328. Let X be the number of hits the player gets during 20 times at bat. Use the central limit theorem to find the approximate probability $P(X \leq k)$ for $k = 1, 3, 6$. Compare with the exact probability for each k.

9.15 Let $X_1, \ldots, X_{30}$ be independent random variables with density

$$f(x) = 3x^2, \quad \text{if } 0 < x < 1.$$

Use the central limit theorem to approximate

$$P(22 < X_1 + \cdots + X_{30} < 23).$$

9.16 Let $X_1, \ldots, X_n$ be an i.i.d. sample from a population with unknown mean μ and standard deviation σ. We take the sample mean $\bar{X} = (X_1 + \cdots + X_n)/n$ as an estimate for μ.

(a) According to Chebyshev's inequality, how large should the sample size n be so that with probability 0.99 the error $|\bar{X} - \mu|$ is less than 2 standard deviations?

(b) According to the central limit theorem, how large should n be so that with probability 0.99 the error $|\bar{X} - \mu|$ is less than 2 standard deviations?

9.17 Let $X_1, \ldots, X_{10}$ be independent Poisson random variables with $\lambda = 1$. Consider $P(\sum_{i=1}^{10} X_i \geq 14)$.

(a) What does Markov's inequality say about this probability?

(b) What does Chebyshev's inequality say?

(c) What does the central limit theorem say?

(d) What does the central limit theorem say with continuity correction?

(e) Find the exact probability.

9.18 Consider a random walk as described in Example 9.13. After one million steps, find the probability that the walk is within 500 steps from the origin.

9.19 Let $X \sim \text{Gamma}(a, \lambda)$, where a is a large integer. Without doing any calculations, explain why $X \approx \text{Norm}(a/\lambda, a/\lambda^2)$.

9.20 Show that

$$\lim_{n \to \infty} \int_0^n \frac{e^{-x} x^{n-1}}{(n-1)!} \, dx = \frac{1}{2}.$$

Hint: Consider an independent sum of n Exponential(1) random variables and apply the central limit theorem.

9.21 A random variable Y is said to have a *lognormal distribution* if $\log Y$ has a normal distribution. Equivalently, we can write $Y = e^X$, where X has a normal distribution.

(a) If $X_1, X_2, \ldots$ is an independent sequence of uniform (0,1) variables, show that the product $Y = \prod_{i=1}^n X_i$ has an approximate lognormal distribution. Show that the mean and variance of $\log Y$ are, respectively, $-n$ and n.

(b) If $Y = e^X$, with $X \sim \text{Norm}(\mu, \sigma^2)$, it can be shown that

$$E[Y] = e^{\mu + \sigma^2/2} \text{ and } V[Y] = \left(e^{\sigma^2 - 1}\right) e^{2\mu + \sigma^2}.$$

Let $X_1, \ldots, X_{100}$ be an independent sequence of uniform (0,1) variables. Estimate

$$P\left(3^{-100} \leq \prod_{i=1}^{100} X_i \leq 2^{-100}\right).$$

(c) Verify the aforementioned results with a simulation experiment in R .

9.22 Consider a *biased* random walk that starts at the origin and that is twice as likely to move to the right as it is to move to the left. After how many steps will the probability be greater than 99% that the walk is to the right of the origin?

Moment-Generating Functions

9.23 A random variable X has a mgf

$$m(t) = pe^{-t} + qe^{t} + 1 - p - q,$$

where $p + q = 1$. Find all the moments of X.

9.24 A random variable X takes values -1, 0, and 2, with respective probabilities 0.2, 0.3, and 0.5. Find the mgf of X. Use the mgf to find the first two moments of X.

9.25 Find the moment-generating function of a geometric distribution with parameter p.

9.26 Find the mgf of a Bernoulli random variable with parameter p. Use this to find the mgf of a binomial random variable with parameters n and p.

9.27 Let $X \sim \text{Unif}(a, b)$. Find the mgf of X and use it to find the variance of X.

9.28 Let X and Y be independent standard normal random variables. Find the mgf of X^2+Y^2. What can you conclude about the distribution of X^2+Y^2? (Hint: See Example 9.19.)

9.29 Let X and Y be independent binomial random variables with parameters (m, p) and (n, p), respectively. Use the mgfs to show that $X+Y$ is a binomial random variable with parameters $m + n$ and p.

9.30 Find the second, third, and fourth moments of the exponential distribution using the mgfs. Give a general expression for the kth moment of the exponential distribution.

9.31 Let X be a random variable with mean μ and standard deviation σ. The *skewness* of X is defined as

$$\text{skew}(X) = \frac{E[(X - \mu)^3]}{\sigma^3}.$$

Skewness is a measure of the asymmetry of a distribution. Distributions that are symmetric about μ have skewness equal to 0. Distributions that are right skewed have positive skewness. Left-skewed distributions have negative skewness.

(a) Show that

$$\text{skew}(X) = \frac{E[X^3] - 3\mu\sigma^2 - \mu^3}{\sigma^3}.$$

(b) Use the mgfs to find the skewness of the exponential distribution with parameter λ.

(c) Use the mgfs to find the skewness of a normal distribution.

9.32 Let X be a random variable with mean μ and standard deviation σ. The *kurtosis* of X is defined as

$$\text{kurtosis}(X) = \frac{E[(X - \mu)^4]}{\sigma^4}.$$

The kurtosis is a measure of the "peakedness" of a probability distribution. Use the mgfs and find the kurtosis of a standard normal distribution.

9.33 Let X be a random variable, not necessarily positive.

(a) Using Markov's inequality, show that for $x > 0$ and $t > 0$,

$$P(X \geq x) \leq \frac{E[e^{tx}]}{e^{tx}} = e^{-tx} m(t), \tag{9.5}$$

assuming that $E[e^{tx}]$ exists, where m is the mgf of X.

(b) For the case when X has a standard normal distribution, give the upper bound in Equation 9.5. Note that the bound holds for all $t > 0$.

(c) Find the value of t that minimizes your upper bound. If $Z \sim$ Norm(0,1), show that for $z > 0$,

$$P(Z \geq z) \leq e^{-z^2/2}. \tag{9.6}$$

The upper bounds in Equations 9.5 and 9.6 are called *Chernoff bounds*.

9.34 Use the mgfs to show that the binomial distribution converges to the Poisson distribution. The convergence is taken so that $pn \to \lambda > 0$. (Hint: Write the p in the binomial distribution as λ/n.)

Simulation and R

9.35 The following code was used to generate the graphs in Figure 9.3. Modify the code to illustrate the strong law of large numbers for an i.i.d. sequence with the following distributions: (i) Pois($\lambda = 5$), (ii) Norm$(-4, 4)$, (iii) Exp($\lambda = 0.01$).

R: STRONG LAW OF LARGE NUMBERS

```
> par(mfrow=c(3,3))
> n = 1000
> p = 1/2
> for (i in 1:9) {
+   seq <- rbinom(n,1,p)   # Distribution
+   avgs <- cumsum(seq)/(1:n)
+     plot(avgs,type="l",xlab="n",ylab="Average")
+     abline(h = p)
+     }
>
```

9.36 See the code in Example 9.13 for generating a simple random walk. Write a function for simulating a biased random walk where the probability of moving left and right is p and $1 - p$, respectively. Graph your function obtaining pictures like Figure 9.8. What do you notice about the behavior of the random walk when $p = 0.60, 0.55, 0.505$?

9.37 (i) Write a function to simulate a random walk in the plane that moves up, down, left, and right with equal probability. Use your function to estimate the average distance from the origin after $n = 1000$ steps. (ii) Modify your function to simulate a three-dimensional random walk that moves in one of six directions with equal probability. Estimate the average distance from the origin after $n = 1000$ steps.

10

ADDITIONAL TOPICS

Not what we have but what we enjoy, constitutes our abundance.

—Epicurus

10.1 BIVARIATE NORMAL DISTRIBUTION

The multivariate normal distribution for random variables $(X_1, \ldots, X_n)$ generalizes the one-dimensional normal distribution to n dimensions. Here we introduce the two-dimensional bivariate normal distribution for X and Y. The distribution is specified by five parameters: μ_X, μ_Y, σ_X^2, σ_Y^2, and ρ: the means and variances of X and Y and their correlation. If $\mu_X = \mu_Y = 0$ and $\sigma_X^2 = \sigma_Y^2 = 1$, this gives the bivariate standard normal distribution.

BIVARIATE STANDARD NORMAL DISTRIBUTION

Random variables X and Y have a *bivariate standard normal distribution with correlation* ρ if the joint density function of (X, Y) is

$$f(x, y) = \frac{1}{2\pi\sqrt{1-\rho^2}} e^{-\frac{x^2-2\rho xy+y^2}{2(1-\rho^2)}}, \qquad (10.1)$$

for $-\infty < x, y < \infty$, where $-1 < \rho < 1$.

Probability: With Applications and R, First Edition. Robert P. Dobrow.

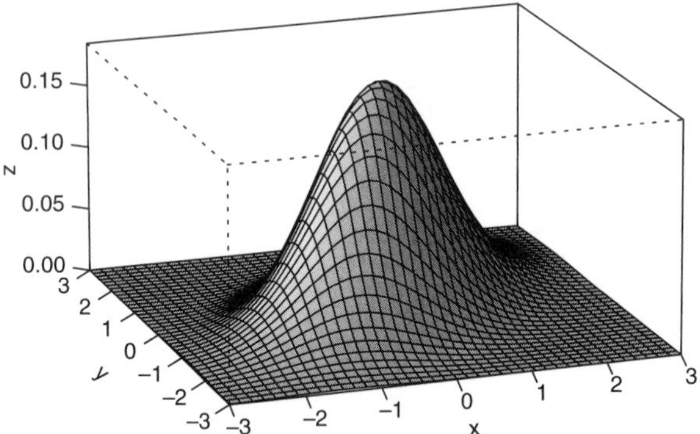

FIGURE 10.1: Bivariate standard normal distribution.

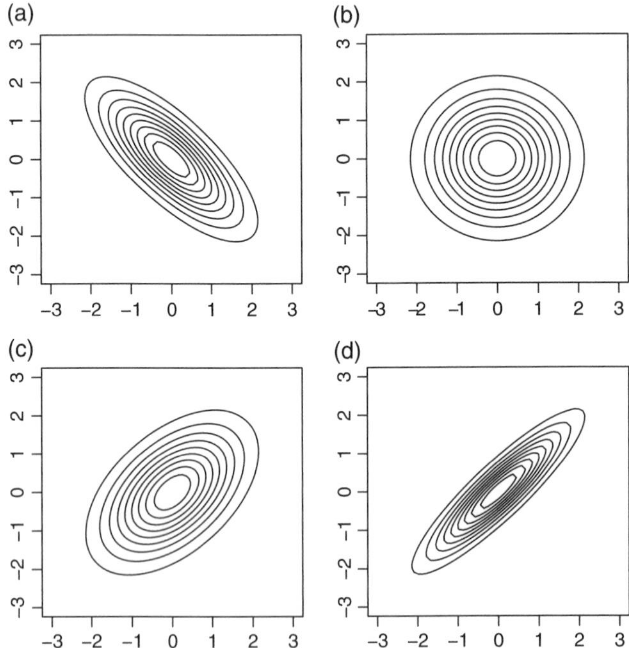

FIGURE 10.2: Contour plots of standard bivariate normal densities with (a) $\rho = -0.75$, (b) $\rho = 0$, (c) $\rho = 0.5$, and (d) $\rho = 0.9$.

See Figure 10.1 for the graph of a bivariate standard normal density. In Figure 10.2 are contour plots for normal distributions with four different values of ρ. The bivariate normal distribution is widely used in science and statistics. In Hollowed et al. (2011), a bivariate normal distribution is used to model the habitat of arrowtooth flounder in Alaska fisheries (see Fig. 10.3).

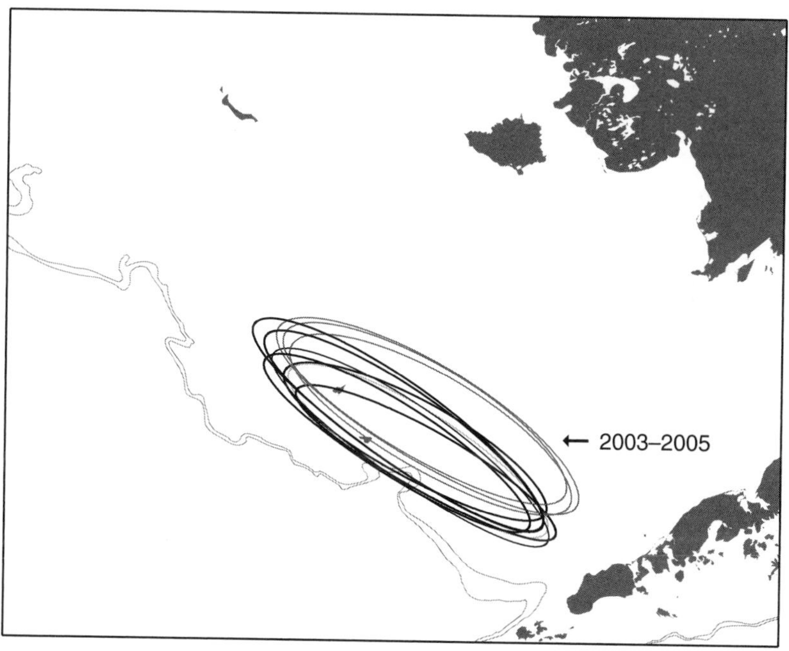

FIGURE 10.3: Distribution of arrowtooth flounder in Alaskan fisheries. Ellipses representing 30% probability contours of bivariate normal distribution fit to EBS survey CPUE data for arrowtooth flounder for the five coldest (black; 1994, 1999, 2008–2010) and warmest (gray; 1996, 1998, 2003–2005) years from 1982 to 2010. Source: Hollowed et al. (2011)

The joint density function of the general bivariate distribution is complicated. Please, do not memorize it! Often it will suffice to work with the standard bivariate density. As in the univariate case, X and Y have a bivariate normal distribution with parameters $\mu_X, \mu_Y, \sigma_X^2, \sigma_Y^2, \rho$ if and only if the standardized variables $(X - \mu_X)/\sigma_X$ and $(Y - \mu_Y)/\sigma_Y$ have a bivariate standard normal distribution.

For completeness, we give the general expression for the bivariate normal density.

BIVARIATE NORMAL DENSITY

Random variables X and Y have a joint bivariate normal distribution with parameters μ_X, μ_Y, σ_X^2, σ_Y^2, and ρ, if the joint density of X and Y is

$$f(x, y) = \frac{1}{\sigma_X \sigma_Y 2\pi\sqrt{1-\rho^2}} \exp\left(-\frac{d(x, y)}{2(1-\rho^2)}\right), \tag{10.2}$$

where

$$d(x,y) = \left(\frac{x-\mu_X}{\sigma_X}\right)^2 - 2\rho\left(\frac{x-\mu_X}{\sigma_X}\right)\left(\frac{y-\mu_Y}{\sigma_Y}\right) + \left(\frac{y-\mu_Y}{\sigma_Y}\right)^2.$$

The parameter constraints are $\sigma_X > 0$, $\sigma_Y > 0$, and $-1 < \rho < 1$.

Example 10.1 Fathers and sons. Sir Francis Galton, one of the "founding fathers" of statistics, introduced the concept of correlation in the late nineteenth century in part based on his study of the relationship between heights of fathers and their adult sons. Galton took 1078 measurements of father–son pairs. From his data, the mean height of fathers is 69 inches, the mean height of their sons is 70 inches, and the standard deviation of height is 2 inches for both fathers and son. The correlation is 0.5. Galton's data are well fit by a bivariate normal distribution (see Fig. 10.4). ■

The bivariate normal distribution has many remarkable properties, including the fact that both marginal and conditional distributions are normal. We, summarize the main properties for the bivariate standard normal distribution next. Results extend naturally to the general case.

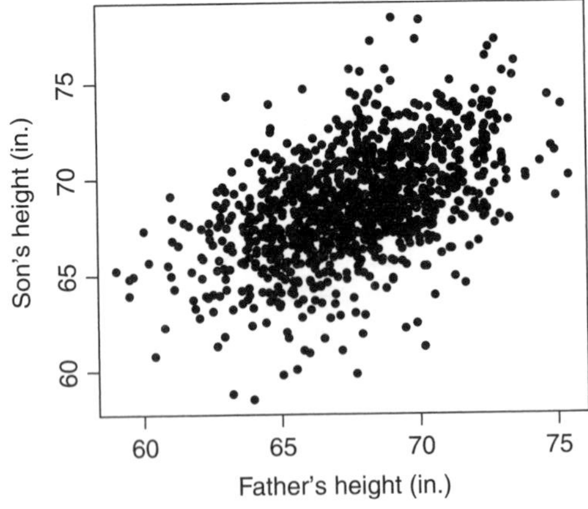

FIGURE 10.4: Galton's height data for fathers and sons are well fit by a bivariate normal distribution with parameters $(\mu_F, \mu_S, \sigma_F^2, \sigma_S^2, \rho) = (69, 70, 2^2, 2^2, 0.5)$.

PROPERTIES OF BIVARIATE STANDARD NORMAL DISTRIBUTION

Suppose random variables X and Y have a bivariate standard normal distribution with correlation ρ. Then the following properties hold.

1. *Marginal distribution:* The marginal distributions of X and Y are each standard normal.

2. *Conditional distribution:* The conditional distribution of X given $Y = y$ is normally distributed with mean ρy and variance $1-\rho^2$. That is, $E[X|Y=y] = \rho y$ and variance $V[X|Y=y] = 1-\rho^2$. Similarly, the conditional distribution of Y given $X = x$ is normal with $E[Y|X = x] = \rho x$ and $V[Y|X=x] = 1-\rho^2$.
3. *Correlation and independence:* If $\rho = 0$, that is, X and Y are uncorrelated, then X and Y are independent random variables.
4. *Transforming X and Y to independent random variables:* Let $Z_1 = X$ and $Z_2 = (Y-\rho X)/\sqrt{1-\rho^2}$. Then Z_1 and Z_2 are independent standard normal random variables.
5. *Linear functions of X and Y:* For nonzero constants a and b, $aX + bY$ is normally distributed with mean 0 and variance $a^2 + b^2 + 2ab\rho$.

We consider each of the five properties.

1. *Marginals:* The marginal density of Y is found by a straightforward calculation, which we leave to the exercises. The key step in the derivation is completing the square in the exponent of e, writing $x^2 - 2\rho xy + y^2 = (x-\rho y)^2 + (1-\rho^2)y^2$. Similarly for the marginal density of X.
2. *Conditional distributions:* The conditional density of Y given $X = x$ is

$$f_{Y|X}(y|x) = \frac{\dfrac{1}{2\pi\sqrt{1-\rho^2}}\exp\left(-\dfrac{x^2-2\rho xy+y^2}{2(1-\rho^2)}\right)}{\dfrac{1}{\sqrt{2\pi}}\exp\left(-\dfrac{x^2}{2}\right)}$$

$$= \frac{1}{\sqrt{2\pi}\sqrt{1-\rho^2}}\exp\left(-\frac{y^2-2\rho xy-\rho^2x^2}{2(1-\rho^2)}\right)$$

$$= \frac{1}{\sqrt{2\pi}\sqrt{1-\rho^2}}\exp\left(-\frac{(y-\rho x)^2}{2(1-\rho^2)}\right),$$

which is the density function of a normal distribution with mean ρx and variance $1-\rho^2$.
3. *Correlation and independence*: Independent random variables always have correlation equal to 0. However, the converse is generally not true. For normal random variables, however, it is. Let $\rho = 0$ in the bivariate joint density function. The density is equal to

$$f(x,y) = \left(\frac{1}{\sqrt{2\pi}}e^{-x^2/2}\right)\left(\frac{1}{\sqrt{2\pi}}e^{-y^2/2}\right),$$

which is the product of the marginal densities of X and Y. Thus, X and Y are independent.

It follows that if X and Y are standard normal random variables that have a joint bivariate normal distribution, and if $E[XY] = E[X]E[Y]$, then X and Y are independent.

4. *Transforming X and Y:* The proof of this property will be given in Section 10.2, where we discuss transformations of two random variables.

   The given transformation $Z_1 = X$ and $Z_2 = (Y - \rho X)/\sqrt{1-\rho^2}$ is a powerful way of transforming bivariate normal random variables into independent normals. Equivalently, if Z_1 and Z_2 are independent normal random variables and $-1 < \rho < 1$, then setting $X = Z_1$ and $Y = (1-\rho^2)Z_2 + \rho Z_1$ gives a pair of random variables (X, Y) that have a bivariate standard normal distribution with correlation ρ. This gives a way to simulate from the bivariate normal distribution.

5. *Linear functions of X and Y:* Property 4 tells us that we can write $X = Z_1$ and $Y = \sqrt{1-\rho^2}Z_2 + \rho Z_1$, where Z_1 and Z_2 are independent standard normals. Thus,

$$aX + bY = aZ_1 + b(\sqrt{1-\rho^2}Z_2 + \rho Z_1) = (a + b\rho)Z_1 + b\sqrt{1-\rho^2}Z_2.$$

   That is, we can write $aX + bY$ as a sum of independent random variables. The result follows.

R: SIMULATING BIVARIATE NORMAL RANDOM VARIABLES

We generate 1000 observations from a bivariate standard normal distribution with correlation $\rho = -0.75$. The resulting plot is shown in Figure 10.5.

```
> n <- 1000
> rho <- -0.75
> xlist <- numeric(n)
> ylist <- numeric(n)
> for (i in 1:n) {
   z1 <- rnorm(1)
   z2 <- rnorm(1)
   xlist[i] <- z1
   ylist[i] <- rho*z1 + sqrt(1-rho^2)*z2
   }
> plot(cbind(xlist,ylist))
```

The properties and results for the bivariate standard normal distribution extend to the general bivariate normal distribution. We summarize the conditional distribution results for the general case.

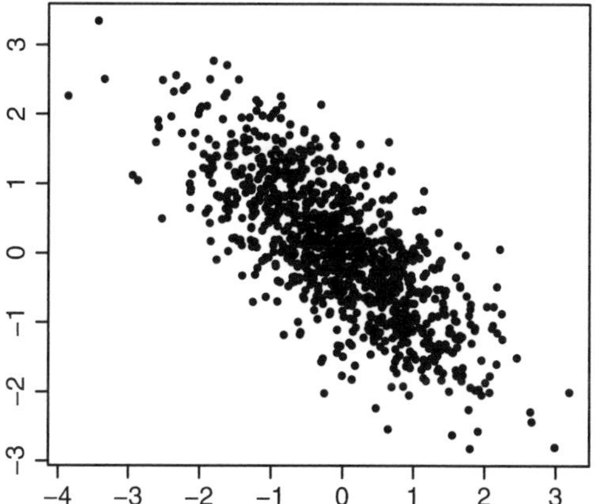

FIGURE 10.5: Plot of 1000 observations from bivariate standard normal distribution with $\rho = -0.75$.

BIVARIATE NORMAL: CONDITIONAL DISTRIBUTION OF Y GIVEN $X = x$

Suppose the distribution of X and Y is bivariate normal with parameters $\mu_X, \mu_Y, \sigma_X^2, \sigma_Y^2, \rho$. Then the conditional distribution of Y given $X = x$ is normal with conditional mean

$$E[Y|X = x] = \mu_Y + \rho \frac{\sigma_Y}{\sigma_X}(x - \mu_X)$$

and conditional variance

$$V[Y|X = x] = \sigma_Y^2(1 - \rho^2).$$

The result is derived with the assistance of Property 4. Write $X = \sigma_X Z_1 + \mu_X$, where $Z_1 \sim$ Norm(0,1). Similarly, write $Y = \sigma_Y Z_2 + \mu_Y$, where $Z_2 \sim$ Norm(0,1). Then

$$\begin{aligned}
E[Y|X = x] &= E[\mu_Y + \sigma_Y Z_2 | \mu_X + \sigma_X Z_1 = x] \\
&= E[\mu_Y + \sigma_Y Z_2 | Z_1 = (x - \mu_X)/\sigma_X] \\
&= \mu_Y + \sigma_Y E[Z_2 | Z_1 = (x - \mu_X)/\sigma_X] \\
&= \mu_Y + \rho \frac{\sigma_Y}{\sigma_X}(x - \mu_X)
\end{aligned}$$

and

$$\begin{aligned} V[Y|X=x] &= V[\mu_Y + \sigma_Y Z_2 | \mu_X + \sigma_X Z_1 = x] \\ &= V[\mu_Y + \sigma_Y Z_2 | Z_1 = (x-\mu_X)/\sigma_X] \\ &= \sigma_Y^2 V[Z_2|Z_1 = (x-\mu_X)/\sigma_X] \\ &= \sigma_Y^2(1-\rho^2). \end{aligned}$$

■ **Example 10.2 Fathers and sons continued.** We use the Galton dataset as the basis of a model for heights for fathers and their adult sons. For a father-son pair, let F denote father's height and S denote son's height. We assume (F, S) has a bivariate normal distribution with a parameters $\mu_F = 69$, $\mu_S = 70$, $\sigma_F = \sigma_S = 2$, and $\rho = 0.50$. (i) Find the probability that a son is taller than his father. (ii) Suppose a father is 67 inches tall. What is the probability that his son will be over 6 feet?

(i) The desired probability is $P(S > F) = P(S-F > 0)$. By Property 5, $S-F$ is normally distributed with mean

$$E[S-F] = \mu_S - \mu_F = 70 - 69 = 1$$

and variance

$$\begin{aligned} V[S-F] &= V[S] + V[F] - 2\text{Cov}(S,F) \\ &= \sigma_S^2 + \sigma_F^2 - 2\rho\sigma_S\sigma_F \\ &= 4 + 4 - 2(0.5)(2)(2) = 4. \end{aligned}$$

Thus $S-F \sim$ Norm(1,4). The desired probability $P(S > F) = P(S-F > 0)$ is found in R .

```
> 1-pnorm(0,1,2)
[1] 0.6914625
```

(ii) The question asks for $P(S > 72|F = 67)$. The conditional distribution of S given $F = f$ is normal with mean

$$E[S|F=f] = 70 + (0.5)\frac{2}{2}(f-69) = 70 + \frac{f-69}{2}.$$

At $f = 67$, the conditional mean is $70 + (67-69)/2 = 69$. The conditional variance is

$$V[S|F=67] = (1-(0.5)^2)4 = 3.$$

The desired probability is gotten in R.

```
> 1- pnorm(72,69,sqrt(3))
[1] 0.04163226
```

■

10.2 TRANSFORMATIONS OF TWO RANDOM VARIABLES

Suppose X and Y are continuous random variables with joint density $f_{X,Y}$. Given functions of two variables $g_1(x,y)$ and $g_2(x,y)$, let $V = g_1(X,Y)$ and $W = g_2(X,Y)$. For example, if $g_1(x,y) = x+y$ and $g_2(x,y) = x-y$, then $V = X+Y$ and $W = X-Y$. In this section, we show how to find the joint density $f_{V,W}$ of V and W from the joint density $f_{X,Y}$ of X and Y.

The method for finding the joint density of a function of two random variables extends how we found the density of a function of one variable. In one dimension, suppose $Z = g(X)$, for some function g. We further assume that g is invertible. Then

$$F_Z(z) = P(Z \le z) = P(g(X) \le z) = P(X \le g^{-1}(z)) = F_X(g^{-1}(z)).$$

Differentiating with respect to z, using the chain rule, gives

$$f_Z(z) = f_X(g^{-1}(z))\frac{d}{dz}g^{-1}(z).$$

We see that the density of Z involves an "inverse piece" and a "derivative piece."

For two variables, suppose g_1 and g_2 are "invertible" in the sense that $v = g_1(x,y)$ and $w = g_2(x,y)$ can be solved uniquely for x and y with $x = h_1(v,w)$ and $y = h_2(v,w)$. This is equivalent to saying that g_1 and g_2 define a one-to-one transformation in the plane.

For instance, if $g_1(x,y) = x+y$ and $g_2(x,y) = x-y$, then solving

$$v = x+y \text{ and } w = x-y$$

gives

$$x = \frac{v+w}{2} \text{ and } y = \frac{v-w}{2},$$

and thus $h_1(v,w) = (v+w)/2$ and $h_2(v,w) = (v-w)/2$.

If h_1 and h_2 have continuous partial derivatives, let J denotes the Jacobian consisting of the determinant of partial derivatives

$$J = \begin{vmatrix} \frac{\partial h_1}{\partial v} & \frac{\partial h_1}{\partial w} \\ \frac{\partial h_2}{\partial v} & \frac{\partial h_2}{\partial w} \end{vmatrix} = \frac{\partial h_1}{\partial v}\frac{\partial h_2}{\partial w} - \frac{\partial h_2}{\partial v}\frac{\partial h_1}{\partial w}.$$

Assume $J \neq 0$. The foregoing gives the joint density function of V and W.

JOINT DENSITY OF V AND W

Under the aforementioned assumptions,

$$f_{V,W}(v,w) = f_{X,Y}(h_1(v,w), h_2(v,w))|J|. \tag{10.3}$$

The factor $f_{X,Y}(h_1(v,w), h_2(v,w))$ is the "inverse piece," and the Jacobian is the "derivative piece." We do not give the proof of the joint density formula. It follows directly from the change of variable formula for multiple integrals in multivariable calculus. Most multivariable calculus textbooks contain a derivation.

In our working example, with $h_1(v,w) = (v+w)/2$ and $h_2(v,w) = (v-w)/2$, The Jacobian is

$$\begin{aligned} J &= \begin{vmatrix} \frac{\partial h_1}{\partial v} & \frac{\partial h_1}{\partial w} \\ \frac{\partial h_2}{\partial v} & \frac{\partial h_2}{\partial w} \end{vmatrix} = \begin{vmatrix} \frac{\partial}{\partial v}\frac{v+w}{2} & \frac{\partial}{\partial w}\frac{v+w}{2} \\ \frac{\partial}{\partial v}\frac{v-w}{2} & \frac{\partial}{\partial w}\frac{v-w}{2} \end{vmatrix} \\ &= \begin{vmatrix} 1/2 & 1/2 \\ 1/2 & -1/2 \end{vmatrix} = (-1/4) - (1/4) = -1/2. \end{aligned}$$

This gives the joint density of V and W in terms of the joint density of X and Y:

$$f_{V,W}(v,w) = \frac{1}{2} f_{X,Y}\left(\frac{v+w}{2}, \frac{v-w}{2}\right).$$

Example 10.3 Suppose X and Y are independent standard normal random variables. Then from the last results, the joint distribution of $V = X + Y$ and $W = X - Y$ is

$$\begin{aligned} f_{V,W}(v,w) &= \frac{1}{2} f_{X,Y}\left(\frac{v+w}{2}, \frac{v-w}{2}\right) \\ &= \frac{1}{2} f\left(\frac{v+w}{2}\right) f\left(\frac{v-w}{2}\right) \\ &= \frac{1}{4\pi} e^{-(v+w)^2/8} e^{-(v-w)^2/8} \\ &= \frac{1}{4\pi} e^{-(v^2+w^2)/4} \\ &= \left(\frac{1}{2\sqrt{\pi}} e^{-v^2/4}\right)\left(\frac{1}{2\sqrt{\pi}} e^{-w^2/4}\right). \end{aligned}$$

This is a product of two normal densities, each with mean $\mu = 0$ and variance $\sigma^2 = 2$. In addition to finding the joint density of $X+Y$ and $X-Y$, we have also established an interesting result for normal random variables. If X and Y are independent, then so are $X+Y$ and $X-Y$. ■

Example 10.4 Sometimes a one-dimensional problem can be approached more easily by working in two-dimensions. Let X and Y be independent exponential random variables with parameter λ. Find the density of $X/(X+Y)$.

Let $V = X/(X+Y)$. We consider a two-dimensional problem letting $W = X$. Then $V = g_1(X,Y)$, where $g_1(x,y) = x/(x+y)$. And $W = g_2(X,Y)$, where $g_2(x,y) = x$. Solving $v = x/(x+y)$ and $w = x$ for x and y gives $x = vw$ and $y = w - vw$. Thus,

$$h_1(v,w) = vw \text{ and } h_2(v,w) = w - vw.$$

The Jacobian is

$$J = \begin{vmatrix} w & v \\ -w & 1-v \end{vmatrix} = w - wv + wv = w.$$

The joint density of X and Y is

$$f_{X,Y}(x,y) = \lambda^2 e^{-\lambda(x+y)}, \quad x > 0, y > 0.$$

The joint density of V and W is thus

$$f_{V,W}(v,w) = f_{X,Y}(vw, w - vw)w = \lambda^2 we^{-\lambda(vw+(w-vw))} = \lambda^2 we^{-\lambda w},$$

for $0 < v < 1$ and $w > 0$.

To find the density of $V = X/(X+Y)$, integrate out the w variable, giving

$$f_V(v) = \int_0^\infty \lambda^2 we^{-\lambda w}\, dw = \lambda\left(\frac{1}{\lambda}\right) = 1, \quad \text{for } 0 < v < 1.$$

That is, the distribution of $X/(X+Y)$ is uniform on (0,1). ■

Example 10.5 Bivariate standard normal density. The bivariate normal distribution was introduced in Section 10.1. Here we derive the bivariate normal distribution and its joint density function.

The derivation starts with a pair of independent standard normal random variables Z_1 and Z_2. Let $-1 < \rho < 1$. We transform (Z_1, Z_2) into a pair of random variables (X, Y) such that (i) marginally X and Y each have standard normal distributions and (ii) the correlation between X and Y is ρ. We then derive the joint density function of X and Y.

Let $X = Z_1$ and $Y = \rho Z_1 + \sqrt{1-\rho^2} Z_2$. Trivially, $X \sim \text{Norm}(0,1)$. Since Y is the sum of two independent normal variables, it follows that Y is normally distributed with mean

$$E[Y] = \rho E[Z_1] + \sqrt{1-\rho^2} E[Z_2] = 0$$

and variance

$$V[Y] = \rho^2 V[Z_1] + (1-\rho^2) V[Z_2] = \rho^2 + 1 - \rho^2 = 1.$$

Since both X and Y have mean 0 and variance 1,

$$\begin{aligned} \text{Corr}(X,Y) = E[XY] &= E[Z_1(\rho Z_1 + \sqrt{1-\rho^2} Z_2)] \\ &= \rho E[Z_1^2] + \sqrt{1-\rho^2} E[Z_1 Z_2] = \rho. \end{aligned}$$

We show that the joint density function of X and Y is the bivariate standard normal density

$$f(x,y) = \frac{1}{2\pi\sqrt{1-\rho^2}} e^{-\frac{x^2 - 2\rho xy + y^2}{2(1-\rho^2)}}.$$

Let $g_1(z_1, z_2) = z_1$ and $g_2(z_1, z_2) = \rho z_1 + \sqrt{1-\rho^2} z_2$. Solving for the inverse functions give

$$h_1(x,y) = x \text{ and } h_2(x,y) = \frac{y - \rho x}{\sqrt{1-\rho^2}}.$$

The Jacobian is

$$J = \begin{vmatrix} 1 & 0 \\ -\rho/\sqrt{1-\rho^2} & 1/\sqrt{1-\rho^2} \end{vmatrix} = \frac{1}{\sqrt{1-\rho^2}}.$$

The joint density of X and Y is thus

$$\begin{aligned} f_{X,Y}(x,y) &= \frac{1}{\sqrt{1-\rho^2}} f_{Z_1,Z_2}\left(x, (y-\rho x)/\sqrt{1-\rho^2}\right) \\ &= \frac{1}{2\pi\sqrt{1-\rho^2}} e^{-(x^2/2 + (y-\rho x)^2/(1-\rho^2)} \\ &= \frac{1}{2\pi\sqrt{1-\rho^2}} e^{-(x^2 - 2\rho xy + y^2)/2(1-\rho^2)}. \end{aligned}$$ ∎

Results for transformations of more than two random variables extend naturally from the two-variable case. Let $(X_1, \ldots, X_k)$ have joint density function $f_{X_1,\ldots,X_k}$. Given functions of k variables $g_1, \ldots, g_k$, let

$$V_i = g_i(X_1, \ldots, X_k), \quad \text{for } i = 1, \ldots, k.$$

Assume the functions $g_1, \ldots, g_k$ define a one-to-one transformation in k dimensions with inverse functions $h_1, \ldots, h_k$ as in the two-variable case. Then the joint density of $(V_1, \ldots, V_k)$ is

$$f_{V_1,\ldots,V_k}(v_1, \ldots, v_k) = f_{X_1,\ldots,X_k}(h_1(v_1, \ldots, v_k), \ldots, h_k(v_1, \ldots, v_k))|J|,$$

where J is the determinant of the $k \times k$ matrix of partial derivatives whose (i, j)th entry is $\frac{\partial h_i}{\partial v_j}$, for $i, j = 1, \ldots, k$.

10.3 METHOD OF MOMENTS

The method of moments is a statistical technique for using data to estimate the unknown parameters of a probability distribution.

Recall that the kth moment of a random variable X is $E[X^k]$. We will also call this the *kth theoretical moment.*

Let $X_1, \ldots, X_n$ be an i.i.d. sample from a probability distribution with finite moments. Think of $X_1, \ldots, X_n$ as representing data from a random sample. The *kth sample moment* is defined as

$$\frac{1}{n}\sum_{i=1}^{n} X_i^k.$$

In a typical statistical context, the values of the X_i's are known (they are the data), and the parameters of the underlying probability distribution are unknown.

In the method of moments one sets up equations that equate sample moments with corresponding theoretical moments. The equation(s) are solved for the unknown parameter(s) of interest. The method is reasonable since if X is a random variable from the probability distribution of interest then by the strong law of large numbers, with probability 1,

$$\frac{1}{n}\sum_{i=1}^{n} X_i^k \to E[X^k], \text{ as } n \to \infty,$$

and thus for large n,

$$\frac{1}{n}\sum_{i=1}^{n} X_i^k \approx E[X^k].$$

For example, suppose a biased die has some unknown probability of coming up 5. An experiment is performed whereby the die is rolled repeatedly until a 5 occurs. Data are kept on the number of rolls until a 5 occurs. The experiment is repeated 10 times. The following data are generated:

$$3 \quad 1 \quad 4 \quad 4 \quad 1 \quad 6 \quad 4 \quad 2 \quad 9 \quad 2.$$

(Thus on the first experiment, the first 5 occurred on the third roll of the die. On the second experiment, a 5 occurred on the first roll of the die, etc.)

The data are modeled as outcomes from a geometric distribution with unknown success parameter p, where p is the probability of rolling a 5. We consider the sample as the outcome of i.i.d. random variables $X_1, \ldots, X_{10}$, where $X_i \sim \text{Geom}(p)$.

Let $X \sim \text{Geom}(p)$. The first theoretical moment is $E[X] = 1/p$. The first sample moment is

$$\frac{1}{10}\sum_{i=1}^{10} X_i = \frac{3+1+4+4+1+6+4+2+9+2}{10} = 3.6.$$

Set the theoretical and sample moments equal to each other. Solving $1/p = 3.6$ gives $p = 1/3.6 = 0.278$, which is the method of moments estimate for p.

Example 10.6 Hoffman (2003) suggests a negative binomial distribution to model water bacteria counts for samples taken from a water purification system. The goal is to calculate an upper control limit for the number of bacteria in future water samples.

Eighteen water samples were taken from a normally functioning water purification system. The data in Table 10.1 are bacteria counts per milliliter.

The negative binomial distribution is a flexible family of discrete distributions with two parameters r and p. It is widely used to model data in the form of counts. A variation of the distribution, described in Section 5.2 and used by R, takes nonnegative integer values starting at 0. An even more general formulation of the distribution allows the parameter r to be real.

TABLE 10.1: Bacteria counts at a water purification system.

Sample	1	2	3	4	5	6	7	8	9
Counts	20	5	23	19	13	12	14	17	2

Sample	10	11	12	13	14	15	16	17	18
Counts	7	30	4	16	2	4	1	2	0

Source: Hoffman (2003).

For a random variable $X \sim \text{NegBin}(r, p)$, the first two theoretical moments are

$$E[X] = \frac{r(1-p)}{p}$$

and

$$E[X^2] = V[X] + E[X]^2 = \frac{r(1-p)}{p^2} + \left(\frac{r(1-p)}{p}\right)^2.$$

The first and second sample moments for these data are

$$\frac{1}{18}\sum_{i=1}^{18} X_i = 10.611 \text{ and } \frac{1}{18}\sum_{i=1}^{18} X_i^2 = 186.833.$$

The two methods of moments equations are thus

$$\frac{r(1-p)}{p} = 10.611$$

and

$$\frac{r(1-p)}{p^2} + \left(\frac{r(1-p)}{p}\right)^2 = 186.833.$$

Substituting the first equality into the second equation gives

$$186.833 = \frac{10.611}{p} + (10.611)^2.$$

Solving gives $p = 0.143$. Back substituting then finds $r = 1.77$.

The data are fit to a negative binomial distribution with $r = 1.77$ and $p = 0.143$. The theoretical distribution is graphed in Figure 10.6.

The vertical line at $x = 27$ marks the *95th percentile* of the distribution. That is, it separates the bottom 95% of the area from the top 5%. For quality control purposes, engineers can use such a model to set an upper limit of bacteria counts for future samples. The top 5% of the distribution might be considered "unusual." Bacteria counts greater than 27 per milliliter could be a warning sign that the purification system is not functioning normally. ■

10.4 RANDOM WALK ON GRAPHS

A *graph* consists of a set of vertices and a set of edges. The graph in Figure 10.7 has four vertices and four edges. We say that two vertices are *neighbors* if there is an edge joining them. Thus, vertex c has two neighbors b and d. And vertex a has only one neighbor b.

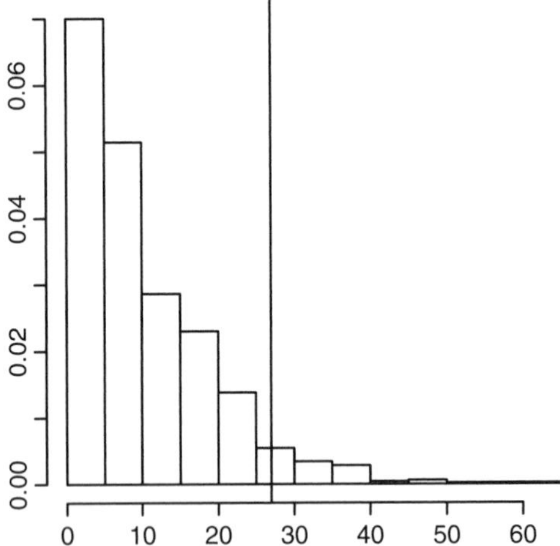

FIGURE 10.6: Negative binomial distribution with parameters $r = 1.77$ and $p = 0.143$ fits water sample data.

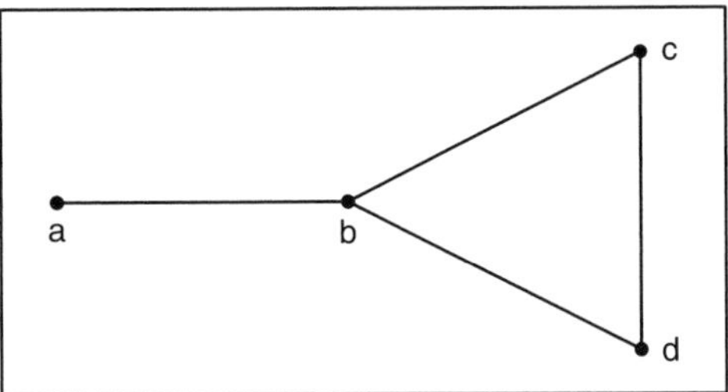

FIGURE 10.7: Graph on four vertices.

Imagine the vertices as lily-pods on a pond, with a frog sitting on one lily-pod. At each discrete unit of time, the frog hops to a neighboring lily-pod with probability proportional to the number of its neighbors.

For instance, if the frog is on lily-pod c, it jumps to b or d with probability 1/2 each. If the frog is on a, it always jumps to b. And if the frog is on b it jumps to a, c, or d, with probability 1/3 each.

As the frog hops from lily-pod to lily-pod, the frog's successive locations form a random walk on the graph. For instance, if the frog starts at a, the random walk sequence might look like $a, b, a, b, c, b, d, c, d, b, a, b, \ldots$.

Define a sequence of random variables $X_0, X_1, X_2, \ldots,$ taking values in the vertex set of the graph. Let X_0 be the frog's initial lily-pod (e.g., vertex). For each $n = 1, 2, \ldots,$ let X_n be the frog's position after n hops. The sequence $X_0, X_1, X_2, \ldots$ forms a *random walk on the graph.*

(Note that we have quietly extended our definition of random variable to allow the variables to take values in the vertex set, not just real numbers. Thus if the frog's random walk starts off as $a, b, a, b, c, \ldots$, then $X_0 = a$, $X_1 = b$, $X_2 = a$, etc.)

For vertices i and j, write $i \sim j$ if i and j are neighbors, that is, if there is an edge joining them. The *degree* of vertex i, written $\deg(i)$, is the number of neighbors of i. In the graph in Figure 10.7 (hereafter called the *frog graph*), $\deg(a) = 1$, $\deg(b) = 3$, and $\deg(c) = \deg(d) = 2$.

A *simple random walk* moves from a vertex to its neighbor with probability proportional to the degree of the vertex. That is, if the frog is on vertex i, then the probability that it hops to vertex j is $1/\deg(i)$, if $i \sim j$, and 0, otherwise. This is true at any time n. That is, for vertices i and j,

$$P(X_{n+1} = j | X_n = i) = \begin{cases} 1/\deg(i), & \text{if } i \sim j \\ 0, & \text{otherwise}, \end{cases} \tag{10.4}$$

for $n = 0, 1, 2 \ldots$. This *transition probability* describes the basic mechanism of the random walk process. The transition probability does not depend on n and thus for all n,

$$P(X_{n+1} = j | X_n = i) = P(X_1 = j | X_0 = i).$$

The transition probabilities can be represented as a matrix T whose ijth entry is the probability of moving from vertex i to vertex j in one step. This matrix is called the *transition matrix* of the random walk. Write T_{ij} for the ijth entry of T. For a random walk on a graph with k vertices labeled $\{1, \ldots, k\}$, T will be a $k \times k$ matrix with

$$T_{ij} = P(X_1 = j | X_0 = i), \quad \text{for } i, j = 1, \ldots, k.$$

For the frog's random walk, the transition matrix is

$$T = \begin{array}{c} \\ a \\ b \\ c \\ d \end{array} \begin{array}{c} \begin{array}{cccc} a & b & c & d \end{array} \\ \begin{pmatrix} 0 & 1 & 0 & 0 \\ 1/3 & 0 & 1/3 & 1/3 \\ 0 & 1/2 & 0 & 1/2 \\ 0 & 1/2 & 1/2 & 0 \end{pmatrix} \end{array}.$$

Observe that each row of a transition matrix sums to 1. If the random walk is at vertex i, then row i gives the probability distribution for the next step of the walk. That is, row i is the conditional distribution of the walk's next position given that it is at vertex i. For each i, the sum of the ith row is

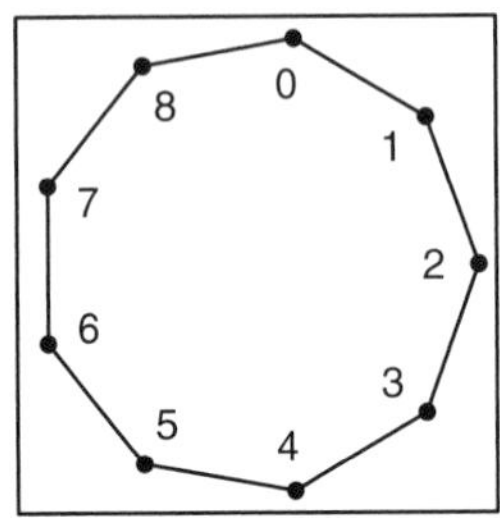

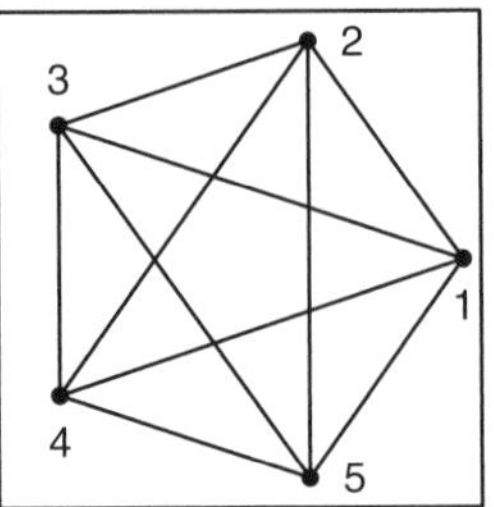

FIGURE 10.8: Left: cycle graph on $k = 9$ vertices. Right: complete graph on $k = 5$ vertices.

$$\sum_{j=1}^{k} T_{ij} = \sum_{j=1}^{k} P(X_1 = j | X_0 = i)$$
$$= \sum_{j=1}^{k} \frac{P(X_0 = i, X_1 = j)}{P(X_0 = i)} = \frac{P(X_0 = i)}{P(X_0 = i)} = 1.$$

■ **Example 10.7 Cycle, complete graph.** Examples of the *cycle graph* and *complete graph* are shown in Figure 10.8. Simple random walk on the cycle graph moves left or right with probability 1/2. Each vertex has degree two. The transition matrix is described using "clock arithmetic":

$$T_{ij} = \begin{cases} 1/2, & \text{if } j = i \pm 1 \\ 0, & \text{otherwise.} \end{cases}$$

In the complete graph, every pair of vertices is joined by an edge. The complete graph on k vertices has $\binom{k}{2}$ edges. Each vertex has degree $k - 1$. The entries of the transition matrix are

$$T_{ij} = \begin{cases} 1/(k-1), & \text{if } i \neq j, \\ 0, & \text{if } i = j. \end{cases}$$

■

■ **Example 10.8 Card shuffling by random transpositions.** Random walks are used to model methods of shuffling cards. Consider the following shuffle. Given a deck of cards pick two distinct positions in the deck at random. Switch the two cards at those positions. This is known as the "random transpositions" shuffle. Successive random transpositions forms a random walk on a graph whose vertices are the set of orderings (permutations) of the deck of cards. For a deck of k cards, this gives a graph with $k!$ vertices. Since there are $\binom{k}{2}$ ways to select two positions in the deck, the degree of each vertex is $\binom{k}{2}$. Letting i and j denote orderings of the deck of cards the transition matrix is defined by

$$T_{ij} = \begin{cases} 1/\binom{k}{2}, & \text{if } i \text{ differs from } j \text{ in exactly two locations} \\ 0, & \text{otherwise.} \end{cases}$$

Observe that the matrix T is symmetric since if the walk can move from i to j by switching two cards, it can move from j to i by switching those same two cards. Here is the transition matrix for random transpositions on a three-card deck with cards labeled 1, 2, and 3.

$$T = \begin{array}{c} \\ 123 \\ 132 \\ 213 \\ 231 \\ 312 \\ 321 \end{array} \begin{array}{c} \begin{array}{cccccc} 123 & 132 & 213 & 231 & 312 & 321 \end{array} \\ \left(\begin{array}{cccccc} 0 & 1/3 & 1/3 & 0 & 0 & 1/3 \\ 1/3 & 0 & 0 & 1/3 & 1/3 & 0 \\ 1/3 & 0 & 0 & 1/3 & 1/3 & 0 \\ 0 & 1/3 & 1/3 & 0 & 0 & 1/3 \\ 0 & 1/3 & 1/3 & 0 & 0 & 1/3 \\ 1/3 & 0 & 0 & 1/3 & 1/3 & 0 \end{array}\right) \end{array}.$$

■

Example 10.9 Hypercube. The *k-hypercube graph* has vertex set consisting of all k-element sequences of zeros and ones. Two vertices (sequences) are connected by an edge if they differ in exactly one coordinate. The graph has 2^k vertices and $\binom{k}{2}$ edges. Each vertex has degree k (see Fig. 10.9).

Random walk on the hypercube can be described as follows. Given a k-element 0–1 sequence, pick one of the k coordinates uniformly at random. Then "flip the bit" at that coordinate. That is, switch the number from 0 to 1, or from 1 to 0. Here is the transition matrix for the 3-hypercube.

$$T = \begin{array}{c} \\ 000 \\ 100 \\ 010 \\ 110 \\ 001 \\ 101 \\ 110 \\ 111 \end{array} \begin{array}{c} \begin{array}{cccccccc} 000 & 100 & 010 & 110 & 001 & 101 & 011 & 111 \end{array} \\ \left(\begin{array}{cccccccc} 0 & 1/3 & 1/3 & 0 & 1/3 & 0 & 0 & 0 \\ 1/3 & 0 & 0 & 1/3 & 0 & 1/3 & 0 & 0 \\ 1/3 & 0 & 0 & 1/3 & 0 & 0 & 1/3 & 0 \\ 0 & 1/3 & 1/3 & 0 & 0 & 0 & 0 & 1/3 \\ 1/3 & 0 & 0 & 0 & 0 & 1/3 & 1/3 & 0 \\ 0 & 1/3 & 0 & 0 & 1/3 & 0 & 0 & 1/3 \\ 0 & 0 & 1/3 & 0 & 1/3 & 0 & 0 & 1/3 \\ 0 & 0 & 0 & 1/3 & 0 & 1/3 & 1/3 & 0 \end{array}\right) \end{array}.$$

■

10.4.1 Long-Term Behavior

Of particular interest in the study of random walk is the long-term behavior of the process. The *limiting distribution* describes this long-term behavior.

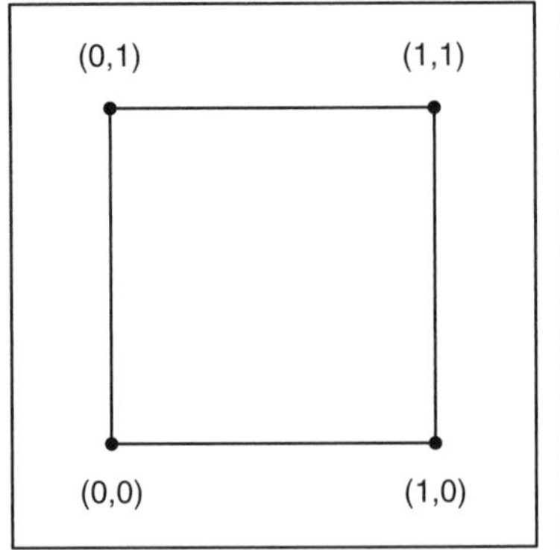

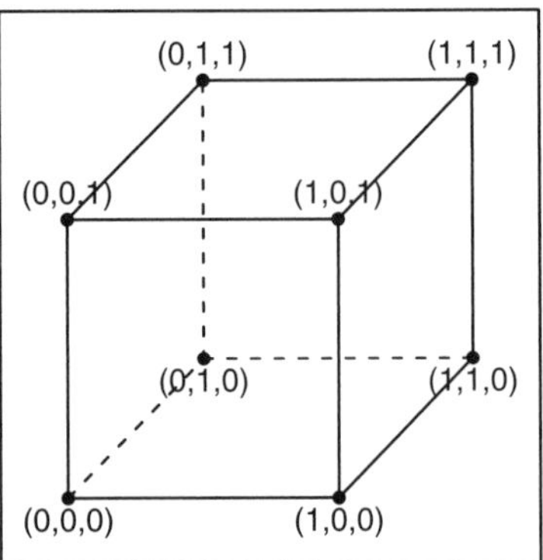

FIGURE 10.9: The k-hypercube graph for $k = 2$ (left) and $k = 3$ (right).

In the foregoing we will write finite probability distributions as vectors. Thus if $\mu = (\mu_1, \ldots, \mu_k)$ is the distribution of a random variable X taking values $\{1, \ldots, k\}$, then $P(X = i) = \mu_i$, for $i = 1, \ldots, k$.

The *initial distribution* of a random walk is the distribution of the starting vertex X_0. A random walk is described by its initial distribution and transition matrix.

If a random walk has a limiting distribution, then as the walk evolves over time the distribution of the walk's position reaches an "equilibrium," which is independent of where the walk began. That is, the effect of the initial distribution "wears off" as the random walk evolves.

LIMITING DISTRIBUTION

Let $X_0, X_1, X_2, \ldots$ be a random walk on a graph. A probability distribution $\pi = (\pi_1, \ldots, \pi_k)$ is the *limiting distribution* of the random walk if for all vertices j,

$$P(X_n = j | X_0 = i) \to \pi_j, \text{ as } n \to \infty,$$

for all initial vertices i.

A consequence of this definition is that $P(X_n = j) \to \pi_j$, as $n \to \infty$, since

$$\begin{aligned}\lim_{n\to\infty} P(X_n = j) &= \lim_{n\to\infty} \sum_{i=1}^{k} P(X_n = j | X_0 = i) P(X_0 = i) \\ &= \sum_{i=1}^{k} \lim_{n\to\infty} P(X_n = j | X_0 = i) P(X_0 = i) \\ &= \sum_{i=1}^{k} \pi_j P(X_0 = i) = \pi_j.\end{aligned}$$

TABLE 10.2: Simple random walk for the cycle graph on nine vertices after n steps. Simulation of X_n for $n = 5, 10, 50, 100$.

Steps	5	6	7	8	0	1	2	3	4
5	0.025	0.06	0.125	0.185	0.21	0.185	0.125	0.06	0.025
10	0.075	0.09	0.12	0.14	0.15	0.14	0.12	0.09	0.075
50	0.11	0.11	0.12	0.10	0.12	0.10	0.12	0.10	0.11
100	0.11	0.11	0.11	0.11	0.11	0.11	0.11	0.11	0.11

To illustrate the limiting distribution, imagine a frog hopping on the cycle graph on nine vertices as in Figure 10.8. Suppose the frog starts at vertex 0. After just a few hops, the frog will tend to be relatively close to 0. However, after the frog has been hopping around for a long time, intuitively the frog's position will tend to be uniformly distributed around the cycle, independent of the frog's starting vertex.

Observe the behavior of random walk on the cycle graph on $k = 9$ vertices in Table 10.2. We simulated the random walk for a fixed number of steps n, and then repeated 100,000 times to simulate the distribution of X_n. The walk starts at vertex 0. After five steps, the walk is close to its starting position with high probability. The probability is almost 60% that $X_5 \in \{8, 0, 1\}$. The distribution of X_5 is strongly dependent on the initial vertex. After 10 steps, there is still bias toward vertex 0 but not as much. Still the random walk is almost twice as likely to be close to 0 than it is to be at the opposite side of the cycle near 4 or 5. By 50 steps, however, the dependency on the initial vertex has almost worn off, and the distribution of X_{50} is close to uniform. By 100 steps, the distribution is uniform to two decimal places. The table suggests that the limiting distribution of the random walk is uniform on the set of vertices.

For the cycle graph, the limiting distribution is uniform. What about for the "frog graph" in Figure 10.7? If the frog has been hopping from lily-pod to lily-pod on this graph for a long time what is the probability that it is now on, say, vertex a?

Based on the structure of the graph, it is reasonable to think that after a long time, the frog is least likely to be at vertex a and most likely to be at vertex b since there are few edges into a and many edges into b.

The limiting distribution for random walk on this graph, and in general, has a nice intuitive description. The long-term probability that the walk is at a particular vertex is proportional to the degree of the vertex.

LIMITING DISTRIBUTION FOR SIMPLE RANDOM WALK ON A GRAPH

Let π be the limiting distribution for simple random walk on a graph. Then

$$\pi_j = \frac{\deg(j)}{\sum_{i=1}^{k} \deg(i)} = \frac{\deg(j)}{2e}, \quad \text{for all vertices } j, \tag{10.5}$$

where e is the number of edges in the graph.

For the frog-hopping random walk, the limiting distribution is

$$\pi = (\pi_a, \pi_b, \pi_c, \pi_d) = (1/8, 3/8, 2/8, 2/8) = (0.125, 0.375, 0.25, 0.25).$$

See the following simulation.

Remarks:

1. For any graph, the sum of the vertex degrees is equal to twice the number of edges since every edge contributes two to the sum of the degrees, one for each endpoint.
2. Not every graph has a limiting distribution. Whether or not a random walk has a limiting distribution depends on the structure of the graph. There are two necessary conditions:
   (i) The graph must be *connected.* This means that for every pair of vertices there is a sequence of edges that forms a path connecting the two vertices.
   (ii) The graph must be *aperiodic*. To understand this condition, consider random walk on the square with vertices labeled (clockwise) 0, 1, 2, and 3. If the walk starts at vertex 0, then after an even number of steps it will always be on either 0 or 2. And after an odd number of steps it will always be on 1 or 3. Thus, the position of the walk depends on the starting state and there is no limiting distribution. The square is an example of a *bipartite graph.* In such a graph, one can color the vertices of the graph with two colors black and white so that every edge has one endpoint that is black and the other endpoint that is white. Random walk on a bipartite graph gives rise to periodic behavior and no limiting distribution exists.

   Every graph that is connected and nonbipartite has a unique limiting distribution. In the foregoing, all the graphs that we discuss will meet this condition.
3. Observe from the limiting distribution formula in Equation 10.5 that if all the vertex degrees of a graph are equal, then the limiting distribution is uniform. A graph with all degrees the same is called *regular.* The complete graph, hypercube graph, and random transpositions graph are all regular and thus simple random walk on these graphs have uniform limiting distributions. The cycle graph is also regular but is bipartite if the number of vertices is even. The cycle graph has a uniform limiting distribution when the number of vertices is odd.
4. A classic problem in probability asks: How many shuffles does it take to mix up a deck of cards? The problem can be studied by means of random walks as in Example 10.8. If the uniform distribution is the limiting distribution of a card-shuffling random walk, then the problem is equivalent to asking: How many steps of a card-shuffling random walk are necessary to get close to the limiting distribution, when the deck of cards is "mixed up?" Such questions study the rate of convergence of random walk to the limiting distribution and form an active area of modern research in probability.

R: RANDOM WALK ON A GRAPH

The script file **Frog.R** simulates random walk on the frog graph for $n = 200$ steps starting from a uniformly random vertex. The 200-step walk is repeated 10,000 times, keeping track of the last position X_{200}.

```
> n <- 200
> trials <- 10000
> simlist <- numeric(trials)
> for (i in 1:trials) {
    k <- 0
    pos <- sample(1:4,1) # initial position
    while (k < n) {
    k <- k+1
    if (pos==1) {pos <- 2; next}
    if (pos==2) {pos <- sample(c(1,3,4),1); next}
    if (pos==3) {pos <- sample(c(2,4),1); next}
    if (pos==4) pos <- sample(c(2,3),1); }
    simlist[i] <- pos  }
> table(simlist)/trials
simlist
     1      2      3      4
0.1248 0.3735 0.2507 0.2510
```

Observe how closely the simulated distribution matches the exact limiting distribution $\pi = (1/8, 3/8, 1/4, 1/4)$.

10.5 RANDOM WALKS ON WEIGHTED GRAPHS AND MARKOV CHAINS

> Let us finish the article and the whole book with a good example of dependent trials, which approximately can be regarded as a simple chain.
>
> —Alexei Andreyevich Markov

Random walks on graphs are a special case of Markov chains. A Markov chain is a sequence of random variables $X_0, X_1, X_2, \ldots$, with the property that for all n, the conditional distribution of X_{n+1} given the past history $X_0, \ldots, X_n$ is equal to the conditional distribution of X_{n+1} given X_n. This is sometimes stated as the distribution of the future given the past only depends on the present. The set of values of the Markov chain is called the *state space*.

Simple random walk on a graph is a Markov chain since the distribution of the walk's position at any fixed time only depends on the last vertex visited and not on the previous locations of the walk. The state space is the vertex set of the graph.

As in the case of random walk on a graph, a Markov chain can be described by its transition matrix T and an initial distribution. The concept of a limiting distribution introduced in the last section extends naturally to general Markov chains.

Markov chains are remarkably useful models. They are used extensively in virtually every applied field to model random processes that exhibit some dependency structure between successive outcomes. A recent Google search of "Markov chain" returned over five million hits.

■ **Example 10.10** Consider a Markov chain model for winter weather in Minnesota. Suppose on any day the weather can be in one of three states: clear, rain, and snow. A meteorologist suggests the following transition matrix:

$$T = \begin{matrix} & \begin{matrix} c & r & s \end{matrix} \\ \begin{matrix} c \\ r \\ s \end{matrix} & \begin{pmatrix} 1/6 & 1/3 & 1/2 \\ 1/8 & 1/8 & 3/4 \\ 1/3 & 1/6 & 1/2 \end{pmatrix} \end{matrix}.$$

The script file **Markov.R** contains the function `markov(matrix,initial, n)` for simulating n steps of a Markov chain with a given transition matrix and initial distribution. Suppose it is clear today. Here are the simulated weather states for the next 25 days (with 1 = clear, 2 = rain, and 3 = snow):

```
> markov(T,1,25)
 [1] 1 3 2 3 1 2 3 2 1 3 2 3 2 3 3 3 3 3 3 3 1 2 3 1 3
```

What is the long-term behavior of the Markov chain? That is, what is the long-term weather according to this model? We ran the Markov chain for $n = 200$ steps (representing days) and repeated 10,000 times with the following results. We also observed the chain's behavior after 300 steps.

R: WEATHER MARKOV CHAIN

```
> simlist <- replicate(10000,markov(T,1,200))
> table(simlist)/10000
simlist
     1      2      3
0.2512 0.2004 0.5484
> simlist<- replicate(10000,markov(T,1,300))
> table(simlist)/10000
simlist
     1      2      3
0.2497 0.2036 0.5467
```

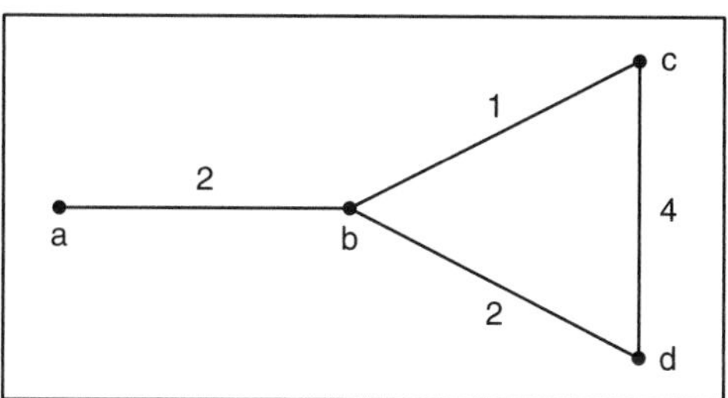

FIGURE 10.10: Weighted graph.

There is little difference in the distribution of X_{200} and X_{300}. The simulated limiting distribution suggests that the long-term weather forecast is approximately 25% chance of clear skies, 20% chance of rain, and 55% chance of snow. ■

General Markov chains can be studied in terms of random walks on graphs if we extend our notions of graph. A *weighted graph* is a graph with a positive number, or weight, assigned to each edge. A *directed graph* is a graph where edges have direction associated with them. Thus for every pair of vertices i and j, one can have an edge from i to j and an edge from j to i. An example of a weighted, but undirected, graph is shown in Figure 10.10.

Example 10.11 PageRank. The PageRank algorithm used by the search engine Google was presented in the book's introduction. When you make an inquiry using Google, it returns an ordered list of sites by assigning a rank to each page. The rank it assigns is essentially the limiting distribution of a Markov chain.

This Markov chain can be described as a random walk on the *web graph*. In the web graph, vertices represent web pages, edges represent hyperlinks. A directed edge joins page i to page j if page i links to page j. Imagine a random walker that visits web pages according to this model moving from a page with probability proportional to the number of "out-links" from that page. The long-term probability that the random walker is at page i is precisely the PageRank of page i.

Write $i \rightsquigarrow j$ if there is a directed edge from i to j. Let $\text{link}(i)$ be the number of directed edges from i. Transition probabilities are defined as follows:

$$T_{ij} = \begin{cases} P(X_1 = j | X_0 = i) = 1/\text{link}(i), & \text{if } i \rightsquigarrow j \\ 0, & \text{otherwise.} \end{cases}$$

There will soon be over a billion websites on the Internet. Computing the PageRank distribution amounts to finding the limiting distribution for a matrix that has hundreds of millions of rows and columns. Using the tools of probability and linear algebra the computation is remarkably fast and efficient. For a detailed treatment, see *The $25,000,000,000 eigenvector* by Bryan and Leise (2006).

For random walk on a weighted graph, transition probabilities are proportional to the sum of the weights. If vertices i and j are neighbors, let $w(i, j)$ denote the weight of the edge joining i and j. Let $w(i) = \sum_{i \sim j} w(i, j)$ be the sum of the weights on all the edges joining i to its neighbors. The transition probabilities, and the transition matrix, for random walk on a weighted graph are defined as

$$T_{ij} = P(X_1 = j | X_0 = i) = \begin{cases} w(i,j)/w(i), & \text{if } i \sim j \\ 0, & \text{otherwise.} \end{cases}$$

For the weighted frog graph, $w(a) = 2$, $w(b) = w(c) = 5$, and $w(d) = 6$. The transition matrix is

$$T = \begin{array}{c} \\ a \\ b \\ c \\ d \end{array} \begin{array}{c} \begin{array}{cccc} a & b & c & d \end{array} \\ \begin{pmatrix} 0 & 1 & 0 & 0 \\ 2/5 & 0 & 1/5 & 2/5 \\ 0 & 1/5 & 0 & 4/5 \\ 0 & 1/3 & 0 & 2/3 \end{pmatrix} \end{array}.$$

We will also allow a random walk to transition from a vertex back to itself in one step. This is done by adding an edge called a "loop" at a vertex. If a vertex has a loop, then it is a neighbor to itself. In the weighted frog graph if we add a loop of weight one at every vertex, then $w(a) = 3$, $w(b) = w(c) = 6$, and $w(d) = 7$. The new transition matrix is

$$T = \begin{array}{c} \\ a \\ b \\ c \\ d \end{array} \begin{array}{c} \begin{array}{cccc} a & b & c & d \end{array} \\ \begin{pmatrix} 1/3 & 2/3 & 0 & 0 \\ 1/3 & 1/6 & 1/6 & 1/3 \\ 0 & 1/6 & 1/6 & 2/3 \\ 0 & 2/7 & 4/7 & 1/7 \end{pmatrix} \end{array}.$$

Every Markov chain can be described as a random walk on a directed, weighted graph. Given a Markov chain with transition matrix T_{ij}, form a graph whose vertex set is the state space of the Markov chain. And define weights $w(i, j) = T_{ij}$ for all i and j. This gives a directed, weighted graph. Similarly, given a directed, weighted graph with weight function $w(i, j)$ define transition probabilities by $T_{ij} = w(i, j)/w(i)$ for all i and j to obtain the transition matrix for a Markov chain. ■

10.5.1 Stationary Distribution

Stationary distributions play an important role in the study of Markov chains with intimate connections to limiting distributions.

STATIONARY DISTRIBUTION

Definition 10.1. *Given a Markov chain with transition matrix* T, *a probability distribution* $\mu = (\mu_1, \ldots, \mu_k)$ *is a stationary distribution of the Markov chain if for all states j,*

$$\mu_j = \sum_{i=1}^{k} \mu_i T_{ij}. \tag{10.6}$$

The reason for the name "stationary" is because a Markov chain that starts in its stationary distribution stays in that distribution. Suppose μ is both a stationary distribution and the initial distribution of a Markov chain. That is, $P(X_0 = j) = \mu_j$, for all j. Consider the distribution of X_1. By conditioning on X_0,

$$P(X_1 = j) = \sum_{i=1}^{k} P(X_1 = j | X_0 = i) P(X_0 = i) = \sum_{i=1}^{k} T_{ij}\mu_i = \mu_j,$$

for all j. Thus, the distribution of X_1 is also given by μ. And if $X_1 \sim \mu$, by the same argument $X_2 \sim \mu$, and so on. Hence, $X_n \sim \mu$ for all n. Thus, the sequence $X_0, X_1, X_2, \ldots$ is identically distributed with common distribution μ. We refer to the *stationary Markov chain* when the process is started in its stationary distribution.

Example 10.12 For the weather chain of Example 10.10, the distribution $\mu = (1/4, 1/5, 11/20)$ is a stationary distribution. We verify that it is so by showing that it satisfies the defining property. Checking for each $j = 1, 2, 3$ gives

$$\sum_{i=1}^{3} \mu_i T_{i1} = (1/4)(1/6) + (1/5)(1/8) + (11/20)(1/3) = 1/4 = \mu_1,$$

$$\sum_{i=1}^{3} \mu_i T_{i2} = (1/4)(1/3) + (1/5)(1/8) + (11/20)(1/6) = 1/5 = \mu_2,$$

$$\sum_{i=1}^{3} \mu_i T_{i3} = (1/4)(1/2) + (1/5)(3/4) + (11/20)(1/2) = 11/20 = \mu_3.$$

■

There is an intimate connection between the stationary distribution of a Markov chain and the limiting distribution. In fact, for a large class of Markov chains they are equal. We show that every limiting distribution is also a stationary distribution.

Suppose the limiting distribution of a Markov chain is π. Then for all states j,

$$\begin{aligned}\sum_{i=1}^{k} T_{ij}\pi_i &= \sum_{i=1}^{k} P(X_{n+1}=j|X_n=i)\lim_{n\to\infty} P(X_n=i)\\ &= \lim_{n\to\infty}\sum_{i=1}^{k} P(X_{n+1}=j|X_n=i)P(X_n=i)\\ &= \lim_{n\to\infty}\sum_{i=1}^{k} P(X_{n+1}=j, X_n=i)\\ &= \lim_{n\to\infty} P(X_{n+1}=j)=\pi_j.\end{aligned}$$

Thus, π is a stationary distribution.

A limiting distribution of a Markov chain is always a stationary distribution. The converse, however, is not necessarily true. Some Markov chains can have many stationary distributions. And some Markov chains with stationary distributions do not have limiting distributions. However, random walk on a weighted graph that is connected and nonbipartite admits a unique positive stationary distribution, which is also the limiting distribution. This result is a fundamental theorem of Markov chains, which we restate without proof.

STATIONARY, LIMITING DISTRIBUTION FOR RANDOM WALK ON WEIGHTED GRAPHS

Suppose a weighted graph G is connected and nonbipartite. Then random walk on the graph has a unique, positive stationary distribution π, which is also the limiting distribution.

The following result is extremely useful for finding stationary distributions.

Theorem 10.2.

Given a Markov chain with transition matrix T, a probability distribution $\mu = (\mu_1, \ldots, \mu_k)$ *is said to satisfy the detailed balance condition if*

$$\mu_i T_{ij} = \mu_j T_{ji}, \text{ for all } i \text{ and } j. \tag{10.7}$$

If a probability distribution satisfies the detailed balance condition, then it is a stationary distribution of the Markov chain.

Proof. Suppose μ satisfies the detailed balance condition. Then for all j,

$$\sum_{i=1}^{k}\mu_i T_{ij} = \sum_{i=1}^{k}\mu_j T_{ji} = \mu_j\sum_{i=1}^{k} T_{ji} = \mu_j.$$

Thus, μ is a stationary distribution. The first equality is because of the detailed balance condition. The last equality is because the jth row of the transition matrix sums to 1.

Apply Theorem 10.2 to find the stationary distribution for random walk on a weighted graph. We search for a candidate π that satisfies the detailed balance condition. Let T be the transition matrix. Then for all i and j, $\pi_i T_{ij} = \pi_j T_{ji}$ and thus

$$\pi_i \frac{w(i,j)}{w(i)} = \pi_j \frac{w(j,i)}{w(j)}.$$

Since $w(i,j) = w(j,i)$ for all i and j, we have that $\pi_i/w(i) = \pi_j/w(j)$ for all vertices. Call this constant c. Hence $\pi_i = cw(i)$ for all i. Summing over i gives $1 = c\sum_{i=1}^{k} w(i)$. And thus $c = 1/\sum_{i=1}^{k} w(i)$. We find that

$$\pi_j = \frac{w(j)}{\sum_{i=1}^{k} w(i)}$$

is the stationary distribution. It is also the limiting distribution.

Observe that simple random walk on a graph can be considered a random walk on a weighted graph where all the weights are equal to 1. In that case, $w(j) = \deg(j)$ and $\pi_j = \deg(j)/\sum_{i=1}^{k} \deg(i)$.

For the random walk on the graph in Figure 10.10, the stationary distribution is

$$\pi = (\pi_a, \pi_b, \pi_c, \pi_d) = (2/18, 5/18, 5/18, 6/18).$$

Example 10.13 Two-state Markov chain and Alexander Pushkin. The general two-state Markov chain on states a and b has transition matrix

$$T = \begin{array}{c} \\ a \\ b \end{array} \begin{array}{c} \begin{array}{cc} a & \quad b \end{array} \\ \begin{pmatrix} 1-p & p \\ q & 1-q \end{pmatrix} \end{array},$$

where $0 \leq p, q, \leq 1$. If p and q are neither 0 nor 1, the Markov chain can be cast as a random walk on a weighted graph with loops as shown in Figure 10.11. This gives $w(a) = q$ and $w(b) = p$.

The stationary distribution probabilities are

$$\pi_a = \frac{w(a)}{w(a) + w(b)} = \frac{q}{p+q} \text{ and } \pi_b = \frac{w(b)}{w(a) + w(b)} = \frac{p}{p+q}.$$

Andrei Andreyevich Markov introduced Markov chains 100 years ago. He first applied a two-state chain to analyze the successive vowels and consonants in Alexander Puskin's poem *Eugéne Onégin.* In 20,000 letters of the poems, Markov

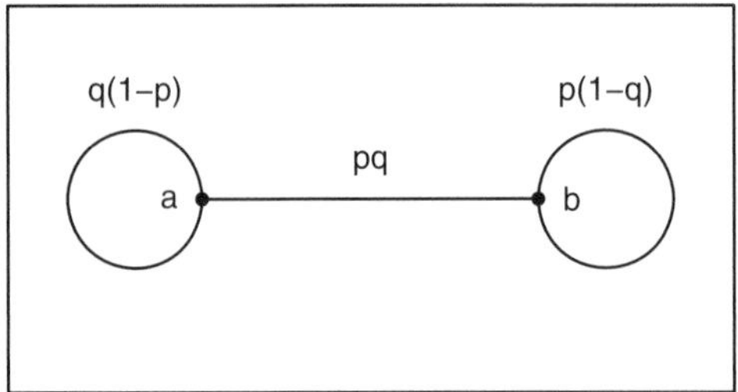

FIGURE 10.11: Weighted graph for the general two-state Markov chain.

found that there are 8638 vowels and 11,362 consonants with 1104 vowel–vowel pairs, 7534 vowel–consonant and consonant–vowel pairs, and 3828 consonant–consonant pairs. In modern notation, the transition matrix is

$$T = \begin{matrix} & v & c \\ v & \\ c & \end{matrix} \begin{pmatrix} 1104/8638 & 7534/8638 \\ 7534/11362 & 3828/8638 \end{pmatrix} = \begin{matrix} v \\ c \end{matrix} \begin{pmatrix} 0.128 & 0.872 \\ 0.663 & 0.337 \end{pmatrix},$$

with stationary distribution

$$\begin{aligned} \pi = (\pi_v, \pi_c) &= \left(\frac{0.663}{0.872 + 0.663}, \frac{0.872}{0.872 + 0.663} \right) \\ &= (0.432, 0.568), \end{aligned}$$

which gives the proportion of vowels and consonants, respectively, in the poem. ■

A Markov chain whose stationary distribution π satisfies the detailed balance condition is said to be *time-reversible*. The descriptive language is used because a time-reversible stationary chain "looks the same" going forward in time as it does going backward.

Suppose a stationary Markov chain is time-reversible with stationary distribution π. For all states i and j,

$$\begin{aligned} P(X_0 = i, X_1 = j) &= P(X_1 = j | X_0 = i) P(X_0 = i) \\ &= T_{ij} \pi_i = \pi_j T_{ji} \\ &= P(X_1 = i, X_0 = j). \end{aligned}$$

More generally, it can be shown that for all n,

$$P(X_0 = i_0, X_1 = i_1, \ldots, X_n = i_n) = P(X_n = i_0, X_{n-1} = i_1, \ldots, X_0 = i_n),$$

for all $i_0, i_1, \ldots, i_n$.

10.6 FROM MARKOV CHAIN TO MARKOV CHAIN MONTE CARLO

> The impact of Gibbs sampling and MCMC was to change our entire method of thinking and attacking problems, representing a *paradigm shift*. Now the collection of real problems that we could solve grew almost without bound. Markov chain Monte Carlo changed our emphasis from "closed form" solutions to algorithms, expanded our impact to solving "real" applied problems and to improving numerical algorithms using statistical ideas, and led us into a world where "exact" now means "simulated." This has truly been a quantum leap in the evolution of the field of statistics, and the evidence is that there are no signs of slowing down. . . . The size of the data sets, and of the models, for example, in genomics or climatology, is something that could not have been conceived 60 years ago, when Ulam and von Neumann invented the Monte Carlo method.
>
> —Christian Robert and George Casella (2011)

We are given a large graph in which the number of vertices is unknown. However, the graph can be described *locally* in the sense that from any vertex i we can easily find the neighbors of i. Thus, we are able to run a simple random walk on the graph. Let V be the number of vertices in the graph. The goal is to sample from the uniform distribution on the vertex set. That is, from the distribution $P(i) = 1/V$, for all i. But, as we said, V is unknown.

At first sight it seems incredible that such a problem can be solved. Yet we will show—using Markov chains—that it can.

An example of the type of graph we have in mind is the web graph introduced in the context of the PageRank algorithm in Example 10.11. The number of vertices in the graph, the total number of web pages on the Internet, is extremely hard to compute and, for all practical purposes, unknown. Yet from any vertex in the graph, from any web page, the number of neighbors is just the number of hyperlinks on that page. We would like to simulate an observation from the uniform distribution on the set of all web pages.

Why might such a simulation be useful? Suppose we would like to know the average number of hyperlinks on a web page on the Internet. Call this μ. One approach to finding μ is to look at every web page on the Internet, count the number of hyperlinks, and take the average. My computer science colleagues tell me that even with a lot of processors it would take between a week and a month to crawl the Internet and collect these data from *every* web page.

On the other hand, suppose we are able to simulate i.i.d. observations $X_1, \ldots, X_n$ from the uniform distribution on the web graph, where X_i represents a uniformly random web page (e.g., vertex). Given a web page i, let $h(i)$ be the number of hyperlinks on i. Then $h(X_k)$ is the number of hyperlinks on a uniformly random web page. A Monte Carlo estimate of μ is

$$\mu \approx \frac{h(X_1) + \cdots + h(X_n)}{n}, \quad \text{for large } n.$$

But in order to make this estimate we need i.i.d. observations from the uniform distribution.

Markov chain Monte Carlo is a powerful method for simulating observations from an unknown probability distribution. The method constructs a Markov chain whose limiting distribution is the unknown distribution of interest.

Let π be the "target" probability distribution of interest. The basic problem is to simulate from π, that is, to generate a random variable X whose distribution is π. Markov chain Monte Carlo constructs a Markov chain sequence $X_0, X_1, X_2, \ldots$ whose limiting distribution is π. Having constructed such a sequence, we can then take $X = X_n$, for n sufficiently large, as an approximate sample from π.

The MCMC algorithm for constructing a Markov chain with a given limiting distribution is called the Metropolis–Hastings algorithm. It is named after Nicholas Metropolis, a physicist, and W. Keith Hastings, a statistician. Metropolis co-authored a 1953 paper that first proposed the algorithm. Hastings later extended the work in 1970.

Metropolis–Hastings algorithm. For simplicity assume that π is finite, taking values $\{1, \ldots, k\}$. Let T be the transition matrix of *any* Markov chain on $\{1, \ldots, k\}$. We assume that we know how to sample from this chain. In particular, if the chain is at state i, we can simulate the next step of the chain according to the ith row of T. The T matrix is called the *proposal* matrix.

We construct a new Markov chain $X_0, X_1, X_2, \ldots$ by describing its transition mechanism. Suppose after m steps, $X_m = i$. The next state X_{m+1} is determined by a two-step procedure.

1. **Propose.** Choose a *proposal state* according to the ith row of T. That is, state j is chosen with probability $T_{ij} = P(X_{m+1} = j | X_m = i)$.
2. **Accept.** Decide whether or not to *accept* the proposal state. Suppose the proposal state is j. Compute the *acceptance function*

$$a(i, j) = \frac{\pi_j T_{ji}}{\pi_i T_{ij}}.$$

Let U be a random variable uniformly distributed on $(0,1)$. Then the next state of the chain is

$$X_{m+1} = \begin{cases} j, & \text{if } U \leq a(i,j) \\ i, & \text{if } U > a(i,j). \end{cases}$$

Theorem 10.3. (MCMC: Metropolis–Hastings algorithm). *Let* X_0 *be an arbitrary initial state. The sequence of random variables* $X_0, X_1, X_2, \ldots$ *constructed by the aforementioned algorithm is a time-reversible Markov chain whose limiting distribution is* π.

Before proving this theorem consider the following toy example. Suppose $\pi = (0.1, 0.2, 0.3, 0.4)$ is the desired target distribution. Let T be the transition matrix for simple random walk on the square, labeled 1, 2, 3, 4. From any vertex the random walk moves left or right with probability 1/2 each.

The algorithm proceeds as follows. Let X_0 be any vertex. Start a simple random walk on the square. If the current state is $X_m = i \in \{1, 2, 3, 4\}$, choose the proposal $j = i \pm 1$ (in clock arithmetic) with probability 1/2 each (e.g., if the walk is at state 1, then choose 4 or 2 with probability 1/2 each). Compute the acceptance function

$$a(i, j) = \frac{\pi_j T_{ji}}{\pi_i T_{ij}} = \frac{\pi_j}{\pi_i}, \quad \text{if } i \sim j,$$

since $T_{ij} = T_{ji} = 1/2$, for $i \sim j$. Let $U \sim$ Unif(0,1). Then, the next step of the walk is set to be j, if $U \leq \pi_j/\pi_i$, and i, otherwise. Following is an implementation.

R: MCMC—A TOY EXAMPLE

The target distribution is $\pi = (0.1, 0.2, 0.3, 0.4)$, with state space $\{1, 2, 3, 4\}$. The function `mcmc(n)` simulates X_n from a Markov chain constructed according to the Metropolis–Hastings algorithm. After 100 steps, X_{100} is output, and repeated 10,000 times. The simulated distribution of X_{100} serves as an approximation of π.

```
> pi <- c(0.1,0.2,0.3,0.4)
> mcmc <- function(n) {
   current_state <- 1
   for (i in 1:n) {
     proposal <-(current_state+sample(c(-1,1),1)) %% 4
     if (proposal==0) proposal <- 4
     accept <- pi[proposal]/pi[current_state]
     if (runif(1) < accept) current_state <- proposal}
   current_state }
> replicate(20,mcmc(100))
 [1] 4 3 4 2 4 2 3 3 4 4 3 3 4 3 3 3 4 2 4 1
> trials <- 10000
> simlist <- replicate(trials,mcmc(100))
> table(simlist)/trials
simlist
     1      2      3      4
0.1032 0.2034 0.2983 0.3951
```

Proof of Metropolis–Hastings algorithm: Let $X_0, X_1, \ldots$ be the sequence constructed by the Metropolis–Hastings algorithm with transition matrix P. By construction, for each m, X_m only depends on the previous state X_{m-1} and not on the full past history of the sequence, and thus the sequence is in fact a Markov chain. Let P be the transition matrix. To prove that the limiting distribution is π, we show that the detailed balance condition is satisfied.

For $i \neq j$, consider $P(X_1 = j|X_0 = i)$, the ijth entry of P. If $X_0 = i$, in order to transition to j, (i) state j must be a proposal state and (ii) j must be accepted. The first event (i) occurs with probability T_{ij}. And (ii) occurs if $U \leq a(i,j)$, where $U \sim \text{Unif}(0,1)$. Observe that

$$P(U \leq a(i,j)) = \begin{cases} a(i,j), & \text{if } \pi_j T_{ji} \leq \pi_i T_{ij} \\ 1, & \text{if } \pi_j T_{ji} > \pi_i T_{ij}. \end{cases}$$

Thus, the transition matrix P is given by

$$P_{ij} = P(X_1 = j|X_0 = i) = \begin{cases} T_{ij} a(i,j), & \text{if } \pi_j T_{ji} \leq \pi_i T_{ij} \\ T_{ij}, & \text{if } \pi_j T_{ji} > \pi_i T_{ij}, \end{cases}$$

for $i \neq j$. The diagonal entries P_{ii} are found so that the rows sum to 1. That is, $P_{ii} = 1 - \sum_{j \neq i} P_{ij}$.

To show that π satisfies the detailed balance condition $\pi_i T_{ij} = \pi_j T_{ji}$, let $i \neq j$ and suppose $\pi_j T_{ji}/\pi_i T_{ij} \leq 1$. Then necessarily, $\pi_i T_{ij}/\pi_j T_{ji} > 1$ and

$$\pi_i P_{ij} = \pi_i T_{ij} a(i,j) = \pi_i T_{ij} \frac{\pi_j T_{ji}}{\pi_i T_{ij}} = \pi_j T_{ji} = \pi_j P_{ji}.$$

The result holds similarly for the case $\pi_j T_{ji}/\pi_i T_{ij} > 1$.

We have shown that $X_0, X_1, \ldots$ is a time-reversible Markov chain with limiting distribution π.

Remarks:

1. The algorithm requires the computation of ratios of the form π_i/π_j. Often, the distribution π is specified up to a proportionality constant, which may be unknown. Since only ratios are required in the algorithm, the proportionality constant is not needed.
2. In the original version of the algorithm, the matrix T was taken to be symmetric, with $T_{ij} = T_{ji}$. In that case, the acceptance function simplifies to $a(i,j) = \pi_j/\pi_i$.
3. There are several accounts of the rich history of Markov chain Monte Carlo. We suggest the papers by Richey (2010) and Robert and Casella (2011).

Example 10.14 We revisit the general problem introduced at the beginning of this section. Let G be a graph where the number of vertices V is both very large and unknown. However, given any vertex i, we are able to compute the degree of i and perform a simple random walk on the graph. Our goal is to sample from the uniform distribution on the set of vertices. To implement MCMC, let T be the transition matrix for simple random walk on the graph. Using T as the proposal matrix, the acceptance function is

$$a(i,j) = \frac{\pi_j T_{ji}}{\pi_i T_{ij}} = \frac{(1/V)(1/\deg(j))}{(1/V)(1/\deg(i))} = \frac{\deg(i)}{\deg(j)}, \quad \text{for } i \sim j.$$

Thus, an MCMC algorithm to simulate from the uniform distribution is based on modifying a simple random walk on G: From vertex i, propose vertex j according to such a random walk. Then compute $a(i,j) = \deg(i)/\deg(j)$. Let $U \sim \text{Unif}(0,1)$. If $U \leq a(i,j)$, then accept j as the next step in the chain. Otherwise, stay at i.

To implement the algorithm on the web graph in order to generate a uniformly random web page, the basic transition mechanism is as follows: from web page i choose a hyperlink uniformly at random to reach a proposal page j. The acceptance function is then

$$a(i,j) = \frac{\text{Number of hyperlinks on page } i}{\text{Number of hyperlinks on page } j}.$$

Let $U \sim \text{Unif(0,1)}$. The Markov chain moves to j if $U < a(i,j)$, or stays at i for the next step of the sequence. Run the chain a "long time" and output X_n as an approximate sample from the uniform distribution. ■

The reader will no doubt wonder after the last example: What is a "long time"? This is perhaps the biggest open question in the study of Markov chain Monte Carlo. In practice there are numerous empirical methods for estimating the number of steps required in order to implement MCMC effectively. There are theoretical results for Markov chains on graphs that exhibit a lot of symmetry, such as the cycle or complete graphs. But these types of highly structured chains are typically not encountered in real-life applications.

Example 10.15 Cryptography

> ahicainqcaqx ic zqcqwbl bwq zwqbj xjustlicz tlhamx ic jyq kbr ho jybj albxx ho jyicmqwx kyh ybgq tqqc qnuabjqn jh mchk chjyicz ho jyq jyqhwr ho dwhtbtilijiqx jybj jyqhwr jh kyiay jyq shxj zlhwihux htpqajx ho yusbc wqxqbway bwq icnqtjqn ohw jyq shxj zlhwihux ho illuxjwbjihcx
>
> —qnzbw bllqc dhq jyq suwnqwx ic jyq wuq shwzuq

This example is based on Diaconis (2008), who gives a compelling demonstration of the power and breadth of MCMC with an application to cryptography.

The coded message above was formed by a simple substitution cipher—each letter standing for one letter in the alphabet. The hidden message can be thought of as a *coding function* from the letters of the coded message to the regular alphabet. For example, given an encrypted message $xoaaoaaoggo$, the function that maps x to m, o to i, a to s, and g to p decodes the message to $mississippi$.

If one keeps tracks of all 26 letters and spaces, then there are $27! \approx 10^{28}$ possible coding functions. The goal is to find the one that decodes the message.

My colleague Jack Goldfeather assisted in downloading the complete works of Jane Austen (about four million characters) from Project Gutenberg's online site (http://www.gutenberg.org) and recorded the number of transitions of consecutive text symbols. For simplicity we ignored case and only kept track of spaces and the letters a to z. The counts are kept in a 27×27 matrix M of transitions indexed by $(a, b, \ldots, z, [\text{space}])$. For example, there are 6669 places in Austen's work where b follows a and thus $M_{12} = 6669$.

The encoded message has 320 characters, denoted $(c_1, \ldots, c_{320})$. For each coding function f associate a *score function*

$$\text{score}(f) = \prod_{i=1}^{319} M_{f(c_1), f(c_{i+1})}.$$

The score is a product over all successive pairs of letters in the decrypted text $(f(c_i), f(c_{i+1}))$ of the number of occurrences of that pair in the reference Austen text. The score is higher when successive pair frequencies in the decrypted message match those of the reference text. Coding functions with high scores are good candidates for decryption. The goal is to find the coding function of maximum score.

A probability distribution proportional to the scores is obtained by letting

$$\pi_f = \frac{\text{score}(f)}{\sum_g \text{score}(g)}. \tag{10.8}$$

From a Monte Carlo perspective, we want to sample from π, with the idea that a sample is most likely to return a value of maximum probability. Of course, the denominator in Equation 10.8 is intractable, being the sum of 27! terms. But the beauty of Metropolis–Hastings is that the denominator is not needed since the algorithm relies on ratios of the form π_f/π_g.

The MCMC implementation runs a random walk on the set of coding functions. Given a coding function f, the transition to a proposal function f^* is made by picking two letters at random and switching the values that f assigns to these two symbols. This method of "random transpositions" gives a symmetric proposal matrix T, simplifying the acceptance function

$$a(f, f^*) = \frac{\pi_{f^*} T_{f^*,f}}{\pi_f T_{f,f^*}} = \frac{\pi_{f^*}}{\pi_f} = \frac{\text{score}(f^*)}{\text{score}(f)}.$$

The algorithm is as follows:

1. Start with any f. For convenience we use the identity function.
2. Pick two letters uniformly at random and switch the values that f assigns to these two symbols. Call this new proposal function f^*.
3. Compute the acceptance function $a(f, f^*) = \text{score}(f^*)/\text{score}(f)$.
4. Let $U \sim \text{Unif}(0,1)$. If $U < a$, accept f^*. Otherwise, stay with f.

We ran this algorithm on the coded message. See the script file **Decode.R**. The results are shown next. At iteration 2658 the message was decoded.

DECODING THE MESSAGE

[100] xlitxisatxau it hatawnc nwa hwand udepocith oclxfu it dra mng ly drnd xcnuu ly dritfawu mrl rnza oaat asexndas dl ftlm tldrith ly dra dralwg ly bwlonoicidiau drnd dralwg dl mrixr dra plud hclwileu lojaxdu ly repnt wauanwxr nwa itsaodas ylw dra plud hclwileu ly icceudwndiltu – ashnw nccat bla dra pewsawu it dra wea plwhea

[500] gsadgaredgen ad wedeoil ioe woeit ntubcladw clsgyn ad the kif sm thit glinn sm thadyeon khs hive ceed erugiter ts ydsk dsthadw sm the thesof sm poscicalataen thit thesof ts khagh the bsnt wlsoasun sczegtn sm hubid oeneiogh ioe adrecter mso the bsnt wlsoasun sm alluntoitasdn – erwio illed pse the buoreon ad the oue bsowue

[1000] goidgimedges id wedenal ane wneat strpclidw clogus id the fak oy that glass oy thiduens fho have ceed emrgatem to udof dothidw oy the theonk oy bnocacilities that theonk to fhigh the post wloniors ocxegts oy hrpad neseangh ane idmectem yon the post wloniors oy illrstnatiods – emwan alled boe the prnmens id the nre ponwre

[1400] goingidenges in beneral are breat stumplinb plogks in the way of that glass of thinkers who have peen edugated to know nothinb of the theory of cropapilities that theory to whigh the most blorious opxegts of human researgh are indepted for the most blorious of illustrations – edbar allen coe the murders in the rue morbue

[1600] goingidenges in ceneral are creat stumplinc plogks in the way of that glass of thinkers who have peen edugated to know nothinc of the theory of bropapilities that theory to whigh the most clorious opxegts of human researgh are indepted for the most clorious of illustrations – edcar allen boe the murders in the rue morcue

[1800] coincidences in general are great stumpling plocks in the way of that class of thinkers who have peen educated to know nothing of the theory of bropapilities that theory to which the most glorious opzects of human research are indepted for the most glorious of illustrations – edgar allen boe the murders in the rue morgue

[2658] coincidences in general are great stumbling blocks in the way of that class of thinkers who have been educated to know nothing of the theory of probabilities that theory to which the most glorious objects of human research are indebted for the most glorious of illustrations – edgar allen poe the murders in the rue morgue

■

Gibbs sampler. The original MCMC algorithm was developed in 1953 motivated by problems in physics. In 1984, a landmark paper by Geman and Geman (1984) showed how the algorithm could be adapted for the high-dimensional problems that arise in Bayesian statistics. The name "Gibbs sampling" was coined in that paper because of connections with Gibbs fields and the work of physicist Josiah Gibbs.

In order to understand Gibbs sampling, we need to broaden our notion of Markov chain and transition matrix to allow for infinite and continuous state spaces. In particular, the transition matrix T in the Metropolis–Hastings algorithm is replaced by a transition *function* $T(i, j)$ that, for fixed i, is a conditional density function, as opposed to a conditional pmf in the discrete case.

In the Gibbs sampler, the target distribution is a joint distribution

$$\pi(\boldsymbol{x}) = \pi(x_1, \ldots, x_k).$$

A Markov chain is constructed whose limiting distribution is π and that takes values in a k-dimensional space. The algorithm generates elements of the form

$$(X_1^{(0)}, \ldots, X_k^{(0)}), (X_1^{(1)}, \ldots, X_k^{(1)}), (X_1^{(2)}, \ldots, X_k^{(2)}), \ldots$$

eventually generating $(X_1^{(n)}, \ldots, X_k^{(n)})$ for large n as a sample from π.

Chain elements are simulated by sampling from conditional distributions. We illustrate with a simple example: simulating from a bivariate standard normal distribution with correlation ρ. Recall that if (X, Y) has a bivariate standard normal distribution, then the conditional distribution of X given $Y = y$ is normal with mean ρy and variance $1 - \rho^2$. Similarly, the conditional distribution of Y given $X = x$ is normal with mean ρx and variance $1 - \rho^2$.

At each step of the Gibbs sampler, each of the two coordinates of the joint distribution are updated by sampling from the conditional distribution of one given the other back and forth as described in the following implementation:

1. Initiate $(X_0, Y_0) = (0, 0)$.
2. At step m, having already simulated (X_{m-1}, Y_{m-1}),
   (a) take X_m as a sample from the conditional distribution of X_m given Y_{m-1}. That is, if $Y_{m-1} = y$, then simulate X_m from a normal distribution with mean ρy and variance $1 - \rho^2$, Then,
   (b) take Y_m as a sample from the conditional distribution of Y_m given X_m. That is, having just simulated $X_m = x$, now simulate Y_m from a normal distribution with mean ρx and variance $1 - \rho^2$.
3. For large n, output (X_n, Y_n) as a sample from the desired bivariate distribution.

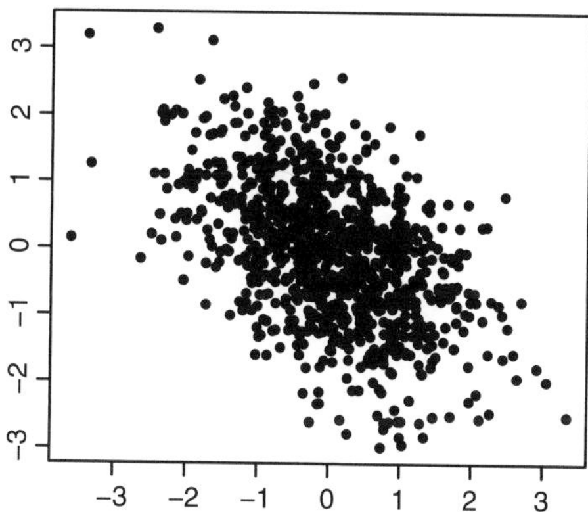

FIGURE 10.12: Gibbs sampler simulation of bivariate standard normal with $\rho = -0.5$.

R: SIMULATION OF BIVARIATE STANDARD NORMAL DISTRIBUTION

This code simulates 1000 observations of a bivariate standard normal distribution with $\rho = -0.5$ using the Gibbs sampler. We take $n = 100$ iterations and output (X_{100}, Y_{100}). See Figure 10.12 for the graph of the output.

```
gibbsnormal <- function(n, rho) {
  x <- 0
  y <- 0
  sd <- sqrt(1-rho^2)
  for (i in 1:n) {
    x <- rnorm(1,rho*y,sd)
    y <- rnorm(1,rho*x,sd) }
  return(c(x,y))}
> simlist <- replicate(1000,gibbsnormal(100,-0.2)
> plot(t(simlist))
```

Why does this work? The Gibbs sampler is actually a special case of the Metropolis–Hastings algorithm, where transitions are based on conditional distributions. Consider one step of the Gibbs sampler. Suppose $\boldsymbol{i} = (x, y)$ is the current state and we are proposing an update for the first coordinate from x to z. Let $\boldsymbol{j} = (z, y)$.

The proposal z is obtained from the conditional distribution of X given $Y = y$. Let $f_{X|Y}(x|y)$ denote the conditional density of X given $Y = y$. Let $f_Y(y)$ denote the marginal density of Y. The acceptance function is thus

$$a(\boldsymbol{i}, \boldsymbol{j}) = \frac{\pi_{\boldsymbol{j}} T_{\boldsymbol{ji}}}{\pi_{\boldsymbol{i}} T_{\boldsymbol{ij}}} = \frac{\pi(z, y) f_{X|Y}(x|y)}{\pi(x, y) f_{Z|Y}(z|y)} = \frac{\pi(z, y)\pi(x, y) f_Y(y)}{\pi(x, y)\pi(z, y) f_Y(y)} = 1.$$

Since the acceptance function is equal to one, we always accept a transition. Similarly, when the second coordinate is updated from (z, y). Thus, Gibbs sampling is a special case of the Metropolis–Hastings algorithm in which we always accept proposals.

More generally, suppose we want to sample from a multivariate probability distribution

$$\pi(\boldsymbol{x}) = \pi(x_1, \ldots, x_k) = P(X_1 = x_1, \ldots, X_k = x_k).$$

We assume that for each i, we can sample from the conditional distribution

$$P(X_i = x_i | X_j = x_j, j \neq i).$$

That is, we can simulate from the conditional distribution of each coordinate of $(X_1, \ldots, X_k)$ given the other $k-1$ coordinates. As shown Earlier, the acceptance function is always equal to 1, and the algorithm proceeds by iteratively choosing coordinates to update, simulating from the conditional distribution given the remaining coordinates.

Example 10.16 The following three-dimensional example is based on Casella and George (1992). Consider the mixed joint density

$$\pi(x, p, n) \propto \binom{n}{x} p^x (1-p)^{n-x} e^{-4} \frac{4^n}{n!}, \tag{10.9}$$

for $x = 0, 1, \ldots, n$, $0 < p < 1$, $n = 0, 1, \ldots$. The p variable is continuous; variables x and n are discrete. The distribution arises from the following application. Conditional on $N = n$ and $P = p$, let X represent the number of successful hatchings from n insect eggs, where each egg has success probability p. Both N and P are random and vary across insects. We seek the expectation and variance of X, the number of successful hatchings among all insects.

There is no simple closed form expression for the marginal distribution of X. Gibbs sampling is used to simulate from the joint density of (X, P, N) and then extract the marginal information.

The key to using the Gibbs sampler is our ability to sample easily from the conditional distributions of each coordinate variable given the other coordinate values. For fixed p and n in Equation 10.9, the conditional distribution of X given $P = p$ and $N = n$ is binomial with parameters n and p. (To see this, observe that anything in the expression that does not involve x can be treated as a constant and thus absorbed in the constant of proportionality.) Similarly, the conditional distribution of P given $X = x$ and $N = n$ is a beta distribution with parameters $x + 1$ and $n - x + 1$. And the conditional pmf of N given $X = x$ and $P = p$, up to proportionality, is

$$\begin{aligned} P(N = n | P = p, X = x) &\propto \frac{1}{(n-x)!}(1-p)^{n-x} 4^n \\ &\propto e^{-4(1-p)} \frac{(4(1-p))^{n-x}}{(n-x)!}, \end{aligned}$$

for $n = x, x+1, \ldots$. This is a "shifted" Poisson distribution. It is the distribution of $Z + x$, where $Z \sim \text{Pois}(4(1-p))$.

As we can simulate from the three conditional distributions, the Gibbs sampler proceeds by cycling through each of the three coordinates of the joint distribution and updating each coordinate with the conditional distribution of that coordinate given the other two coordinate values.

R: GIBBS SIMULATION OF TRIVARIATE DISTRIBUTION

The Gibbs sampler is run for 500 iterations and outputs $(X_{500}, P_{500}, N_{500})$ as an approximate sample from the joint distribution. We then repeat 10,000 times storing the output in the 3×500 matrix `simmat`. The first row `simmat[1,]` is taken as a sample from the marginal distribution of X. See the histogram of the distribution in Figure 10.13.

```
> gibbsthree <- function(trials) {
  x <- 1
  p <- 1/2
  n <- 2
> for (i in 1:trials) {
  x <- rbinom(1,n,p)
  p <- rbeta(1,x+1,n-x+1)
  n <- x + rpois(1,4*(1-p)) }
  return(c(x,p,n))  }
> simmat <- replicate(10000,gibbsthree(500))
> marginal <- simmat[1,]
> mean(marginal)
[1] 1.9964
> var(marginal)
[1] 3.286116
> hist(marginal)
```

∎

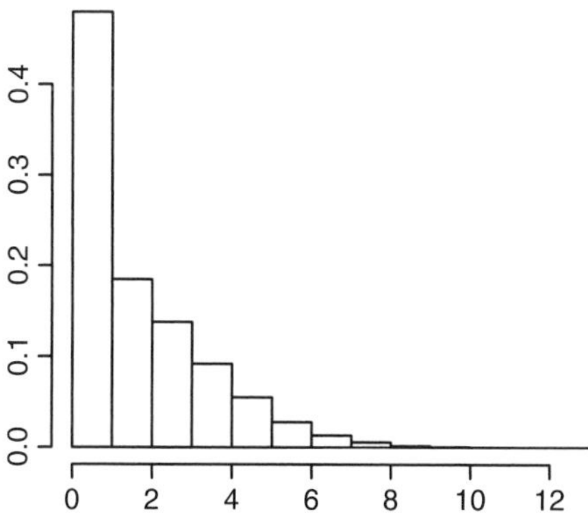

FIGURE 10.13: Distribution of the marginal distribution of X from 10,000 runs of the Gibbs sampler. Each sampler is run for 500 iterations.

10.7 SUMMARY

The first section of this chapter introduces the bivariate normal distribution, one of the most widely used distributions in statistics and applied fields. The distribution generalizes the one-dimensional normal distribution. It has several special properties, such as that marginal and conditional distributions are normal. Also, if bivariate normal random variables X and Y are uncorrelated ($E[XY] = E[X]E[Y]$), then the variables are independent.

Transformations of two random variables are presented in the second section. The formula for the probability density function is a consequence of the change of variable formula in multivariable calculus. Among the examples is included a derivation of the joint density function of the bivariate normal distribution.

The method of moments is included in the third section. This technique, which originates in statistics, is used to estimate the unknown parameters of a probability distribution based on data gotten from a random sample.

The final three sections of the book are devoted to random walk on graphs and Markov chains, culminating in the final treatment of Markov chain Monte Carlo (MCMC). Simple random walk and random walk on weighted graphs are introduced. The treatment gives a gentle introduction to Markov chains, which does not require knowledge of linear algebra. Several examples of the power and scope of MCMC are presented. Both the Metropolis–Hastings algorithm and the Gibbs sampler are introduced.

- **Bivariate standard normal density:** X and Y have a bivariate standard normal distribution with correlation ρ, if the joint density function is

$$f(x, y) = \frac{1}{2\pi\sqrt{1-\rho^2}} e^{-\frac{x^2-2\rho xy+y^2}{2(1-\rho^2)}}.$$

- **Properties of bivariate standard normal distribution:** If X and Y have a bivariate standard normal distribution with correlation ρ, then
  1. Marginal distributions of X and Y are standard normal.
  2. Conditional distribution of X given $Y = y$ is normal with mean ρy and variance $1 - \rho^2$. Similarly for Y given $X = x$.
  3. If $\rho = 0$, X and Y are independent.
  4. Let $Z_1 = X$ and $Z_2 = (Y - \rho X)/\sqrt{1-\rho^2}$. Then Z_1 and Z_2 are independent standard normal variables.
  5. For nonzero constants a and b, $aX + bY$ is normally distributed with mean 0 and variance $a^2 + b^2 + 2ab\rho$.
- **Transformation of two random variables—joint density:** Let X and Y have joint density $f_{X,Y}$. Suppose $V = g_1(X, Y)$ and $W = g_2(X, Y)$ with $X = h_1(V, W)$ and $Y = h_2(V, W)$. Let J be the Jacobian $J = \frac{\partial h_1}{\partial v}\frac{\partial h_2}{\partial w} - \frac{\partial h_2}{\partial v}\frac{\partial h_1}{\partial w}$. Then

$$f_{V,W}(v, w) = f_{X,Y}\left(h_1(v, w), h_2(v, w)\right)|J|.$$

- **Sample moments:** Given i.i.d. random variables $X_1, \ldots, X_n$, the kth sample moment is $(1/n) \sum_{X_i^k}$.
- **Method of moments:** A technique for estimating the unknown parameters of a probability distribution. For sufficiently many k, equate sample moments with theoretical moments $E[X^k]$ and solve for unknown parameters.
- **Graph:** A set of vertices together with a set of edges.
- **Neighbors and degrees:** For vertices i and j, say i is a neighbor of j if there is an edge between them. The degree of a vertex is the number of its neighbors.
- **Simple random walk on a graph:** The random walk moves from a given vertex by choosing one of its neighbors uniformly at random.
- **Transition matrix:** For a graph on k vertices, this is a $k \times k$ matrix whose ijth entry $P(X_1 = j | X_0 = i)$ is the probability of moving to vertex j given that the walk is on vertex i.
- **Initial distribution:** The distribution of X_0, the initial vertex of the random walk.
- **Limiting distribution:** A probability distribution π on the vertex set with the property that for all initial vertices i, $P(X_n = j | X_0 = i) \rightarrow \pi_j$, as $n \rightarrow \infty$.
- **Limiting distribution for simple random walk on a graph:** Under suitable conditions on the graph, $\pi_j = \deg(j) / \sum_{i=1}^{k} \deg(i)$.
- **Weighted graph:** A graph with a positive number assigned to each edge.
- **Random walk on weighted graph:** The random walk moves from a given vertex i by choosing a neighbor j with probability $w(i,j)/w(i)$, where $w(i,j)$ is the weight of the edge joining i and j, and $w(i)$ is the sum of the weights on the edges that join i to all of its neighbors.
- **Markov chain:** A sequence of $X_0, X_1, \ldots$ with the property that for all n, the conditional distribution of X_{n+1} (the "future") given $X_0, \ldots, X_n$ (the "past") only depends on X_n (the "present").
- **Stationary distribution:** A probability distribution $\mu = (\mu_1, \ldots, \mu_k)$ on the states of a Markov chain that satisfies $\mu_j = \sum_{i=1}^{k} \mu_i T_{ij}$ for all j.
- **Detailed balance condition:** A probability distribution μ satisfies this condition if $\mu_i T_{ij} = \mu_j T_{ji}$ for all i and j. If a probability distribution satisfies this condition, it is a stationary distribution of the Markov chain.
- **Markov chain Monte Carlo:** A technique for simulating from a given probability distribution by running a Markov chain whose limiting distribution is the distribution of interest.
- **Metropolis–Hastings algorithm:** Given a probability distribution π, this algorithm constructs a Markov chain whose limiting distribution is π.
- **Gibbs sampler:** An MCMC method for simulating from a multivariate joint distribution $\pi = (\pi_1, \ldots, \pi_k)$. The algorithm is based on the ability to sample from the conditional distribution of each variable given the other variables.

EXERCISES

Bivariate normal distribution

10.1 Let X and Y be independent and identically distributed normal random variables. Show that $X + Y$ and $X - Y$ are independent.

10.2

(a) Let U and V have a bivariate standard normal distribution with correlation ρ. Find $E[UV]$.

(b) Let X and Y have a bivariate normal distribution with parameters μ_X, μ_Y, σ_X^2, σ_Y^2, ρ. Find $E[XY]$.

10.3 Suppose that math and reading SAT scores have a bivariate normal distribution with the mean of both scores 500, the standard deviation of both scores 100, and correlation 0.70. For someone who scores 650 on the reading SAT, find the probability that they score over 700 on the math SAT.

10.4 Let (X, Y) have a bivariate standard normal distribution with correlation ρ. Using results for the conditional distribution of Y given $X = x$, illustrate the law of total variance and find $V[Y]$ with a bare minimum of calculation.

10.5 Let X and Y have joint density

$$f(x, y) = \frac{1}{\pi\sqrt{3}} e^{-\frac{2}{3}(x^2 - xy + y^2)},$$

for real x, y.

(a) Identify the distribution of X and Y and parameters.

(b) Identify the conditional distribution of X given $Y = y$.

(c) Use R to find $P(X > 1|Y = 0.5)$

10.6 Let X and Y have a bivariate normal distribution with parameters $\mu_X = -1$, $\mu_y = 4$, $\sigma_X^2 = 1$, $\sigma_Y^2 = 25$, and $\rho = -0.75$.

(a) Find $P(3 < Y < 6|X = 0)$.

(b) Find $P(3 < Y < 6)$.

10.7 Let X and Y have a bivariate standard normal distribution with correlation $\rho = 0$. That is, X and Y are independent. Let (x, y) be a point in the plane. The rotation of (x, y) about the origin by angle θ gives the point

$$(u, v) = (x \cos\theta - y \sin\theta, x \sin\theta + y \cos\theta).$$

Show that the joint density of X and Y has *rotational symmetry* about the origin. That is, show that $f(x, y) = f(u, v)$.

10.8 Let X and Y have a bivariate standard normal distribution with correlation ρ. Find $P(X > 0, Y > 0)$ by the following steps.

(a) Write $X = Z_1$ and $Y = \rho Z_1 + \sqrt{1-\rho^2} Z_2$, where (Z_1, Z_2) are independent standard normal random variables. Rewrite the probability in terms of Z_1 and Z_2.

(b) Think geometrically. Express the event as a region in the plane.

(c) See the last Exercise 10.7. Use rotational symmetry and conclude that

$$P(X > 0, Y > 0) = \frac{1}{4} + \frac{\sin^{-1} \rho}{2\pi}.$$

10.9 If X and Y have a joint bivariate normal distribution, show that

$$\rho^2 = \frac{V[E[Y|X]]}{V[Y]}.$$

Transformations of random variables

10.10 Suppose X and Y are independent exponential random variables with parameter λ. Find the joint density of $V = X/Y$ and $W = X + Y$. Use the joint density to find the marginal distributions.

10.11 Let X and Y be jointly continuous with density $f_{X,Y}$. Let (R, Θ) be the polar coordinates of (X, Y).

(a) Give a general expression for the joint density of R and Θ.

(b) Suppose X and Y are independent with common density function $f(x) = 2x$, for $0 < x < 1$. Use your result to find the probability that (X, Y) lies inside the circle of radius one centered at the origin.

10.12 Suppose X and Y have joint density

$$f(x, y) = 4xy, \quad \text{for } 0 < x < 1, 0 < y < 1.$$

Find the joint density of $V = X$ and $W = XY$. Find the marginal density of W.

10.13 Recall that the density function of the Cauchy distribution is

$$f(x) = \frac{1}{\pi(1 + x^2)}, \quad \text{for all } x.$$

Show that the ratio of two independent standard normal random variables has a Cauchy distribution by finding a suitable transformation of two variables.

10.14 Let X and Y be independent standard normal random variables. Let $V = X^2 + Y^2$ and $W = \tan^{-1}(Y/X)$.

(a) Show that V and W are independent with $V \sim \text{Exp}(1/2)$ and $W \sim \text{Unif}(0, 2\pi)$. (Hint: $x = \sqrt{v} \cos w$ and $y = \sqrt{v} \sin w$.)

(b) Show that this gives the following method for simulating a pair of standard normal random variables given a pair of independent uniforms. Let $U_1, U_2 \sim \text{Unif}(0, 1)$. Let

$$X = \sqrt{-2 \ln U_1} \cos(2\pi U_2), \quad Y = \sqrt{-2 \ln U_1} \sin(2\pi U_2).$$

(c) Implement this method and plot 1000 pairs of points.

10.15 (This exercise requires knowledge of three-dimensional determinants.) Let (X, Y, Z) be independent standard normal random variables. Let (Φ, Θ, R) be the corresponding spherical coordinates. The correspondence between rectangular and spherical coordinates is given by

$$x = r \sin \phi \cos \theta, \quad y = r \sin \phi \sin \theta, \quad z = r \cos \phi,$$

for $0 \leq \phi \leq \pi$, $0 \leq \theta \leq 2\pi$, and $r > 0$. Find the joint density of (Φ, Θ, R) and the marginal density of R.

Method of Moments

10.16 Following are data from an i.i.d. sample taken from a Poisson distribution with unknown parameter λ.

$$2 \quad 3 \quad 0 \quad 7 \quad 2 \quad 2 \quad 3 \quad 5 \quad 2 \quad 2 \quad 2 \quad 0.$$

Find the method of moments estimate for λ.

10.17 Let $X_1, \ldots, X_{25}$ be an i.i.d. sample from a binomial distribution with parameters n and p. Suppose n and p are unknown. Write down the method of moments equations that would need to be solved to estimate n and p.

10.18 Let $X_1, \ldots, X_n$ be an i.i.d. sample from a normal distribution with mean μ and variance σ^2. Find the general method of moments estimators for μ and σ^2.

Random walk on graphs and Markov chains

10.19 The star graph on k vertices contains one center vertex and $k-1$ other vertices called leaves. Between each leaf and the center vertex there is one edge. Thus the graph has k edges. Find the stationary distribution of random walk on the star graph.

10.20 The *lollipop graph* on $2k - 1$ vertices is defined as follows: a complete graph on k vertices is joined with a path on k vertices by identifying one of the

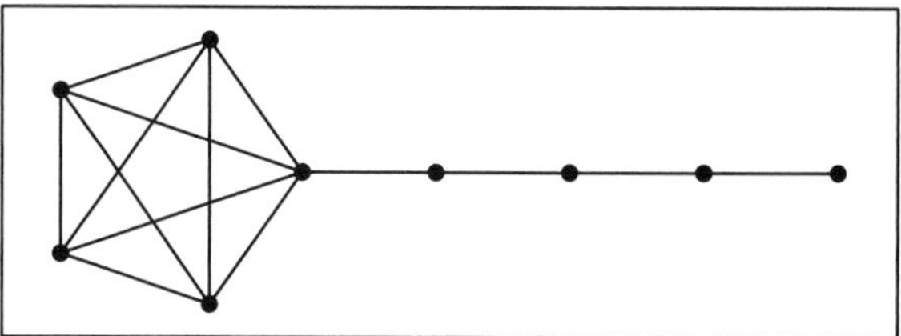

FIGURE 10.14: Lollipop graph on nine vertices.

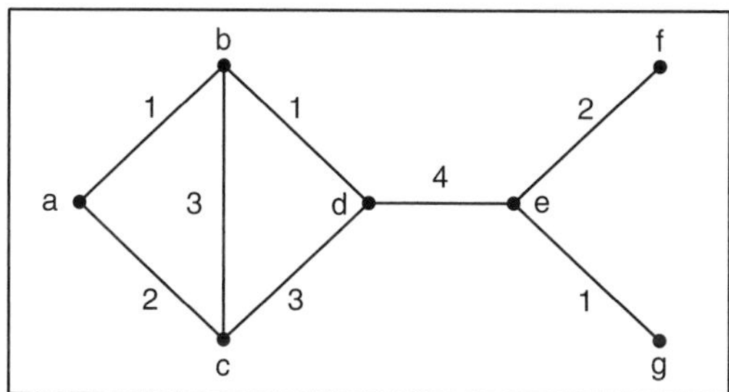

FIGURE 10.15: Weighted graph.

endpoints of the path with one of the vertices of the complete graph (see Fig 10.14). Find the limiting distribution for simple random walk on the lollipop graph.

10.21 Find the stationary distribution for random walk on the weighted graph in Figure 10.15.

10.22 The rows of a Markov chain transition matrix sum to one. A matrix is called *doubly stochastic* if its columns also sum to one. If a Markov chain has a doubly stochastic transition matrix, show that its stationary distribution is uniform.

10.23 Show that if a transition matrix for a Markov chain is symmetric, that is, if $T_{ij} = T_{ji}$ for all i and j, then the Markov chain is time-reversible.

10.24 A lone king on a chessboard conducts a random walk by moving to a neighboring square with probability proportional to the number of neighbors. The walk defines a simple random walk on a graph consisting of 64 vertices. Find the value of the stationary distribution at a corner square of the chessboard.

10.25 The weather Markov chain of Example 10.10 has stationary distribution is $\pi = (1/4, 1/5, 11/20)$. Determine whether or not the Markov chain is time-reversible.

10.26 Suppose a time-reversible Markov chain has transition matrix P and stationary distribution π. Show that the Markov chain can be regarded as a random walk on a weighted graph with edge weights $w(i, j) = \pi_i T_{ij}$ for all states i and j.

10.27 A Markov chain has transition matrix

$$T = \begin{array}{c} \\ a \\ b \\ c \\ d \end{array} \begin{array}{c} \begin{array}{cccc} a & b & c & d \end{array} \\ \begin{pmatrix} 0 & 1/4 & 1/4 & 1/2 \\ 1/4 & 0 & 0 & 3/4 \\ 1/2 & 0 & 0 & 1/2 \\ 1/3 & 1/2 & 1/6 & 0 \end{pmatrix} \end{array}.$$

(a) Show that the stationary distribution is $\pi = (1/4, 1/4, 1/8, 3/8)$.

(b) The Markov chain can be regarded as a random walk on a weighted graph. Determine the graph and the weights.

10.28 Suppose a Markov chain with unique positive stationary distribution π starts at state i. The expected number of steps until the chain revisits i is called the *expected return time of state* i. We state without proof that the expected return time of state i is equal to $1/\pi_i$. Find the expected return time of the center vertex in the lollipop graph of Figure 10.14.

10.29 See Exercise 10.28. A lone knight performs a random walk on a chessboard. From any square, the knight looks at the squares that it can legally move to in chess, and picks one uniformly at random to move to. If the knight starts this random walk at one of the corner squares of the chessboard, find the expected number of steps until the knight returns to its starting square.

Markov chain Monte Carlo

10.30 See the toy example following Theorem 10.3. Find the transition matrix for the Markov chain constructed by the Metropolis–Hastings algorithm. Show that $\pi = (0.1, 0.2, 0.3, 0.4)$ is the stationary distribution and that the detailed-balance condition is satisfied.

10.31 Use the Metropolis–Hastings algorithm to simulate a Poisson random variable with parameter λ. Let T be the (infinite) matrix that describes a simple random walk on the integers. From an integer i, the walks moves to $i - 1$ or $i + 1$ with probability 1/2 each. Show how to simulate a Poisson random variable. Then implement your algorithm and give evidence that it works.

10.32 Modify the simulation code for a bivariate standard normal distribution to simulate a bivariate normal distribution with parameters $\mu_X = 20, \sigma_X^2 = 100$, $\mu_Y = -14, \sigma_Y^2 = 4$, and $\rho = -0.8$ using the Gibbs sampler.

10.33 Research project: Implement the MCMC cryptography algorithm as described in Example 10.15. Study ways to adjust parameters in order to improve the algorithm for more efficient decoding.

APPENDIX A

GETTING STARTED WITH R

There are many excellent R primers and tutorials on the web. To learn more about R , the best starting place is the homepage of the R Project for Statistical Computing at `http://www.r-project.org/`. Go there to download the software. The site contains many links to books, manuals, demonstrations, and other resources. For this introduction, we assume that you have R up and running on your computer.

1. **R as a calculator**
   When you bring up R , the first window you see is the R console. You can type directly on the console using R like a calculator.

```
> 1+1
[1] 2
> 2-2
[1] 0
> 3*3
[1] 9
> 4/4
[1] 1
> 5^5
[1] 3125
```

   R has many built-in math functions. Their names are usually intuitive.

Probability: With Applications and R, First Edition. Robert P. Dobrow.

```
> pi
[1] 3.141593
> cos(pi)
[1] -1
> exp(1)
[1] 2.718282
> abs(-6)
[1] 6
> factorial(4)
[1] 24
> factorial(52)
[1] 8.065818e+67
> sqrt(9)
[1] 3
```

2. **Navigating the keyboard**
   Perhaps the most important keystrokes to learn are the up and down cursor/arrow keys. Each command you submit in R is stored in history. The up arrow (↑) key scrolls back along this history; the down arrow (↓) scrolls forward.

   Suppose you want to calculate $\log 8$ and you type

```
> Log(8)
Error: could not find function "Log"
```

   You get an error because R is case-sensitive. All commands start with lowercase letters. Rather than retype the command, navigate back with the up arrow key and fix.

```
> log(8)
[1] 2.079442
```

3. **Asking for help**
   Prefacing any R command with ? will produce a help file in another window.

```
> ?log
```

   Reading the help for the `log` function, we learn that the log function has the form `log(x, base = exp(1))`. The second `base` argument defaults to $e = \exp(1)$. For logarithms in other bases, give a second argument.

```
> log(8,2)
[1] 3
> log(100,10)
[1] 2
```

4. **Vectors**

The real power of R lies in its ability to work with lists of numbers, called vectors. The command `a:b` creates a vector of numbers from a to b.

```
> 1:10
[1]  1  2  3  4  5  6  7  8  9 10
> -3:5
[1] -3 -2 -1  0  1  2  3  4  5
> -5:5
 [1] -5 -4 -3 -2 -1  0  1  2  3  4  5
```

For more control with sequence generation, use the command `seq`.

```
> seq(1,10)
[1]  1  2  3  4  5  6  7  8  9 10
> seq(1,20,2)
[1]  1  3  5  7  9 11 13 15 17 19
> seq(20,1,-4)
[1] 20 16 12  8  4
```

To create a vector and assign it to a variable, type

```
> x <- c(2,3,5,7,11)
> x
[1]  2  3  5  7 11
```

The assignment operator is `<-` . The `c` stands for *concatenate*. If we want to extend the vector `x` to include two more elements, type

```
> x <- c(x,13,17)
> x
[1]  2  3  5  7 11 13 17
> c(x,x)
 [1]  2  3  5  7 11 13 17  2  3  5  7 11 13 17
```

Elements of vectors are accessed with `[ ]`.

```
> x[4]
[1] 7
> x[1:3]
[1] 2 3 5
> x[c(2,4,6,8)]
[1]  3  7 13 NA
```

The vector `x` has seven elements. When we asked for the eighth element, R returned an `NA`.

Vectors can consist of numbers or characters or even strings of characters.

```
> c("d","o","g")
[1] "d" "o" "g"
> y <- c("Probability", "is", "very", "very", "cool")
> y[c(1,2,5)]
[1] "Probability" "is"          "cool"
```

With numeric vectors you can add, subtract, and do all the usual mathematical operations, treating the entire vector as a single object.

```
> dog
[1]  1  3  5  7  9 11 13 15
> dog+1
[1]  2  4  6  8 10 12 14 16
> dog*3
[1]  3  9 15 21 27 33 39 45
> 2^dog
[1]     2     8    32   128   512  2048  8192 32768
> cat<-dog+1
> cat
[1]  2  4  6  8 10 12 14 16
> dog/cat
[1] 0.5000000 0.7500000 0.8333333 0.8750000
    0.9000000
[6] 0.9166667 0.9285714 0.9375000
> dog+cat
[1]  3  7 11 15 19 23 27 31
> dog*cat
[1]   2  12  30  56  90 132 182 240
```

Notice that when we add a single number to a vector that number gets added to each element of the vector. But when two vectors of the same length are added to each other, then the corresponding elements are added together. This applies to most binary operations.

5. **Operations on vectors**
Some common operations on vectors are (i) sum the elements, (ii) take the average (mean), and (iii) find the length.

```
> tiger <- 1:1000
> sum(tiger)
[1] 500500
> mean(tiger)
[1] 500.5
```

```
> length(tiger)
[1] 1000
```

The `sample` command is one of many commands for generating random numbers. The following picks one number at random from the integers 1 to 10.

```
> sample(1:10,1)
[1] 8
```

To simulate 10 rolls of a six-sided die with labels $\{1, 2, 3, 4, 5, 6\}$, type

```
> roll <- sample(1:6,10,replace=TRUE)
> roll
 [1] 6 3 6 1 3 6 5 3 2 1
```

When working with vectors we may want to find elements of the vector that satisfy some condition. For instance, in the 10 dice rolls, how many times did we get a 2? Use the logical operators $<$, $>$, and $==$, for less than, greater than, and equal to. Typing

```
> roll == 2
 [1] FALSE FALSE FALSE FALSE FALSE FALSE FALSE
 [8] FALSE TRUE FALSE
```

creates a TRUE–FALSE vector identifying where the twos are in the original vector. We can treat this TRUE–FALSE vector as if it consists of zeros and ones. By summing that vector, we count the number of 2's.

```
> sum(roll==2)
[1] 1
```

The following one-line command simulates rolling 600 dice and counting the number of 2's that appear. After you type the command once, hit the up arrow key to repeat. By repeating over and over again, you begin to get a feel for the typical behavior of the number of 2's in 600 dice rolls.

```
> sum(2==sample(1:6,600,replace=TRUE))
[1] 99
> sum(2==sample(1:6,600,replace=TRUE))
[1] 107
> sum(2==sample(1:6,600,replace=TRUE))
[1] 115
> sum(2==sample(1:6,600,replace=TRUE))
[1] 97
```

6. **Plots and graphs**
   To graph functions, and plot data, use `plot`. This workhorse command has enormous versatility. Here is a simple plot comparing two vectors.

```
> radius <- 1:20
> area <- pi*radius^2
> plot(radius,area, main="Area as function of
  radius")
```

   To graph smooth functions you can also use the `curve` command.

```
> curve(pi*x^2,1,20,xlab="radius",ylab="area",
+ main="Area as function of radius")
```

   The syntax is `curve(expr,from=,to=)`, where `expr` is some expression that is a function of x.

7. **Script files**
   When you have many R commands to manage, it is extremely useful to keep your work in a *script file.* A script file is a text file that contains your commands and that can be executed in the R console, saved, edited, and executed again later. Most of the R examples in this book are contained in script files that you can download, execute, and modify.

   In the Mac environment, click on **File: New Document** to bring up a new window for typing your R commands. To execute a portion of your code, highlight the text that you want to execute and then hit **Command-Return**. The highlighted code will appear in the R console and be executed.

   In the PC environment, click on **File: New Script** to bring up a new window for a script file. To execute a portion of your code, highlight the text. Under the task bar, press the **Run line or selection** button (third from the left) to execute your code.

   Open the script file **ScriptSample.R**. The # symbol is for comments. Everything after the # symbol will be ignored by R . The script file contains two blocks of code. In the first block, we plot the area of a circle as a function of radius for radii from 1 to 20. Highlight the three lines of R code in the first part of the file. Execute the code and you should see a plot like the one in Figure A.1.

```
# ScriptSample.R
# Area of circle
radius <- 1:20
area <- pi*radius^2
plot(radius,area, main="Area as function of
  radius")
```

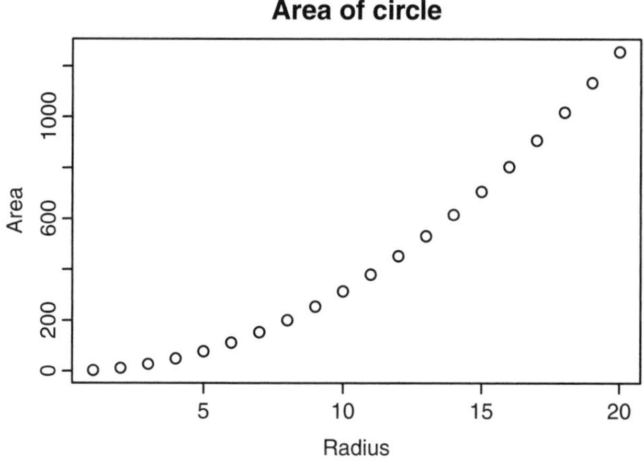

FIGURE A.1

```
####################################
# Coin flips
n <- 1000 # Number of coin flips
coinflips <- sample(0:1,n,replace=TRUE)
coinflips
heads <- cumsum(coinflips)
prop <- heads/(1:n)
plot(1:n,prop,type="l",xlab="Number of coins",
  ylab="Running average",
  main="Proportion of heads in 1000 coin flips")
abline(h=0.5)
```

In the second block of code, we flip 1000 coins and plot the running proportion of heads from 1 to 1000. Let zero represent tails and one represent heads. A 1000-element coin flip vector of zeros and ones is stored in the variable `coinflips`. The `cumsum` command generates a cumulative sum and stores the resulting vector in `heads`. The kth element of `heads` is equal to the number of ones in the first k positions of `coinflips`, that is, the number of heads in the first k coin flips. Then `prop` stores the proportion of heads in the first k coin flips. Finally, the running proportions are plotted. We see that the proportion of heads appears to converge to one-half (Figure A.2 and Figure A.3).

8. **Write your own functions**
   You can create your own functions in R . The syntax is

```
> name <- function(arg1, arg2, . . . ) expression
```

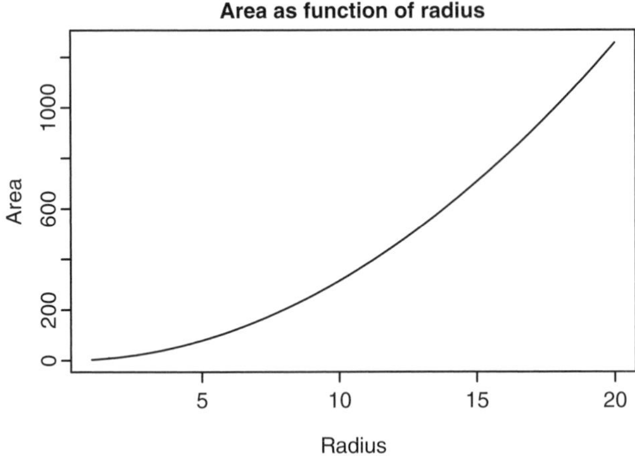

FIGURE A.2

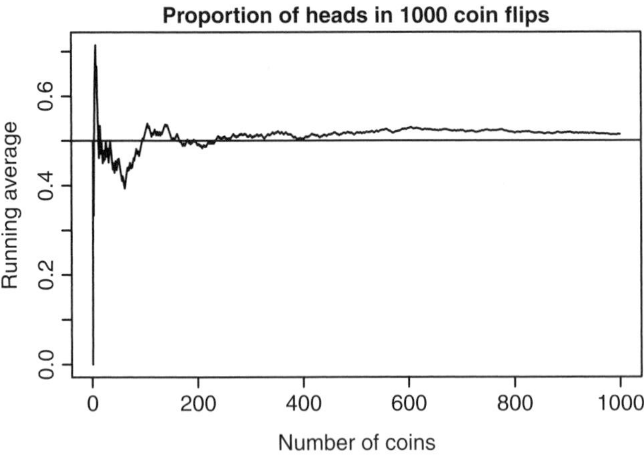

FIGURE A.3

A function can have one, several, or no arguments. Here is a simple function to find the area of a circle of given radius.

```
> area <- function(radius) pi*radius^2
```

The name of the function is `area`. It is a function of one variable.

```
> area(1)
[1] 3.141593
> area(7.5)
[1] 176.7146
```

The function cone computes the volume of a cone of height h and circular base of radius r.

```
> cone <- function(r,h) (1/3)*pi*r^2*h
> cone(1,1)
[1] 1.047198
> cone(3,10)
[1] 94.24778
```

If the function definition contains more than one line of code, enclose the code in curly braces { }. The following function sums accepts a vector vec as its input and outputs the sum, sum of squares, and sum of cubes of the elements of vec.

```
> sums <- function(vec) {
 s1 <- sum(vec)
 s2 <- sum(vec^2)
 s3 <- sum(vec^3)
 c(s1,s2,s3) }
> sums(1:5)
[1]  15  55 225
> sums(-5:5)
[1]   0 110   0
```

9. **Other useful commands**

```
> table(c(0,1,1,1,1,1,2,2,6,6,6,6,6,6,6))
   # tabulates the number each distinct entry in a
   vector
0 1 2 6
1 5 2 7
>
> sort(c(3,10,-5,3,8,4,-2)) # sort in increasing
  order
[1] -5 -2  3  3  4  8 10

> min(0:20)    # minimum element
[1] 0

> max(0:20)    # maximum element
[1] 20

> round(pi,3)  # rounds to a given number of decimal
  places
[1] 3.142
```

The `replicate` command is very useful. The syntax is `replicate (n,expr)`. The expression `expr` is replicated n times.

```
> c(replicate(2,0),replicate(4,1),replicate(6,2))
 [1] 0 0 1 1 1 1 2 2 2 2 2 2
```

Choose a number at random from 1 to 20. Then find the area of a circle whose radius is equal to that number. Suppose we now want to replicate this five times. Using our `area` function, the command

```
> area(sample(1:20,1))}
[1] 530.9292
```

finds the area of a circle with a random radius. The experiment is repeated five times with

```
> replicate(5,area(sample(1:20,1)))}
 [1]  706.85835   78.53982 1017.87602   78.53982
      530.92916
```

Exercises: Now it is your turn. Keep your R code in a script file and execute commands from the script.

A.1 Collect a small amount of data from any source (between 10 and 50 numbers). Enter the data and assign to a vector named `mydata`. Compute the following "summary statistics" on your data:
   (a) Sum.
   (b) Average.
   (c) Minimum value.
   (d) Maximum value.
   (e) The *geometric mean* of n numbers is the nth root of their product. The R command `prod` finds the product of the elements of a vector. Use `prod` to find the geometric mean of your data.

A.2 Let $n = 10$. Create an n-element vector `vec` defined as

$$\texttt{vec}_k = k + \frac{2\pi k}{n}, \quad \text{for } k = 1, \ldots, n.$$

   (a) Find the sine and cosine of the elements of `vec`.
   (b) Plot the points $(\cos(k), \sin(k)),\ \text{for } k = 1, \ldots, n.$
   (c) Repeat the earlier steps for $n = 20, 50$, and 100.

A.3 Write a function that takes as input the lengths of two sides of a right triangle and returns the length of the hypotenuse.

A.4 A tetrahedron die is a four-sided die with labels $\{1, 2, 3, 4\}$. Write a function that rolls 10 tetrahedron die and computes the average. Now use the `replicate` command to repeat this 1000 times. Take the grand average of your 1000 averages.

A.5 Open the **ScriptSample.R** file and execute the coin-flipping portion of the code. Generate a similar graph for 10 coin flips, 100 coin flips, and 10,000 coin flips. Replace the second line of the code block with the command

```
coinflips <- sample(0:1,n,replace=TRUE,
prob=c(.49,51))
```

This will flip slightly "biased" coins, where the probability of heads is 51% instead of 50%. What happens to the resulting plot?

APPENDIX B

PROBABILITY DISTRIBUTIONS IN R

There are four commands for working with probability distributions in R. The commands take the root name of the probability distribution (see Table B.1) and prefix the root with `d`, `p`, `q`, or `r`. These give continuous density or discrete probability mass function (`d`), cumulative distribution function (`p`), quantile (`q`), and random variable simulation (`r`).

TABLE B.1: Probability distributions in R.

Distribution	Root	Distribution	Root
Beta	beta	Log-normal	lnorm
Binomial	binom	Multinomial	multinom
Cauchy	cauchy	Negative binomial	nbinom
Chi-squared	chisq	Normal	norm
Exponential	exp	Poisson	pois
F	f	Student's t	t
Gamma	gamma	Uniform	unif
Geometric	geom	Weibull	weibull
Hypergeometric	hyper		

Probability: With Applications and R, First Edition. Robert P. Dobrow.

APPENDIX C

SUMMARY OF PROBABILITY DISTRIBUTIONS

Discrete Distributions

Distribution	PMF, Expectation, and Variance
Uniform$(1,\ldots,n)$	$P(X=k)=\frac{1}{n}, \quad k=1,\ldots,n$
	$E[X]=\frac{n+1}{2} \qquad V[X]=\frac{n^2-1}{12}$
Binomial	$P(X=k)=\binom{n}{k}p^k(1-p)^{n-k}, \quad k=0,1,\ldots,n$
	$E[X]=np \qquad V[X]=np(1-p)$
Poisson	$P(X=k)=\frac{e^{-\lambda}\lambda^k}{k!}, \quad k=0,1,2,\ldots$
	$E[X]=\lambda \qquad V[X]=\lambda$

Probability: With Applications and R, First Edition. Robert P. Dobrow.

Geometric	$P(X=k)=(1-p)^{k-1}p, \quad k=1,2,\ldots$ $E[X]=\frac{1}{p} \qquad V[X]=\frac{1-p}{p^2}$ $P(X>t)=(1-p)^t$
Negative binomial	$P(X=k)=\binom{k-1}{r-1}p^r(1-p)^{k-1},$ $k=r,r+1,\ldots, \quad r=1,2,\ldots$ $E[X]=\frac{r}{p} \qquad V[X]=\frac{r(1-p)}{p^2}$
Hypergeometric	$P(X=k)=\frac{\binom{D}{k}\binom{N-D}{n-k}}{\binom{N}{n}}, \quad k=0,1,\ldots,n$ $E[X]=\frac{nD}{N} \qquad V[X]=\frac{nD(N-D)}{N^2}\left(1-\frac{n-1}{N-1}\right)$

CONTINUOUS DISTRIBUTIONS

Distribution	Density, CDF, and Expectation, Variance
Uniform(a,b)	$f(x)=\frac{1}{b-a}, \quad a<x<b$ $F(x)=\frac{x-a}{b-a}, \quad a<x<b$ $E[X]=\frac{a+b}{2} \qquad V[X]=\frac{(b-a)^2}{12}$
Exponential	$f(x)=\lambda e^{-\lambda x}, \quad x>0$ $F(x)=1-e^{-\lambda x}, \quad x>0$ $E[X]=\frac{1}{\lambda} \qquad V[X]=\frac{1}{\lambda^2}$

Normal	$f(x) = \frac{1}{\sigma\sqrt{2\pi}} \exp\left[\frac{-(x-\mu)^2}{2\sigma^2}\right], \quad -\infty < x < \infty$ $E[X] = \mu \qquad V[X] = \sigma^2$
Gamma	$f(x) = \lambda e^{-\lambda x} \frac{(\lambda x)^{r-1}}{\Gamma(r)}, \quad x > 0$ where $\Gamma(r) = \int_0^\infty t^{r-1} e^{-t}\, dt$ $E[X] = \frac{r}{\lambda} \qquad V[X] = \frac{r}{\lambda^2}$
Beta	$f(x) = \frac{1}{B(\alpha,\beta)} x^{\alpha-1}(1-x)^{\beta-1}, \quad 0 < x < 1$ where $B(\alpha,\beta) = \frac{\Gamma(\alpha)\Gamma(\beta)}{\Gamma(\alpha+\beta)}.$ $E[X] = \frac{\alpha}{\alpha+\beta} \qquad V[X] = \frac{\alpha\beta}{(\alpha+\beta+1)(\alpha+\beta)^2}$

APPENDIX D

REMINDERS FROM ALGEBRA AND CALCULUS

Exponents

$$e^{a+b} = e^a e^b$$

$$[e^a]^b = e^{ab}$$

Logarithms

$$\log ab = \log a + \log b$$

$$\log a^r = r \log a$$

$$\log x = y \text{ if and only if } x = e^y$$

Calculus

Limits

$$\lim_{x \to \infty} \left(1 + \frac{a}{x}\right)^x = e^a$$

Probability: With Applications and R, First Edition. Robert P. Dobrow.

Derivatives

$$f'(x) = \lim_{h \to 0} \frac{f(x+h) - f(x)}{h}$$

Definite Integral

$$\int_a^b f(x)\,dx = \lim_{n \to \infty} \frac{b-a}{n} \sum_{i=1}^{n} f(x_i), \quad a = x_1 < x_2 < \cdots < x_n = b$$

Chain Rule

$$\frac{d}{dx} f(g(x)) = f'(g(x))g'(x)$$

Fundamental Theorem of Calculus

$$\frac{d}{dx} \int_a^x f(t)\,dt = f(x)$$

Integration by Parts for definite integrals

$$\int_a^b u\,dv = uv\big|_a^b - \int_a^b v\,du$$

Series

$$\sum_{k=0}^{\infty} \frac{x^k}{k!} = e^x, \quad \text{for all } x$$

$$\sum_{k=0}^{\infty} x^k = \frac{1}{1-x}, \quad \text{for } |x| < 1$$

$$\sum_{k=0}^{n} x^k = \frac{1-x^{n+1}}{1-x}, \quad \text{for } x \neq 1$$

APPENDIX E

MORE PROBLEMS FOR PRACTICE

Following are additional problems for self-study. They can be used for midterm and final exam preparation. Solutions to all problems are given at the end of the book.

E.1. A four-sided tetrahedron die is labeled with numbers $\{1, 2, 3, 4\}$. A die is rolled twice.

- **(a)** Give the sample space.
- **(b)** Let X be the maximum of the two rolls. Find the pmf of X.
- **(c)** Find $E[1/X]$.

E.2. Suppose A, B, C are independent events with respective probabilities 1/6, 1/4, and 1/2.

- **(a)** What is the probability that at least one of the events occurs?
- **(b)** What is the probability that A does not occur given that B and C both occur?
- **(c)** Find the probability that A and B occur given that A or B occur.

E.3. There are three coins in a box. One is two-headed. One is fair. And one is biased to come up heads three-fourths of the time. A coin is selected at random, flipped, and shows heads. What is the probability that it was the two-headed coin?

E.4. A hotel has 100 rooms. In each room the number of phone calls made by the guest has a Poisson distribution with $\lambda = 2$. Find the probability that in at least three rooms more than six calls will be made by the guest. Use R .

Probability: With Applications and R, First Edition. Robert P. Dobrow.

E.5. A bag of Scrabble tiles contains two each of the letters r, a, n, d, o, and m (12 in all). Six letters are picked without replacement and then placed from left to right on a Scrabble rack.

(a) Find the probability of spelling $random$.

(b) Find the probability of picking both r's.

(c) Find the probability that no m is picked.

(d) Find the probability that the ordered letters $rando$ are in your rack.

E.6. A random variable X has probability mass function

$$P(X=k)=\begin{cases}0.1, & \text{if } k=0\\ 0.2, & \text{if } k=1\\ 0.3, & \text{if } k=2\\ 0.4, & \text{if } k=3.\end{cases}$$

(a) Find $E[X]$.

(b) Find $V[X]$.

(c) Find $E[e^X]$.

E.7. Use indicator random variables to find the expected number of Zodiac signs that are shared by at least four students in a class of 30.

E.8. Suppose X and Y have joint pmf

$$P(X=x,Y=y)=c(x+y),\quad x,y=0,1,2.$$

(a) Find c.

(b) Find $P(X \leq 1, Y \geq 1)$.

(c) Find $E[XY]$.

(d) $P(X < Y)$.

(e) Find the marginal distributions of X and Y.

(f) Find $E[Y|X=2]$.

E.9. A bag contains three red, five green, and seven blue balls. A sample of three balls is picked. Let X be the number of reds in the sample. Let Y be the number of greens in the sample. Find the joint pmf of X and Y.

E.10. Let X have density function proportional to $x(1-x)$ on (0,1).

(a) Find the constant of proportionality.

(b) Find $E[X]$ and $V[X]$.

(c) Find, and plot, the CDF.

(d) Find $P(0.1 < X < 0.3)$ in two ways: (i) using the CDF and (ii) using the density function.

(e) Find the density of $Y=2X$.

(f) Find the expectation of Y, two ways: (i) using the density of Y in and (ii) not using the density of Y.

(g) Find $E[e^X/(1-X)]$.

E.11. Let $X_1, \ldots, X_{10}$ be an i.i.d. sequence with common distribution given by the density function $f(x) = 3/x^4$, $x > 1$.

(a) Let $M = \max(X_1, \ldots, X_n)$. Derive the density.

(b) Show how to use the inverse transform method and R to simulate M.

E.12. Let $X_1, \ldots, X_n$ be an i.i.d. sequence of independent uniform (0,1) random variables. Find the expectation of the minimum of the X_i's.

E.13. Let X have density function $f(x) = xe^{-x}$ for $x > 0$.

(a) Find the density of $Y = cX$, where $c > 0$.

(b) Let Z have the same distribution as X and be independent of X. Find the density of $X + Z$.

E.14. Let X have density function $f(x) = ce^{-x}$, for $x > 2$.

(a) Find the constant of proportionality.

(b) Let $Z \sim$ Exp(1). Show that X has the same distribution as $Z + 2$.

(c) Find the mean and variance of X. (Think about what is the best way to do this.)

(d) Find the density function of $Y = e^{-X}$.

E.15. $X \sim$ Exp$(1/2)$ and $Y \sim$ Exp(1/3). The random variables are independent.

(a) Find the joint density of X and Y.

(b) Find the joint CDF of X and Y.

(c) Find Cov(X, Y) and Corr(X, Y).

(d) Find the density of $X + Y$.

(e) Find $E[e^{-(X+3Y)}]$.

E.16. Suppose X and Y are uniformly distributed on (0,1) and are independent. Show how to use geometric probability to find the CDF and density function of $Z = Y/X$. (Hint: there are two cases to consider.)

E.17. Without using calculus solve the integral

$$\int_0^\infty e^{-2x^2}\, dx.$$

E.18. Suppose X is uniformly distributed on $(0, 2)$. If $X = x$, then Y is uniformly distributed on $(0, x)$. Find the joint, marginal, and conditional distributions of X and Y.

E.19. X and Y have joint density $f(x, y) = 3y$, for $0 < x < y < 1$.

(a) Find the marginal densities of X and Y.

(b) Find Cov(X, Y).

(c) Find the conditional density functions of Y given $X = x$ and X given $Y = y$. Give a qualitative description of the conditional distribution of X given $Y = y$.

(d) Find $P(X < 1/2|Y = 3/4)$.

(e) Find $E[Y|X = 1/2]$.

(f) Find $E[X|Y]$.

(g) Illustrate the law of total expectation in this example. Show $E[E]X|Y]] = E[X]$.

(h) Find $V[X]$ in two ways: (i) using the density of X and (ii) using the law of total variance.

E.20. Halloween problem: A boy is lost in a haunted house. He is in a room with three doors. If he takes door A, he will find his way out of the house in 20 minutes. If he takes door B, he will find his way out of the house in 40 minutes. If he takes door C, he will wander around the house for 30 minutes and then wind up in the same room as he started in. Whenever he is in that room he picks doors A, B, and C with respective probabilities 0.1, 0.3, and 0.6. How long will it take, on average, to get out of the house?

E.21. Suppose $Z \sim \text{Geom}(p)$. Let $X_1, \ldots, X_n$ be an i.i.d. sequence of normally distributed random variables with mean 2 and variance 9. Consider the random sum

$$Y = \sum_{i=1}^{Z} X_i = X_1 + \cdots + X_Z.$$

Find $E[Y]$ and $V[Y]$.

E.22. For the following situations describe the distribution, including parameters, of the given random variables. Give the most reasonable distribution for the situation.

(a) Tom is waiting for a text message. Typically he receives about five texts per hour. Let X be the time he will wait.

(b) Every day there is a 10% chance that Danny will receive no mail. Let X be the number of times he receives mail over the next week.

(c) Joe is playing craps at the casino. Recall that the probability of winning craps is about 0.49. He will keep playing until he wins. Let X be the number of times he will play.

(d) Lisa is examining the flowers in her rose garden. Petal lengths tend to be about 12 mm long with standard deviation 2 mm. Let X be the length of a rose petal.

(e) Laura is biking from her house to another city, 30 miles away. She is equally likely to be anywhere between the two. Let X be her position.

(f) Hurricanes hit this state at the average rate of five per year. Let X be the number of hurricanes next year.

E.23. Suppose $X_1, \ldots, X_{25}$ are i.i.d. exponential random variables with $\lambda = 0.1$. Consider $P(\sum_{i=1}^{25} X_i > 350)$. What can be said about this probability using

(a) Markov's inequality

(b) Chebyshev's inequality

(c) Central limit theorem

(d) The exact distribution

E.24. Use Monte Carlo integration to solve

$$I = \int_0^4 \sin\left(x^x\right)\, dx.$$

E.25. Let $X \sim \text{Norm}(\mu_x, \sigma_x^2)$ and $Y \sim \text{Norm}(\mu_y, \sigma_y^2)$ be independent. Use moment generating functions to show that $X + Y \sim \text{Norm}(\mu_x + \mu_y, \sigma_x^2 + \sigma_y^2)$.

SOLUTIONS TO EXERCISES

SOLUTIONS FOR CHAPTER 1

1.3 (i) Choosing a topping; (ii) set of possible toppings. Let n, p, r, denote pineapple, peppers, and pepperoni, respectively. Then

$$\Omega = \{\emptyset, n, p, r, np, nr, pr, npr\}.$$

(iii) X is the number of toppings.

1.5 (i) Harvesting tomatoes; (ii) set of all possibilities for bad and good tomatoes among the 100; (iii) {At most five tomatoes are bad}; (iv) Let X be the number of bad tomatoes among the 100.

1.7 (b) $\{R = 1, B = 2\}$; (c) $\{R + B = 4\}$; (d) $\{R = 2B\}$.

1.9 $P(\omega_4) = 1/41$.

1.11 Show

$$\sum_{\omega} P(\omega) = 1.$$

1.13 $p = 0$ or 1.

1.15 $1/16$.

Probability: With Applications and R, First Edition. Robert P. Dobrow.

1.17 (a) $1/6^4$; (b) 0.598; (c) 0.093.

1.19 $1/(n \cdot n-1 \cdots n-k+1)$.

1.21 (a) 0.90

1.21 (b) 0.

1.23 (i) $A^c = \{X=1\} \bigcup \{X \geq 5\}$; (ii) $B^c = \{1 \leq X \leq 3\}$; (iii) $AB = \{X = 4\}$; and (iv) $A \bigcup B = \{X \geq 2\}$.

1.25 0.80.

1.27 (i) $P(\text{HHHT, HHTH}) = 2/16$; (ii) $3/8$; (iii) $P(\text{HHHH,HHTT})$.

1.29 (a)

$$\Omega = \{pp, pn, pd, pq, np, nn, nd, nq, dp, dn, dd, dq, qp, qn, qd, qq\}.$$

$$\{X=1\} = \{pp\}, \ \ \{X=5\} = \{pn, np, nn\}$$

$$\{X=10\} = \{pd, nd, dd, dn, dp\}, \ \ \{X=25\} = \{pq, nq, dq, qq, qd, qnqp\}.$$

1.29 (c) 3/8.

1.31 (a)

$$\sum_{k=0}^{\infty} \frac{2}{3^{k+1}} = \frac{2}{3} \sum_{k=0}^{\infty} \frac{1}{3^k} = \frac{2}{3} \frac{1}{1-1/3} = 1.$$

1.31 (b) 1/27.

1.33 (a) $A \cup B \cup C$; (b) BA^cC^c; (c) $A^cB^cC^c \cup AB^cC^c \cup BA^cC^c \cup CA^cB^c$; (d) ABC; (e) $A^cB^cC^c$.

1.35 1/2.

1.37 0.40.

SOLUTIONS FOR CHAPTER 2

2.3 $(2p_1 - p_2)/p_1$.

2.5 (a) 0; (b) 1; (c) $P(A)/P(B)$; (d) 1.

2.7 (a) False. For instance, consider tossing two coins. Let A be the event that both coins are heads. Let B be the event that the first toss is heads. Then $P(A|B) + P(A|B^c) = 1/2 + 0 = 1/2 \neq 1$.

(b) True.

$$P(A|B) = \frac{P(AB)}{P(B)} = \frac{P(B) - P(A^cB)}{P(B)} = 1 - P(A^c|B).$$

2.9 0.00198.

2.11 (a) $p_1 = P(AB|A) = P(AB)/P(A)$; (b) $p_2 = P(AB|A \cup B) = P(AB)/P(A \cup B)$, since AB implies $A \cup B$; (c) Since $A \subseteq A \cup B$, $P(A) \leq P(A \cup B)$ and thus $p_1 \geq p_2$.

2.13
$$\begin{aligned} P(A \cup B|C) &= \frac{P((A \cup B)C)}{P(C)} = \frac{P(AC \cup BC)}{P(C)} \\ &= \frac{P(AC) + P(BC) - P(ABC)}{P(C)} \\ &= P(A|C) + P(B|C) - P(AB|C) \end{aligned}$$

2.15 Apply the birthday problem. Approximate probability is 0.63.

2.17 (ii)

$$P(G) = P(G|1)P(1) + P(G|2)P(2) = \frac{3}{5}\left(\frac{1}{2}\right) + \frac{2}{6}\left(\frac{1}{2}\right) = \frac{7}{15}.$$

2.19 $P(A|B^c) = \dfrac{P(A) - P(AB)}{1 - P(B)}.$

2.23 Solve $0.20 = 0.35(0.30) + x(0.70)$ for x.

2.25 (a) 0.32; (b) 76/77 = 98.7% reliable.

2.27 0.10.

2.29
```
d1 <- function() {
   sample(c(3,5,7),1,replace=T)    }
d2 <- function() {
   sample(c(2,4,9),1,replace=T)    }
d3 <- function() {
   sample(c(1,6,8),1,replace=T)    }
simlist1 <- replicate(10000,d1() > d2())
mean(simlist1)
[1] 0.5638
simlist2 <- replicate(10000,d1() > d2())
mean(simlist2)
[1] 0.562
simlist3 <- replicate(10000,d1() > d2())
mean(simlist3)
[1] 0.5585
```

SOLUTIONS FOR CHAPTER 3

3.1 If A and B are independent,

$$\begin{aligned}P(A^cB^c) &= P\left((A \cup B)^c\right) = 1 - P(A \cup B)\\ &= 1 - [P(A) + P(B) - P(AB)] = P(A^c) - P(B) + P(A)P(B)\\ &= P(A^c) - P(B)[1 - P(A)] = P(A^c) - P(A^c)P(B)\\ &= P(A^c)P(B^c).\end{aligned}$$

3.3 0.25.

3.5 0.04914.

3.7 $P(X = 0) = 0.729; P(X = 1) = 0.243;$

$P(X = 2) = 0.027; P(X = 3) = 0.001.$

3.9

2	3	4	5	6	7	8
0.01	0.04	0.10	0.20	0.25	0.24	0.16

3.11

3	4	5	6	7	8	9
1/27	3/27	6/27	7/27	6/27	3/27	1/27

3.13 (a) $P(4432) = 0.2155$. $P(4333) = 0.1054$. (b) $P(5332) = 0.1552$; $P(5431) = 0.1293$; $P(5422) = 0.1058$.

3.15 9.10947×10^{-6}.

3.17 (b) Hint: Use the ballot problem.

3.19 Suppose we want to pick an k element subset from $S = \{1, \ldots, n\}$. One way to choose the set is to pick k numbers from S. This can be done in $\binom{n}{k}$ ways. Another way to choose the set is to pick the $n - k$ numbers from S that will not be in the set. This can be done in $\binom{n}{n-k}$ ways.

3.21 (a) 0.126; (b) 0.048; (c) 0.22; (d) 0.999568.

3.23 (b) Yes. $n = 7$, $p = 0.25$; (b) No fixed number of trials; (c) Sampling without replacement. No independence or fixed probability; (d) No fixed probability; (e) Yes, assuming independence for pages.

3.25 (a) 0.2778; (b) 0.03845; (c) 0.365; (d) 0.058.

3.27 0.0384. It is very unlikely (less than a 4% chance that the drug is ineffective).

3.29 $\lambda = 4$ and $P(X = 3) = 0.1954$.

3.31 $\lambda = -\ln 0.10$ and the probability of hatching at least two eggs is 0.6697.

3.33 0.1018.

3.35 $(1 - e^{-2\lambda})/2$.

3.37 (a) $P(1 \leq X \leq 3) = 0.7228997$; (b) $P(1 \leq X \leq 3) \approx 0.7228572$.

3.39 Let X have a Poisson distribution with $\lambda = 1$. The series gives the probability that X is even.

3.43
```
> lambda = 2
> sim <- rpois(100000,lambda)
> sum( sim%%2 ==1)/100000
[1] 0.49048
> (1-exp(-2*2))/2
[1] 0.4908422
```

SOLUTIONS FOR CHAPTER 4

4.1 1.722.

4.3 (a) $P(W = 0) = 3/6$; $P(W = 10) = 2/6$; $P(W = 24) = 1/6$: (b) $E[W] = 7.33$.

4.5 \$11.

4.9 $E[X!] = e^{-\lambda}/(1 - \lambda)$, if $0 < \lambda < 1$.

4.11 (a) $P(T = t, E = z) = e^{-6}4^t 2^z/(t!z!)$; (b) e^{-6}; (c) 0.54; (d) 0.95

4.13 (a) $c = 6/(n(n+1)(n+2))$; (b) 0.30

4.15 (a) 1/24. (b) $(e+1)(e+2)(e+3)/24$.

4.17 (a) X is uniform on $\{1, \ldots, n\}$; (b) $1/2$.

4.19 (a) $P(X = 1) = 3/8$; $P(X = 2) = 5/8$ and $P(Y = 1) = 1/4$; $P(Y = 2) = 3/4$; (b) No; (c) 1/2; (d) 7/8.

4.21 Let $I_k = 1$ if the kth ball in the sample is red. $E[R] = nr/(r + g)$.

4.23 7.54.

4.25 2.61.

4.27 $V[X] = 1.25n$.

4.29 (a) 61; (b) 75.

4.31 $a(1 - a) + b(1 - b) + 2(c - ab)$.

4.33 (a) $E[S] = n(2p - 1)$; (b) $V[S] = 4np(1 - p)$.

4.35 (a) $c = 1/15$; (b) $P(X = 0) = 6/15 = 1 - P(X = 1)$; $P(Y = 1) = 3/15$; $P(Y = 2) = 5/15$; $P(Y = 3) = 7/15$. (c) $\text{Cov}(X, Y) = -2/75$.

4.38 If $X = 12$, then necessarily two 6's were rolled and thus $Y = 0$. Thus X and Y are not independent. Show $E[XY] = E[D_1^2 - D_2^2] = 0 = E[X]E[Y]$ to show X and Y are uncorrelated.

4.40 (a) 101; (b) 1410; (c) 3; (d) 1266.

4.45 $\text{Cov}(X, Y) = \text{Cov}(X, aX + b) = a\text{Cov}(X, X) = aV[X]$.

4.49 $P(A|B)$.

4.51 (a) False. (b) True. (c) False.

4.53 $P(X + Y = k) = \begin{cases} (k-1)/n^2, & \text{for } k = 2, \ldots, n+1 \\ (2n - k + 1)/n^2, & \text{for } k = n+2, \ldots, 2n. \end{cases}$

SOLUTIONS FOR CHAPTER 5

5.1 694.167.

5.3 $V[X] = (p-1)/p^2$.

5.5 Send out 11 applications.

5.7 $\dfrac{\sum_{k=1}^{n}(k-1)n}{(n+1-k)^2}$.

5.9 $\dfrac{E[2^X] = 2p}{(1 - 2(1-p))}$, for $1/2 < p < 1$.

5.11 (a) 0.082; (b) 0.254; (c) $P(Y = k) = P(X = k + r) = \binom{k+r-1}{r-1}p^r(1-p)^k$, for $k = 0, 1, \ldots$.

5.13 $P(X + Y = k) = (k-1)p^2(1-p)^{k-2}$, $k = 2, 3, \ldots$.

5.15 0.183.

5.17 $P(\text{Andy wins}) = 0.1445$. Andy gets \$14.45 and Beth gets \$85.55.

5.19 $X \sim \text{Geom}(p^2)$.

5.21 (a) Hypergeometric; (b) $P(X = k) = \binom{4}{k}\binom{4}{4-k}/8choose4$.

5.25 (a) $E[G] = 200$, $V[G] = 160$. (b) $E[G + U] = 300$, $V[G + U] = 160 + 90 - 40 = 210$.

5.27 (a) 0.074; (b) 0.155.

5.29 Apply the multinomial theorem to expand $(1 + 1 + 1 + 1)^4$.

5.33 Let C, D, H, S, be the number of clubs, diamonds, hearts, and spades, respectively, in the drawn cards. Then (C, D, H, S) has a multinomial distribution with parameters $(8, 1/4, 1/4, 1/4, 1/4)$. Desired probability is 0.038.

5.35 0.54381.

SOLUTIONS FOR CHAPTER 6

6.1 (a) $c = 1/(e^2 - e^{-2}) = 0.13786$. (b) $P(X < -1) = 0.032$; (c) $E[X] = 1.0746$.

6.3 (a) 0.0988; (b) $(\pi - 2)/2$.

6.5 (a) 1/4; (b) $\dfrac{\int_{-\infty}^{\infty} x}{(1+x^2)}\, dx$ does not exist.

6.7 $E[X^k] = \dfrac{(b^{k+1} - a^{k+1})}{((k+1)(b-a))}$.

6.9 $0 \le a \le 1$ and $b = 1 - a$.

6.11 $E[aX + b] = \int_{-\infty}^{\infty} (ax+b) f(x)\, dx = a \int_{-\infty}^{\infty} x f(x)\, dx + b \int_{-\infty}^{\infty} f(x)\, dx$
$= aE[X] + b$.

6.13 Not true. $P(X > s + t | X > t) = P(X > s)$, not $P(X > s + t)$.

6.15 $V[X] = \dfrac{1}{\lambda^2}$.

6.17 0.988.

6.19 0.95.

6.21 $Y \sim \text{Exp}(\lambda/c)$.

6.23 $e - 1$.

6.25 For all real x,

$$P(1/X \le t) = P(X \ge 1/t) = 1 - P(X < 1/t).$$

Density of $1/X$ is thus

$$f(t) = f_X(1/t)/t^2 = f_X(t).$$

6.27 For $ma + n < y < mb + n$,

$$\begin{aligned} P(Y \le y) &= P(mX + n \le y) = P(X \le (y-n)/m) \\ &= \frac{(y-n)/m - a}{b - a} = \frac{y - (ma+n)}{mb - ma}, \end{aligned}$$

which is the cdf of a uniform distribution on $(ma + n, mb + n)$.

6.29 $Y \sim \text{Geom}(1 - e^{-\lambda})$.

6.31 (a) $P(X \le x, Y \le y) = (1 - e^{-x})(1 - e^{-2y}), x > 0, y > 0$; (b) $P(X \le x) = 1 - e^{-x}, x > 0$; (c) 1/3.

6.33 $X \sim \text{Exp}(6)$ and $Y \sim \text{Unif}(1,4)$

6.37 $f(x) = \lambda^3 x^2 e^{-\lambda x}/2,\ x > 0.$

6.39 $F(X) \sim \text{Unif}(0,1).$

6.41 $\frac{1}{3}$

6.44 $F(x) = P(X/Y \leq x) = x/(x+1),\ x > 0.\ P(X/Y < 1) = 1/2.$

6.45 $f_X(x) = (2/\pi)\sqrt{1-x^2},\ -1 < x < 1$; not independent.

6.51 (a) $\frac{\sqrt{2}}{3}$; (b) $\frac{7}{12}$.

6.53 $\frac{2\sqrt{3}}{3\pi} \approx 0.37.$

SOLUTIONS FOR CHAPTER 7

7.1 (a) About 0.025; (b) 68%; (c) 0.16.

7.3 (a) 87.1; (b) 0.0003; (c) 0.0067.

7.5 Amy: 0.943; Bill: 0.868; Carrie: 0.691

7.7 (a) $E[T] = 1500$; $SD[T] = 315.18$; (b) 80th percentile is 1765; 90th percentile is 1904; (c) 0.171.

7.9 $x = \mu \pm \sigma.$

7.11 (a) 0.68; (b) 0.997.

7.15 $\frac{f_{Z^2}(t) = e^{-t/2}}{\sqrt{2\pi t},\ t > 0}.$

7.17 $\frac{f(x) = \sqrt{x}e^{-x/2}}{\sqrt{2\pi},\ x > 0}.$

7.19 $V[X] = a/\lambda^2.$

7.21 (a) $E[N_{15}] = V[N_{15}] = 30$; (b) 0.542; (c) 0.476; (d) 0.152; (e) 9:03 and 30 s.

7.23 (a) $e^{-4\lambda\pi}$; (b) $f(x) = 2\pi\lambda x e^{-\lambda\pi x^2},\ x > 0.$

7.25 $e^{-\lambda(s-t)}(\lambda(s-t))^{k-n}/(k-n)!$

7.27 $\lambda s.$

7.29 (a) 1320; (b) $\mu = 1/3$ and $\sigma^2 = 2/117$; (c) 0.1133.

7.31 (a) 0.1314; (b) $\mu = 0.9406$ and $\sigma^2 = 0.00055$; (c) 99/101.

7.33 $f(x) = \left(\frac{1}{t-s}\right)\frac{\Gamma(a+b)}{\Gamma(a)\Gamma(b)}\left(\frac{x-s}{t-s}\right)^{a-1}\left(\frac{t-x}{t-s}\right)^{b-1}, \; s < x < t.$

7.35 $1 - X \sim \text{Beta}(b, a)$.

7.37 (a) $E[X] = am/(a-1)$ for $a > 1$; (b) $m^2a/((a-1)^2(a-2))$ for $a > 2$.

7.39 The probabilities for $k = 3, 4, 5, 6$ are $0.008, 0.0047, 0.003, 0.0021$, respectively.

7.43 $a = 1.0044$.

SOLUTIONS FOR CHAPTER 8

8.1 (a) $f_X(x) = 12x(1-x)^2, \; 0 < x < 1.$

$$f_Y(y) = 6y^2 - 4y^3, \; 0 < y < 1.$$

(b) $f_{Y|X}(y|x) = \dfrac{1}{1-x}, \; x < y < 1.$

The conditional distribution is uniform on $(x, 1)$. (c) 0.3125.

8.3 1/3.

8.5 $f_{X+Y|X}(t|x) = \lambda e^{-\lambda(t-x)}, \; t > x.$

Let $Z \sim \text{Exp}(\lambda)$. Then the desired conditional distribution is the same as the distribution of $Z + x$.

8.11 (a) $f_{X|Y=y}(x|y) = (y+1)e^{-x(y+1)}, \; x > 0,$

which is an exponential density with parameter $y + 1$.

(b) $E[X|Y = y] = 1/(y+1)$ and $E[X|Y] = 1/(Y+1)$.

(c) (i) $E[X] = E[E[X|Y]] = E[1/(Y+1)] = \int_0^{e-1} \frac{1}{(y+1)^2}\, dy = \frac{e-1}{e}.$

(ii) $E[X] = \int_0^{\infty} x \frac{e^{-x} - e^{-ex}}{x}\, dx = \frac{e-1}{e}.$

8.13 $E[X|Y = 0] = 9/12$ and $E[X|Y = 1] = 8/12$.

8.15 (a) As per hint,

$$P(M \leq m, N > n) = 2P(X \leq m, Y > n) = 2m(1-n).$$
$$P(M \leq m, N \leq n) = P(M \leq m) - P(M \leq m, N > n)$$
$$= 2m - m^2 - 2m(1-n).$$

Thus $f(m,n) = 2, \;$ for $0 < m < n < 1$.

(b) $f_{N|M}(n|m) = 1/(1-m), \; m < n < 1$. The conditional distribution is uniform on $(m, 1)$.

8.17 (a) 5/2; (b) $S/2$; (c)

$$E[M|X_1 = x] = \begin{cases} 5/2, & \text{if } x = 1 \\ 11/4, & \text{if } x = 2 \\ 13/4, & \text{if } x = 3 \\ 4, & \text{if } x = 4. \end{cases}$$

(d) $5X_1/2$.

8.19 4.5.

8.23 (a) $f_{XY}(t) = \int_0^\infty \frac{1}{y} f_X(t/y) f_Y(y)\, dy, \;\; t > 0$
(c) For real t,

$$f_{X-Y}(t) = \int_t^\infty f_X(t+y) f_Y(y)\, dy.$$

8.25 (a) $E[X] = p\lambda$; (b) $V[X] = p\lambda$.

8.27 $\mu = \$250$, $\sigma = \$25.98$.

8.29 $E[D] = 3/2$; $V[D] = 9/4$.

8.31 (a) $(X_1 + (n-1)\mu)/n$; (b) $(n-1)\sigma^2/n^2$.

SOLUTIONS FOR CHAPTER 9

9.3 (i) $P(S \geq 380) \leq 0.921$; (ii) $P(S \geq 380) \leq 0.324$.

9.5 (a) $P(X \geq 60) \leq 0.833$; (b) $P(X \geq 60) \leq 0.25$.

9.7 $P(\log X \geq c) = P(X \geq e^c) \leq \mu/e^c$.

9.9 (a) Let $U_1, \ldots, U_n$ be i.i.d. uniform (0,1) variables. Let $f(x) = \sin(x)e^{-x^2}$. Then $I \approx (f(U_1) + \cdots + f(U_n))/n$.

(c) Let $X_1, \ldots, X_n$ be i.i.d. normal variables with mean 0 and variance 1/2. Let $f(x) = \log(x^2)\sqrt{\pi}$. Then $I \approx (f(X_1) + \cdots + f(X_n))/n$.

(e) Let $X_1, \ldots, X_n$ be i.i.d. Poisson variables with mean one. Let $f(x) = e \cos \cos x$. Then $S \approx (f(X_1) + \cdots + f(X_n))/n$.

9.11 0.508.

9.13 0.9869.

9.15 0.363.

9.17 Let $p = P(\sum_{i=1}^{10} X_i \geq 14)$. (a) $p \leq 0.714$; (b) $p \leq 0.625$; (c) $p \approx 0.103$; (d) $p \approx 0.134$; (e) $p = 0.1355$.

9.19 Consider a sum of a independent exponential random variables with $\mu = \sigma = \lambda$.

9.21 (a) $\log Y = \sum_{i=1}^{n} \log X_i$, a sum of i.i.d. random variables. By clt, $\log Y$ is approximately normal.

$$\mu = E[\log X_1] = \int_0^1 \log x \, dx = -1,$$

$$E[(\log X_1)^2] = \int_0^1 (\log x) \, dx = 2$$

and $\sigma^2 = 1$. The result follows.
(b) 0.837.

9.23 $E[X^k] = \begin{cases} q - p, & \text{if } k \text{ is odd} \\ q + p, & \text{if } k \text{ is even.} \end{cases}$

9.25 $m(t) = \dfrac{pe^t}{1 - (1-p)e^t}$.

9.27 $m(t) = \dfrac{e^{tb} - e^{ta}}{t(b-a)}$.

$V[X] = (b-a)^2/12$.

9.29 $m_X(t)m_Y(t) = (1-p+pe^t)^m(1-p+pe^t)^n(1-p+pe^t)^{m+n} = m_{X+Y}(t)$.

SOLUTIONS FOR CHAPTER 10

10.1 Show $E[(X+Y)(X-Y)] = E[X+Y]E[X-Y]$.

10.3 0.092.

10.5 (a) Bivariate standard normal distribution with $\rho = 1/2$.

(b) Normal with (conditional) mean $y/2$ and variance 3/4.

(c) 0.193.

10.10 $f_{V,W}(v,w) = \frac{w}{(1+v)^2}\lambda^2 e^{-\lambda w}$, for $v > 0, w > 0$.

10.12 $f_{V,W}(v,w) = \frac{4w}{v}$, for $0 < w < v < 1$.

$f_W(w) = -4w \log w$, for $0 < w < 1$.

10.15 $f(\theta, \phi, r) = \frac{r^2 \sin\phi}{(2\pi)^{3/2}} e^{-r^2/2}$, $0 \le \theta \le 2\pi$, $0 \le \phi \le \pi$, $r > 0$.

$f_R(r) = \sqrt{\frac{2}{\pi}} r^2 e^{-r^2/2}$, $r > 0$.

10.17 $np = (X_1 + \cdots + X_n)/n$ and $np(1-p) + (np)^2$
$= (X_1^2 + \cdots + X_n^2)/n$.

10.19 $P(\text{Center}) = 1/2$ and $P(\text{leaf}) = 1/(2(k-1))$.

10.21 $P(a) = 3/34$; $P(b) = 5/34$; $P(c) = 8/34$; $P(d) = 8/34$; $P(e) = 7/34$; $P(f) = 2/34$; $P(g) = 1/34$.

10.23 If the transition matrix is symmetric, it is doubly stochastic, and thus the stationary distribution is uniform. Then $\pi_i T_{ij} = T_{ij}/k = T_{ji}/k = \pi_j T_{ji}$.

10.25 Not time-reversible.

10.29 Expected return time for the knight is 168 steps.

SOLUTIONS FOR APPENDIX E

E.1 (a) $\Omega = \{11, 12, \ldots, 43, 44\}$.

(b) $P(X = k) = \begin{cases} 1/16, & \text{if } k = 1 \\ 3/16, & \text{if } k = 2 \\ 5/16, & \text{if } k = 3 \\ 7/16, & \text{if } k = 4. \end{cases}$

(c) $E[1/X] = 1(1/16) + (1/2)(3/16) + (1/3)(5/16) + (1/4)(7/16) = 0.370$.

E.2 (a) $1 - P(A^c B^c C^c) = 1 - P(A^c)P(B^c)P(C^c) = 11/16$.

(b) $P(A^c|BC) = P(A^c) = 5/6$.

(c) $P(AB|A \cup B) = P(AB)/P(A \cup B) = (1/24)/(9/24) = 1/9$.

E.3 In suggestive notation, using Bayes formula,

$$P(2H|H) = \frac{P(H|2H)P(2H)}{P(H|2H)P(2H) + P(H|F)P(F) + P(H|B)P(B)} = \frac{4}{9}.$$

E.4 Let X be the number of rooms in which more than six calls are made. Then X has a binomial distribution with parameters n and $p = P(Y > 6)$, where $Y \sim \text{Pois}(2)$. Thus $p = 1 - \texttt{ppois}(5, 2) = \texttt{0.01656}$. The desired probability is $P(X \geq 3) = 1 - P(X \leq 2) = 1 - \texttt{pbinom}(100, 0.01656) = \texttt{0.23}$.

E.5 (a) $\frac{2^6}{12 \cdot 11 \cdots 7} = 0.0000962$.

(b) $\binom{2}{2}\binom{10}{4}/\binom{12}{6} = \frac{5}{22} = 0.227$.

(c) $\binom{2}{0}\binom{10}{6}/\binom{12}{6} = \frac{5}{22} = 0.227$.

(d) $\frac{2 \times 2^5 \times 7}{12 \cdots 7} = 0.00067$.

E.6 (a) $E[X] = 0(0.1) + 1(0.2) + 2(0.3) + 3(0.4) = 2.0$

(b) $E[X^2] = 0(0.1) + 1(0.2) + 4(0.3) + 9(0.4) = 5$,

giving $V[X] = E[X^2] - E[X]^2 = 5 - 4 = 1$.

(c) $E[e^X] = e^0(0.1) + e^1(0.2) + e^2(0.3) + e^3(0.4) = 10.8946$.

E.7 Let X be the number of Zodiac signs shared by at least four students. For $k = 1, \ldots, 12$, let $I_k = 1$, if the kth Zodiac sign is shared by at least four students in the class, and 0, otherwise. Then $X = I_1 + \cdots + I_{12}$ and

$$\begin{aligned} E[X] &= E[I_1 + \cdots + I_{12}] = E[I_1] + \cdots + E[I_{12}], \\ &= \sum_{k=1}^{12} P(\text{At least four students share Zodiac sign } k). \end{aligned}$$

Let Z be the number of students in the class whose sign is, say, Sagittarius. Then Z has a binomial distribution with parameters $n = 30$ and $p = 1/12$. And

$$\begin{aligned} &P(\text{At least four students are Sagittarius}) = P(Z \geq 4) \\ &= \sum_{k=4}^{\infty} \binom{30}{k} (1/12)^k (11/12)^{30-k} = 0.10. \end{aligned}$$

It follows that $E[X] = 12(0.1) = 1.2$.

E.8 (a) $c = 1/18$.

(b) $P(X \leq 1, Y \geq 1) = 4/9$.

(c) $E[XY] = 30/18 = 15/9$.

(d) $P(X < Y) = 6/18 = 1/3$.

(e) The distributions are the same. The common pmf is $P(X = 0) = 1/6$, $P(X = 1) = 2/6$, $P(X = 2) = 3/16$.

(f) $E[Y|X = 2] = 0(2/9) + 1(3/9) + 2(4/9) = 11/9$.

E.9 $P(X = x, Y = y) = \binom{3}{x}\binom{5}{y}\binom{7}{3-x-y} \Big/ \binom{15}{3}$.

E.10 (a) $c = 6$; (b) $E[X] = 1/2$; $V[X] = (3/10) - (1/4) = 1/20$.

(c) $F(x) = 3x^2 - 2x^3, \;\; 0 < x < 1$. (d)

$$\begin{aligned}
\text{(i) } P(0.1 < X < 0.3) &= F(0.3) - F(0.1) \\
&= (3(0.3)^2 - 2(0.3)^3) - (3(0.1)^2 - 2(0.1)^3) \\
&= 0.188.
\end{aligned}$$

(ii) $\displaystyle\int_{0.1}^{0.3} 6x(1 - x)\,dx = 0.188$.

(e) $P(2X \leq x) = P(X \leq x/2)$. Thus

$$f_{2X}(x) = (1/2)f_{x/2} = 3(x/2)(1 - x/2), \; 0 < x < 2.$$

(f) $E[Y] = \int_0^2 y3(y/2)(1 - y/2)\,dy = 1 = 2E[X] = 2(1/2) = 1$.

(g) $E\left[\dfrac{e^X}{1-X}\right] = \int_0^1 \dfrac{e^x}{1-x} x(1 - x)\,dx = \int_0^1 xe^x\,dx = 1$.

E.11 (a) For $x > 1$,

$$\begin{aligned}
P(M \leq m) &= P(\max(X_1, \ldots, X_n) \leq m) \\
&= P(X_1 \leq m, \ldots, X_{10} \leq m) \\
&= P(X_1 \leq m)^{10} = \left(1 - \frac{1}{m^3}\right)^{10},
\end{aligned}$$

giving

$$f_M(m) = \frac{30}{m^2}\left(1 - \frac{1}{m^3}\right)^9.$$

(b) The inverse of the original cdf is $F^{-1}(x) = 1/(1-x)^{1/3}$. Let $U \sim \text{Unif}(0,1)$. In R, type

```
> max(replicate(10,1/(1-runif(1))^(1/3)))
```

E.12 For the density of the minimum M, consider

$$P(M > m) = P(X_1 > m, \ldots, X_n > m) = P(X_1 > m)^n = (1-m)^n,$$

for $0 < m < 1$, and $f_M(m) = n(1-m)^{n-1}$ with expectation

$$\int_0^1 mn(1-m)^{n-1}\,dm = \frac{1}{n+1}.$$

E.13 (a) For $y > 0$, $P(Y \leq y) = P(cX \leq y) = P(X \leq y/c)$ giving

$$f_Y(y) = (1/c)f_X(y/c) = \frac{y}{c^2}e^{-y/c}.$$

(b) The convolution formula gives the density as

$$g(t) = \int_{-\infty}^{\infty} f(t-y)f(y)\,dy = \int_0^t (t-y)e^{-(t-y)}ye^{-y}\,dy = \frac{e^{-t}t^3}{6}.$$

E.14 (a) $c = e^2$; (b) For $x > 2$,

$$P(Z+2 \leq x) = P(Z \leq x-2) = \int_0^{x-2} e^{-t}\,dt = \int_0^x e^{-(t-2)}\,dt,$$

which is the cdf of X. (c) $E[X] = E[Z+2] = E[Z]+2 = 1+2 = 3$. $V[X] = V[Z+2] = V[Z] = 1$.

(d) For $0 < y < e^{-2}$,

$$\begin{aligned} P(Y \leq y) &= P(e^{-X} \leq y) = P(-X \leq \ln y) \\ &= P(X \geq -\ln y) = 1 - P(X < -\ln y) \end{aligned}$$

and

$$f_Y(y) = (1/y)f_X(-\ln y) = (1/y)e^2 y = e^2.$$

Thus Y is uniformly distributed on $(0, e^{-2})$.

E.15 (a) $f(x, y) = f_X(x)f_Y(y) = (1/6)e^{-(x/2+y/3)}$.

(b) $P(X \leq x, Y \leq y) = P(X \leq x)P(Y \leq y) = (1 - e^{-x/2})(1 - e^{-y/2})$.

(c) By independence, $\text{Cov}(X, Y) = \text{Corr}(X, Y) = 0$.

(d) For $t > 0$¡

$f_{X+Y}(t) = \int_0^t (1/2)e^{-(t-y)/2}(1/3)e^{-y/3}\, dy = e^{-t/2}(e^{t/6} - 1)$.

(e) $E[e^{-X+3Y}] = E[e^{-X}]E[e^{-3Y}] = (1/3)(1/10) = 1/30$.

E.16 For $t > 0$, to find $P(Y/X \leq t) = P(Y \leq tX)$. Consider the line $y = tx$ in the square $[0, 1] \times [0, 1]$. If $t < 1$, the line from the origin meets the right side of the square, and the area under the line is $t/2$. If $t > 1$, the line from the origin meets the top side of the square, and the area under the line is $1 - 1/2t$. Differentiating gives

$$f_{Y/X}(t) = \begin{cases} 1/2, & \text{if } t \leq 1 \\ 1/(2t^2), & \text{if } t > 1. \end{cases}$$

E.17 The integrand is proportional to a normal density with mean 0 and variance 1/4. The "missing" constant is $2/\sqrt{2\pi}$. The density is symmetric about zero and the limits of integration are zero to infinity. The integral is equal to $\sqrt{2\pi}/4$.

E.18 $f(x, y) = \frac{1}{2x},\ 0 < y < x < 2$.

$f_X(x) = 1/2,\ 0 < x < 2$.

$f_Y(y) = \int_y^2 \frac{1}{2x}\, dx = \frac{\ln(y/2)}{2}\ 0 < y < 2$.

$f_{Y|X}(y|x) = 1/x,\ 0 < y < x$.

$f_{X|Y}(x|y) = \frac{f(x,y)}{f_Y(y)} = \frac{\ln(y/2)}{x},\ y < x < 2$.

E.19 (a) $f_X(x) = \int_x^1 3y\, dy = \frac{3}{2}(1 - x^2),\ 0 < x < 1$.

$f_Y(y) = \int_0^y 3y\, dx = 3y^2,\ 0 < y < 1$.

(b) $E[XY] = \int_0^1 \int_0^y xy(3y)\, dx\, dy = \frac{3}{10}$.

$E[X] = \int_0^1 x\frac{3}{2}(1 - x^2)\, dx = \frac{3}{8}$.

$E[Y] = \int_0^1 3y^3\, dy = \frac{3}{4}$.

$\mathrm{Cov}(X, Y) = (3/10) - (3/8)(3/4) = \frac{3}{160}$.

(c) $f_{Y|X}(y|x) = \frac{3y}{(3/2)(1-x^2)} = \frac{2y}{1-x^2}$, $x < y < 1$.

$f_{X|Y}(x|y) = \frac{3y}{3y^2} = \frac{1}{y}$, $0 < x < y$.

The conditional distribution of X given $Y = y$ is uniform on $(0, y)$.

(d) $P(X < 1/2|Y = 3/4) = \int_0^{1/2} \frac{1}{3/4}\, dx = \frac{2}{3}$.

(e) $E[Y|X = 1/2] = \int_{1/2}^1 y \frac{2y}{1-(1/2)^2}\, dy = \frac{7}{9}$.

(f) $E[X|Y] = Y/2$.

(g) $E[E[X|Y]] = E[Y/2] = (1/2)(3/4) = 3/8$.

(h) (i) $E[X^2] = \int_0^1 x^2(3/2)(1 - x^2)\, dx = \frac{1}{5}$,

giving

$V[X] = (1/5) - (3/8)^2 = \frac{19}{320}$.

(ii) $V[X] = V[E[X|Y]] + E[V[X|Y]] = V[Y/2] + E[Y^2/12]$

$$= \frac{1}{4}V[Y] + \frac{1}{12}E[Y^2] + \frac{1}{4}\frac{3}{80} + \frac{1}{12}\frac{3}{5} = \frac{19}{320}.$$

E.20 Let T be the time it takes to get out of the house. Then

$$E[T] = E[T|A](0.1) + E[T|B](0.3) + E[T|C](0.6)$$
$$= 20(0.1) + 40(0.3) + (E[T] + 30)(0.6) = 32 + 6E[T]/10.$$

Thus $E[T] = 32(10)/4 = 80$ minutes.

E.21 By results for random sums of random variables, $E[Y] = (1/p)(2) = 2/p$. $V[Y] = 9/p + (2)^2(1 - p)/p^2 = (5p + 4)/p^2$.

E.22 (a) $X \sim \mathrm{Exp}(5)$.

(b) $X \sim \mathrm{Binom}(7, 0.90)$.

(c) $X \sim \mathrm{Geom}(0.49)$.

(d) $X \sim \text{Norm}(12, 4)$.

(e) $X \sim \text{Unif}(0, 30)$.

(f) $X \sim \text{Pois}(5)$.

E.23 (a) $P(\sum_{i=1}^{25} X_i > 350) \leq 250/350 = 0.714$.

(b) $P(\sum_{i=1}^{25} X_i > 350) = P(|\sum_{i=1}^{25} X_i - 250| > 100) \leq 2500/100^2 = 0.25$.

(c) $P\left(\frac{\sum_{i=1}^{25} X_i - 250}{\sqrt{2500}} > \frac{350-250}{\sqrt{2500}}\right) \approx P(Z > 2) = 0.02275$.

(d) $\sum_{i=1}^{25} X_i \sim \text{Gamma}(25, 0.1)$ = `1 - pgamma(350,25,0.1)` = `0.03237`.

E.24 The integral is equal to $4E[f(X)]$, where $f(x) = \sin(x^x)$ and $X \sim \text{Unif}(0, 4)$. In R, type

```
> f <- function(x) sin(x^x)
> 4*mean(f(runif(10000,0,4)))
[1] 1.225753
```

E.25 First find the moment-generating function of X. Write $X = \sigma_x Z + \mu_x$, where Z has a standard normal distribution. Then

$$\begin{aligned} m_X(t) &= E[e^{tX}] = E[e^{t(\sigma_x Z + \mu_x)}] \\ &= e^{t\mu_x} E[e^{(t\sigma_x)Z}] = e^{t\mu_x} m_Z(t\sigma_x) \\ &= e^{t\mu_x} e^{t^2\sigma_x^2/2}. \end{aligned}$$

The moment-generating function of $X + Y$ is

$$\begin{aligned} m_{X+Y}(t) &= m_X(t) m_Y(t) = e^{t\mu_x} e^{t^2\sigma_x^2/2} e^{t\mu_y} e^{t^2\sigma_y^2/2} \\ &= e^{t(\mu_x+\mu_y)} e^{(\sigma_x^2+\sigma_y^2)t^2/2}, \end{aligned}$$

which is the moment generating function of a normal random variable with mean $\mu_x + \mu_y$ and variance $\sigma_x^2 + \sigma_y^2$.

REFERENCES

Chrisna Aing, Sarah Halls, Kiva Ohen, Robert Dobrow, and John Fieberg. A Bayesian hierarchical occupancy model for track surveys conducted in a series of linear, spatially correlated, sites. *Journal of Applied Ecology*, 48(6):1508–1517, 2011.

John Aldrich. Earliest uses of symbols in probability and statistics. http://jeff560.tripod.com/stat.html. Accessed on May 14, 2003.

Avi Arampatzis and Jaap Kamps. A study of query length. In *Proceedings of the 31st Annual International ACM SIGIR Conference on Research and Development in Information Retrieval*, SIGIR '08, pp. 811–812. ACM, 2008.

Card trick defies the odds. http://news.bbc.co.uk/2/hi/uk_news/50977.stm. Accessed on May 14, 2003.

Frank Benford. The law of anomalous numbers. *Proceedings of the American Philosophical Society*, 78:551–572, 1938.

Gunnar Blom, Lars Holst, and Dennis Sandell. *Problems and Snapshots from the World of Probability*. New York:Springer-Verlag, 1991.

College Board. 2011 College-Bound Seniors. Total Group Profile Report, 2011.

Kurt Bryan and Tanya Leise. The $25,000,000,000 eigenvector. The linear algebra behind Google. *SIAM Review*, 48(3):569–581, 2006.

Lewis Carroll. *The Mathematical Recreations of Lewis Carroll: Pillow Problems and a Tangled Tale*. New York: Dover Publications, 1958.

George Casella and Edward I. George. Explaining the Gibbs sampler. *American Statistician*, 46(3):167–174, 1992.

Probability: With Applications and R, First Edition. Robert P. Dobrow.

E. Chia and M. F. Hutchinson. The beta distribution as a probability model for daily cloud duration. *Agricultural and Forest Meteorology*, 56(3–4):195–208, 1991.

B. Dawkins. Siobhan's problem: the coupon collector revisited. *American Statistician*, 45: 76–82, 1991.

Keith Devlin. *The Unfinished Game: Pascal, Fermat, and the Seventeenth-Century Letter that Made the World Modern*. New York: Basic Books, 2008.

Persi Diaconis. Dynamical bias in coin tossing. *SIAM Review*, 49(2):211–235, 2007.

Persi Diaconis. The Markov chain Monte Carlo revolution. *Bulletin of the American Mathematical Society*, 46(2):179–205, 2008.

Persi Diaconis and Frederick Mosteller. Methods for studying coincidences. *Journal of the American Statistical Association*, 84(408):853–861, 1989.

William Dorsch, Tom Newland, David Jassone, Samuel Jymous, and David Walker. A statistical approach to modeling the temporal patterns in ocean storms. *Journal of Coastal Research*, 24(6):1430–1438, 2008.

Cindy Durtschi, William Hillison, and Carl Pacini. The effective use of Benford's law to assist in detecting fraud in accounting data. *Journal of Forensic Accounting*, 5(1):17–34, 2004.

Sherwood F. Ebey and John J. Beauchamp. Larval fish, power plants, and Buffon's needle problem. *American Mathemartical Monthly*, 84(7):534–541, 1977.

Roger Eckhardt. Stan Ulam, John von Neumann, and the Monte Carlo method. *Los Alamos Science*, 15:131–137, 1987.

Andrew Ehrenberg. The pattern of consumer purchases. *Applied Statistics*, 8(1):26–41, 1959.

William Feller. *An Introduction to Probability Theory and Its Applications*. New York Wiley, 1968.

Thomas S. Ferguson. Who solved the secretary problem? *Statistical Science*, 4(3):282–296, 1989.

Mark Finkelstein, Howard G. Tucker, and Jerry Alan Veeh. Confidence intervals for the number of unseen types. *Statistics and Probability Letters*, 37:423–430, 1998.

Sir Ronald A. Fisher. *The Design of Experiments*. Edinburg Oliver and Boyd, 1935.

David Freedman, Robert Pisani, and Robert Purves. *Statistics*. New York W.W. Norton and Company, 2007.

S. Geman and D. Geman. Stochastic relaxation, Gibbs distributions, and the Bayesian restoration of images. *IEEE Transactions on Pattern Analysis and Machine Intelligence*, 6:721–741, 1984.

Arjun K. Gupta and Saralees Nadarajah, editors. *Handbook of Beta Distribution and Its Applications*. New York: Marcel Dekker, 2004.

Robert Harris. Reliability applications of a bivariate exponential distribution. *Operations Research*, 16(1):18–27, 1968.

David Hoffman. Negative binomial control limits for count data with extra-Poisson variation. *Pharmaceutical Statistics*, 2:127–132, 2003.

Anne B. Hollowed, Teresa Amar, Steven Barbeau, Nicholas Bond, James N. Ianelli, Paul Spencer, and Thomas Wilderbuer. Integrating ecosystem aspects and climate change forecasting into stock assessments. *Alaska Fisheries Science Center Quarterly Report Feature Article*, July/September 2011.

Michael Huber and Andrew Glen. Modeling rare baseball events—are they memoryless. *Journal of Statistics Education*, 15(1), 2007.

Darrel Huff. *How to take a chance*. New York: W.W. Norton and Company, 1964.

Steven Krantz. *Mathematical Apocrypha Redux*. More stories and anecdotes of Mathematics and Mathematical Washington, DC: Mathematical Association of America, 2005.

B.S. Van Der Laan and A.S. Louter. A statistical model for the costs of passenger car traffic accidents. *Journal of the Royal Statistical Society, Series D*, 35(2):163–174, 1986.

Kevin Z. Leder, Saverio E. Spagniole, and Stefan M. Wild. Probabilistically optimized airline overbooking strategies, or "Anyone willing to take a later flight?!". *Journal of Undergraduate Mathematics and Its Applications*, 23(3):317–338, 2002.

Robert E. Leiter and M. A. Hamdan. Some bivariate probability models applicable to traffic accidents and fatalities. *International Statistical Review*, 41(1):87–100, 1973.

Dana Mackenzie. Compressed sensing makes every pixel count. http://www.ams.org/samplings/math-history/hap7-pixel.pdf. Accessed on May 13, 2013.

Eamonn B. Mallon and Nigel R. Franks. Ants estimate area using buffon's needle. *Proceedings of the Royal Society B Biological Sciences*, 267(1445):765–770, 2000.

J.S. Marshall and W.M. Palmer. The distribution of raindrops with size. *Journal of Meteorology*, 5:165–166, 1948.

Gely P. Masharin, Amy N. Langville, and Valeriy A. Naumov. The life and work of A. A. Markov. *Linear Algebra and its Applications*, 386:3–26, 2004.

Norm Matloff. *From Algorithms to Z-scores: Probabilistic and Statistical Modeling in Computer Science*. Orange Grove Texts Plus, 2009.

Webb Miller et al. Polar and brown bear genomes reveal ancient admixture and demographic footprints of past climate change. *Proceedings of the National Academy of Sciences*, 109(36): E2382–E2390 2012.

Chris Mulligan. Births by day of year. http://chmullig.com/2012/06/births-by-day-of-year/. Accessed on May 13, 2013.

Daniel B. Murray and Scott W. Teare. Probability of a tossed coin landing on edge. *Physical Review E (Statistical Physics, Plasmas, Fluids, and Related Interdisciplinary Topics)*, 48(4):2547–2552, 1993.

Michael W. Nachman and Susan L. Crowell. Estimate of the mutation rate per nucleotide in humans. *Genetics*, 156:297–304, 2000.

Mej Newman. Power laws, Pareto distributions, and Zipf's law. *Contemporary Physics*, 46(5):323–351, 2005.

Mark J. Nigrini. *Benford's Law*: Applications for Forensic Accounting, and Fraud Detection. Hoboken:Wiley, 2012.

Opte project. http://opte.org/. Accessed on May 13, 2013.

T. Parsons and E.L. Geist. Tsunami probability in the Caribbean region. *Pure and Applied Geophysics*, 165(11–12):2089–2116, 2008.

Project Gutenberg. http://www.gutenberg.org. Accessed on May 13, 2013.

J.F. Ramaley. Buffon's noodle problem. *American Mathematical Monthly*, 76(8):916–918, 1969.

Matthew Richey. Evolution of Markov chain Monte Carlo methods. *American Mathematical Monthly*, 117(5):383–413, 2010.

Christian Robert and George Casella. Short history of Markov chain Monte Carlo: subjective recollections from incomplete data. *Statistical Science*, 26(1):102–115, 2011.

Kenneth A. Ross. Benford's law, a growth industry. *American Mathematical Monthly*, 118:571–583, 2011.

Sheldon Ross. *A First Course in Probability*. Upper Saddle River:Prentice Hall, 2012.

Zvi Schechner, Jonathan J. Kaufman, and Robert S. Siffert . A Poisson process model for hip fracture risk. *Medical and Biological Engineering and Computing*, 48(8):799–810, 2010.

Hal Stern. Shooting darts. *Chance*, 10(3):16–19, 1997.

Stephen M. Stigler. Isaac Newton as a probabilist. *Statistical Science*, 21(3):400–403, 2006.

Lajos Tákacs. The problem of coincidences. Archive for History of Exact Sciences, 21(3): 229–244, 1980.

K. Troyer, T. Gilroy, and B. Koeneman. A nine STR locus match between two apparent unrelated individuals using AmpFISTR Profiler Plus™and COfiler™. *Proceedings of the Promega 12th International Symposium on Human Identification*, 2001.

Marilyn vos Savant. Game show problem. http://marilynvossavant.com/game-show-problem/. Accessed on May 13, 2013.

Thayer Watkins. Raindrop size. http://www.sjsu.edu/faculty/watkins/raindrop.htm.

Malcolm Williams. Can we measure homelessness? A critical evaluation of 'capture-recapture'. *Methodological Innovations*, 5(2):49–59, 2010.

Marcel Zwahlen, Beat E. Neuenschwander, André Jeannin, Francoise Dubois-Arber, and david Vlahov. HIV testing and retesting for men and women in Switzerland. *European Journal of Epidemilogy*, 16(2):123–133, 2000.

INDEX

Probability: With Applications and R, First Edition. Robert P. Dobrow.